Remote Sensing of Target Object Detection and Identification II

Remote Sensing of Target Object Detection and Identification II

Editor

Paolo Tripicchio

MDPI

Basel • Beijing • Wuhan • Barcelona • Belgrade • Novi Sad • Cluj • Manchester

Editor
Paolo Tripicchio
Scuola Superiore Sant'Anna
Pisa
Italy

Editorial Office
MDPI AG
Grosspeteranlage 5
4052 Basel, Switzerland

This is a reprint of articles from the Special Issue published online in the open access journal *Remote Sensing* (ISSN 2072-4292) (available at: https://www.mdpi.com/journal/remotesensing/special_issues/GDWU5I62TK).

For citation purposes, cite each article independently as indicated on the article page online and as indicated below:

Lastname, A.A.; Lastname, B.B. Article Title. *Journal Name* **Year**, *Volume Number*, Page Range.

ISBN 978-3-7258-2119-8 (Hbk)
ISBN 978-3-7258-2120-4 (PDF)
doi.org/10.3390/books978-3-7258-2120-4

Cover image courtesy of Paolo Tripicchio

Contents

About the Editor . . . vii

Paolo Tripicchio
Remote Sensing of Target Object Detection and Identification II
Reprinted from: *Remote Sens.* **2024**, *16*, 3106, doi:10.3390/rs16163106 . . . 1

Tianqi Zhao, Yongcheng Wang, Zheng Li, Yunxiao Gao, Chi Chen, Hao Feng and Zhikang Zhao
Ship Detection with Deep Learning in Optical Remote-Sensing Images: A Survey of Challenges and Advances
Reprinted from: *Remote Sens.* **2024**, *16*, 1145, doi:10.3390/rs16071145 . . . 8

Le Xia, Fulai Wang, Chen Pang, Nanjun Li, Runlong Peng, Zhiyong Song and Yongzhen Li
An Identification Method of Corner Reflector Array Based on Mismatched Filter through Changing the Frequency Modulation Slope
Reprinted from: *Remote Sens.* **2024**, *16*, 2114, doi:10.3390/rs16122114 . . . 48

Weishan Zhao, Lijia Huang, Haitian Liu and Chaobao Yan
Scattering-Point-Guided Oriented RepPoints for Ship Detection
Reprinted from: *Remote Sens.* **2024**, *16*, 933, doi:10.3390/rs16060933 . . . 70

Yang Tian, Xuan Wang, Shengjie Zhu, Fang Xu and Jinghong Liu
LMSD-Net: A Lightweight and High-Performance Ship Detection Network for Optical Remote Sensing Images
Reprinted from: *Remote Sens.* **2023**, *15*, 4358, doi:10.3390/rs15174358 . . . 89

Shilong Jing, Hengyi Lv, Yuchen Zhao, Hailong Liu and Ming Sun
MVT: Multi-Vision Transformer for Event-Based Small Target Detection
Reprinted from: *Remote Sens.* **2024**, *16*, 1641, doi:10.3390/rs16091641 . . . 115

Xiaozhen Wang, Chengshan Han, Jiaqi Li, Ting Nie, Mingxuan Li, Xiaofeng Wang and Liang Huang
Multiscale Feature Extraction U-Net for Infrared Dim- and Small-Target Detection
Reprinted from: *Remote Sens.* **2024**, *16*, 643, doi:10.3390/rs16040643 . . . 136

Meihui Li, Yuxing Wei, Bingbing Dan, Dongxu Liu and Jianlin Zhang
Infrared Small Dim Target Detection Using Group Regularized Principle Component Pursuit
Reprinted from: *Remote Sens.* **2024**, *16*, 16, doi:10.3390/rs16010016 . . . 156

Xuying Hao, Xianyuan Liu, Yujia Liu, Yi Cui and Tao Lei
Infrared Small-Target Detection Based on Background-Suppression Proximal Gradient and GPU Acceleration
Reprinted from: *Remote Sens.* **2023**, *15*, 5424, doi:10.3390/rs15225424 . . . 177

Xi'ai Chen, Zhen Wang, Kaidong Wang, Huidi Jia, Zhi Han and Yandong Tang
Multi-Dimensional Low-Rank with Weighted Schatten p-Norm Minimization for Hyperspectral Anomaly Detection
Reprinted from: *Remote Sens.* **2024**, *16*, 74, doi:10.3390/rs16010074 . . . 200

Wenrong Yue, Feng Xu and Juan Yang
Tracking-by-Detection Algorithm for Underwater Target Based on Improved Multi-Kernel Correlation Filter
Reprinted from: *Remote Sens.* **2024**, *16*, 323, doi:10.3390/rs16020323 . . . 219

Lei Zhang, Peng Rao, Yang Hong, Xin Chen and Liangjie Jia
Infrared Dim Star Background Suppression Method Based on Recursive Moving Target Indication
Reprinted from: *Remote Sens.* **2023**, *15*, 4152, doi:10.3390/rs15174152 . . . . **235**

Yueqi Su, Xin Chen, Gaorui Liu, Chen Cang and Peng Rao
Implementation of Real-Time Space Target Detection and Tracking Algorithm for Space-Based Surveillance
Reprinted from: *Remote Sens.* **2023**, *15*, 3156, doi:10.3390/rs15123156 . . . . **253**

Getachew Nadew Wedajew and Sendren Sheng-Dong Xu
SE-RRACycleGAN: Unsupervised Single-Image Deraining Using Squeeze-and-Excitation-Based Recurrent Rain-Attentive CycleGAN
Reprinted from: *Remote Sens.* **2024**, *16*, 2642, doi:10.3390/rs16142642 . . . . **287**

About the Editor

Paolo Tripicchio

Paolo Tripicchio is an Assistant Professor at the Department of Excellence in Robotics and AI, Mechanical Intelligence Institute, Sant'Anna School of Superior Studies, Pisa, Italy. His main research areas cover different technological topics, such as studies on robotic perception, applications and theories of computer vision, AI systems and their applications, human–machine interaction with a particular focus on haptics, the fast-growing industrial robotics sector, field robotics applications, and the use of virtual and augmented reality for visualization, interaction, and simulation. He was the recipient of the Best Paper Award at the International Conference on Systems, Analysis and Automatic Control of the 11th International Multi-Conference on Systems, Signals and Devices 2014, a Best Paper Award Finalist at the 2019 IEEE International Conference on Real-Time Computing and Robotics, and received the One Star Innovation Award at MBDA Innovation Awards 2020 and the 2020 Researcher Prize in Artificial Intelligence from the Department of Excellence in Robotics and AI of the Scuola Superiore Sant'Anna.

remote sensing

MDPI

Editorial

Remote Sensing of Target Object Detection and Identification II

Paolo Tripicchio

Institute of Mechanical Intelligence, Department of Excellence in Robotics and AI, Scuola Superiore Sant'Anna, 56124 Pisa, Italy; paolo.tripicchio@santannapisa.it

The ability to detect and identify target objects from remote images and acquisitions is paramount in remote sensing systems for the proper analysis of territories. The field of applying such a technology spans environmental [1] and urban [2] monitoring, hazard and disaster management, and defense and military applications. The existing literature has taken advantage of the large amounts of data acquired by sensors mounted on satellite, airborne, and unmanned aerial vehicle (UAV) platforms. While satellite imaging is still the foremost source of data, as also confirmed by the contributions collected in this Special Issue, UAV platforms have had exponential growth in recent years [3] and, given their low-cost effectiveness, this has allowed, and will allow in future, the acquisition and coverage of a wide range of environments exploiting customized setups and coverage algorithms [4]. Research applications exploit different phenomena and technologies, which include synthetic aperture radar (SAR) [5] imaging, multispectral and hyperspectral imaging, and images (or videos) acquired in the visible and near-infrared (VNIR) wavelength ranges. With the recent improvements in the sensing technologies regarding their spatial resolution and spectral content, and with the rapid development of artificial intelligence techniques that exploit convolutional neural networks (CNNs) or deep neural networks (DNNs), the results that novel approaches will achieve in the near future are promising.

The articles belonging to this Special Issue provide a comprehensive overview of the advancements, challenges, and future trends in object detection and tracking, with a particular focus on remote sensing applications. They discuss a wide range of topics, including different types of targets (e.g., ships, small targets), imaging modalities (e.g., optical, SAR, infrared), image processing techniques, and deep learning algorithms.

This editorial attempts to summarize the novelties and drawbacks of the methods and studies presented by the contributors in the context of current research trends, and also considering future developments.

A group of articles discusses different aspects of ship detection in remote sensing images, including challenges, advancements, and datasets. These sources specifically focus on ship detection in SAR images, which poses unique challenges due to the presence of speckle noise and the need for robust algorithms that can handle different ship sizes and orientations. Another group addresses the problem of detecting small targets in infrared images, which is a complex task due to the small size of the targets, low contrast with the background, and the presence of noise and clutter. A third group focuses on target tracking in image sequences, which involves estimating the trajectory of a target over time. This is particularly useful in applications such as surveillance and navigation. All contributions refer to the use of various image processing techniques to either enhance the quality of images or extract meaningful information. Examples of these techniques include background suppression, edge enhancement, and Hough transformation [6]. Many of the sources discuss the use of deep learning algorithms, particularly convolutional neural networks (CNNs) and Transformers [7], for object detection and tracking tasks, and several of them highlight the importance of evaluating the performance of detection and tracking algorithms using appropriate metrics and datasets.

A detailed summary of the novelty and drawbacks of each contribution, whose list can be found at the end of this editorial, is provided to introduce and discuss the current challenges and future development in remote sensing object detection and tracking.

Citation: Tripicchio P. Remote Sensing of Target Object Detection and Identification II. *Remote Sens.* **2024**, *16*, 3106. https://doi.org/10.3390/rs16163106

Received: 6 August 2024
Accepted: 20 August 2024
Published: 22 August 2024

The first contribution is a survey that offers a comprehensive overview of the technologies, challenges, and prospects of ship detection in optical images obtained through remote sensing. The article examines various ship detection technologies in chronological order, dividing them into traditional methods, methods based on convolutional neural networks (CNNs), and methods based on Transformers. The advantages and disadvantages of each category of method are analyzed. The article particularly focuses on the challenges that arise from ship detection in optical images obtained from remote sensing, which are mainly: complex marine environments, in which images can be influenced by factors such as light, weather, and the presence of objects other than ships, making identification difficult; the presence of insufficient discriminating features, since ships often occupy very small areas in images, and this makes it difficult to extract distinctive features; the problems of large scale variations, density distribution, and rotation, since ships can have very different sizes in images, can be very close to each other, and are probably oriented in different directions; the presence of large aspect ratios, since ships often have an elongated shape; and the problem of imbalance between positive and negative samples, since images usually contain many more background areas than ships. For each challenge, the article reviews and analyzes the solutions proposed in the literature, mainly based on CNNs, and highlights their advantages and disadvantages. In addition, the article presents a collection of public datasets of optical images for ship detection, offering detailed information on their data distribution, such as the number of ships and their size in pixels. The performance of different detection models on these datasets is compared, and the effects of the different optimization strategies to address the challenges of ship detection are analyzed. Finally, the article explores the application of Transformers in ship detection, comparing their feature extraction capability with that of CNNs. The results show that Transformers, thanks to their ability to model long-range dependencies, have great potential in this field.

Xia et al. (Contribution 2) present a novel method for differentiating corner reflector arrays from ships in anti-ship scenarios. This distinction is crucial in naval warfare, as corner reflector arrays are often deployed as decoys to confuse enemy radar systems. Their method leverages the distinct scattering characteristics exhibited by corner reflector arrays and ships. These characteristics become more pronounced when processed using a mismatched filter with an adjusted frequency modulation slope. By modifying the frequency modulation slope of the LFM signal within the filter, the main lobe of the signal output is broadened. This broadening reduces the compression level, as compared to a matched filter, thereby accentuating the differences in scattering characteristics between ships and corner reflector arrays. Two key features are extracted from the two-dimensional range-Doppler image obtained after applying the mismatched filter. These are the variance of the width, and the variance of the intervals of regions with normalized amplitude within a specific range. These features effectively capture the differences in the spatial distribution of scattering points between the two target types, aiding in their discrimination. The extracted features are then used to train a Support Vector Machine (SVM) classifier with a Gaussian kernel. This classifier demonstrates high efficacy in distinguishing between ships and corner reflector arrays. It has to be noted, however, that the method's performance is susceptible to degradation in the presence of noise, particularly at low signal-to-noise ratios (SNRs). This vulnerability arises because the extracted features rely on the distribution of scattering points, which noise can distort. The effectiveness of the method also hinges on the selection of specific parameters, such as the Doppler factor range and the point extraction range. While the method has shown resilience to minor variations in these parameters, optimizing their selection is paramount for achieving optimal performance. The authors propose potential solutions to mitigate these limitations, including the implementation of noise suppression techniques and the refinement of parameter selections. They also suggest future research directions, such as validating the method using real-world data in operational scenarios and integrating information from other domains, such as polarization, to enhance identification accuracy.

The third contribution describes a method for detecting oriented ships in Synthetic Aperture Radar (SAR) images based on RepPoints, which are representation points that capture the object's shape and orientation, which is based on an anchor-free detection architecture consisting of two main components: Scattering-Point-Guided Adaptive Sample Selection(SPG-ASS) and SPG learning. The improved sample selection method integrates the scattering point location information to select higher-quality samples during training, thus preventing model degradation caused by low-quality samples. The SPG learning mechanism improves the quality of RepPoints in the initialization stage, enabling the network to learn more refined representations of ships' electromagnetic characteristics, while reducing land clutter interference in complex nearshore environments. The method has shown good generalization and reliability across different datasets with varying characteristics, suggesting its adaptability to practical application scenarios. Furthermore, ablation experiments demonstrate the effectiveness of the individual components, namely SPG-ASS and SPG learning, in improving detection performance. However, both the adaptive sample selection scheme and the adaptive learning part rely on extracting the scattering points from the target. If the area occupied by ships is limited, or if the scattering from the ships is weak, resulting in fewer or no corner points being extracted, the method might fail. The authors have suggested that, in the future, they will explore redesigning the scattering point extraction part and introducing more efficient and advanced network structures for scattering feature extraction and fusion.

Tian et al. (contribution 4) present a new lightweight model (LMSD-Net) for ship detection, specifically small ships, in optical remote sensing imagery. The model is designed to address challenges posed by variations in ship size, background clutter, and the limited capabilities of embedded systems. A fog simulation method is used to augment the training dataset with more foggy images. This method simulates the effect of fog on scene radiance, improving the model's robustness in adverse weather conditions. A new feature extraction module, called Efficient Layer Aggregation of C3 (ELA-C3), is introduced for more efficient information aggregation. ELA-C3 enhances feature learning without significantly increasing the number of model parameters. A feature fusion method is proposed to fuse features extracted at different scales. It utilizes learnable weights for channels during bidirectional fusion, allowing the model to focus on the most relevant information and reducing the number of parameters, as compared to the original architecture it is based on. A Contextual Transformer (CoT) block is added to the detection head to improve its localization accuracy. The CoT block combines the global relationship modeling capability of transformers with the computational efficiency of convolutional neural networks. Finally, an improved version of the CIoU loss function, called V-CIoU, is proposed to address the issue of slow convergence when the aspect ratios of the ground truth box and the predicted box are similar. V-CIoU introduces a penalty term based on the variance of aspect ratios, improving detection performance for small ships.

Contribution 5 introduces a new method for detecting small objects in remote sensing images. The method, called Multi-Vision Transformer (MVT), is based on a Transformer-like neural network, and proposes the first remote sensing dataset based on event cameras, called the Event Object Detection Dataset (EOD Dataset). This dataset consists of over 5000 event streams, and includes six object categories: cars, buses, pedestrians, bicycles, boats, and ships. MVT consists of three modules: a downsampling Module, a Channel Spatial Attention Module (CSA), and a Global Spatial Attention Module (GSA). The CSA focuses on short-range dependencies within feature maps, improving the representation of channel- and spatial location-level features. The GSA module, consisting of Window-Attention and Grid-Attention, considers long-range dependencies in the feature maps, capturing the global information and long-distance connections in a single operation. Finally, a novel cross-scale attention mechanism (Cross Deformable Attention (CDA)) that progressively merges high-level features with low-level features is proposed, reducing the computational complexity of the Transformer encoder and the entire network while preserving the original performance. The authors suggest that, in order to improve possible loss of details due to event camera captures, a possible solution could be to combine data from the event cameras with those from traditional cameras to exploit the advantages of both technologies.

The Multiscale Feature Extraction U-Net (MFEU-Net), presented in the sixth contribution, is a convolutional neural network designed for infrared small and dim target detection (ISDT). The network's architecture is based on the U-Net structure, which enables the fusion of multiscale information through skip connections. This allows the network to have different receptive fields at different levels, improving its ability to detect targets of varying sizes. In particular, MFEU-Net utilizes a combination of Residual U-block (RSU) blocks and Inception modules to extract multiscale information. Moreover, it incorporates a multidimensional attention mechanism, which operates on both channels and space. This mechanism enables the network to focus on the important areas of the image, enhancing detection in complex scenarios and reducing false alarms. The results show superior performance compared to other ISDT detection algorithms, achieving higher detection rates, lower false alarm rates, and higher IoU values on various datasets.

Contribution 7 presents another method aimed at ISDT detection, called the Group Regularized Principle Component Pursuit (GPCP), which is a group-regularized low-rank and sparsity decomposition model. This method addresses the limitations of traditional patch-based models, such as Infrared Patch Image (IPI), which are often sensitive to strong edges and background clutter due to their failure to consider the diversity of data structure. Unlike traditional methods that utilize a single low-rank constraint for the entire background component, GPCP employs a group low-rank constraint for background estimation. This approach allows for the use of different singular value thresholds for the low-rank decomposition of image groups corresponding to different complexities. Consequently, GPCP can better explore the local structure of the image and achieve a more accurate decomposition result. By dividing image data into groups based on brightness and clutter level, GPCP can more effectively suppress background clutter, particularly in areas with strong edges. This capability is demonstrated by experimental results on various detection scenes, where the GPCP achieves higher background suppression factors, as compared to other methods. Although GPCP utilizes Singular Value Decomposition (SVD) [8], in the same way as other patch-based models, its computational complexity is lower than its baseline model, IPI. This is attributed to the grouping strategy that divides image data into smaller groups, reducing the overall computational cost of SVD decomposition. By integrating group low-rank regularization with the sparsity constraint for background and target separation, GPCP improves the detection accuracy and overcomes the limitations of traditional decomposition-based methods. However, further research is needed to optimize the grouping criteria, further reduce the computational complexity, and explore more efficient background modeling methods.

On the same topic, the Background-Suppression Proximal Gradient (BSPG) method (contribution 8) enhances detection accuracy and computational efficiency in patch-based methods, particularly in suppressing strong background edges. It incorporates a novel continuation strategy during the alternating update of low-rank and sparse components, so as to suppress strong edges that are often mistaken as targets. This strategy retains more components during the low-rank matrix update while reducing the sparse matrix's update speed, enabling the model to mitigate the influence of strong edges. Approximate Partial SVD (APSVD) is employed to expedite the resolution of the low-rank sparse decomposition problem. This approach is more efficient than the full SVD because it leverages the fact that the soft-thresholding operation utilizes only a portion of the singular values. To further enhance processing speed, BSPG employs GPU multi-thread parallelism strategies to accelerate the construction and reconstruction of patch images, which can be divided into repetitive and independent subtasks. While efforts have been made to reduce computational complexity and exploit computation parallelism, further research may enhance the time performance and possible limitations due to data dependency.

Chen et al. (contribution 9) present a novel approach for anomaly detection in hyperspectral images (HSI), named the Multi-Dimensional Low-Rank (MDLR) method. Unlike previous tensor-based methods that mainly focused on the low dimensionality of the spectral dimension, MDLR considers low dimensionality along all dimensions of HSI: width, height, and spectrum. This three-dimensional analysis allows for more comprehensive background information extraction, improving the separation between background and anomalies. To impose low-rank constraints on the background tensor, MDLR utilizes Weighted Schatten p-norm Minimization (WSNM) on the slices of the f-diagonal tensor obtained through t-SVD decomposition. This approach allows for better preservation of the low-rank structure of the background, as compared to traditional nuclear norm minimization. MDLR utilizes a norm that penalizes the anomaly tensor, promoting joint sparsity in both the spectral and spatial domains. This takes into account the fact that anomalies tend to be spatially localized and exhibit low spectral density. Finally, the authors suggest that dimensionality reduction techniques could be integrated in the future to mitigate the computational complexity due to the t-SVD.

The authors of contribution 10 describe a tracking algorithm for sonar detection, called the Improved Multi-Kernel Correlation Filter (IMKCF), which is designed to detect and track weak underwater targets in complex marine environments. This problem is particularly challenging in environments with a low signal-to-reverberation ratio, where reverberation interference can make it difficult to distinguish the target from the background noise. Although the kernel correlation filter algorithm has been successful in visual tracking, it has not previously been applied to underwater target tracking. Using weighted information from historical samples to solve the coefficients of multiple nonlinear kernels adaptively, this method addresses the problem of limited robustness in tracking single features, taking full advantage of multiple complementary features. In particular, when a tracking result is deemed unreliable, a redetection module uses the historical reliability tracking results to drive a Kalman filter, which predicts the location of the target candidate. The use of multiple features, an adaptive update of the kernel coefficients, and the inclusion of a redetection module improve the method's performance over traditional tracking algorithms.

Contribution 11 proposes a method for suppressing faint/dim background stars in infrared, based on recursive moving target indication, to enhance the detection of space targets in optical image sequences. The suppression of stars with a low signal-to-noise ratio (SNR) has been largely ignored by previous research, but can negatively impact accuracy and real-time performance, particularly for time-before-space (TBS) detection methods. Unlike other TBS methods, which are closely tied to their corresponding target detection methods, the proposed method is versatile, and can be utilized as an efficient pre-processing step for most target detection and tracking methods. Additionally, a multi-frame adaptive threshold segmentation method is put forward to create an accurate star mask, enabling the real-time suppression of dim stars.

Contribution 12 presents a multi-stage joint detection and tracking model (MJDTM) for the real-time detection of space targets, such as space debris and satellites, using optical image sequences. The authors argue that although space target surveillance is critical for aerospace safety, it is becoming increasingly difficult, due to increasingly complex space environments. This article addresses the limitations of existing approaches that struggle to suppress background noise and mostly focus on single tasks, such as detection or tracking. The model uses an improved local contrast method to extract potential small space targets in optical image sequences. It uses a star target suppression method that exploits the differences in motion relative to Earth and real-time satellite attitude data to distinguish between space and star targets. The model is implemented on a specialized heterogeneous multi-core processing platform based on FPGA and DSP to meet real-time processing requirements. The authors note that although the proposed model shows improvements in detection accuracy while maintaining real-time processing speed, it may not perform well for targets with a low SNR. Additionally, the model relies on real-time satellite attitude data, which could be a limitation.

The SE-RRACycleGAN algorithm (contribution 13) introduces several improvements for single image deraining in an unsupervised manner. It contains an innovative Recurrent Rain-Attentive Module, designed to enhance the detection of rain-related information by concurrently considering both rainy and clean images, by not only incorporating spatial and channel attention blocks, but also employing an LSTM unit to capture spatiotemporal dependencies within images, facilitating the modeling of complex rain streak patterns that interact with the scene. The addition of Squeeze-and-Excitation (SE) blocks to the generator enables the model to learn discriminative features, facilitating the capture of intricate rain patterns and the representation of the underlying image structure. This capability is particularly significant for deraining tasks requiring both local and global features. Finally, to enhance the visual similarity between the generated image and the input image, the algorithm's loss function includes the content loss. These improvements allow SE-RRACycleGAN to surpass most state-of-the-art unsupervised methods, particularly on the Rain12 dataset and real rainy images, making it highly competitive compared to supervised techniques.

Conflicts of Interest: The author declares no conflicts of interest.

List of Contributions

1. Zhao, T.; Wang, Y.; Li, Z.; Gao, Y.; Chen, C.; Feng, H.; Zhao, Z. Ship Detection with Deep Learning in Optical Remote-Sensing Images: A Survey of Challenges and Advances. *Remote Sens.* **2024**, *16*, 1145.
2. Xia, L.; Wang, F.; Pang, C.; Li, N.; Peng, R.; Song, Z.; Li, Y. An Identification Method of Corner Reflector Array Based on Mismatched Filter through Changing the Frequency Modulation Slope. *Remote Sens.* **2024**, *16*, 2114.
3. Zhao, W.; Huang, L.; Liu, H.; Yan, C. Scattering-Point-Guided Oriented RepPoints for Ship Detection. *Remote Sens.* **2024**, *16*, 933.
4. Tian, Y.; Wang, X.; Zhu, S.; Xu, F.; Liu, J. LMSD-Net: A Lightweight and High-Performance Ship Detection Network for Optical Remote Sensing Images. *Remote Sens.* **2023**, *15*, 4358.
5. Jing, S.; Lv, H.; Zhao, Y.; Liu, H.; Sun, M. MVT: Multi-Vision Transformer for Event-Based Small Target Detection. *Remote Sens.* **2024**, *16*, 1641.
6. Wang, X.; Han, C.; Li, J.; Nie, T.; Li, M.; Wang, X.; Huang, L. Multiscale Feature Extraction U-Net for Infrared Dim- and Small-Target Detection. *Remote Sens.* **2024**, *16*, 643.
7. Li, M.; Wei, Y.; Dan, B.; Liu, D.; Zhang, J. Infrared Small Dim Target Detection Using Group Regularized Principle Component Pursuit. *Remote Sens.* **2024**, *16*, 16.
8. Hao, X.; Liu, X.; Liu, Y.; Cui, Y.; Lei, T. Infrared Small-Target Detection Based on Background-Suppression Proximal Gradient and GPU Acceleration. *Remote Sens.* **2023**, *15*, 5424.
9. Chen, X.; Wang, Z.; Wang, K.; Jia, H.; Han, Z.; Tang, Y. Multi-Dimensional Low-Rank with Weighted Schatten p-Norm Minimization for Hyperspectral Anomaly Detection. *Remote Sens.* **2024**, *16*, 74.
10. Yue, W.; Xu, F.; Yang, J. Tracking-by-Detection Algorithm for Underwater Target Based on Improved Multi-Kernel Correlation Filter. *Remote Sens.* **2024**, *16*, 323.
11. Zhang, L.; Rao, P.; Hong, Y.; Chen, X.; Jia, L. Infrared Dim Star Background Suppression Method Based on Recursive Moving Target Indication. *Remote Sens.* **2023**, *15*, 4152.
12. Su, Y.; Chen, X.; Liu, G.; Cang, C.; Rao, P. Implementation of Real-Time Space Target Detection and Tracking Algorithm for Space-Based Surveillance. *Remote Sens.* **2023**, *15*, 3156.
13. Wedajew, G.N.; Xu, S.S.-D. SE-RRACycleGAN: Unsupervised Single-Image Deraining Using Squeeze-and-Excitation-Based Recurrent Rain-Attentive CycleGAN. *Remote Sens.* **2024**, *16*, 2642.

References

1. Yuan, Q.; Shen, H.; Li, T.; Li, Z.; Li, S.; Jiang, Y.; Xu, H.; Tan, W.; Yang, Q.; Wang, J.; et al. Deep learning in environmental remote sensing: Achievements and challenges. *Remote Sens. Environ.* **2020**, *241*, 111716. [CrossRef]
2. Jat, M.K.; Garg, P.K.; Khare, D. Monitoring and modelling of urban sprawl using remote sensing and GIS techniques. *Int. J. Appl. Earth Obs. Geoinf.* **2008**, *10*, 26–43. [CrossRef]
3. Pajares, G. Overview and current status of remote sensing applications based on unmanned aerial vehicles (UAVs). *Photogramm. Eng. Remote Sens.* **2015**, *81*, 281–330. [CrossRef]

4. Tripicchio, P.; Unetti, M.; D'Avella, S.; Avizzano, C.A. Smooth Coverage Path Planning for UAVs with Model Predictive Control Trajectory Tracking. *Electronics* **2023**, *12*, 2310. [CrossRef]
5. Liu, C.A.; Chen, Z.X.; Yun, S.H.A.O.; Chen, J.S.; Hasi, T.; Pan, H.Z. Research advances of SAR remote sensing for agriculture applications: A review. *J. Integr. Agric.* **2019**, *18*, 506–525. [CrossRef]
6. Illingworth, J.; Kittler, J. A survey of the Hough transform. *Comput. Vision, Graph. Image Process.* **1988**, *44*, 87–116. [CrossRef]
7. Aleissaee, A.A.; Kumar, A.; Anwer, R.M.; Khan, S.; Cholakkal, H.; Xia, G.S.; Khan, F.S. Transformers in remote sensing: A survey. *Remote Sens.* **2023**, *15*, 1860. [CrossRef]
8. Andrews, H.; Patterson, C.L.I.I.I. Singular value decomposition (SVD) image coding. *IEEE Trans. Commun.* **1976**, *24*, 425–432. [CrossRef]

remote sensing

MDPI

Review

Ship Detection with Deep Learning in Optical Remote-Sensing Images: A Survey of Challenges and Advances

Tianqi Zhao [1,2], Yongcheng Wang [1,*], Zheng Li [1,2], Yunxiao Gao [1,2], Chi Chen [1,2], Hao Feng [1,2] and Zhikang Zhao [1,2]

[1] Changchun Institute of Optics, Fine Mechanics and Physics, Chinese Academy of Sciences, Changchun 130033, China; zhaotianqi22@mails.ucas.ac.cn (T.Z.); lizheng20@mails.ucas.ac.cn (Z.L.); gaoyunxiao19@mails.ucas.ac.cn (Y.G.); chenchi21@mails.ucas.ac.cn (C.C.); fenghao21@mails.ucas.ac.cn (H.F.); zhaozhikang20@mails.ucas.ac.cn (Z.Z.)

[2] University of Chinese Academy of Sciences, Beijing 100049, China

* Correspondence: wangyc@ciomp.ac.cn

Abstract: Ship detection aims to automatically identify whether there are ships in the images, precisely classifies and localizes them. Regardless of whether utilizing early manually designed methods or deep learning technology, ship detection is dedicated to exploring the inherent characteristics of ships to enhance recall. Nowadays, high-precision ship detection plays a crucial role in civilian and military applications. In order to provide a comprehensive review of ship detection in optical remote-sensing images (SDORSIs), this paper summarizes the challenges as a guide. These challenges include complex marine environments, insufficient discriminative features, large scale variations, dense and rotated distributions, large aspect ratios, and imbalances between positive and negative samples. We meticulously review the improvement methods and conduct a detailed analysis of the strengths and weaknesses of these methods. We compile ship information from common optical remote sensing image datasets and compare algorithm performance. Simultaneously, we compare and analyze the feature extraction capabilities of backbones based on CNNs and Transformer, seeking new directions for the development in SDORSIs. Promising prospects are provided to facilitate further research in the future.

Keywords: ship detection; deep learning; optical remote-sensing images; convolutional neural network; transformer

Citation: Zhao, T.; Wang, Y.; Li, Z.; Gao, Y.; Chen, C.; Feng, H.; Zhao, Z. Ship Detection with Deep Learning in Optical Remote-Sensing Images: A Survey of Challenges and Advances. *Remote Sens.* **2024**, *16*, 1145. https://doi.org/10.3390/rs16071145

Academic Editor: Paolo Tripicchio

Received: 2 February 2024
Revised: 18 March 2024
Accepted: 19 March 2024
Published: 25 March 2024

1. Introduction

Ship detection has important applications in areas such as fisheries management, maritime patrol, and maritime rescue. It contributes to ship traffic management and the maintenance of maritime safety. Therefore, ship detection has broad application prospects in civil and military fields [1]. The core objective is to determine the position of ships and identify their categories.

Optical remote-sensing images are captured via imaging distant ground surfaces using electro-optical sensors on aerial platforms and artificial Earth satellites [2]. With the rapid development of remote sensing, the resolution of optical remote-sensing images has continuously improved. They can provide more details, such as color and texture, as well as a comprehensive database for ship detection. Therefore, how to effectively utilize the existing favorable conditions to maximize the application benefits is an urgent issue to be solved.

Ship-detection methods have experienced two stages of development: rule-based classification and deep learning. In the early methods, the sliding window method was employed to systematically judge all potential areas. It relies on fixed-pattern approaches, such as geometric elements and manually designed features to extract ship features. However, the early methods may generate large amounts of redundant computations, which significantly impact detection speed. Additionally, the manually designed features lack the robustness to resist the interference from complex backgrounds. Therefore, early approaches struggled to meet the requirements of both performance and efficiency.

Compared with traditional methods, deep learning can extract features with stronger semantic information, and enable autonomous learning. In recent years, deep learning has developed rapidly. It has gradually migrated and innovated in the field of ship detection, achieving good results in ship detection in optical remote-sensing images (SDORSIs). However, influenced by factors such as complex maritime environments and ship characteristics, the results of SDORSIs based on deep learning still need improvement. Furthermore, achieving a balance between accuracy and speed is also one of the significant challenges.

At present, some reviews have been published in ship detection. Er et al. [3] collated a large number of popular datasets and reviewed the existing object-detection models. Joseph et al. [4] and Li et al. [5] systematically analyzed the typical methods at each stage of SDORSIs. Kanjir et al. [6] conducted a detailed analysis of the impact of environmental factors on SDORSIs. Li et al. [7] summarized the ship-detection techniques in synthetic aperture radar (SAR) images, along with their advantages and disadvantages.

Different from existing reviews, this paper primarily focuses on the challenges associated with SDORSIs. It aims to establish a refined classification system that progresses from the main problems to solutions, and provides readers with a comprehensive understanding of this field. Specifically, according to the characteristics of optical remote-sensing images and ships, we summarize the challenges as follows: complex marine environments, insufficient discriminative features, large scale variations, dense and rotated distributions, large aspect ratios, and imbalances between positive and negative samples, as shown in Figure 1. We take the problems as the driving force and conduct an in-depth analysis for each one. We comprehensively summarize the corresponding solutions and analyze the advantages and disadvantages of the respective solutions. In addition, we chronologically summarize ship-detection technologies, including methods based on manual feature extraction, convolutional neural networks (CNN) and Transformer. Finally, for the first time, we separate and aggregate ship information from comprehensive datasets. We also summarize and analyze the performance improvement effects of existing solutions, as well as compare the feature extraction capabilities of CNNs and Transformer. It is worth noting that the ship-detection methods and datasets discussed in this paper are only for nadir imagery.

To summarize, the main contributions are as follows:

- We systematically review ship-detection technologies in chronological order, including traditional methods, CNN-based methods, and Transformer-based methods.
- Guided by ship characteristics, we classify and outline the existing challenges in SDORSIs. based on CNNs and analyze their advantages and disadvantages.
- We summarize ship datasets and evaluation metrics. Furthermore, we are the first to separate and aggregate ship information from comprehensive datasets. At the same time, we compare and analyze performance improvement of the solutions and the feature extraction abilities of different backbones.
- Prospects of SDORSIs are presented.

The remaining components of this review are as follows: Section 2 chronologically reviews ship-detection technologies. Section 3 sorts out SDORSI challenges, summarizing improvement methods and their pros and cons. Section 4 summarizes ship datasets and evaluation metrics, comparing the performance of existing algorithms. Section 5 discusses the future development trends. Finally, Section 6 provides a summary of this paper. A research content diagram of this paper is shown in Figure 2.

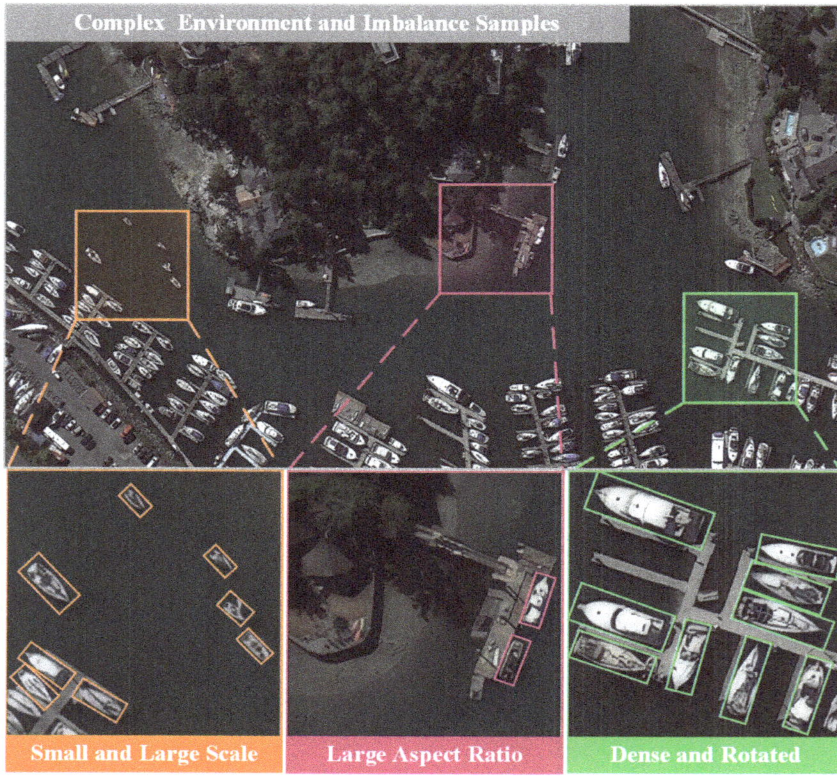

Figure 1. Main challenges in SDORSIs.

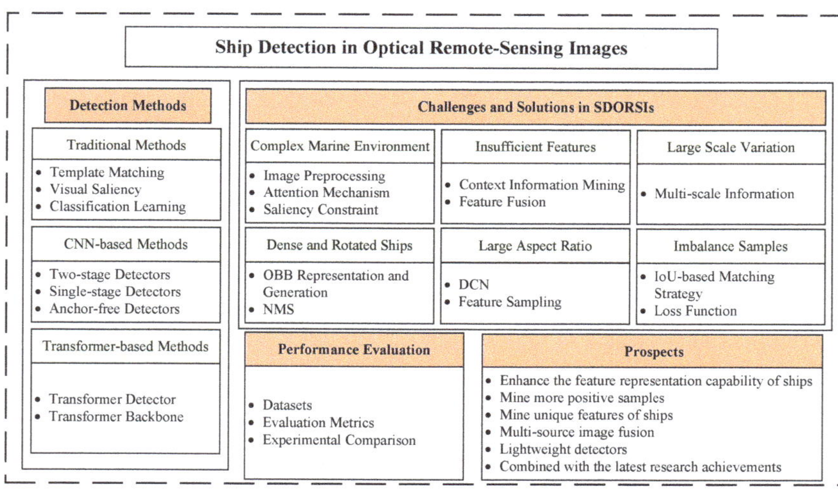

Figure 2. The research content of the paper.

2. Methods

Ship detection is an important research topic. In this section, we chronologically review the methods of ship-detection technologies, including traditional methods, CNN-based methods, and Transformer-based methods. The timeline of ship-detection methods is shown in Figure 3.

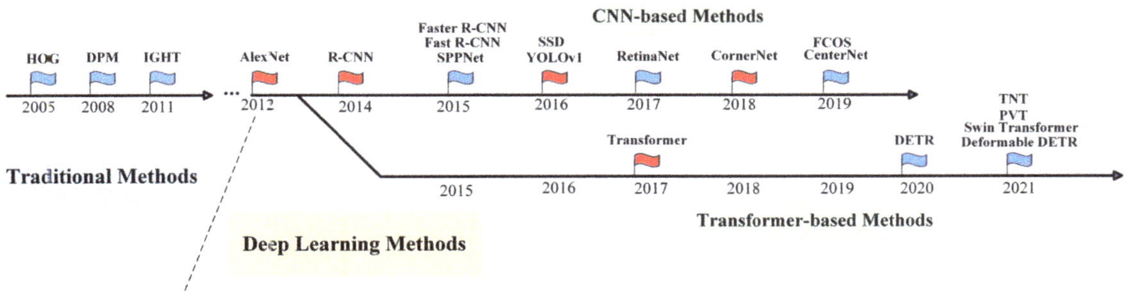

Figure 3. The timeline of ship-detection methods.

2.1. Traditional Methods

Most traditional ship-detection methods rely on geometric elements and manually designed features to locate ships within the background. Furthermore, they achieve good detection results in specific scenarios. The traditional methods are as follows: template matching, visual saliency, and classification learning.

2.1.1. Template-Matching-Based Method

Template-matching-based methods initially collect ship templates from various angles and environments. Then, they calculate the similarity between the templates and input images to determine the presence of ships. The methods primarily include global template matching, local template matching and feature-point matching. They are simple to operate and exhibit good detection performance in specific scenarios.

Xu et al. [8] proposed a method based on an invariant generalized Hough transform. It exhibited invariance to translation, scaling, and rotation transformation to extract ship shapes. Harvey et al. [9] performed rotational transformation on ship samples to increase the diversity of the templates. The method enhanced the generalization capability of the detector. He et al. [10] proposed a new method based on pose-weighted voting. It is robust in template matching. It further improved the performance.

Template-matching-based methods achieve good results in traditional ship detection. However, they require a lot of prior knowledge to build a template database and are sensitive to the environment, leading to a poor generalization capability.

2.1.2. Visual-Saliency-Based Method

The visual-saliency-based method prioritizes detector focus on regions with visually prominent features by analyzing image characteristics. The method first utilizes saliency detection algorithms to calculate the contrast between a certain region and its surrounding areas. Subsequently, it accomplishes the extraction of ship regions according to the results. The method achieves good results in ship detection.

Xu et al. [11] proposed a saliency model with adaptive weights for extracting candidate ships. The method can identify ships and suppress the interference from complex backgrounds effectively. Nie et al. [12] proposed a method that combined extended wavelet transform with phase saliency regions. It effectively achieved the extraction of regions of interests (ROIs) from complex backgrounds. Qi et al. [13] utilized the phase spectrum of Fourier transform to measure saliency, resulting in better identification of ships.

Bi et al. [14] employed a visual attention algorithm to highlight the positions of ships and provided their approximate regions.

The visual-saliency-based method finds extensive application in traditional ship detection. However, it has higher requirements for image quality. When ships are disturbed by cloud or the ship areas are large, it is difficult to obtain ideal results.

2.1.3. Classification-Learning-Based Method

Supervised machine learning is utilized in traditional ship detection. Thus, it is necessary to design suitable classifiers. The network trains classifiers by extracting ship features and labels to predict ships, and then establishes the relationship between ship features and ship categories. The main features include Scale Invariant Feature Transform (SIFT) features [15], histogram of oriented gradients (HOG) features [16], shape and texture features, etc. The commonly used classifiers are SVM, logistic regression, and AdaBoost.

Corbane et al. [17] utilized Radon transform and wavelet transform to extract ship features. Subsequently, the features were combined, employing logistic regression to accomplish ship detection. Song et al. [18] combined shape features with HOG features to construct a feature vector independent of size. Then, the method detected ships through AdaBoost.

However, the above manually designed features only utilize the low-level visual information, and cannot accurately express the complex high-level semantic information in the image. Moreover, because of the large amount of calculation in classifier detection, it is difficult to meet the application requirements of a real-time system.

2.1.4. Summary

In addition to the aforementioned methods, nearshore ship–land segmentation [19–22] and grayscale information [23] are also common traditional ship-detection methods. They have achieved some good results in specific scenarios. However, they are vulnerable to complex environment and heavily rely on prior knowledge. Additionally, the features are manually designed, and lack good robustness and generalization ability in traditional methods.

2.2. *CNN-Based Methods*

The CNN-based AlexNet [24] won the first prize in the 2012 ImageNet competition, marking the advent of the CNN era. Since then, CNN-based ship-detection technologies have developed rapidly and achieved excellent results. Compared with traditional methods, CNNs can automatically extract ship features without manual design. The features possess more advanced semantic information, contributing to the improvement of detection results. CNN-based methods are mainly divided into anchor-based methods and anchor-free methods, in which anchor-based methods include a two-stage detector and a single-stage detector.

2.2.1. Two-Stage Detector

The anchor-based detector locates ships by defining a set of anchor boxes. Anchor boxes are a set of rectangular bounding boxes with different sizes and aspect ratios, and evenly distributed at each pixel position in the image. The network predicts and adjusts the positions of anchor boxes to precisely cover the ships. Then, by further judging the category of ship, the network completes the detection. Anchor-based detectors include a two-stage detector and a single-stage detector. The two-stage detector divides the ship detection into two stages. The network first predicts all proposed regions containing ships in the first stage, and then modifies these regions to accurately locate and classify ships in the second stage, as shown in Figure 4. The two-stage detector has high accuracy and robustness. However, due to the refinement process of the proposed regions, the detection efficiency still needs further improvement.

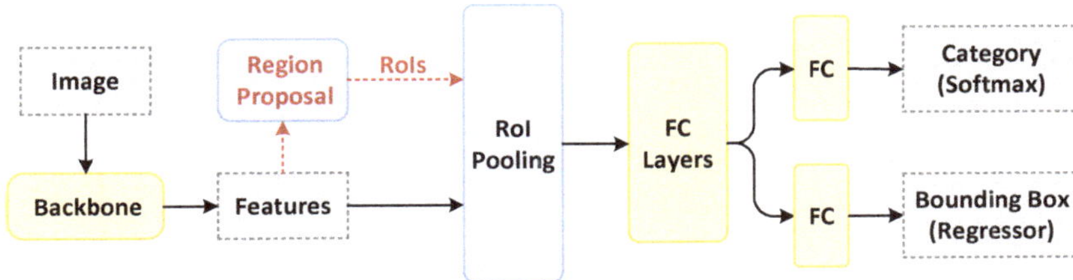

Figure 4. Schematic diagram of two-stage detector.

R-CNN: Girshick et al. [25] proposed R-CNN in 2014, marking the first attempt to incorporate deep learning into object detection. It significantly improves the results of detection. R-CNN uses the deep semantic features extracted by a CNN to replace the original shallow features (HOG, SIFT, etc.), further enhancing the discriminability of ships. Specifically, R-CNN first employs the Selective Search (SS) algorithm to divide the input image into approximately 2000 proposed regions, aiming to comprehensively cover the ships. Then, the network utilizes a CNN to extract features of each proposed region in turn, and sends them into the SVM classifier to obtain the detection results. At the same time, the network uses the regressor to adjust the positions of these proposed regions to accurately represent the ships.

SPPNet: Due to the size requirements of the classifier, R-CNN needs to standardize the sizes of proposed regions. It leads to the distortion and deformation of ships. To this end, He et al. [26] proposed SPPNet in 2015 which introduced spatial pyramid pooling (SPP). SPP divides the feature map into a fixed number of grids, and then performs max pooling for each grid. As a result, it can convert feature maps of arbitrary size into fixed-size feature vectors. Furthermore, compared with R-CNN, SPPNet significantly improves detection speed.

Fast R-CNN: In order to enable end-to-end learning for object detection and further improve the training speed, Girshick et al. [27] proposed Fast R-CNN in 2015. The network no longer needs to extract features for each proposed region separately; instead, it cleverly maps the regions to the feature map of the input image. At the same time, Fast R-CNN innovatively proposed ROI pooling, which can adapt the proposed regions of different sizes to a unified size to fit into the subsequent fully connected network. Fast R-CNN replaces the SVM classifier with a softmax layer. Furthermore, by designing a multi-task loss, the network is unified into a whole to train and optimize. Fast R-CNN greatly reduces training costs.

Faster R-CNN: Ren et al. [28] proposed Faster R-CNN, in which a region proposal network (RPN) replaced the SS algorithm for extracting ROIs. RPN proposed anchor boxes for the first time and it greatly improved the detection speed. Anchor boxes are evenly distributed at each pixel position of the feature map and fully cover it. Specifically, in the first stage, Faster R-CNN predicts the foreground and background probability of anchor boxes and performs rough boundary adjustments. Then, it maps anchor boxes to the feature map to support predictions in the second stage.

R-CNN improvement: Following the concept of R-CNN, some detectors improved from R-CNN have been successively proposed, such as Mask R-CNN [29], Cascade R-CNN [30], Libra R-CNN [31], Grid R-CNN [32], etc. These detectors improve Faster R-CNN from different aspects, aiming to meet the application requirements in various scenarios and achieving excellent detection results.

A two-stage detector achieves high precision and robustness in ship detection. For example, Guo et al. [33] proposed rotational Libra R-CNN to accurately predict the position of rotated ships. Li et al. [34] introduced the hierarchical selective filtering layer into Faster R-CNN to generate more accurate prediction boxes. Nie et al. [35] proposed a nearshore ship-detection method based on Mask R-CNN which introduced Soft-NMS to reduce the occurrence of missed detection.

2.2.2. Single-Stage Detector

In the single-stage detector, the results can be directly output after passing through a deep network, eliminating the time-consuming aspect of region proposals, as shown in Figure 5. Compared with the two-stage detector, the single-stage detector trades off the accuracy and efficiency. It is suitable for applications that require high real-time accuracy and high efficiency.

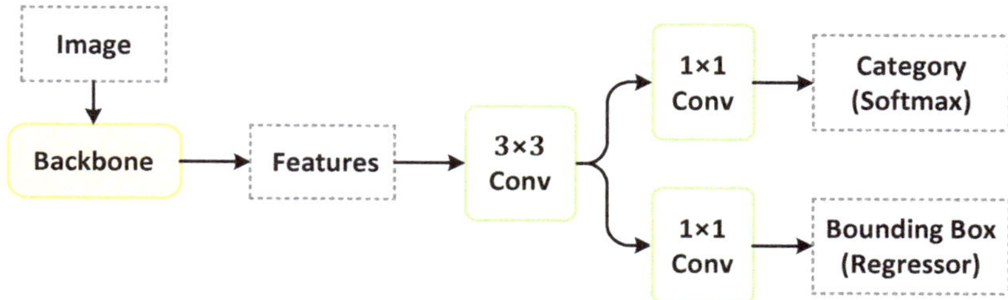

Figure 5. Schematic diagram of single-stage detector.

YOLO: Redmon et al. [36] first proposed the representative of single-stage detectors in 2016, known as You Only Look Once (YOLO). The image only passes through the CNN, and the ship category and location can be generated directly. Specifically, YOLOv1 divides the input image into 7×7 grids, and each grid generates two prediction boxes to predict the ship category and location. YOLO reduces the complexity of the algorithm and increases the detection speed. However, YOLOv1 can only detect one ship per grid, resulting in poor detection performance for dense ships. Therefore, many researchers have made a series of improvements on the basis of YOLOv1, including data preprocessing, feature extraction, and anchor box generation [37–41]. These methods have elevated the accuracy of single-stage detectors to a new level while maintaining YOLO's high detection speed, achieving further balance in performance. To date, the latest algorithm in the YOLO series, YOLOv8, has been published in GitHub. It incorporates innovative improvements over YOLOv5, including backbone, decoupling detection head, loss function, and sets the algorithm in an anchor-free form. YOLOv8 has the advantages of light weight and high efficiency.

SSD: Liu et al. [42] combined the regression concept of YOLO with the anchor mechanism of Faster R-CNN, proposing the SSD in 2016. SSD sets anchor boxes with different aspect ratios at each pixel of the feature map for predicting the classification and regression of ships. At the same time, multi-scale detection technology is introduced in SSD. By setting up six scale feature maps, the model gains the capability to detect ships at multiple scales, especially small ones. SSD provides a new approach for the design of single-stage detectors by incorporating the anchor mechanism, which can achieve effective coverage of ships.

RetinaNet: During the training process, anchor mechanisms may lead the model to excessively focus on the background regions where negative samples are located, thereby affecting detection performance. For this reason, Lin et al. [43] proposed RetinaNet in 2017, and Focal Loss effectively addresses the issues of positive and negative samples imbalance as well as difficulty imbalance. By utilizing Focal Loss, the network achieves weighted positive samples through balanced cross-entropy, enhancing the ability to detect positive samples. Simultaneously, the network maps the confidence of each category to a weight coefficient added to the loss, improving the ability of the network to detect difficult samples. The proposal of RetinaNet makes it possible to imagine that the single-stage detector can compete with the two-stage detector in detection accuracy.

There are strict limitations on the detection speed due to the real-time requirements of monitoring the sea situation. Therefore, more and more researchers are committed to deep development of single-stage detectors to meet the requirements of ship detection. For example, Patel et al. [44] compared the detection capabilities of different versions of the

YOLO algorithm. Gong et al. [45] integrated the shallow features of SSD and introduced context information, improving the detection accuracy. Wu et al. [46] employed RetinaNet as the backbone and proposed the hierarchical atrous spatial pyramid to obtain larger receptive fields.

In summary, anchor-based detectors include two-stage detectors and single-stage detectors. Anchor boxes fully cover the image per pixel, significantly enhancing detection accuracy. However, the drawbacks of the anchor mechanism are as follows: Firstly, the ship regions occupy only a small portion of an image, resulting in the majority of anchor boxes being assigned to irrelevant backgrounds. Therefore, the massive tiling of anchor boxes introduces redundant computations. Secondly, anchor boxes require setting hyperparameters, and unreasonable configurations may lead to performance degradation. Finally, predefined aspect ratios result in poor performance when matching irregularly shaped ships, causing the detector to lack generalization.

2.2.3. Anchor-Free Detector

The anchor-free detector breaks limitations of the anchor-based detector, providing a new reference path for ship detection. The anchor-free detector uses keypoints instead of anchor boxes to detect ships, which enhances the ability to process ships of different shapes, as shown in Figure 6. It improves the generalization of the model.

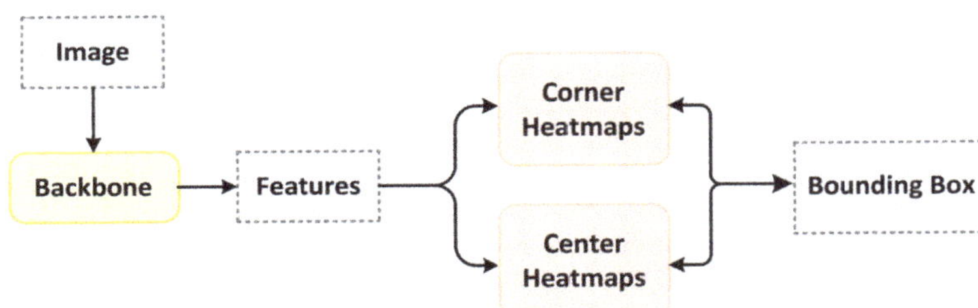

Figure 6. Schematic diagram of anchor-free detector.

CornerNet: Law et al. [47] proposed CornerNet which was the first to implement the anchor-free detector in 2018. It adopts the keypoint detection method and proposes corner pooling. By predicting the top-left and bottom-right points, Corner pooling generates prediction boxes to determine the ship positions. It significantly reduces the amount of calculation and improves the speed of detection.

CenterNet: Inspired by CornerNet, Zhou et al. [48] proposed CenterNet in 2019. CenterNet takes the peak points of the heatmap generated by the image as the center points of ships. Then, it regresses the width, height, weight, and other information of ships based on the center points to generate prediction boxes.

FCOS: Tian et al. [49] proposed an anchor-free detector using pixel prediction in 2019, named FCOS. It introduces center-ness to measure the distance between predicted pixels and the actual center of ships. Center-ness effectively inhibits the generation of low-quality prediction boxes.

An anchor-free detector has the obvious advantage of alleviating the imbalance between positive and negative samples. Therefore, it achieves excellent performance in ship detection. For example, Yang et al. [50] improved the weight assignment method of center-ness in FCOS, making it better aligned with the shape of ships. It more effectively suppressed the generation of low-quality prediction boxes. Zhuang et al. [51] proposed CMDet based on FCOS to detect rotated ships. Zhang et al. [52] introduced the recall-priority branch based on CenterNet to alleviate the occurrence of missed detection.

However, due to the lack of anchor boxes, the capability of ship detection completely depends on the recognition of keypoints. Anchor-free detector exhibits poor performance for ships with ambiguous keypoints. Moreover, it cannot effectively handle overlapping or occluded ships.

2.2.4. Summary

Compared with traditional ship-detection methods, CNN-based methods demonstrate superior robustness and accuracy. Currently, CNN-based methods have become the primary methods for ship detection. According to the specific requirements, different detectors are adopted in different ship detections. For high-precision detection, two-stage detectors are considered more suitable. Furthermore, single-stage detectors are more suitable for scenes with high requirements for real-time performance. In addition, anchor-free detectors can effectively address problems such as imbalance between positive and negative samples, and redundant calculations in anchor-based detectors.

2.3. Transformer-Based Methods

Vaswani et al. [53] proposed a simple network architecture, Transformer, and implemented efficient natural language processing (NLP) in 2017. Transformer abandons traditional recurrent and convolutional structures, adopting an encoder–decoder structure based on multi-head self-attention mechanism, as shown in Figure 7a,b. In this process, the encoder maps input sequences into a continuous representative sequence through global attention operations. Furthermore, the decoder is auto-regressive. It is able to better capture long-range contextual relationships by interacting with the output of the encoder during sequence generation. Furthermore, the parallel computing capability of Transformer greatly enhances training speed. Benefiting from the satisfactory performance in NLP, researchers are attempting to explore its applications in computer vision. In recent years, Transformer has been extended to object detection and has made great contributions. According to differences in model design, it can be divided into Transformer-based detector and Transformer-based backbone.

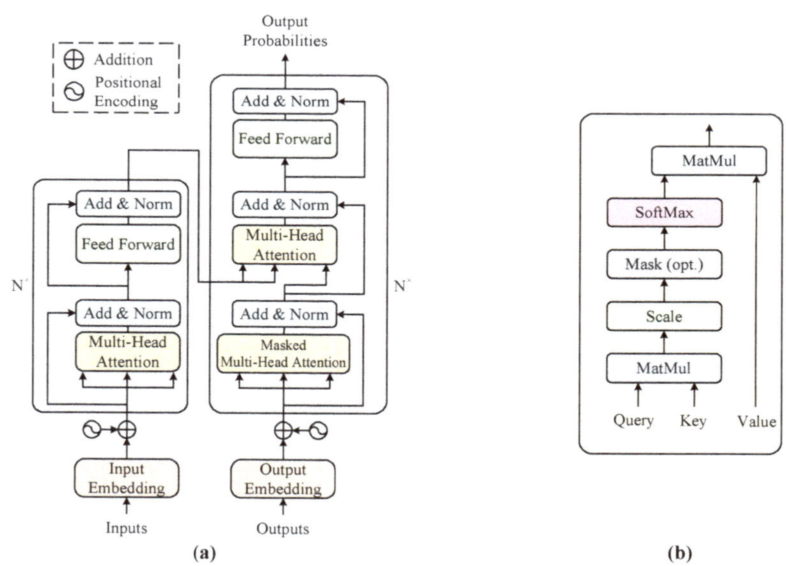

Figure 7. Schematic diagram of Transformer. (**a**) Encoder–decoder structure. (**b**) Self-attention mechanism.

2.3.1. Transformer-Based Detector

DETR: Carion et al. [54] proposed DETR, which first applied Transformer to object detection in 2020. DETR views ship detection as a set prediction problem. Specifically, DETR first extracts feature maps using CNN. Then, they are converted into one-dimensional vectors and fed into the encoder along with positional codes. Afterward, the encoder sends the output vectors into the decoder along with object queries. Finally, the decoder sends the output to a shared feed-forward network to obtain the detection result. DETR matches the predicted object queries with ships, seeking an optimal matching scheme with the lowest cost. Therefore, DETR circumvents the NMS procedure and achieves end-to-end detection.

Deformable DETR: The high computational cost and spatial complexity of the self-attention mechanism result in a slow convergence speed of DETR. The resolution that DETR can process is limited, and it is not ideal for detecting small ships. To address it, Zhu et al. [55] incorporated the concepts of deformable convolution and multi-scale features into DETR, proposing Deformable DETR. Furthermore, the deformable attention module was designed to replace the traditional attention module. It allows each reference point to focus only on a set of sampling points in its neighborhood, and the positions of these sampling points are learnable. It reduces the computational burden in irrelevant regions and decreases training time. At the same time, the introduction of multi-scale feature maps realizes the hierarchical processing for ships of different sizes. Deformable DETR is capable of effectively performing detection tasks of different scales.

2.3.2. Transformer-Based Backbone

Swin Transformer: Liu et al. [56] proposed Swin Transformer, attempting to combine the prior knowledge of a CNN with Transformer. Swin Transformer employs the idea of the local context in a CNN, where the model calculates self-attention only within each local window. It significantly reduced the sequence length and improved computational efficiency. Swin Transformer also introduced the idea of translational invariance from CNNs. The shifted window approach facilitates information interaction between adjacent windows, achieving the goal of global information extraction. It first demonstrated that Transformer can be used as a general backbone in computer vision.

PVT: Wang et al. [57] proposed a Transformer backbone suitable for dense object detection, named PVT. By incorporating the pyramid structure from CNN, PVT can extract better multi-scale feature information. Meanwhile, compared with traditional multi-head attention, spatial reduction attention ensures that PVT can obtain high-resolution feature maps while reducing computational cost.

TNT: Transformer struggles to capture the correlation within patches, which leads to the omission of small objects. To this end, Han et al. [58] proposed a Transformer in Transformer (TNT) architecture. TNT further divides each patch and then computes self-attention within each patch. As a result, TNT cannot only model global information, but also better capture local information, extracting more detailed features.

2.3.3. Summary

The issues of high parameters and computational consumption in Transformer greatly restrict its practical application scenarios. Furthermore, the high data requirements make it challenging to achieve satisfactory results on small datasets. These factors limit its development in ship detection. However, compared to CNN-based methods, Transformer can thoroughly explore long-range dependencies in targets, and effectively capture global features. It increases the identifiable information of ships from a global perspective. Transformer has significant potential for development in ship detection. However, there is currently a lack of research on the optimization of ship characteristics, which may be a key hindrance to the development of this field. Therefore, addressing the above issues and fully leveraging the advantages of Transformer in ship detection require more efforts in the future. Furthermore, in order to facilitate the comparison of the three methods, we summarize them and their advantages and disadvantages in Table 1.

Table 1. Methods of ship detection and main advantages and disadvantages.

Methods		Advantages	Disadvantages	References
Traditional Methods	Template Matching	It is simple to operate.	It requires a lot of prior knowledge and is sensitive to the environment.	[8–10]
	Visual Saliency	It calculates the contrast between a certain region and its surrounding areas to extract regions.	It has higher requirements for image quality.	[11–14]
	Classification Learning	It establishes the relationship between ship features and ship categories.	The manually designed features only utilize the low-level visual information and cannot express the complex semantic information.	[17,18]
CNN-based Methods	Two-stage Detector	It divides the ship detection into two stages and has high accuracy and robustness.	Detection efficiency of two-stage detector may be lower than single-stage detector.	[25–32]
	Single-stage Detector	It is suitable for the applications that require high real-time accuracy and high efficiency.	Detection accuracy of single-stage detector may be lower than two-stage detector.	[36–43]
	Anchor-free Detector	It uses keypoints instead of anchor boxes to detect ships which improves the generalization of the model.	It exhibits poor performance for ships with ambiguous keypoints.	[47–49]
Transformer Methods	Detector Backbone	It can explore long-range dependencies in targets, and effectively capture global features.	The high data requirements make it challenging to achieve satisfactory results on small datasets.	[54–58]

3. Challenges and Solutions in Ship Detection

Due to the significant differences between optical remote-sensing images and natural images, and variations in the features of ships compared with other targets, applying classical object detection algorithms directly results in low detection accuracy and missed detection. Therefore, this section summarizes the reasons for the low accuracy in SDORSIs, including complex marine environments, insufficient discriminative features, large scale variations, dense and rotated distributions, large aspect ratios, and imbalances between positive and negative samples. Furthermore, the corresponding solutions based on CNNs and their advantages and disadvantages are analyzed in detail. Challenges and solutions are shown in Figure 8.

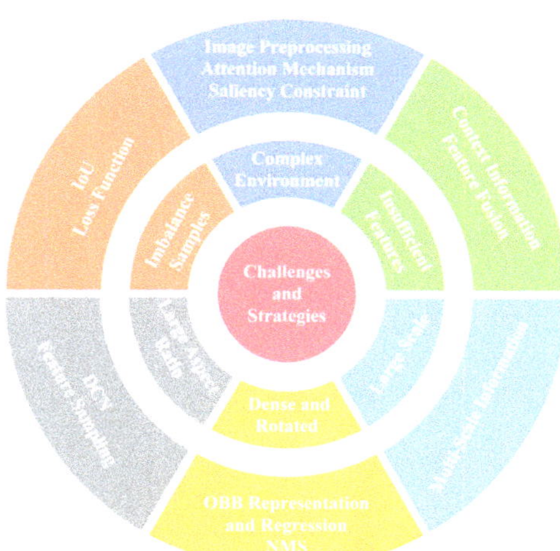

Figure 8. Challenges and solutions for improvement.

3.1. Complex Marine Environments

Optical remote-sensing images can provide rich information, but they are susceptible to factors such as light and weather. These adverse background factors bring significant interference to ship detection, resulting in missed or false detection. At the same time, there are usually only a few ships in remote-sensing images of the sea, while the background occupies the majority of the area. The extreme imbalance phenomenon causes the detector to overly focus on background regions, but ignores the effective extraction of ships. Therefore, it is a necessary processing strategy to guide the network to pay more attention to ships and ignore irrelevant background in SDORSIs. At present, there are several main solutions for complex backgrounds: image preprocessing, attention mechanisms, and salience constraints.

3.1.1. Image-Preprocessing-Based Method

Image preprocessing is one of the feasible methods to deal with complex background. It primarily suppresses the expression of background through prior information during the image preparation stage to reduce the contribution of the background, allowing the model to focus on learning ship features. Through the method of active guidance, image preprocessing greatly reduces the impact of complex background in SDORSIs.

Yu et al. [59] developed an embedded cascade structure. It removes the majority of irrelevant background in advance, and selects regions containing ships for training. The method alleviates the imbalance of the foreground and background, and reduces the interference of the background. Zheng et al. [60], Song et al. [61], and Yang et al. [62] designed image dehazing algorithms to restore images, addressing the issues of cloud occlusion in ocean scenes. Dehazing algorithms improve the image quality and are beneficial for enhancing detection accuracy. However, Li et al. [63] argued that existing dehazing algorithms did not distinguish between blurry and clear images. Excessive deblurring of clear images could lead to degrading image quality. Therefore, they proposed the blurred classification and deblurring module which obtained clear images and improved detection accuracy.

However, it should be noted that some image preprocessing methods require processing images independently based on prior knowledge, lacking generalization ability. Furthermore, some methods may introduce more convolutional layers which require additional training for the network.

3.1.2. Attention-Mechanism-Based Method

Due to the bottleneck in information processing, human cognitive systems always tend to selectively focus on important information and ignore secondary information. The core idea is to weight different parts of the input sequence according to the importance of features, and enhance the contrast between ships and the background at the feature level. Without human intervention, the attention mechanism operates end-to-end. Attention-mechanism-based methods generate prominent feature maps, which effectively highlight ship regions and suppress the expression of irrelevant background regions. Therefore, introducing attention mechanism is one of the effective methods to deal with complex background issues.

Li et al. [64] introduced the channel attention mechanism, as shown in Figure 9b, into multiple receptive field fusion modules to suppress irrelevant background information. Wang et al. [65] attached the channel attention mechanism to the backbone to enhance the capability of extracting ship features in complex backgrounds. Hu et al. [66] and Qin et al. [67] incorporated both a spatial attention mechanism, as shown in Figure 9a, and a channel attention mechanism to highlight the ships. Chen et al. [68] designed a coordinate attention module. It effectively combines spatial attention and channel attention to enhance the ability of ship feature representation. Qu et al. [69] added a convolutional attention module to YOLOv3, as shown in Figure 9c, highlighting ship features and improving detection accuracy.

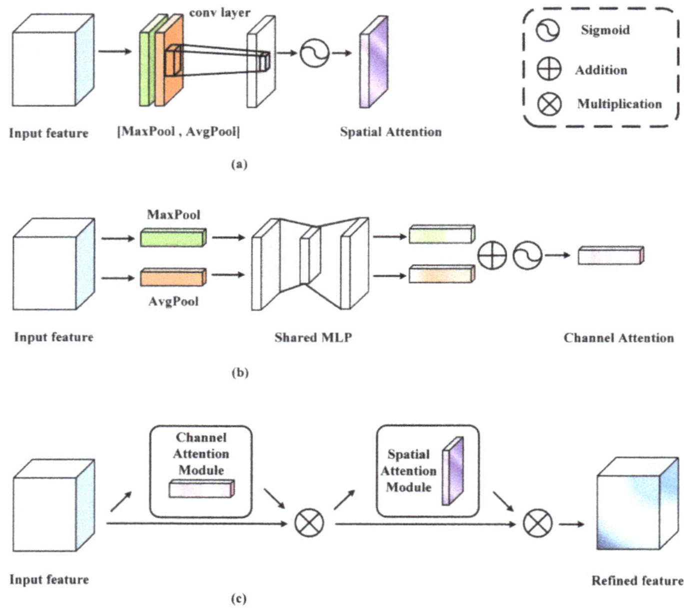

Figure 9. Schematic diagram of attention mechanisms. (**a**) Spatial attention mechanism. (**b**) Channel attention mechanism. (**c**) Convolutional block attention module.

However, an attention mechanism increases the complexity of network computing. Furthermore, if the network overly relies on it in SDORSIs, it may lead to a decreased ability to generalize.

3.1.3. Saliency-Constraint-Based Method

The saliency-constraint-based method adopts the idea of multi-task learning, constraining the network to focus on ships by designing the loss function, as shown in Figure 10. Firstly, the method utilizes prior information to create significance maps as labels. The values on labels reflect the importance of pixel positions, and the higher value indicates the higher attention of the ship. Then, a saliency prediction branch is added to output the predicted saliency maps. Through pixel-level loss constraints, the model pays more attention to ship regions during the training phase, thereby suppressing the impact of the background. The method enables the network to prioritize focusing on saliency regions with obvious visual features, and ignore the irrelevant background. It can narrow down the detection range and enhance detection efficiency.

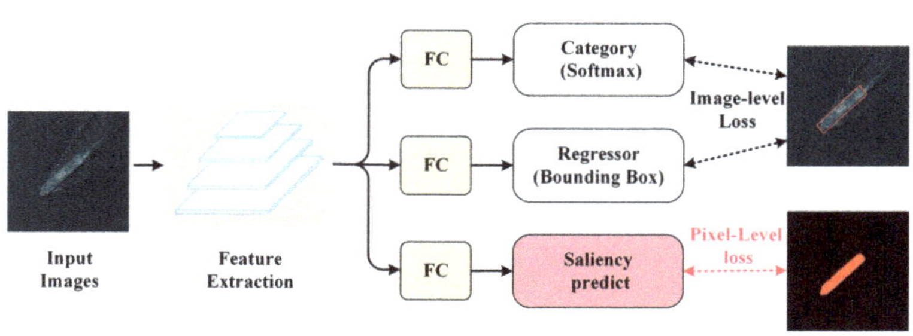

Figure 10. Schematic diagram of saliency constraint, the red square is the saliency constraint.

Ren et al. [70] added a saliency prediction branch to introduce saliency information with stronger foreground expression ability in SDORSIs. It improves the ship detection capability in complex environments. Chen et al. [71] designed a degradation reconstruction enhancement network. By selective degradation, the network obtains "pseudo saliency maps". Then, the maps are used to guide the network to focus more on ship information and ignore the irrelevant background in the training stage.

Visual saliency employs pixel-level supervision to guide the network and greatly addresses the challenge of complex backgrounds in SDORSIs. However, the generation of saliency maps requires clearer spatial distribution, which has demands on the details and resolution of feature maps. Furthermore, the weight of multi-task loss needs to be adjusted manually.

3.1.4. Summary

Complex environmental interference is one of the main challenges for the difficult improvement of SDORSI results. The existing research indicates that optimization strategies such as image preprocessing, attention mechanisms, and saliency constraints contribute to improving detection performance. The essence of these methods is to highlight ships and make the network focus on ship features. However, the methods are inevitably associated with some disadvantages. Simple methods are not suitable for more complex environments. Furthermore, paying too much attention to the background of a specific dataset leads to overfitting, hindering the network from generalizing. In order to provide readers with a more intuitive understanding, the methods and the main advantages and disadvantages in complex marine environments are shown in Table 2.

Table 2. Methods and main advantages and disadvantages of complex marine environments.

Methods		Advantages	Disadvantages	References
Image Preprocessing	Exclude Background	It filters out untargeted images in advance.	Introducing convolutional layers requires additional training for the network.	[59]
	Dehazing Algorithm	It improves the quality of the image by eliminating the impact of clouds and fog.	Excessive dehazing may result in information loss. Simple algorithms are not suitable for complex scenes.	[60–63]
Attention Mechanism	Channel Attention Mechanism	It adjusts channel weights dynamically to focus on ships.	It has limitations in extracting global information.	[64–67]
	Spatial Attention Mechanism	It highlights important information in the image to focus on ships.	It may excessively focus on local structures, leading to a decreased ability to generalize.	[66,67]
	Convolutional Attention Module	It adjusts convolutional kernel weights dynamically at different positions to focus on ships.	Introducing additional computation.	[68,69]
Saliency Constraint	Saliency Constraint	It uses the concept of multi-task learning and pixel-level supervision to focus on ships.	It has a high requirement for the resolution of the images. The weight needs to be adjusted manually.	[70,71]

3.2. *Insufficient Discriminative Features*

Unlike occupying a large proportion in natural images, ships usually cover only a few dozen pixels in optical remote-sensing images, which makes them challenging to detect. As a deep network continuously compresses and extracts features, the crucial information of small ships is easily suppressed. Therefore, insufficient discriminative features of small ships are the main reason for missed detection. It remains a challenge in ship detection, and has not been effectively solved. Currently, context information mining and feature fusion are effective methods to improve the accuracy of small ship detection. These methods focus on extracting effective information from the surroundings or inside of ships to enhance the feature expression ability.

3.2.1. Context Information Mining-Based Method

Context information mining refers to enhancing the information processing ability of the network by obtaining the environment information around the ship. The information is closely related to ships and helps to identify small ships with network uncertainty, thereby improving the accuracy and robustness. When detecting small ships, exploring contextual information that is closely connected with the ship can help obtain contents conducive to detection. It can alleviate the issue of insufficient discriminative features of small ships and improve the detection accuracy.

Ship-wake-based method: Ships navigating at sea usually occupy only a few dozen pixels in optical remote-sensing images, but their wake often reaches hundreds of pixels, as shown in Figure 11a. Wake refers to the visual trace created by the movement of ships, such as waves or disturbances on the sea. It is closely associated with ships and provides crucial contextual information, which can be used to enhance ship detection performance. Liu et al. [72], Xue et al. [73], Liu et al. [74], Liu et al. [75], and Liu et al. [76] introduced wake as contextual information. By employing a cascaded method of ships and wake, the network achieved excellent performance.

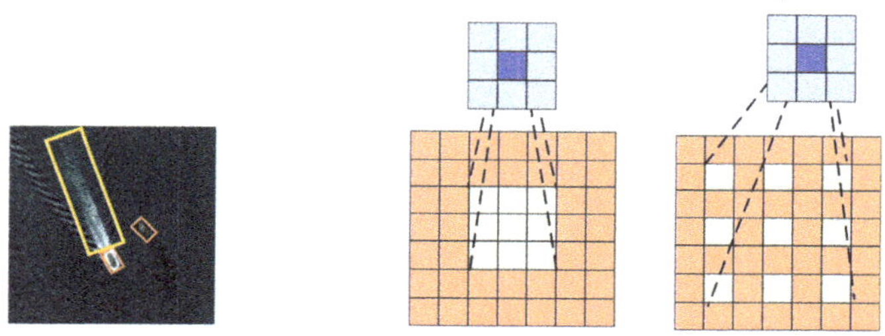

(a) (b)

Figure 11. Schematic diagram of context information mining. (**a**) Comparison between the ship and its wake. (**b**) Comparison between standard convolution (kernel size = 3, rate = 1) and dilated convolution (kernel size = 3, rate = 2).

Dilated-convolution-based method: Increasing the receptive fields while maintaining resolution can help obtain more contextual information, helping the network to detect small ships better. Using a large kernel to extract information is regarded as an effective method for increasing the receptive fields. However, the parameters of it increase the computational burden. Therefore, the dilated convolution is developed as the context information mining method, as shown in Figure 11b. Xu et al. [77], Chen et al. [78], and Zhou et al. [79] used dilated convolution instead of regular convolution to extract ship features. Dilated convolution can capture more context information without bringing too many parameters, introducing more references in SDORSIs.

It is worth noting that the extraction of context information requires a balance, as introducing irrelevant information may harm the performance. Furthermore, because of gaps in the dilated convolution kernel, the feature extraction may result in discontinuity of information. Therefore, the network needs to stack multiple dilated convolutions to ensure the integrity of feature.

3.2.2. Feature-Fusion-Based Method

A CNN has a hierarchical structure, and generates features with multiple resolutions. Shallow features contain more detailed information, such as ship boundary, which is beneficial for ship localization. Furthermore, deep features contain more semantic information, such as the discriminant parts of the ship, which is more conducive to ship classification. Feature fusion can obtain rich semantic information and localization information on a feature map to enhance the discriminative features of small ships.

Liu et al. [80] integrated three feature maps of different sizes in the same channel dimension, enhancing discriminative features. Li et al. [81] first proposed a pooling-based method to integrate features, fully leveraging the advantages of features with different resolutions in ship detection. Tian et al. [82] designed a dense feature reconstruction module. By integrating high-resolution detailed information with low-resolution semantic information, small ship features were enhanced. Qin et al. [83] aggregated features based on residual network to improve the accuracy of ship detection. Han et al. [84] proposed a dense feature fusion network. It effectively integrated information without consuming additional memory space. Wen et al. [85] proposed a method of cross-skip connection to flexibly fuse information.

Feature fusion is an effective method to detect insufficient discriminative features of small ships. However, it increases the computation and model complexity, which are detrimental to detection speed. Furthermore, improper fusion methods may result in loss or confusion of information.

3.2.3. Summary

Insufficient discriminative features are a major challenge in SDORSIs, and enhancing the feature representation ability of ships is a key technology to alleviate this problem. The experiments indicate that methods of context information mining and feature fusion can enhance the discriminative ability of small ships, further improving the detection effect. However, the significant performance gap between small and large ships indicates that there is still considerable room for improvement. Specifically, the unfairness in Intersection over Union (IoU) evaluation and the indifference in regression loss contribute to the disregard of small ships in detection. Therefore, in order to effectively address this challenge, increasing the attention of small ships detection is the key point for future work. The methods and the main advantages and disadvantages of insufficient discriminative feature are shown in Table 3.

Table 3. Methods and main advantages and disadvantages of insufficient discriminative feature.

Methods		Advantages	Disadvantages	References
Context Information Mining	Ship Wake	The wake is closely related to the ship and can provide additional discriminative information.	Excessive context information may compromise detection performance.	[72–76]
	Dilated Convolution	It enhances the receptive field without introducing additional parameters while maintaining resolution.	There are gaps in the dilated convolution kernel, which leads to information discontinuity.	[77–79]
Feature Fusion	Feature Fusion	Integrating information from feature maps with different resolutions can extract rich semantic information and localization information to enhance information interaction capabilities.	Improper fusion methods may result in loss or confusion of information.	[80–85]

3.3. Large Scale Variation

Compared with natural images, the scale variation of ships in optical remote-sensing images is larger. With the down sampling of optical remote-sensing images, the spatial resolution decreases. The information of small ships may vanish in deep features, causing the detector to fail to identify crucial discriminative features. Therefore, only relying on single-scale information for detecting ships of various scales cannot achieve desirable results. The current research challenge lies in achieving satisfactory detection results for ships with different scales using the same network. At present, the introduction of multi scale information is an effective method to address this issue. The essence is to perform hierarchical processing for large, medium, and small ships.

3.3.1. Multi-Scale Information-Based Method

Due to the absence of excessive down sampling in shallow features, important high frequency information can be preserved, such as texture, color, and edges. The information helps with the prediction of small ships. After multiple down samplings, the deep features can obtain larger receptive fields, which is helpful for the prediction of large ships. Therefore, utilizing multi-scale feature maps can better complete the fine-grained detection of different scales. However, they are independent from each other in the early prediction of ships, lacking mutual correlation. Then, a multi-scale information-based method based on feature fusion is proposed to alleviate this problem. It enhances the information interaction ability of different scale feature maps, and is widely applied in ship detection.

Feature Pyramid Network (FPN) [86] is a representative method that uses feature fusion to enhance multi-scale information. Through the lateral connection and the top-down pathway, a high-level feature transfers downward and fuses with a low-level feature as shown in Figure 12. It combines the semantic information and positional information of feature maps, improving the representational ability of multi-scale information. Therefore, FPN can more comprehensively detect multi-scale ships. Tian et al. [87] and Ren et al. [70] proposed a multi-node feature fusion method based on FPN. It fully integrates information from feature maps at different scales, and improves the detection ability of multi-scale ships. Si et al. [88] and Yan et al. [89] used an improved bidirectional FPN to enhance the interactive ability of multi-scale features. Li et al. [90] and Yang et al. [50] improved FPN using the Network Architecture Search algorithm (NAS). It can learn features adaptively and choose more suitable fusion paths to enrich information. Chen et al. [91] combined FPN with the recursive mechanism to further enhance the representational capacity of multi-scale information. Xie et al. [92] proposed an adaptive pyramid network. It can enhance important features, improving detection accuracy. Zhang et al. [93] proposed SCM, which addresses the issue of channel imbalance during the feature fusion. Guo et al. [33] proposed Balanced Feature Pyramid (BFP). It adjusts multi-scale feature maps to the same medium size by interpolation and down sampling. Then, the balanced semantic features are generated by scaling and refining the features. The method alleviates the impact of different size feature maps during fusion. Guo et al. [94] improved BFP and proposed Adaptive Balanced Feature Integration (ABFI). The module can assign different weights to the different feature maps during feature fusion, enabling more accurate detection.

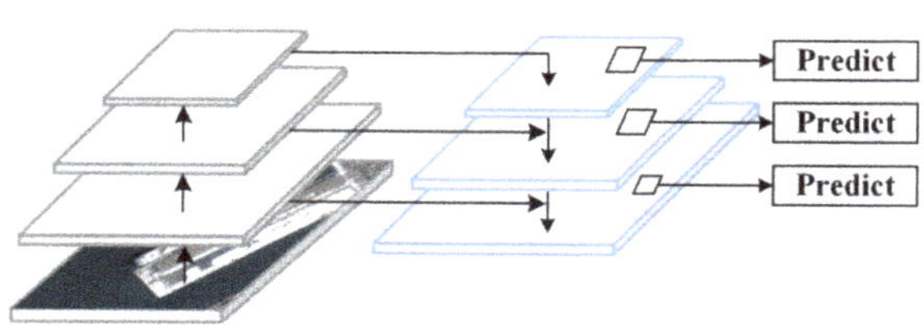

Figure 12. Schematic diagram of FPN.

In conclusion, addressing the multi-scale challenge in SDORSIs requires a comprehensive consideration of factors such as scale differences, algorithm design, FPN construction, and so on. It is essential to ensure that the network can accurately and efficiently detect ships of different scales.

3.3.2. Summary

The large-scale variation in SDORSIs is one of the key factors limiting the improvement of performance. Introducing multi-scale information is one of the commonly used methods. Simultaneously, the key factor contributing to the low detection accuracy of large-scale targets detection is the poor performance in small vessels. In the future, enhancing the feature representation capability of small ships and designing multi-branch detection networks are also strategies to address this issue. Furthermore, the research trend lies in how to enhance the light weight of the network and reduce the application threshold in portable mobile devices while ensuring the accuracy of multi-scale ship detection.The methods and the main advantages and disadvantages of large scale variation are shown in Table 4.

Table 4. Methods and main advantages and disadvantages of large scale variation.

Methods		Advantages	Disadvantages	References
Multi-Scale Information	FPN and Improvements	It enables the model to handle ships of different scales through the pyramid structure and the feature fusion is used to enhance the information interaction ability to improve the detection accuracy.	By introducing the pyramid structure, it increases the computational complexity and training time.	[33,50,70,86–94]

3.4. Dense Distribution and Rotated Ships

Due to the arbitrary orientation of ships in optical remote-sensing images, using horizontal bounding boxes (HBBs) cannot accurately represent the orientation of ships, and also introduce excessive background information. At the same time, ships often exhibit a trend of dense and rotated distribution in areas such as nearshore docks. Excessive overlap between bounding boxes leads to the suppression of correct boxes, which further exacerbates the phenomenon of low recall. Therefore, achieving accurate detection of ships with a dense rotated distribution is a challenge in optical remote-sensing images. Currently, employing arbitrary orientation bounding boxes (OBBs) is an effective strategy for detecting rotated ships. OBBs accurately represent the position and orientation information of ships while effectively reducing the introduction of background information. Additionally, improved methods for Non-Maximum Suppression (NMS) alleviate the issue that detection results are incorrectly suppressed in densely distributed ships to a certain extent.

3.4.1. OBB Representation and Regression-Based Method

OBBs introduce angle information based on HBBs. The angle information can effectively represent the sailing direction of the ship. Therefore, OBBs can better highlight the position and orientation information. OBBs also effectively reduce the introduction of background information and separate the densely distributed ships. Accurately representing and generating arbitrary OBBs to locate ships holds higher application value in optical remote-sensing images.

Representation with five parameters: The method with five parameters is one of the classical representations of OBBs, represented by (x, y, w, h, θ). Specifically, (x, y) represents the center point, (w, h) represents the width and height, and θ represents the rotated angle. The representation of 90° cycle defines the height as a rectangular edge that forms an acute angle with the x-axis, and the range of values for θ is $[0°, 90°)$, as shown in Figure 13a. However, the defined width and height are exchanged when the rotated angle exceeds 90°, as shown in Figure 14a. It affects the convergence effectiveness of the network. The representation of 180° cycle defines the long side of a rectangular box as the height, and the range of values for θ is $[-90°, 90°)$, as shown in Figure 13b. It can effectively avoid the issue of exchanging width and height. However, there is a value difference when there is an overlap of $-90°$ and $90°$ at the boundary, which produces the boundary discontinuity problem, as shown in Figure 14b. It results in a sharp increase in loss at the boundary, affecting the detection performance. Liu et al. [95], Ouyang et al. [96], and Ma et al. [97] used OBBs represented as (x, y, w, h, θ) to locate ships.

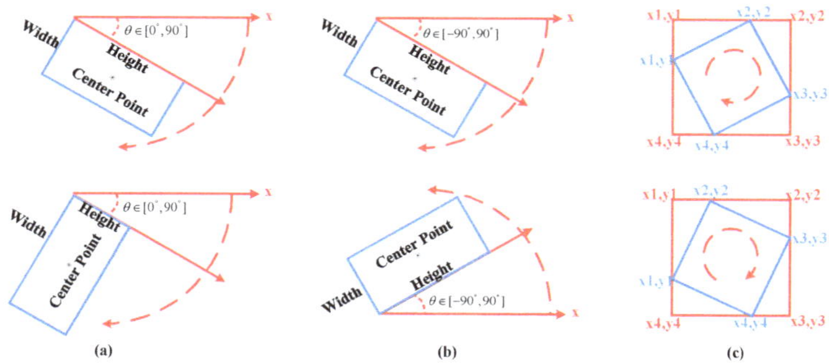

Figure 13. Schematic diagram of classical representations. (**a**) Five parameters (90° cycle). (**b**) Five parameters (180° cycle). (**c**) Eight parameters.

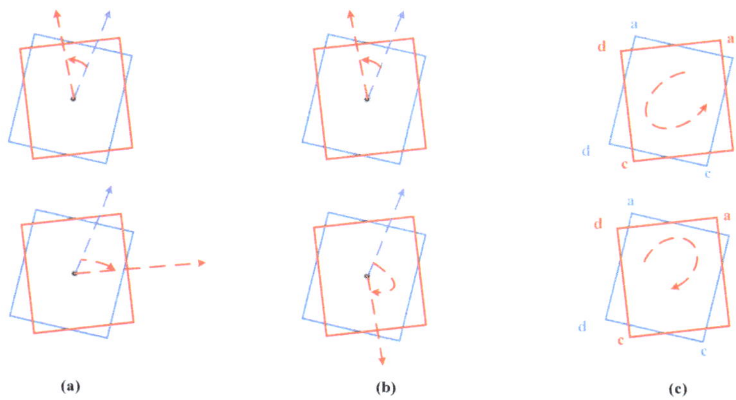

Figure 14. Schematic diagram of the issues of classical representations. The ground truth boxes are shown in red, and the bounding boxes are shown in blue. (**a**) Five parameters (90° cycle). (**b**) Five parameters (180° cycle). (**c**) Eight parameters.

Representation with eight parameters: The method with eight parameters is another classical representation for OBBs, represented by $(x_1, y_1, x_2, y_2, x_3, y_3, x_4, y_4)$. Specifically, (x_n, y_n) represents the coordinates of the four vertices of OBBs, as shown in Figure 13c. The method determines a unique direction by artificially setting the reference point, rather than representing angle values. However, it also exhibits an issue of loss discontinuity during the regression process. As shown in Figure 14c, the ideal regression process from the blue bounding box to the red ground truth box should be $(a \rightarrow a)$, $(b \rightarrow b)$, $(c \rightarrow c)$, $(d \rightarrow d)$. However, the actual regression process is $(a \rightarrow b)$, $(b \rightarrow c)$, $(c \rightarrow d)$, $(d \rightarrow a)$. At the same time, the representation requires more parameters, increasing the learning burden of the network. Zhang et al. [98] used OBBs represented as $(x_1, y_1, x_2, y_2, x_3, y_3, x_4, y_4)$ to locate ships.

Others: The issue of loss discontinuity, calculated by representations with five parameters and eight parameters, significantly impacts the convergence effectiveness of the model. Therefore, proposing new representations to alleviate this problem is the focus of current research. Su et al. [99] proposed the method represented by (x, y, w, h, OH, OV) to locate ships, as shown in Figure 15a. OH and OV were normalized horizontal and vertical distance. The method fundamentally addressed the boundary issue of angle regression. Zhou et al. [100] proposed an ellipse method, represented by $(x, y, |u|, |v|, m, \alpha)$, as shown in Figure 15b, where $\alpha = 0$ represents that the ship belongs to the second and fourth quadrants; $\alpha = 1$ represents that the ship belongs to the first and third quadrants. Furthermore, m is the difference between the length of the major axis and the focal vector. It uses vectors to represent angles, avoiding the issue of loss discontinuity caused by direct angle prediction. Yang et al. [101] and Zhang et al. [93] converted the representation with five parameters into a 2D Gaussian distribution, as shown in Figure 15c. It abandons angle representation, avoiding the issue of discontinuity in angles.

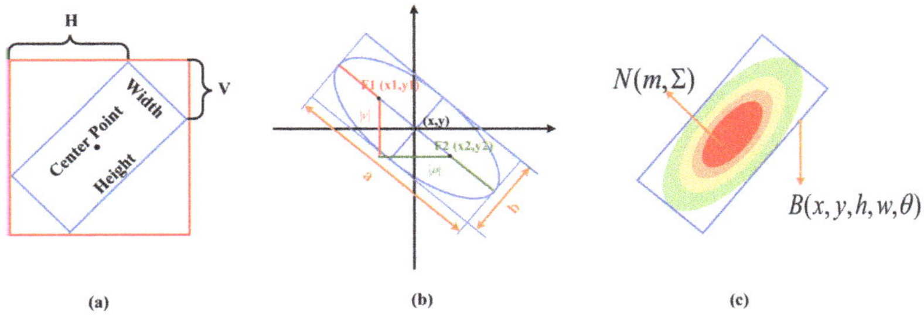

Figure 15. Schematic diagram of others. (**a**) Six parameters represented by (x, y, w, h, OH, OV). (**b**) An ellipse method represented by $(x, y, |u|, |v|, m, \alpha)$. (**c**) Gaussian distribution, and the confidence is highest in the red area.

Anchor-based regression: It is a common method to use the anchor-based detector to generate OBBs. The detector first presets a set of rotated anchor boxes and overlays the input image with pixel-wise prediction. Then, the detector regresses parameters of the rotated angle, center position, width, and height of positive samples by a predefined method to generate OBBs. For example, KOO et al. [102] used the width or height distance projection to predict the angle and generate OBBs. Ouyang et al. [96] first preset a series of horizontal anchor boxes. Then, the rotated proposal regions were generated by bilinear interpolation. Furthermore, through fully connected layers, OBBs were generated. Li et al. [64] proposed the boundary regression module, which achieved more accurate regression by predicting the offset values for the four edges of each bounding box.

Anchor-free regression: The method of generating OBBs using the anchor-free detector is not constrained by anchor boxes. It usually uses keypoints or segmentation techniques to directly generate the OBBs of ships. Furthermore, compared with the anchor-based detector, it reduces hyperparameters and demonstrates greater generalization. Zhang et al. [93] converted the ship detection into a binary semantic segmentation based on the anchor-free detector. The method generates OBBs directly by selecting pixels above the set threshold. Chen et al. [103] used the network to detect three keypoints: the bow, the stern, and the center. Furthermore, they combined the bow and stern to generate a series of prediction boxes. Then, OBBs were generated using the center points and angle information. Zhang et al. [104] used the bow and the center points to determine the orientation and generate OBBs. Cui et al. [105] used the anchor-free detector to predict the center point and shape of ships for accurately generating OBBs.

Using OBBs in rotated ship detection alleviates the issues introduced by HBBs and achieves good results. However, there are certain limitations in OBBs. The loss discontinuity of classical representations seriously impacts efficiency. Currently, some representations solve this problem, but the calculations are complex. Furthermore, the predefined dimensions, aspect ratio, and angles of anchor boxes are closely related to the dataset. The design of different hyperparameters affects the performance of detection. However, the prior knowledge of anchor boxes is crucial. Their absence may cause the detection accuracy to decrease.

3.4.2. NMS-Based Method

Due to the dense distribution of ships, the use of OBBs for close ship detection may also produce the significant overlap. When the IoU between different ships exceeds the predefined parameter, traditional NMS retains only one bounding box with the highest confidence, and completely discards the other. The operation may lead to the suppression of a correct prediction, resulting in the instance of a missed detection. Therefore, in order to eliminate redundant prediction boxes while maximally preserving correct predictions, the improvement methods of NMS have been proposed.

Bodla et al. [106] proposed Soft-NMS, which considers both the confidence and the overlap of different bounding boxes. It weights the overlapping bounding boxes to reduce their scores, rather than simply removing them with non-maximum confidence. Nie et al. [34] and Zhang et al. [107] employed Soft-NMS instead of traditional NMS, improving the recall in ship detection. Inspired by Soft-NMS, Cui et al. [105] proposed Soft-Rotate-NMS. It combines Soft-NMS with rotated features, making it more suitable for ships with arbitrary orientations.

It is important to note that the setting of the IoU threshold has a significant impact on NMS, requiring constantly manual adjustment to find the optimal threshold during the training process. Therefore, an adaptive threshold NMS algorithm is more in line with the current environment.

3.4.3. Summary

The dense and rotated distribution of ships is one of the challenges in SDORSIs. Existing research indicates that the generation of arbitrary OBBs and the improvement methods of NMS have positive effects. OBBs can more accurately locate the position and orientation of rotated ships. Furthermore, the improvement methods of NMS greatly alleviate the problem of missed detection of dense ships. Solving the issue of boundary discontinuity caused by OBBs has significant research value in the future. However, current OBB representations introduce additional parameters, and require a balance between detection accuracy and speed in practical applications. The methods and the main advantages and disadvantages of dense distribution and rotated ships are shown in Table 5.

Table 5. Methods and main advantages and disadvantages of dense distribution and rotated ships.

Methods		Advantages	Disadvantages	References
OBB Representation	Five Parameters	It is represented by (x, y, w, h, θ) and more accurately represents the position and orientation information of ships.	At the angle boundary, angle change leads to a sharp increase in loss.	[95–97]
	Eight Parameters	It is represented by $(x_1, y_1, x_2, y_2, x_3, y_3, x_4, y_4)$ and does not use angles to represent direction.	It produces loss discontinuity and a large number of parameters.	[98]
	Others	It can alleviate the problem of loss discontinuity.	Some methods increase the computational complexity and the training time.	[93,99–101]
OBB Regression	Anchor-Based	It utilizes predefined anchor boxes for the OBB's more accurate regression.	The performance is greatly influenced by hyperparameters, which are related to sizes and aspect ratios of predefined anchor boxes.	[64,96,102]
	Anchor-Free	It is not constrained by sizes and aspect ratios of anchor boxes, reducing hyperparameters.	Due to the absence of prior information provided by anchor boxes, the results are sometimes lower than anchor-based methods.	[93,103–105]
NMS	Soft-NMS	It alleviates the problem of missed dense ships by weighting overlapping bounding boxes.	It is not combined with rotated feature of the ship.	[35,106,107]
	Soft-Rotate -NMS	It combines rotated features with Soft-NMS, making it more suitable for ship detection.	The IoU threshold has a significant impact on NMS.	[105]

3.5. Large Aspect Ratio of Ships

The large aspect ratio is one of the most crucial features of ships. The standard convolution struggles to adapt to the geometric shapes in feature extraction. It inevitably leads to insufficient feature extraction and carries redundant information. Traditional ROI pooling usually extracts square-shaped features during the feature sampling stage. It leads to an uneven distribution of feature samples in two directions, affecting the detection performance. Therefore, it is important to design effective processing methods according to the geometric shapes of ships. Currently, the Deformable Convolutional Network (DCN) and improved methods of feature sampling are effective strategies. These methods aim to adapt to the geometric shapes of ships with large aspect ratios, and enhance the ability to extract irregular features.

3.5.1. DCN-Based Method

DCN [108] achieves the effect of random sampling by adding the offset variable to each sampling point. Moreover, by dynamically adjusting offsets, DCN can adaptively extract feature information from irregularly shaped ships, as shown in Figure 16a. Therefore, compared with the standard convolution, DCN is better able to adapt to geometric deformations such as the shape and size of the ship. It can extract ship features adequately while reducing the introduction of background information.

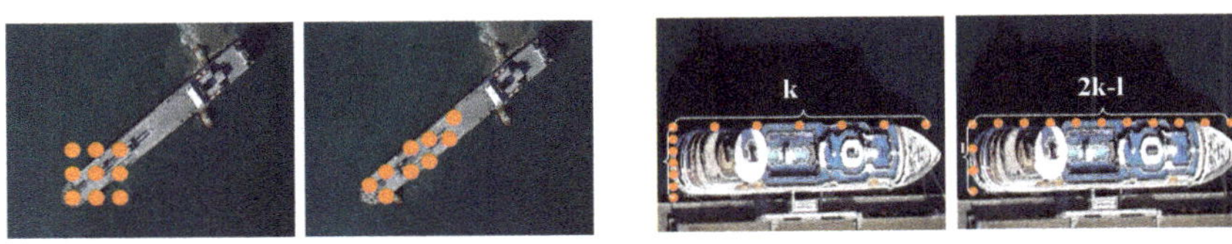

(a) (b)

Figure 16. Schematic diagram of methods for large aspect ratios, orange indicates sampling points. (**a**) Comparison between standard convolution and deformable convolution, and the latter is deformable convolution. (**b**) Comparison between standard sampling and improved sampling, and the latter better matches the shape.

Su et al. [99] and Chai et al. [109] utilized DCN instead of standard convolution to extract features, enhancing the ability to capture irregular ship features. Guo et al. [94] and Cui et al. [110] integrated DCN into FPN to better adapt to the geometric features of ships. Zhang et al. [52] employed DCN for up sampling, which ensured the robust convolutional process and improved the detection ability for ships with various shapes.

However, it is worth noting that the offsets entirely rely on the compensatory predictions of the network. It may result in unstable performance at the beginning of training. Furthermore, DCN consumes more memory compared to the standard convolution.

3.5.2. Feature Sampling-Based Method

Feature sampling refers to the operation of using ROI pooling or ROI align to obtain the fixed-size feature map. However, traditional feature sampling outputs the same number of feature samples along the width and height directions. It leads to a dense distribution of feature samples in the short side, but a sparse distribution in the long side, significantly impacting detection performance. Therefore, it is necessary to propose a new feature sampling method that adapts to ship geometric shapes. The improved method can match ship shapes and extract feature samples uniformly in both directions, as shown in Figure 16b.

Different from the typical ROI pooling, Li et al. [81] designed a shape-adaptive pooling. It obtains uniformly distributed feature samples in both length and width according to the shapes of ships. Then, it combines these samples into a fixed-size feature map. Guo et al. [111] designed a shape-aware rotated ROI align. It alleviates the problem of uneven feature distribution caused by the typical square-shaped sampling approach.

Furthermore, it achieves more accurate feature representations with fewer parameters. Zhang et al. [112] performed three different shape-aware ROI align operations on each ROI. It captures information more accurately for ships with large aspect ratios.

The improved method is an effective approach to enhance the detection result of ships with large aspect ratios. However, it maps multiple feature points to one feature point, which may cause some degree of information loss.

3.5.3. Summary

The large aspect ratio is one of the key factors which constrains the development in SDORSIs. Furthermore, enhancing the ability of network to extract irregular features is a critical technology for alleviating this issue. The experiments show that using DCN to extract features and improving the feature sampling methods are effective strategies. These methods can better adapt to ship shapes and uniformly extract feature samples. However, when extracting features from large images, DCN tends to heavily consume memory which limits application scenarios. Simultaneously, feature sampling maps multiple feature samples to a single feature point, which may cause a certain degree of information loss and calculation errors. The large aspect ratio is the essential distinction between ships and other targets. Therefore, exploring more detection methods designed for the large aspect ratio is one of the future development trends. The methods and the main advantages and disadvantages of large aspect ratio of ships are shown in Table 6.

Table 6. Methods and main advantages and disadvantages of large aspect ratio of ships.

Methods		Advantages	Disadvantages	References
DCN	DCN	It can adaptively extract feature information for irregularly shaped ships by randomly sampling.	The offset of sampling points entirely relies on the prediction of network and DCN consumes more memory compared to the standard convolution.	[52,94,99,109,110]
Feature Sampling	ROI Pooling ROI Align	It adapts to the ship geometry of the large aspect ratio, and extracts features uniformly in different directions.	It maps multiple feature points to one feature point, which may cause some degree of information loss and computational error.	[81,111,112]

3.6. *Imbalance between Positive and Negative Samples*

Ships usually occupy only a small portion in optical remote-sensing images, generating a large number of negative samples [113]. Meanwhile, due to the shapes of ships with large aspect ratios and rotated distribution, IoU-based matching strategy imposes stricter constraint. Even a slight angular deviation between the detection boxes seriously disrupts the calculation of IoU, as shown in Figure 17, resulting in insufficient positive samples. The imbalance between positive and negative samples significantly impacts the training of the network. Therefore, it is important to alleviate this problem for the development of SDORSIs. At present, improving the calculation method of IoU and loss function are effective strategies. These methods aim to explore more positive samples to mitigate the impact of insufficient positive samples.

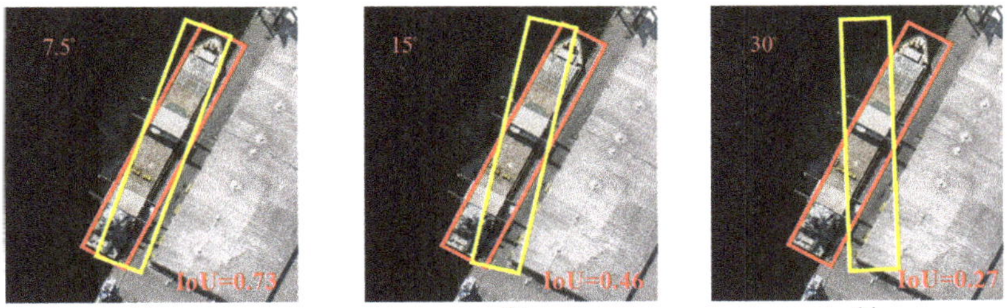

(a) (b) (c)

Figure 17. Schematic diagrams of IoU at different angles. The ground truth boxes are shown in red, and the bounding boxes are shown in yellow. (**a**) The angle difference is 7.5°, the IoU is 0.73. (**b**) The angle difference is 15°, the IoU is 0.46. (**c**) The angle difference is 30°, the IoU is 0.27.

3.6.1. IoU-Based Matching Methods

There is a certain deviation between the prediction box and ground truth box, and IoU is sensitive to angular changes. Even a small angular deviation leads to a large change in the IoU value. Meanwhile, the traditional hard-threshold sample matching strategy also severely limits the selection of positive samples, leading only a small number of high-quality positive samples to meet the filtering criteria. However, these positive samples are insufficient to support the training, constraining the performance of the network. Therefore, improving the calculation method of IoU and dynamically adjusting the IoU threshold are effective strategies to alleviate the imbalance of positive and negative samples.

Zhang et al. [114] and Li et al. [115] proposed a dynamic soft label assignment method, which adjusts the IoU threshold dynamically according to aspect the ratios of ships. It ensures that ships with extreme aspect ratios can still retain sufficient positive samples for training. Song et al. [116] used Skew IoU to calculate the overlapping area between the prediction box and ground truth box. Ma et al. [97] designed a ship orientation classification network. The network first roughly predicts the angular range of each ship. Then, several more precise angles are established within this range. It limits the angular difference to a smaller range, mitigating the impact of angular factors on IoU. Li et al. [81] proposed the orientation-agnostic IoU. The prediction box aligns with the label in orientation, assisting the network in obtaining more positive samples.

The method can better adapt to the features of ships, achieving the exploration of more positive samples. However, improving the calculation method of IoU may introduce additional computation. Furthermore, dynamical threshold requires designing a suitable threshold mapping function and constraining the range of the threshold. Inappropriate mapping ranges may introduce interfering samples.

3.6.2. Loss-Function-Based Method

There is the fact that ships usually occupy a small area in optical remote-sensing images. The number of negative samples is larger than positive samples. It results in the imbalance between positive and negative samples during training. Furthermore, the traditional cross-entropy loss function tends to focus on more negative samples, seriously affecting the detection performance. Therefore, proposing the loss function that can assign more weight to positive samples is an important way to alleviate this problem.

Focal Loss [43] introduced a weighting factor before each category in the loss function to balance the cross-entropy loss:

$$FL(p_t) = -\alpha_t(1 - p_t)^{\gamma} \log(p_t) \tag{1}$$

It can alleviate the imbalance of the network. Liu et al. [117] applied Focal Loss in ship detection and it also enabled to focus more on hard samples, enhancing the robustness of the model. Chen et al. [103] assigned the higher weight to pixels near keypoints when calculating loss. It effectively addressed the imbalance caused by the smaller number of keypoints compared with the total pixels in the image.

The method mitigates the impact of the imbalance between positive and negative samples by increasing the contribution of positive samples during training. However, it is worth noting that the weighting factor requires constant manual search for the optimal value.

3.6.3. Summary

The imbalance between positive and negative samples seriously impacts the performance and constrains the development in SDORSIs. The existing research shows that the improvements of the loss function and IoU are the primary ways to alleviate this problem. Improving the calculation method of IoU and dynamically adjusting IoU threshold aim to explore more positive samples during training. Furthermore, the improved loss function aims to assign more weight to positive samples, preventing the model from focusing more on the larger quantity of negative samples. However, the method of dynamically adjusting the IoU threshold relies on the choice of the dataset. The same network may behave differently on different datasets. Furthermore, there is a certain difficulty in selecting hyperparameters for the loss function. Therefore, alleviating the imbalance of samples has great development potential. The methods and the main advantages and disadvantages of imbalance between positive and negative samples are shown in Table 7.

Table 7. Methods and main advantages and disadvantages of imbalance between positive and negative samples.

Methods		Advantages	Disadvantages	References
IoU	Improved IoU Calculation	It can obtain more positive samples to participate in training by improving the calculation method of IoU.	It introduces additional computation and increases the complexity of the network.	[81,97,116]
	Dynamical IoU Threshold	It dynamically adjusts the threshold based on the shape of the ship to obtain more positive samples.	It requires designing a suitable threshold mapping function and constraining the range of threshold. Inappropriate mapping ranges may introduce interfering samples.	[114,115]
Loss Function	Improved Loss Function	It assigns more weight to positive samples during loss calculation, and improves their contribution in training.	It relies on hyperparameters tuning, and requires constant manual search for the optimal value.	[43,103,117]

4. Datasets, Evaluation Metrics, and Experiments

High-quality datasets are the foundation for the successful development of deep learning and play a crucial role in ship detection. In this section, we summarize the publicly available ship datasets of optical remote-sensing images and evaluation metrics. It is worth noting that we separated ship information from comprehensive datasets to provide more detailed data for the development of SDORSIs. Furthermore, we meticulously recorded the number of ships and the approximate distribution of ship sizes for each dataset, enabling readers to gain a more intuitive understanding of the data distribution. In addition, we compared and analyzed some representative models on different datasets. Furthermore, we summarized the improvement effects of optimization strategies for ship detection challenges. Finally, by analyzing the feature extraction capabilities of different backbones, we provided new insights into the development of SDORSIs.

4.1. Datasets

For the first time, we separated ships from comprehensive datasets and compiled specific ship information from seven commonly used optical remote sensing image datasets, as shown in Table 8. We used a box diagram to depict the pixel distribution of ships in each dataset. As shown in Figure 18a, ShipRSImageNet and HRRSD-ship exhibit larger variations in ship scales, which can be effectively alleviated by introducing multi-scale information during detection. The pixels of ships in DIOR-ship and LEVIR-ship are smaller, and focusing on small targets can effectively improve detection accuracy. Additionally, we visually represented the number of ships in each dataset using a bar chart. As shown in Figure 18b, the number of ships in DIOR-ship and DOTA-ship is higher than in others.

Table 8. Summary of public optical remote sensing image ship datasets.

Dataset	Year	Image	Category	Instance	Resolution	Image Size	Label
HRSC2016 [118]	2016	1070	4	2976	0.4–2 m	300 × 300–1500 × 900	HBB, OBB
DOTA-ship [119]	2017	434	1	37,028	0.5 m	800 × 800–4000 × 4000	HBB, OBB
DIOR-ship [120]	2018	2702	1	62,400	0.5–30 m	800 × 800	HBB
HRRSD-ship [121]	2019	2165	1	3886	0.5–1.2m	270 × 370–4000 × 5500	HBB
FGSD2021 [104]	2021	636	20	5274	1 m	1202 × 1205	OBB
ShipRSImageNet [122]	2021	3435	50	17,573	0.12–6 m	930 × 930	HBB, OBB
LEVIR-ship [71]	2021	3896	1	3119	16m	512 × 512	HBB

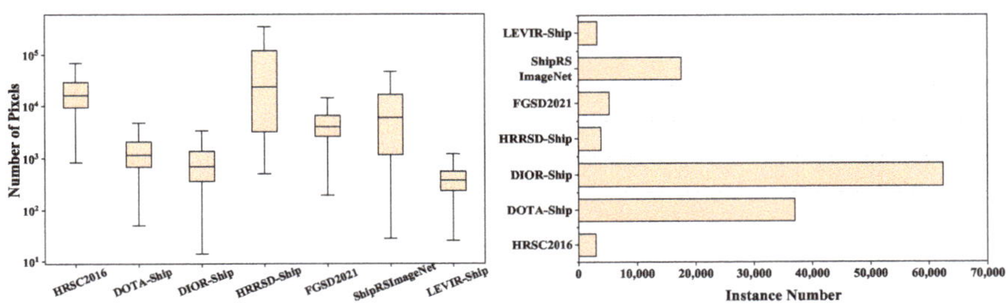

(**a**) (**b**)

Figure 18. Statistical chart of specific ship information. (**a**) Box diagram of ship pixel distribution. (**b**) Bar chart of instance numbers.

HRSC2016: The HRSC2016 [118] dataset was published by Northwestern Polytechnical University in 2016. The dataset consists of 1070 images from six different ports and 2976 labeled ships from Google Earth. The image size ranges from 300 × 300 to 1500 × 900 pixels, and the resolution from 0.4 m to 2 m. It is labeled with HBB and OBB.

DOTA-ship: The DOTA-ship dataset is collected from the DOTA [119] dataset. It includes 434 ship images and 37028 ships. The image size ranges from 800 × 800 to 4000 × 4000 pixels, and the resolution from 0.1m to 1m. It is labeled with HBB and OBB.

DIOR-ship: The DIOR-ship dataset is collected from the DIOR [120] dataset. It includes 2702 ship images and 62,400 ships. The image size is 800 × 800, and the resolution ranges from 0.5 m to 30 m. It is labeled with HBB.

HRRSD-ship: The HRRSD-ship dataset is collected from the HRRSD [121] dataset. It includes 2165 ship images and 3886 ships. The image size ranges from 270 × 370 to 4000 × 5500 pixels, and the resolution from 0.5 m to 1.2 m. It is labeled with HBB.

FGSD2021: Zhang et al. [104] introduced an FGSD2021 dataset at a ground sample distance in 2021. The dataset consists of 636 images from Google Earth and the HRSC2016 dataset. It includes 5274 labeled ships and 20 categories. The average size is 1202 × 1205 pixels, and the resolution is 1m. It is labeled with OBB.

ShipRSImageNet: The ShipRSImageNet [122] dataset is composed of 3435 images from the xView dataset, HRSC2016 dataset, FGSD dataset, Airbus Ship Detection Challenge, and Chinese satellites. It includes 17,573 ships and 50 categories. The size of most original images is 930 × 930 pixels, and the resolution ranges from 0.12 m to 6 m. It is labeled with HBB and OBB.

LEVIR-ship: Chen et al. [71] introduced a LEVIR-ship dataset in 2021, which is a medium-resolution ship dataset. The images were captured from GaoFen-1 and GaoFen-6 satellites. It includes 3896 ship images and 3119 ships. The image size is 512 × 512 pixels, and the resolution is 16 m. It is labeled with HBB.

4.2. Evaluation Metrics

IoU: IoU [123] is a metric used to measure the overlap between the prediction box and the ground truth box. In general, positive samples are filtered by setting the IoU threshold, defined as follows:

$$IoU = \frac{area(Proposal \cap GroundTruth)}{area(Proposal \cup GroundTruth)} \tag{2}$$

However, IoU lacks consideration for the distance between the prediction box and the ground truth box, failing to accurately reflect their spatial relationship. Therefore, metrics such as GIoU [124] and DIoU [125] were introduced. Based on IoU, GIoU introduces geometric factors to calculate the distance between two bounding boxes. Furthermore, DIoU calculates the distance between the centers of two bounding boxes on the basis of GIoU.

Accuracy, Precision, and Recall: First, we define as follows: true positives (TP) indicate that the prediction is positive and the ground truth is also positive; false positives (FP) indicate that the prediction is positive but the ground truth is negative; false negatives (FN) indicate that the prediction is negative but the ground truth is positive; true negatives (TN) indicate that the prediction is negative and the ground truth is also negative. Then, the definitions of accuracy rate, precision rate, and recall rate are given as follows: Accuracy rate represents the proportion of all correctly predicted samples out of the total samples:

$$Accuracy = \frac{TP + TN}{TP + TN + FP + FN} \tag{3}$$

Precision rate represents the proportion of correctly predicted positive samples out of all predicted positive samples:

$$Precision = \frac{TP}{TP + FP} \tag{4}$$

Recall rate represents the proportion of correctly predicted positive samples out of all actual positive samples:

$$Recall = \frac{TP}{TP + FN} \tag{5}$$

Average precision (AP) and mean average precision (mAP): The curve plotted with the recall rate as the horizontal axis and the precision rate as the vertical axis is called the precision recall curve (PRC). Furthermore, the area under the PRC is called AP. AP is used to characterize the detection accuracy for a single category:

$$AP = \int_0^1 P(R)dR \tag{6}$$

Each category corresponds to an AP value, and the average AP value across all categories is called mAP. The mAP is used to evaluate the overall accuracy of the dataset. Furthermore, a higher mAP value indicates better performance of the detector:

$$mAP = \frac{1}{C}\sum_{i=1}^{C} AP = \frac{1}{C}\sum_{i=1}^{C}\int_0^1 P_i(R_i)dR_i \tag{7}$$

Frames Per Second (FPS): The speed is as important as the accuracy of detection when measuring the effect of a model. Furthermore, a commonly used metric to evaluate the detection speed is FPS, which represents the number of images recognized per second.

4.3. Experimentation and Analysis

4.3.1. Algorithm Performance Comparison and Analysis

To visually demonstrate the progress in SDORSIs, we compiled some representative models in recent years and listed them in Tables 9 and 10. According to the data in Table 9, it can be observed that, for the simple ship category datasets, such as HRSC2016, the mAP reaches more than 90%, and the performance is generally saturated since 2023. 3WM-AugNet achieves 90.69% on the HRSC2016 dataset, demonstrating a leading performance.

Table 9. The performance of each algorithm on HRSC2016 datasets. mAP refers to the mAP computed on the PASCAL VOC2007. The optimal results are shown in bold, and sub-optimal results are shown in underline.

Method	Year	Publication	Backbone	Input_Size	mAP
Anchor-based (Two-stage)					
R^2CNN [126]	2017	ICPR	ResNet-101	800 × 800	73.07
RRPN [127]	2018	TMM	ResNet-101	800 × 800	79.08
RoI_Trans [128]	2019	CVPR	ResNet-101	512 × 800	86.20
Gliding Vertex [129]	2021	TPAMI	ResNet-101	512 × 800	88.20
OPLD [130]	2021	JSTAR	ResNet-50	1024 × 1333	88.44
Oriented R-CNN [131]	2021	ICCV	ResNet-101	1333 × 800	<u>90.50</u>
Anchor-based (One-stage)					
DAL [132]	2021	AAAI	ResNet-101	416 × 416	88.95
R^3Det [133]	2021	AAAI	ResNet-101	800 × 800	89.26
DLAO [99]	2022	GRSL	DCNDarknet25	800 × 800	88.28
RIDet-Q [134]	2022	GRSL	ResNet-101	800 × 800	89.10
CFC-Net [135]	2022	TGRS	ResNet-101	800 × 800	89.70
S^2A-Net [136]	2022	TGRS	ResNet-101	512 × 800	90.17
DSA-Net [67]	2022	GRSL	CSPDarknet-53	608 × 608	90.41
DAL-BCL [137]	2023	TGRS	CSPDarknet-53	800 × 800	89.70
3WM-AugNet [63]	2023	TGRS	ResNet-101	512 × 512	**90.69**
Anchor-free					
Axis Learning [138]	2020	RS	ResNet-101	800 × 800	78.15
TOSO [139]	2020	ICASSP	ResNet-101	800 × 800	79.29
SKNet [105]	2021	TGRS	Hourglass-104	511 × 511	88.30
BBAVectors [140]	2021	WACV	ResNet-101	608 × 608	88.60
CHPDet [104]	2022	TGRS	DLA-34	512 × 512	88.81
LCNet [141]	2022	GRSL	RepVGG-A1	416 × 416	89.50
CMDet [51]	2023	GRSL	ResNet-50	640 × 640	90.20
AEDet [100]	2023	JSTAR	CSPDarknet-53	800 × 800	90.45

In contrast, FGSD2021 includes more ship categories and quantities, making it more challenging in SDORSIs. According to the data in Table 10, compared with single-stage detectors, the mAP of two-stage detectors is improved by about 5–10%, meaning that two-stage detectors have the advantage of higher accuracy. Furthermore, compared with anchor-based detectors, the real-time performance of anchor-free detectors is improved by approximately 20–30 FPS. At the same time, it also can achieve satisfactory accuracy. GF-CSL achieves 88.5%, exceeding other algorithms. CenterNet-Rbb demonstrates the best real-time performance. In the 20 categories of FGSD2021, the accuracy of Ave, Sub, and Oth is significantly lower than others. Therefore, it is helpful to design a classification algorithm with stronger discrimination ability to improve the overall detection performance of the model.

Table 10. The performance of each algorithm on FGSD2021 datasets. The short name of the class is defined as (abbreviation-full name): AIR-AIRCRAFT CARRIERS, WAS-WASP CLASS, TAR-TARAWA CLASS, AUS-AUSTIN CLASS, WHI-WHIDBEY ISLAND CLASS, SAN-SAN ANTONIO CLASS, NEW-NEWPORT CLASS, TIC-TICONDEROGA CLASS, BUR-ARLEIGH BURKE CLASS, PER-PERRY CLASS, LEW-LEWIS CLARK CLASS, SUP-SUPPLY CLASS, KAI-HENRY J. KAISER CLASS, HOP-BOB HOPE CLASS, MER-MERCY CLASS, FRE-FREEDOM CLASS, IND-INDEPENDENCE CLASS, AVE-AVENGER CLASS, SUB-SUBMARINE, and OTH-OTHER. mAP refers to the mAP computed on the PASCAL VOC2007. The optimal results are shown in bold, and sub-optimal results are shown in underline.

Method	Backbone	Air	Was	Tar	Aus	Whi	San	New	Tic	Bur	Per	Lew	Sup	Kai	Hop	Mer	Fre	Ind	Ave	Sub	Oth	mAP	FPS
Anchor-based (Two-stage)																							
R^2CNN [126]	Resnet50	89.9	80.9	80.5	79.4	87.0	87.8	44.2	89.0	89.6	79.5	**80.4**	47.7	81.5	87.4	**100**	82.4	**100**	66.4	50.9	57.2	78.1	10.3
RoI_Trans [128]	Resnet50	90.9	88.6	87.2	89.5	78.5	88.8	81.8	89.6	89.8	90.4	71.7	74.7	73.7	81.6	78.6	**100**	75.6	78.4	68.0	66.9	83.5	19.2
Oriented R-CNN [131]	Resnet50	90.9	89.7	81.5	81.1	79.6	88.2	98.9	89.8	90.6	87.8	60.4	73.9	81.8	86.7	**100**	60.0	**100**	79.4	66.9	63.7	82.5	27.4
DEA-Net [142]	Resnet50	90.4	**91.4**	84.6	93.5	88.7	94.5	92.1	90.7	**92.4**	88.9	60.6	81.6	85.4	90.3	99.7	83.1	98.5	76.6	68.5	69.2	86.0	12.1
SCRDet [143]	Resnet50	77.3	90.4	87.4	89.8	78.8	90.9	54.5	88.3	89.6	74.9	68.4	59.2	90.4	77.2	81.8	73.9	**100**	43.9	43.8	57.1	75.9	9.2
ReDet [144]	ReResnet50	90.9	90.6	80.3	81.5	89.3	88.4	81.8	88.8	90.3	**90.5**	78.1	76.0	90.7	87.0	98.2	84.4	90.9	74.6	**85.3**	71.2	85.4	13.8
Anchor-based (One-stage)																							
Retinanet [43]	Resnet50	89.7	89.2	78.2	87.3	77.0	86.9	62.7	81.5	83.3	70.6	46.8	69.9	80.2	83.1	**100**	80.6	89.7	61.5	42.5	9.1	73.5	35.6
CSL [145]	Resnet50	89.7	81.3	77.2	80.2	71.4	77.2	52.7	87.7	87.7	74.2	57.1	**97.2**	77.6	80.5	**100**	72.7	**100**	32.6	37.0	40.7	73.7	10.4
R^3Det [133]	Resnet50	90.9	80.9	81.5	90.1	79.3	87.5	29.5	77.4	89.4	69.7	59.9	67.3	80.7	76.8	72.7	83.3	90.9	38.4	23.1	40.0	70.5	14.0
DCL [146]	Resnet50	89.9	81.4	78.6	80.7	78.0	87.9	49.8	78.7	87.2	76.1	60.6	76.9	90.4	80.0	78.8	77.9	**100**	37.1	31.2	45.6	73.3	10.0
RSDet [147]	Resnet50	89.8	80.4	75.8	77.3	78.6	88.8	26.1	84.7	87.6	75.2	55.1	74.4	89.7	89.3	**100**	86.4	**100**	27.6	37.6	50.6	73.7	15.4
S^2A-Net [136]	Resnet50	90.9	81.4	73.3	89.1	80.9	89.9	81.2	89.2	90.7	88.9	60.5	75.9	81.6	89.2	**100**	68.6	90.9	61.3	55.7	64.7	80.2	33.1
Anchor-free																							
BBAVectors [140]	Resnet50	**99.5**	90.9	75.9	94.3	90.9	52.9	88.5	90.0	80.4	72.2	76.9	88.2	**99.6**	**100**	94.0	**100**	74.5	58.9	63.1	**81.1**	83.6	18.5
CHPDet [104]	DLA34	90.9	90.4	89.6	89.3	89.6	**99.1**	**99.4**	90.2	90.2	90.3	70.7	87.9	89.2	96.5	**100**	85.1	**100**	84.4	68.5	56.9	87.9	41.7
CenterNet [48]	DLA34	67.2	77.9	79.2	75.5	66.8	79.8	76.8	83.1	89.0	77.7	54.5	72.6	77.4	**100**	**100**	60.8	74.8	46.5	44.1	6.8	70.5	**48.5**
RepPoint [148]	Resnet50	91.2	89.2	85.6	89.3	87.6	93.1	94.2	**91.5**	88.7	83.3	71.4	81.1	89.4	91.5	95.6	82.6	**100**	86.6	64.7	57.5	85.7	36.7
GF-CSL [149]	Resnet50	92.6	90.3	86.6	90.5	88.2	95.3	97.9	89.8	91.2	86.9	69.7	85.6	92.7	92.5	99.7	85.1	98.6	**86.7**	79.4	70.4	**88.5**	40.3
DARDet [150]	Resnet50	90.9	89.2	69.7	89.6	88.0	81.4	90.3	89.5	90.5	79.7	62.5	87.9	90.2	89.2	**100**	68.9	81.8	66.3	44.3	56.2	80.3	31.9
DDMNet [151]	DDRNet39	98.2	89.8	**92.5**	**97.1**	**91.6**	94.9	90.9	90.0	90.5	79.0	80.2	91.7	90.0	93.6	**100**	93.2	**100**	74.8	48.7	69.4	87.3	43.8

4.3.2. Performance of Optimization Strategies Comparison and Analysis

The mAP intuitively proves that a series of optimization strategies for ship characteristics are effective in Table 11. Specifically, attention mechanism is the primary method used to address complex background issues. It can enhance the contrast between ships and the background. Compared with the baseline model, the mAP of the algorithm employing this strategy is improved by about 1 to 4%. As one of the primary methods of multi-scale feature representation, FPN is widely applied in SDORSIs. It can enhance the information interaction ability of feature maps, and effectively identify ships with significant scale variations. The improved methods of FPN can enhance the ability to detect multi-scale ships. Table 11 shows an improvement in mAP of approximately 1 to 6%. Furthermore, OBB representation and regression address the issue of loss discontinuity associated with rotation angles. The mAP in Table 11 is improved by about 0.5 to 7%, confirming its effectiveness. DCN and feature sampling are more adaptive to large aspect ratios. They can reduce the introduction of irrelevant information while adequately extracting ship features. The mAP of the algorithm using this strategy is improved by about 1 to 8%.

Table 11. The performance of optimization strategies on HRSC2016 datasets. The improve values are shown in bold.

Challenges	Strategies	Methods	Year	mAP
Complex environment	Attention Mechanism	AM [45]	2021	82.67 **(+1.81)**
		CDA [64]	2021	87.20 **(+0.70)**
		CLM [67]	2022	86.18 **(+1.13)**
		GCM [67]	2022	87.75 **(+2.70)**
		DFAM [84]	2022	78.65 **(+3.70)**
	Image Preprocessing	De_haze [61]	2023	95.27 **(+1.59)**
	Saliency Constraint	SPB * [70]	2022	86.51 **(+0.99)**
Large Aspect Ratio	Feature Sampling	AP [81]	2021	89.20 **(+0.80)**
		OP [105]	2021	88.30 **(+1.80)**
	DCN	DCN [99]	2022	86.42 **(+8.46)**
		DRoI [67]	2022	89.21 **(+0.61)**
Dense and Rotated ship	OBB Representation	Gaussian-Mask [93]	2021	88.38 **(+0.87)**
		Six Parameters [99]	2022	88.28 **(+3.55)**
		ICR-Head [67]	2022	89.17 **(+0.57)**
		MDP-RGH [152]	2023	89.69 **(+4.75)**
		DAL [137]	2023	89.70 **(+0.20)**
	OBB Regression	EL [50]	2021	87.70 **(+1.92)**
		BR [64]	2021	87.40 **(+2.00)**
		OAC [98]	2023	91.07 **(+6.89)**
		KLD [68]	2023	89.87 **(+3.94)**
Large Scale Variation	Multi-scale Information	SCM [93]	2021	88.43 **(+0.92)**
		FFM [45]	2021	83.34 **(+2.48)**
		NASFCOS-FPN [50]	2021	88.20 **(+2.42)**
		FES * [70]	2022	87.01 **(+1.49)**
		DFF [84]	2022	74.95 **(+2.63)**
		FE-FPN [98]	2023	84.11 **(+6.05)**
		AF-OSD [152]	2023	89.69 **(+1.80)**
		RFF-Net [68]	2023	83.91 **(+3.96)**

* means that the model used only partial data.

4.3.3. Exploration of Transformer Application

The performance of some competitive detection models are listed in Tables 9 and 10. It can be observed that most algorithms prioritize the classical CNN models as the primary choice for feature extraction networks. However, the rate of performance growth is slowing down in recent years, indicating that the development of CNN-based algorithms is approaching maturity. To address this, it is an urgent need to break through the bottleneck of algorithmic development to further enhance detection capabilities. In view of the strong performance advantages of Transformer in other computer vision domains, we attempted to explore the feature extraction capability of Transformer for SDORSIs. We compared the detection performance of two representative CNN-based backbones (ResNet and ResNext) and two representative Transformer-based backbones (Swin Transformer and PV Transformer) on the HRSC2016 dataset. At the same time, to ensure the robustness of the results, we chose two detection networks (RetinaNet and RoI_Trans) as baselines. We selected mAP, GFlops, and Parameters as the objective criteria for performance evaluation, as shown in Tables 12 and 13. Furthermore, in order to intuitively demonstrate the relationship between the parameters' count and performance of different backbones, we drewthe experimental results in a line chart, as shown in Figure 19.

Table 12. The performance of different backbones for RetinaNet on HRSC2016 datasets. The optimal results are shown in bold, and sub-optimal results are shown in underline.

Backbones	Params(M)	GFLOPs(G)	mAP
ResNet-18 [153]	11.02	38.07	73.55
ResNet-50 [153]	23.28	86.10	81.07
ResNet-101 [153]	42.28	163.99	82.57
ResNext-50-32 × 4d [154]	22.77	89.25	82.93
ResNext-101-32 × 4d [154]	41.91	167.83	83.73
ResNext-101-64 × 4d [154]	81.00	324.99	84.45
Swin-tiny [56]	27.50	95.36	84.32
Swin-small [56]	48.79	188.10	85.22
Swin-base [56]	86.68	334.16	85.70
PVT-tiny [57]	9.24	32.40	85.15
PVT-small [57]	17.65	63.51	85.62
PVT-Medium [57]	41.07	108.96	**85.93**

Table 13. The performance of different backbones for RoI_Trans on HRSC2016 datasets. The optimal results are shown in bold, and sub-optimal results are shown in underline.

Backbones	Params(M)	GFLOPs(G)	mAP
ResNet-18 [153]	11.02	38.07	72.35
ResNet-50 [153]	23.28	86.10	87.24
ResNet-101 [153]	42.28	163.99	88.62
ResNext-50-32 × 4d [154]	22.77	89.25	88.26
ResNext-101-32 × 4d [154]	41.91	167.83	89.61
ResNext-101-64 × 4d [154]	81.00	324.99	89.40
Swin-tiny [56]	27.50	95.36	90.23
Swin-small [56]	48.79	188.10	90.41
Swin-base [56]	86.68	334.16	**90.49**
PVT-tiny [57]	9.24	32.40	89.69
PVT-small [57]	17.65	63.51	90.04
PVT-Medium [57]	41.07	108.96	90.23

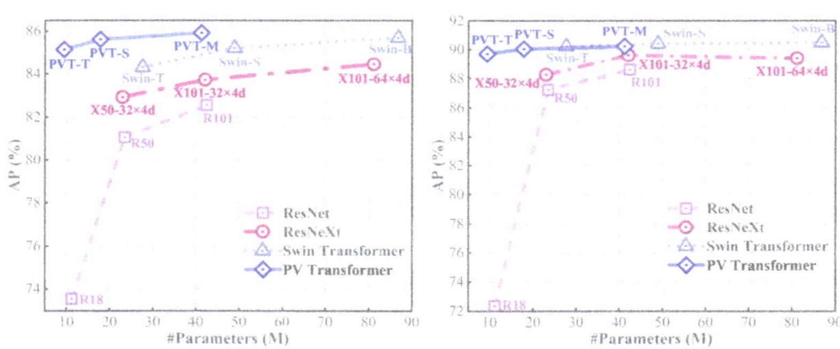

(**a**) (**b**)

Figure 19. The performance for different backbones. (**a**) Line chart of performance for RetinaNet. (**b**) Line chart of performance for RoI_Trans.

It can be observed that under the same parameter level, the feature extraction capabilities of Transformer-based backbones are generally higher than those of CNN-based backbones. In Table 12, PVT-Medium achieves the best mAP of 85.93% when choosing RetinaNet as the baseline. Compared to ResNet-101 and ResNext-101 with the same parameter level, PVT-Medium improves by 3.36% and 2.20%, while significantly reducing GFlops. Swin Transformer also takes a leading position in competition with ResNext at the same parameter level. Specifically, under three model parameters (tiny, small, and base), Swin Transformer improves mAP by 1.39%, 1.49%, and 1.25%. In Table 13, Swin-base achieves the highest mAP when RoI_Trans is selected as the baseline. Furthermore, compared to ResNet-18, PVT-tiny improves the mAP by 17.34%. As shown in Figure 19, it is concluded that under the same parameter level, Transformer exhibits stronger feature extraction capability than CNN, leading to better network performance. This is because Transformer can effectively capture dependencies between targets over long distances, building the ability of global information awareness, while CNNs can only extract information within a small window, and the information is quite limited. Exploring the connections between ship and ship or ship and ocean from a global perspective can provide important clues for SDORSIs. Therefore, Transformer has great potential in SDORSIs. Furthermore, further research is important to explore optimization strategies for Transformers based on the characteristics of ships.

We visualize feature heatmaps of each backbone at the low, middle, and high levels to compare the differences in feature extraction capabilities between CNNs and Transformer. The feature heatmaps for RetinaNet and RoI_Trans are, respectively, presented in Figures 20 and 21. According to the figures, as the network depth increases, CNN-based backbones (ResNet and ResNext) gradually pay more attention to ship regions. This is because the receptive field of deep-layer features increases, resulting in the feature collecting a wider range of information, so that the network can learn the comprehensive features of ships. However, the convolution is still a locally sliding feature extraction operation, and the extracted features are only concerned with the local scenes. Transformer-based backbones (Swin and PVT) process information from a global perspective, and the core self-attention operation can capture correlations between all pixels. For ship detection, the network can gather all ship-related clues to assist in prediction, avoiding the limitations of feature extraction confined to local windows. As shown in Figures 20 and 21, the feature heatmaps of PVT focus on the edge details of ships at shallow feature levels, while the deep-level features establish global dependencies, thereby activating more associated regions to assist ship detection. Furthermore, in order to reduce the computational burden, Swin Transformer limits self-attention within a window and realizes the interaction between windows through sliding operations. The heatmaps in figures also indicate that attention is more concentrated within certain windows.

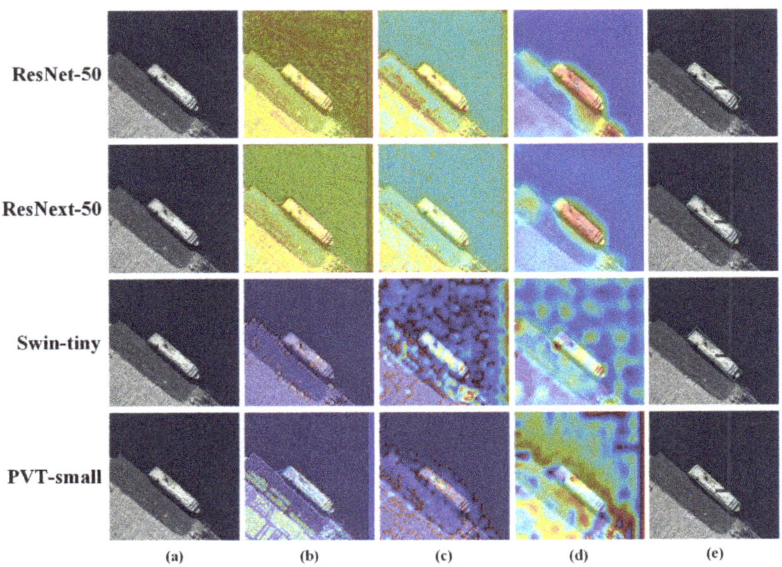

Figure 20. Feature heatmaps of each backbone for RetinaNet. (**a**) Inputs. (**b**) Shallow feature heatmaps. (**c**) Intermediate feature heatmaps. (**d**) Deep feature heatmaps. (**e**) Predicted boxes and confidence scores.

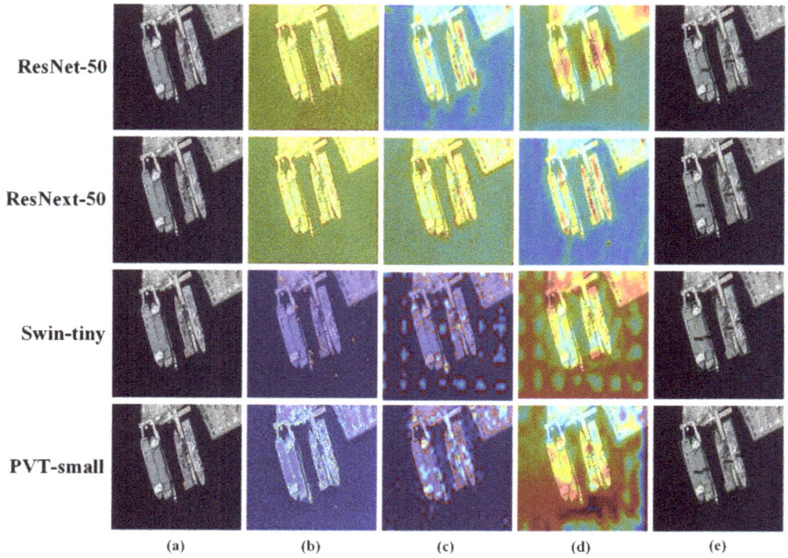

Figure 21. Feature heatmaps of each backbone for RoI_Trans. (**a**) Inputs. (**b**) Shallow feature heatmaps. (**c**) Intermediate feature heatmaps. (**d**) Deep feature heatmaps. (**e**) Predicted boxes and confidence scores.

5. Discussions and Prospects

The rapid development of deep learning has led to significant progress in SDORSIs. However, there is still a considerable gap to reach mature applications, due to the six factors summarized in this paper that constrain the development of SDORSIs. Therefore, we discuss and prospect the future development directions in this section:

1. Utilizing super-resolution and other feature enhancement methods to selectively enhance the feature representation ability of small-scale ships, which improve the recall for small ships when the scale variation is extensive. It contributes to further enhancing the overall detection accuracy.
2. To address the challenge of imbalance between positive and negative samples, supplementing the quantity of positive samples, such as methods of mining samples from the ignored set and using adaptive IoU thresholds, are helpful to increase the contribution of positive samples during network training.
3. Directly transferring common object detection networks to ship detection often fails to produce satisfactory results. Therefore, it is one of the future trends to mine the inherent features of ships, such as the wake of moving ships, large aspect ratios and so on, and design targeted ship detection networks.
4. Utilizing image fusion methods of different modalities, such as spatial information and frequency domain information, optical remote-sensing images and SAR images, enables the advantageous complementarity of information. Therefore, It helps to improve the detection accuracy of ships with cloud and fog cover and small-scale ships.
5. Designing compact and efficient detection models is more in line with the needs of applications. Therefore, the research on lightweight models, such as knowledge distillation, network pruning, and NAS, is an important strategy for deploying models to embedded devices.
6. By comparing the feature extraction capabilities of CNNs and Transformer, this paper preliminarily verifies that the global modeling concept of Transformer is helpful to improve the detection accuracy of the network. Therefore, drawing inspiration from the latest research achievements in computer vision is the direction for future development.

6. Conclusions

Ship detection in optical remote-sensing images has broad application prospects in both civilian and military domains, and is the focal point in object detection. However, a comprehensive and systematic survey that addresses the challenges faced by SDORSIs in realistic scenarios is lacking. To address this gap, this paper based on the characteristics and challenges of ships, systematically reviews the development and current research status in SDORSIs. Specifically, this paper provides a systematic review of object detection methods, including both traditional and deep learning-based methods. Furthermore, the analysis of the application scenarios of these methods is conducted in SDORSIs. Secondly, we analyze the challenges faced in detection based on the characteristics of ships, including complex marine environments, insufficient discriminative features, large scale variations, dense and rotated distributions, large aspect ratios, and imbalances between positive and negative samples. The improvement strategies for these six issues are summarized in detail. Then, we firstly compile ship information from comprehensive datasets and compare the performance of representative models. We explore the application prospects of Transformer in SDORSIs through experiments. Finally, we put forward the prospects for the development trends in SDORSIs.

We hope that this review can promote development in SDORSIs. In the future, we will continue to monitor the latest technologies in ship detection. Furthermore, we are eager to successful deploy ship detectors into embedded devices and achieve high-precision real-time detection.

Author Contributions: T.Z. wrote the manuscript; Y.W. gave professional guidance and edited; Z.L. gave advice and edited; Y.G. and Z.Z. gave advice; C.C. and H.F. revised the manuscript. All authors have read and agreed to the published version of the manuscript.

Funding: This research received no external funding.

Data Availability Statement: The data presented in this study are available in the article.

Conflicts of Interest: The authors declare no conflict of interest.

References

1. Abileah, R. Surveying coastal ship traffic with LANDSAT. In Proceedings of the OCEANS 2009, Biloxi, MS, USA, 26–29 October 2009; pp. 1–6. [CrossRef]
2. Li, Z.; Wang, Y.; Zhang, N.; Zhang, Y.; Zhao, Z.; Xu, D.; Ben, G.; Gao, Y. Deep Learning-Based Object Detection Techniques for Remote Sensing Images: A Survey. *Remote Sens.* **2022**, *14*, 2385. [CrossRef]
3. Er, M.J.; Zhang, Y.; Chen, J.; Gao, W. Ship detection with deep learning: A survey. *Artif. Intell. Rev.* **2023**, *56*, 11825–11865. [CrossRef]
4. Iwin Thanakumar Joseph, S.; Sasikala, J.; Sujitha Juliet, D. Ship detection and recognition for offshore and inshore applications: A survey. *Int. J. Intell. Unmanned Syst.* **2019**, *7*, 177–188.
5. Bo, L.; Xiaoyang, X.; Xingxing, W.; Wenting, T. Ship detection and classification from optical remote sensing images: A survey. *Chin. J. Aeronaut.* **2021**, *34*, 145–163.
6. Kanjir, U.; Greidanus, H.; Oštir, K. Vessel detection and classification from spaceborne optical images: A literature survey. *Remote Sens. Environ.* **2018**, *207*, 1–26. [CrossRef]
7. Li, J.; Xu, C.; Su, H.; Gao, L.; Wang, T. Deep learning for SAR ship detection: Past, present and future. *Remote Sens.* **2022**, *14*, 2712. [CrossRef]
8. Xu, J.; Fu, K.; Sun, X. An Invariant Generalized Hough Transform Based Method of Inshore Ships Detection. In Proceedings of the 2011 International Symposium on Image and Data Fusion, Tengchong, China, 9–11 August 2011; pp. 1–4. [CrossRef]
9. Harvey, N.R.; Porter, R.; Theiler, J. Ship detection in satellite imagery using rank-order grayscale hit-or-miss transforms. In *Proceedings of the Visual Information Processing XIX*; SPIE: Bellingham, WA, USA, 2010; Volume 7701, pp. 9–20.
10. He, H.; Lin, Y.; Chen, F.; Tai, H.M.; Yin, Z. Inshore Ship Detection in Remote Sensing Images via Weighted Pose Voting. *IEEE Trans. Geosci. Remote Sens.* **2017**, *55*, 3091–3107. [CrossRef]
11. Xu, F.; Liu, J.; Sun, M.; Zeng, D.; Wang, X. A hierarchical maritime target detection method for optical remote sensing imagery. *Remote Sens.* **2017**, *9*, 280. [CrossRef]
12. Nie, T.; He, B.; Bi, G.; Zhang, Y.; Wang, W. A method of ship detection under complex background. *ISPRS Int. J. Geo-Inf.* **2017**, *6*, 159. [CrossRef]

13. Qi, S.; Ma, J.; Lin, J.; Li, Y.; Tian, J. Unsupervised Ship Detection Based on Saliency and S-HOG Descriptor from Optical Satellite Images. *IEEE Geosci. Remote Sens. Lett.* **2015**, *12*, 1451–1455. [CrossRef]
14. Bi, F.; Zhu, B.; Gao, L.; Bian, M. A Visual Search Inspired Computational Model for Ship Detection in Optical Satellite Images. *IEEE Geosci. Remote Sens. Lett.* **2012**, *9*, 749–753. [CrossRef]
15. Lowe, D. Object recognition from local scale-invariant features. In Proceedings of the Seventh IEEE International Conference on Computer Vision, Kerkyra, Greece, 20–27 September 1999; Volume 2, pp. 1150–1157. [CrossRef]
16. Dalal, N.; Triggs, B. Histograms of oriented gradients for human detection. In Proceedings of the 2005 IEEE Computer Society Conference on Computer Vision and Pattern Recognition (CVPR'05), San Diego, CA, USA, 20–25 June 2005; Volume1, pp. 886–893. [CrossRef]
17. Corbane, C.; Najman, L.; Pecoul, E.; Demagistri, L.; Petit, M. A complete processing chain for ship detection using optical satellite imagery. *Int. J. Remote Sens.* **2010**, *31*, 5837–5854. [CrossRef]
18. Song, Z.; Sui, H.; Wang, Y. Automatic ship detection for optical satellite images based on visual attention model and LBP. In Proceedings of the 2014 IEEE Workshop on Electronics, Computer and Applications, Ottawa, ON, USA, 8–9 May 2014; pp. 722–725. [CrossRef]
19. Zhu, C.; Zhou, H.; Wang, R.; Guo, J. A Novel Hierarchical Method of Ship Detection from Spaceborne Optical Image Based on Shape and Texture Features. *IEEE Trans. Geosci. Remote Sens.* **2010**, *48*, 3446–3456. [CrossRef]
20. Liu, G.; Zhang, Y.; Zheng, X.; Sun, X.; Fu, K.; Wang, H. A New Method on Inshore Ship Detection in High-Resolution Satellite Images Using Shape and Context Information. *IEEE Geosci. Remote Sens. Lett.* **2014**, *11*, 617–621. [CrossRef]
21. Antelo, J.; Ambrosio, G.; Gonzalez, J.; Galindo, C. Ship detection and recognitionin high-resolution satellite images. In Proceedings of the 2009 IEEE International Geoscience and Remote Sensing Symposium, Cape Town, South Africa, 12–17 July 2009; Volume 4, pp. IV–514–IV–517. [CrossRef]
22. Xu, J.; Sun, X.; Zhang, D.; Fu, K. Automatic Detection of Inshore Ships in High-Resolution Remote Sensing Images Using Robust Invariant Generalized Hough Transform. *IEEE Geosci. Remote Sens. Lett.* **2014**, *11*, 2070–2074. [CrossRef]
23. Zhu, L.; Xiong, G.; Guo, D.; Yu, W. Ship target detection and segmentation method based on multi-fractal analysis. *J. Eng.* **2019**, *2019*, 7876–7879. [CrossRef]
24. Krizhevsky, A.; Sutskever, I.; Hinton, G.E. Imagenet classification with deep convolutional neural networks. *Adv. Neural Inf. Process. Syst.* **2012**, *25*. [CrossRef]
25. Girshick, R.; Donahue, J.; Darrell, T.; Malik, J. Rich Feature Hierarchies for Accurate Object Detection and Semantic Segmentation. In Proceedings of the 2014 IEEE Conference on Computer Vision and Pattern Recognition, Columbus, OH, USA, 23–28 June 2014; pp. 580–587. [CrossRef]
26. He, K.; Zhang, X.; Ren, S.; Sun, J. Spatial Pyramid Pooling in Deep Convolutional Networks for Visual Recognition. *IEEE Trans. Pattern Anal. Mach. Intell.* **2015**, *37*, 1904–1916. [CrossRef] [PubMed]
27. Girshick, R. Fast r-cnn. In Proceedings of the IEEE International Conference on Computer Vision, Santiago, Chile, 7–13 December 2015; pp. 1440–1448.
28. Ren, S.; He, K.; Girshick, R.; Sun, J. Faster R-CNN: Towards Real-Time Object Detection with Region Proposal Networks. *IEEE Trans. Pattern Anal. Mach. Intell.* **2017**, *39*, 1137–1149. [CrossRef] [PubMed]
29. He, K.; Gkioxari, G.; Dollár, P.; Girshick, R. Mask R-CNN. *IEEE Trans. Pattern Anal. Mach. Intell.* **2020**, *42*, 386–397. [CrossRef]
30. Cai, Z.; Vasconcelos, N. Cascade R-CNN: Delving Into High Quality Object Detection. In Proceedings of the 2018 IEEE/CVF Conference on Computer Vision and Pattern Recognition, Salt Lake City, UT, USA, 18–23 June 2018; pp. 6154–6162. [CrossRef]
31. Pang, J.; Chen, K.; Shi, J.; Feng, H.; Ouyang, W.; Lin, D. Libra R-CNN: Towards Balanced Learning for Object Detection. In Proceedings of the 2019 IEEE/CVF Conference on Computer Vision and Pattern Recognition (CVPR), Long Beach, CA, USA, 15–20 June 2019; pp. 821–830. [CrossRef]
32. Lu, X.; Li, B.; Yue, Y.; Li, Q.; Yan, J. Grid R-CNN. In Proceedings of the 2019 IEEE/CVF Conference on Computer Vision and Pattern Recognition (CVPR), Long Beach, CA, USA, 15–20 June 2019; pp. 7355–7364. [CrossRef]
33. Guo, H.; Yang, X.; Wang, N.; Song, B.; Gao, X. A Rotational Libra R-CNN Method for Ship Detection. *IEEE Trans. Geosci. Remote Sens.* **2020**, *58*, 5772–5781. [CrossRef]
34. Li, Q.; Mou, L.; Liu, Q.; Wang, Y.; Zhu, X.X. HSF-Net: Multiscale Deep Feature Embedding for Ship Detection in Optical Remote Sensing Imagery. *IEEE Trans. Geosci. Remote Sens.* **2018**, *56*, 7147–7161. [CrossRef]
35. Nie, S.; Jiang, Z.; Zhang, H.; Cai, B.; Yao, Y. Inshore Ship Detection Based on Mask R-CNN. In Proceedings of the IGARSS 2018—2018 IEEE International Geoscience and Remote Sensing Symposium, Valencia, Spain, 22–27 July 2018; pp. 693–696. [CrossRef]
36. Redmon, J.; Divvala, S.; Girshick, R.; Farhadi, A. You Only Look Once: Unified, Real-Time Object Detection. In Proceedings of the 2016 IEEE Conference on Computer Vision and Pattern Recognition (CVPR), Las Vegas, NV, USA, 27–30 June 2016; pp. 779–788. [CrossRef]
37. Redmon, J.; Farhadi, A. YOLO9000: Better, Faster, Stronger. In Proceedings of the 2017 IEEE Conference on Computer Vision and Pattern Recognition (CVPR), Honolulu, HI, USA, 21–26 July 2017; pp. 6517–6525. [CrossRef]
38. Redmon, J.; Farhadi, A. Yolov3: An incremental improvement. *arXiv* **2018**, arXiv:1804.02767.
39. Bochkovskiy, A.; Wang, C.Y.; Liao, H.Y.M. Yolov4: Optimal speed and accuracy of object detection. *arXiv* **2020**, arXiv:2004.10934.

40. Li, C.; Li, L.; Jiang, H.; Weng, K.; Geng, Y.; Li, L.; Ke, Z.; Li, Q.; Cheng, M.; Nie, W.; et al. YOLOv6: A single-stage object detection framework for industrial applications. *arXiv* **2022**, arXiv:2209.02976.
41. Wang, C.Y.; Bochkovskiy, A.; Liao, H.Y.M. YOLOv7: Trainable bag-of-freebies sets new state-of-the-art for real-time object detectors. In Proceedings of the IEEE/CVF Conference on Computer Vision and Pattern Recognition, Vancouver, BC, Canada, 17–24 June 2023; pp. 7464–7475.
42. Liu, W.; Anguelov, D.; Erhan, D.; Szegedy, C.; Reed, S.; Fu, C.Y.; Berg, A.C. Ssd: Single shot multibox detector. In Proceedings of the Computer Vision–ECCV 2016: 14th European Conference, Amsterdam, The Netherlands, 11–14 October 2016; Proceedings, Part I 14; Springer: Berlin/Heidelberg, Germany, 2016; pp. 21–37.
43. Lin, T.Y.; Goyal, P.; Girshick, R.; He, K.; Dollár, P. Focal Loss for Dense Object Detection. *IEEE Trans. Pattern Anal. Mach. Intell.* **2020**, *42*, 318–327. [CrossRef] [PubMed]
44. Patel, K.; Bhatt, C.; Mazzeo, P.L. Deep learning-based automatic detection of ships: An experimental study using satellite images. *J. Imaging* **2022**, *8*, 182. [CrossRef]
45. Gong, W.; Shi, Z.; Wu, Z.; Luo, J. Arbitrary-oriented ship detection via feature fusion and visual attention for high-resolution optical remote sensing imagery. *Int. J. Remote Sens.* **2021**, *42*, 2622–2640. [CrossRef]
46. Wu, J.; Pan, Z.; Lei, B.; Hu, Y. LR-TSDet: Towards tiny ship detection in low-resolution remote sensing images. *Remote Sens.* **2021**, *13*, 3890. [CrossRef]
47. Law, H.; Deng, J. Cornernet: Detecting objects as paired keypoints. In Proceedings of the European Conference on Computer Vision (ECCV), Munich, Germany, 8–14 September 2018; Springer: Berlin/Heidelberg, Germany, 2018; pp. 734–750.
48. Zhou, X.; Wang, D.; Krähenbühl, P. Objects as points. *arXiv* **2019**, arXiv:1904.07850.
49. Tian, Z.; Shen, C.; Chen, H.; He, T. FCOS: Fully Convolutional One-Stage Object Detection. In Proceedings of the 2019 IEEE/CVF International Conference on Computer Vision (ICCV), Seoul, Republic of Korea, 27 October–2 November 2019; pp. 9626–9635. [CrossRef]
50. Yang, Y.; Pan, Z.; Hu, Y.; Ding, C. CPS-Det: An anchor-free based rotation detector for ship detection. *Remote Sens.* **2021**, *13*, 2208. [CrossRef]
51. Zhuang, Y.; Liu, Y.; Zhang, T.; Chen, H. Contour Modeling Arbitrary-Oriented Ship Detection From Very High-Resolution Optical Remote Sensing Imagery. *IEEE Geosci. Remote Sens. Lett.* **2023**, *20*, 6000805. [CrossRef]
52. Zhang, Y.; Sheng, W.; Jiang, J.; Jing, N.; Wang, Q.; Mao, Z. Priority branches for ship detection in optical remote sensing images. *Remote Sens.* **2020**, *12*, 1196. [CrossRef]
53. Vaswani, A.; Shazeer, N.; Parmar, N.; Uszkoreit, J.; Jones, L.; Gomez, A.N.; Kaiser, Ł.; Polosukhin, I. Attention is all you need. *arXiv* **2023**, arXiv:1706.03762. https://doi.org/10.48550/arXiv.1706.03762.
54. Carion, N.; Massa, F.; Synnaeve, G.; Usunier, N.; Kirillov, A.; Zagoruyko, S. End-to-end object detection with transformers. In Proceedings of the European Conference on Computer Vision, Tel Aviv, Israel, 23–27 October 2020; Springer: Berlin/Heidelberg, Germany, 2020; pp. 213–229.
55. Zhu, X.; Su, W.; Lu, L.; Li, B.; Wang, X.; Dai, J. Deformable detr: Deformable transformers for end-to-end object detection. *arXiv* **2020**, arXiv:2010.04159.
56. Liu, Z.; Lin, Y.; Cao, Y.; Hu, H.; Wei, Y.; Zhang, Z.; Lin, S.; Guo, B. Swin Transformer: Hierarchical Vision Transformer using Shifted Windows. In Proceedings of the 2021 IEEE/CVF International Conference on Computer Vision (ICCV), Montreal, QC, Canada, 10–17 October 2021; pp. 9992–10002. [CrossRef]
57. Wang, W.; Xie, E.; Li, X.; Fan, D.P.; Song, K.; Liang, D.; Lu, T.; Luo, P.; Shao, L. Pyramid vision transformer: A versatile backbone for dense prediction without convolutions. In Proceedings of the IEEE/CVF International Conference on Computer Vision, Montreal, BC, Canada, 11–17 October 2021; pp. 568–578.
58. Han, K.; Xiao, A.; Wu, E.; Guo, J.; Xu, C.; Wang, Y. Transformer in transformer. *Adv. Neur. In.* **2021**, *34*, 15908–15919.
59. Yu, Y.; Yang, X.; Li, J.; Gao, X. A Cascade Rotated Anchor-Aided Detector for Ship Detection in Remote Sensing Images. *IEEE Trans. Geosci. Remote Sens.* **2022**, *60*, 5600514. [CrossRef]
60. Zheng, Y.; Su, J.; Zhang, S.; Tao, M.; Wang, L. Dehaze-AGGAN: Unpaired Remote Sensing Image Dehazing Using Enhanced Attention-Guide Generative Adversarial Networks. *IEEE Trans. Geosci. Remote Sens.* **2022**, *60*, 5630413. [CrossRef]
61. Song, R.; Li, T.; Li, T. Ship detection in haze and low-light remote sensing images via colour balance and DCNN. *Appl. Ocean Res.* **2023**, *139*, 103702. [CrossRef]
62. Yang, Y.; Wang, C.; Liu, R.; Zhang, L.; Guo, X.; Tao, D. Self-augmented Unpaired Image Dehazing via Density and Depth Decomposition. In Proceedings of the 2022 IEEE/CVF Conference on Computer Vision and Pattern Recognition (CVPR), Orleans, LA, USA, 18–24 June 2022; pp. 2027–2036. [CrossRef]
63. Ying, L.; Miao, D.; Zhang, Z. 3WM-AugNet: A Feature Augmentation Network for Remote Sensing Ship Detection Based on Three-Way Decisions and Multigranularity. *IEEE Trans. Geosci. Remote Sens.* **2023**, *61*, 1001219. [CrossRef]
64. Li, L.; Zhou, Z.; Wang, B.; Miao, L.; An, Z.; Xiao, X. Domain adaptive ship detection in optical remote sensing images. *Remote Sens.* **2021**, *13*, 3168. [CrossRef]
65. Wang, Q.; Shen, F.; Cheng, L.; Jiang, J.; He, G.; Sheng, W.; Jing, N.; Mao, Z. Ship detection based on fused features and rebuilt YOLOv3 networks in optical remote-sensing images. *Int. J. Remote Sens.* **2021**, *42*, 520–536. [CrossRef]
66. Hu, J.; Zhi, X.; Shi, T.; Zhang, W.; Cui, Y.; Zhao, S. PAG-YOLO: A portable attention-guided YOLO network for small ship detection. *Remote Sens.* **2021**, *13*, 3059. [CrossRef]

67. Qin, C.; Wang, X.; Li, G.; He, Y. An Improved Attention-Guided Network for Arbitrary-Oriented Ship Detection in Optical Remote Sensing Images. *IEEE Geosci. Remote Sens. Lett.* **2022**, *19*, 6514805. [CrossRef]
68. Chen, Y.; Wang, J.; Zhang, Y.; Liu, Y. Arbitrary-oriented ship detection based on Kullback–Leibler divergence regression in remote sensing images. *Earth Sci. Inform.* **2023**, *16*, 3243–3255. [CrossRef]
69. Qu, Z.; Zhu, F.; Qi, C. Remote sensing image target detection: Improvement of the YOLOv3 model with auxiliary networks. *Remote Sens.* **2021**, *13*, 3908. [CrossRef]
70. Ren, Z.; Tang, Y.; He, Z.; Tian, L.; Yang, Y.; Zhang, W. Ship Detection in High-Resolution Optical Remote Sensing Images Aided by Saliency Information. *IEEE Trans. Geosci. Remote Sens.* **2022**, *60*, 5623616. [CrossRef]
71. Chen, J.; Chen, K.; Chen, H.; Zou, Z.; Shi, Z. A Degraded Reconstruction Enhancement-Based Method for Tiny Ship Detection in Remote Sensing Images With a New Large-Scale Dataset. *IEEE Trans. Geosci. Remote Sens.* **2022**, *60*, 5625014. [CrossRef]
72. Liu, Y.; Zhang, R.; Deng, R.; Zhao, J. Ship detection and classification based on cascaded detection of hull and wake from optical satellite remote sensing imagery. *GIScience Remote Sens.* **2023**, *60*, 2196159. [CrossRef]
73. Xue, F.; Jin, W.; Qiu, S.; Yang, J. Rethinking Automatic Ship Wake Detection: State-of-the-Art CNN-Based Wake Detection via Optical Images. *IEEE Trans. Geosci. Remote Sens.* **2022**, *60*, 5613622. [CrossRef]
74. Liu, Y.; Zhao, J.; Qin, Y. A novel technique for ship wake detection from optical images. *Remote Sens. Environ.* **2021**, *258*, 112375. [CrossRef]
75. Liu, Z.; Xu, J.; Li, J.; Plaza, A.; Zhang, S.; Wang, L. Moving Ship Optimal Association for Maritime Surveillance: Fusing AIS and Sentinel-2 Data. *IEEE Trans. Geosci. Remote Sens.* **2022**, *60*, 5635218. [CrossRef]
76. Liu, Y.; Deng, R.; Zhao, J. Simulation of Kelvin wakes in optical images of rough sea surface. *Appl. Ocean Res.* **2019**, *89*, 36–43. [CrossRef]
77. Xu, Q.; Li, Y.; Shi, Z. LMO-YOLO: A Ship Detection Model for Low-Resolution Optical Satellite Imagery. *IEEE J. Sel. Topics Appl. Earth Observ. Remote Sens.* **2022**, *15*, 4117–4131. [CrossRef]
78. Chen, L.; Shi, W.; Deng, D. Improved YOLOv3 based on attention mechanism for fast and accurate ship detection in optical remote sensing images. *Remote Sens.* **2021**, *13*, 660. [CrossRef]
79. Zhou, L.; Li, Y.; Rao, X.; Liu, C.; Zuo, X.; Liu, Y. Ship Target Detection in Optical Remote Sensing Images Based on Multiscale Feature Enhancement. *Comput. Intell. Neurosci.* **2022**, *2022*, 2605140. [CrossRef] [PubMed]
80. Liu, W.; Ma, L.; Chen, H. Arbitrary-Oriented Ship Detection Framework in Optical Remote-Sensing Images. *IEEE Geosci. Remote Sens. Lett.* **2018**, *15*, 937–941. [CrossRef]
81. Li, L.; Zhou, Z.; Wang, B.; Miao, L.; Zong, H. A Novel CNN-Based Method for Accurate Ship Detection in HR Optical Remote Sensing Images via Rotated Bounding Box. *IEEE Trans. Geosci. Remote Sens.* **2021**, *59*, 686–699. [CrossRef]
82. Tian, L.; Cao, Y.; He, B.; Zhang, Y.; He, C.; Li, D. Image enhancement driven by object characteristics and dense feature reuse network for ship target detection in remote sensing imagery. *Remote Sens.* **2021**, *13*, 1327. [CrossRef]
83. Qin, P.; Cai, Y.; Liu, J.; Fan, P.; Sun, M. Multilayer Feature Extraction Network for Military Ship Detection From High-Resolution Optical Remote Sensing Images. *IEEE J. Sel. Topics Appl. Earth Observ. Remote Sens.* **2021**, *14*, 11058–11069. [CrossRef]
84. Han, Y.; Yang, X.; Pu, T.; Peng, Z. Fine-Grained Recognition for Oriented Ship Against Complex Scenes in Optical Remote Sensing Images. *IEEE Trans. Geosci. Remote Sens.* **2022**, *60*, 5612318. [CrossRef]
85. Wen, G.; Cao, P.; Wang, H.; Chen, H.; Liu, X.; Xu, J.; Zaiane, O. MS-SSD: Multi-scale single shot detector for ship detection in remote sensing images. *Appl. Intell.* **2023**, *53*, 1586–1604. [CrossRef]
86. Lin, T.Y.; Dollár, P.; Girshick, R.; He, K.; Hariharan, B.; Belongie, S. Feature Pyramid Networks for Object Detection. In Proceedings of the 2017 IEEE Conference on Computer Vision and Pattern Recognition (CVPR), Honolulu, HI, USA, 21–26 July 2017; pp. 936–944. [CrossRef]
87. Tian, Y.; Wang, X.; Zhu, S.; Xu, F.; Liu, J. LMSD-Net: A Lightweight and High-Performance Ship Detection Network for Optical Remote Sensing Images. *Remote Sens.* **2023**, *15*, 4358. [CrossRef]
88. Si, J.; Song, B.; Wu, J.; Lin, W.; Huang, W.; Chen, S. Maritime Ship Detection Method for Satellite Images Based on Multiscale Feature Fusion. *IEEE J. Sel. Topics Appl. Earth Observ. Remote Sens.* **2023**, *16*, 6642–6655. [CrossRef]
89. Yan, Z.; Li, Z.; Xie, Y.; Li, C.; Li, S.; Sun, F. ReBiDet: An Enhanced Ship Detection Model Utilizing ReDet and Bi-Directional Feature Fusion. *Appl. Sci.* **2023**, *13*, 7080. [CrossRef]
90. Li, J.; Li, Z.; Chen, M.; Wang, Y.; Luo, Q. A new ship detection algorithm in optical remote sensing images based on improved R3Det. *Remote Sens.* **2022**, *14*, 5048. [CrossRef]
91. Chen, W.; Han, B.; Yang, Z.; Gao, X. MSSDet: Multi-Scale Ship-Detection Framework in Optical Remote-Sensing Images and New Benchmark. *Remote Sens.* **2022**, *14*, 5460. [CrossRef]
92. Xie, X.; Li, L.; An, Z.; Lu, G.; Zhou, Z. Small Ship Detection Based on Hybrid Anchor Structure and Feature Super-Resolution. *Remote Sens.* **2022**, *14*, 3530. [CrossRef]
93. Zhang, X.; Wang, G.; Zhu, P.; Zhang, T.; Li, C.; Jiao, L. GRS-Det: An Anchor-Free Rotation Ship Detector Based on Gaussian-Mask in Remote Sensing Images. *IEEE Trans. Geosci. Remote Sens.* **2021**, *59*, 3518–3531. [CrossRef]
94. Guo, H.; Bai, H.; Yuan, Y.; Qin, W. Fully deformable convolutional network for ship detection in remote sensing imagery. *Remote Sens.* **2022**, *14*, 1850. [CrossRef]
95. Liu, Q.; Xiang, X.; Yang, Z.; Hu, Y.; Hong, Y. Arbitrary Direction Ship Detection in Remote-Sensing Images Based on Multitask Learning and Multiregion Feature Fusion. *IEEE Trans. Geosci. Remote Sens.* **2021**, *59*, 1553–1564. [CrossRef]

96. Ouyang, L.; Fang, L.; Ji, X. Multigranularity Self-Attention Network for Fine-Grained Ship Detection in Remote Sensing Images. *IEEE J. Sel. Topics Appl. Earth Observ. Remote Sens.* **2022**, *15*, 9722–9732. [CrossRef]
97. Ma, J.; Zhou, Z.; Wang, B.; Zong, H.; Wu, F. Ship detection in optical satellite images via directional bounding boxes based on ship center and orientation prediction. *Remote Sens.* **2019**, *11*, 2173. [CrossRef]
98. Zhang, D.; Wang, C.; Fu, Q. OFCOS: An Oriented Anchor-Free Detector for Ship Detection in Remote Sensing Images. *IEEE Geosci. Remote Sens. Lett.* **2023**, *20*, 6004005. [CrossRef]
99. Su, N.; Huang, Z.; Yan, Y.; Zhao, C.; Zhou, S. Detect Larger at Once: Large-Area Remote-Sensing Image Arbitrary-Oriented Ship Detection. *IEEE Geosci. Remote Sens. Lett.* **2022**, *19*, 6505605. [CrossRef]
100. Zhou, K.; Zhang, M.; Zhao, H.; Tang, R.; Lin, S.; Cheng, X.; Wang, H. Arbitrary-Oriented Ellipse Detector for Ship Detection in Remote Sensing Images. *IEEE J. Sel. Topics Appl. Earth Observ. Remote Sens.* **2023**, *16*, 7151–7162. [CrossRef]
101. Yang, X.; Yan, J.; Ming, Q.; Wang, W.; Zhang, X.; Tian, Q. Rethinking rotated object detection with gaussian wasserstein distance loss. In Proceedings of the International Conference on Machine Learning, PMLR, Virtual Event, 18–24 July 2021; pp. 11830–11841.
102. Koo, J.; Seo, J.; Jeon, S.; Choe, J.; Jeon, T. RBox-CNN: Rotated bounding box based CNN for ship detection in remote sensing image. In Proceedings of the 26th ACM SIGSPATIAL International Conference on Advances in Geographic Information Systems, Seattle, WA, USA, 6–9 November 2018; pp. 420–423.
103. Chen, J.; Xie, F.; Lu, Y.; Jiang, Z. Finding Arbitrary-Oriented Ships From Remote Sensing Images Using Corner Detection. *IEEE Geosci. Remote Sens. Lett.* **2020**, *17*, 1712–1716. [CrossRef]
104. Zhang, F.; Wang, X.; Zhou, S.; Wang, Y.; Hou, Y. Arbitrary-Oriented Ship Detection Through Center-Head Point Extraction. *IEEE Trans. Geosci. Remote Sens.* **2022**, *60*, 5612414. [CrossRef]
105. Cui, Z.; Leng, J.; Liu, Y.; Zhang, T.; Quan, P.; Zhao, W. SKNet: Detecting Rotated Ships as Keypoints in Optical Remote Sensing Images. *IEEE Trans. Geosci. Remote Sens.* **2021**, *59*, 8826–8840. [CrossRef]
106. Bodla, N.; Singh, B.; Chellappa, R.; Davis, L.S. Soft-NMS–improving object detection with one line of code. In Proceedings of the IEEE International Conference on Computer Vision, Venice, Italy, 22–29 October 2017; pp. 5561–5569.
107. Zhang, Y.; Guo, L.; Wang, Z.; Yu, Y.; Liu, X.; Xu, F. Intelligent ship detection in remote sensing images based on multi-layer convolutional feature fusion. *Remote Sens.* **2020**, *12*, 3316. [CrossRef]
108. Dai, J.; Qi, H.; Xiong, Y.; Li, Y.; Zhang, G.; Hu, H.; Wei, Y. Deformable Convolutional Networks. In Proceedings of the 2017 IEEE International Conference on Computer Vision (ICCV), Venice, Italy, 22–29 October 2017; pp. 764–773. [CrossRef]
109. Chai, B.; Nie, X.; Gao, H.; Jia, J.; Qiao, Q. Remote Sensing Images Background Noise Processing Method for Ship Objects in Instance Segmentation. *J. Indian Soc. Remote Sens.* **2023**, *51*, 647–659. [CrossRef]
110. Cui, Z.; Sun, H.M.; Yin, R.N.; Jia, R.S. SDA-Net: A detector for small, densely distributed, and arbitrary-directional ships in remote sensing images. *Appl. Intell.* **2022**, *52*, 12516–12532. [CrossRef]
111. Guo, B.; Zhang, R.; Guo, H.; Yang, W.; Yu, H.; Zhang, P.; Zou, T. Fine-Grained Ship Detection in High-Resolution Satellite Images With Shape-Aware Feature Learning. *IEEE J. Sel. Topics Appl. Earth Observ. Remote Sens.* **2023**, *16*, 1914–1926. [CrossRef]
112. Zhang, J.; Huang, R.; Li, Y.; Pan, B. Oriented ship detection based on intersecting circle and deformable RoI in remote sensing images. *Remote Sens.* **2022**, *14*, 4749. [CrossRef]
113. Li, Z.; Wang, Y.; Zhang, Y.; Gao, Y.; Zhao, Z.; Feng, H.; Zhao, T. Context Feature Integration and Balanced Sampling Strategy for Small Weak Object Detection in Remote-Sensing Imagery. *IEEE Geosci. Remote Sens. Lett.* **2024**, *112*, 102966. [CrossRef]
114. Zhang, C.; Xiong, B.; Li, X.; Kuang, G. Aspect-Ratio-Guided Detection for Oriented Objects in Remote Sensing Images. *IEEE Geosci. Remote Sens. Lett.* **2022**, *19*, 8024805. [CrossRef]
115. Li, Y.; Bian, C.; Chen, H. Dynamic Soft Label Assignment for Arbitrary-Oriented Ship Detection. *IEEE J. Sel. Topics Appl. Earth Observ. Remote Sens.* **2023**, *16*, 1160–1170. [CrossRef]
116. Song, Z.; Wang, L.; Zhang, G.; Jia, C.; Bi, J.; Wei, H.; Xia, Y.; Zhang, C.; Zhao, L. Fast Detection of Multi-Direction Remote Sensing Ship Object Based on Scale Space Pyramid. In Proceedings of the 2022 18th International Conference on Mobility, Sensing and Networking (MSN), Guangzhou, China, 4–16 December 2022; pp. 1019–1024. [CrossRef]
117. Liu, M.; Chen, Y.; Ding, D. AureNet: A Real-Time Arbitrary-oriented and Ship-based Object Detection. In Proceedings of the 2023 IEEE 2nd International Conference on Electrical Engineering, Big Data and Algorithms (EEBDA), Changchun, China, 24–26 February 2023; pp. 647–652. [CrossRef]
118. Liu, Z.; Wang, H.; Weng, L.; Yang, Y. Ship Rotated Bounding Box Space for Ship Extraction From High-Resolution Optical Satellite Images With Complex Backgrounds. *IEEE Geosci. Remote Sens. Lett.* **2016**, *13*, 1074–1078. [CrossRef]
119. Xia, G.S.; Bai, X.; Ding, J.; Zhu, Z.; Belongie, S.; Luo, J.; Datcu, M.; Pelillo, M.; Zhang, L. DOTA: A Large-Scale Dataset for Object Detection in Aerial Images. In Proceedings of the 2018 IEEE/CVF Conference on Computer Vision and Pattern Recognition, Salt Lake City, UT, USA, 18–23 June 2018; pp. 3974–3983. [CrossRef]
120. Li, K.; Wan, G.; Cheng, G.; Meng, L.; Han, J. Object detection in optical remote sensing images: A survey and a new benchmark. *ISPRS J. Photogramm. Remote Sens.* **2020**, *159*, 296–307. [CrossRef]
121. Zhang, Y.; Yuan, Y.; Feng, Y.; Lu, X. Hierarchical and Robust Convolutional Neural Network for Very High-Resolution Remote Sensing Object Detection. *IEEE Trans. Geosci. Remote Sens.* **2019**, *57*, 5535–5548. [CrossRef]
122. Zhang, Z.; Zhang, L.; Wang, Y.; Feng, P.; He, R. ShipRSImageNet: A Large-Scale Fine-Grained Dataset for Ship Detection in High-Resolution Optical Remote Sensing Images. *IEEE J. Sel. Topics Appl. Earth Observ. Remote Sens.* **2021**, *14*, 8458–8472. [CrossRef]

123. Yu, J.; Jiang, Y.; Wang, Z.; Cao, Z.; Huang, T. Unitbox: An advanced object detection network. In Proceedings of the 24th ACM international conference on Multimedia, Amsterdam, The Netherlands, 15–19 October 2016; pp. 516–520.
124. Rezatofighi, H.; Tsoi, N.; Gwak, J.; Sadeghian, A.; Reid, I.; Savarese, S. Generalized Intersection Over Union: A Metric and a Loss for Bounding Box Regression. In Proceedings of the 2019 IEEE/CVF Conference on Computer Vision and Pattern Recognition (CVPR), Long Beach, CA, USA, 15–20 June 2019; pp. 658–666. [CrossRef]
125. Zheng, Z.; Wang, P.; Liu, W.; Li, J.; Ye, R.; Ren, D. Distance-IoU loss: Faster and better learning for bounding box regression. In Proceedings of the AAAI Conference on Artificial Intelligence, New York, NY, USA, 7–12 February 2020; Volume 34, pp. 12993–13000.
126. Jiang, Y.; Zhu, X.; Wang, X.; Yang, S.; Li, W.; Wang, H.; Fu, P.; Luo, Z. R2CNN: Rotational region CNN for orientation robust scene text detection. *arXiv* **2017**, arXiv:1706.09579.
127. Ma, J.; Shao, W.; Ye, H.; Wang, L.; Wang, H.; Zheng, Y.; Xue, X. Arbitrary-Oriented Scene Text Detection via Rotation Proposals. *IEEE Trans. Multimedia* **2018**, *20*, 3111–3122. [CrossRef]
128. Ding, J.; Xue, N.; Long, Y.; Xia, G.S.; Lu, Q. Learning RoI Transformer for Oriented Object Detection in Aerial Images. In Proceedings of the 2019 IEEE/CVF Conference on Computer Vision and Pattern Recognition (CVPR), Long Beach, CA, USA, 15–20 June 2019; pp. 2844–2853. [CrossRef]
129. Xu, Y.; Fu, M.; Wang, Q.; Wang, Y.; Chen, K.; Xia, G.S.; Bai, X. Gliding Vertex on the Horizontal Bounding Box for Multi-Oriented Object Detection. *IEEE Trans. Pattern Anal. Mach. Intell.* **2021**, *43*, 1452–1459. [CrossRef] [PubMed]
130. Song, Q.; Yang, F.; Yang, L.; Liu, C.; Hu, M.; Xia, L. Learning Point-Guided Localization for Detection in Remote Sensing Images. *IEEE J. Sel. Topics Appl. Earth Observ. Remote Sens.* **2021**, *14*, 1084–1094. [CrossRef]
131. Xie, X.; Cheng, G.; Wang, J.; Yao, X.; Han, J. Oriented R-CNN for Object Detection. In Proceedings of the 2021 IEEE/CVF International Conference on Computer Vision (ICCV), Montreal, QC, Canada, 10–17 October 2021; pp. 3500–3509. [CrossRef]
132. Ming, Q.; Zhou, Z.; Miao, L.; Zhang, H.; Li, L. Dynamic anchor learning for arbitrary-oriented object detection. In Proceedings of the AAAI Conference on Artificial Intelligence, Washington DC, USA, 7–14 February 2021; Volume 35, pp. 2355–2363.
133. Yang, X.; Yan, J.; Feng, Z.; He, T. R3det: Refined single-stage detector with feature refinement for rotating object. In Proceedings of the AAAI conference on artificial intelligence, Washington DC, USA, 7–14 February 2021; Volume 35, pp. 3163–3171.
134. Ming, Q.; Miao, L.; Zhou, Z.; Yang, X.; Dong, Y. Optimization for Arbitrary-Oriented Object Detection via Representation Invariance Loss. *IEEE Geosci. Remote Sens. Lett.* **2022**, *19*, 8021505. [CrossRef]
135. Ming, Q.; Miao, L.; Zhou, Z.; Dong, Y. CFC-Net: A Critical Feature Capturing Network for Arbitrary-Oriented Object Detection in Remote-Sensing Images. *IEEE Trans. Geosci. Remote Sens.* **2022**, *60*, 5605814. [CrossRef]
136. Han, J.; Ding, J.; Li, J.; Xia, G.S. Align Deep Features for Oriented Object Detection. *IEEE Trans. Geosci. Remote Sens.* **2022**, *60*, 5602511. [CrossRef]
137. Pan, C.; Li, R.; Liu, W.; Lu, W.; Niu, C.; Bao, Q. Remote Sensing Image Ship Detection Based on Dynamic Adjusting Labels Strategy. *IEEE Trans. Geosci. Remote Sens.* **2023**, *61*, 4702621. [CrossRef]
138. Xiao, Z.; Qian, L.; Shao, W.; Tan, X.; Wang, K. Axis learning for orientated objects detection in aerial images. *Remote Sens.* **2020**, *12*, 908. [CrossRef]
139. Feng, P.; Lin, Y.; Guan, J.; He, G.; Shi, H.; Chambers, J. TOSO: Student's-T Distribution Aided One-Stage Orientation Target Detection in Remote Sensing Images. In Proceedings of the ICASSP 2020—2020 IEEE International Conference on Acoustics, Speech and Signal Processing (ICASSP), Barcelona, Spain, 4–8 May 2020; pp. 4057–4061. [CrossRef]
140. Yi, J.; Wu, P.; Liu, B.; Huang, Q.; Qu, H.; Metaxas, D. Oriented Object Detection in Aerial Images with Box Boundary-Aware Vectors. In Proceedings of the 2021 IEEE Winter Conference on Applications of Computer Vision (WACV), Waikoloa, HI, USA, 3–8 January 2021; pp. 2149–2158. [CrossRef]
141. Deng, G.; Wang, Q.; Jiang, J.; Hong, Q.; Jing, N.; Sheng, W.; Mao, Z. A Low Coupling and Lightweight Algorithm for Ship Detection in Optical Remote Sensing Images. *IEEE Geosci. Remote Sens. Lett.* **2022**, *19*, 6513505. [CrossRef]
142. Liang, D.; Geng, Q.; Wei, Z.; Vorontsov, D.A.; Kim, E.L.; Wei, M.; Zhou, H. Anchor Retouching via Model Interaction for Robust Object Detection in Aerial Images. *IEEE Trans. Geosci. Remote Sens.* **2022**, *60*, 5619213. [CrossRef]
143. Yang, X.; Yang, J.; Yan, J.; Zhang, Y.; Zhang, T.; Guo, Z.; Sun, X.; Fu, K. SCRDet: Towards More Robust Detection for Small, Cluttered and Rotated Objects. In Proceedings of the 2019 IEEE/CVF International Conference on Computer Vision (ICCV), Seoul, Republic of Korea, 27 October–2 November 2019; pp. 8231–8240. [CrossRef]
144. Han, J.; Ding, J.; Xue, N.; Xia, G.S. ReDet: A Rotation-equivariant Detector for Aerial Object Detection. In Proceedings of the 2021 IEEE/CVF Conference on Computer Vision and Pattern Recognition (CVPR), Nashville, TN, USA, 20–25 June 2021; pp. 2785–2794. [CrossRef]
145. Yang, X.; Yan, J. Arbitrary-oriented object detection with circular smooth label. In *Proceedings of the Computer Vision–ECCV 2020: 16th European Conference, Glasgow, UK, 23–28 August 2020*; Proceedings, Part VIII 16; Springer: Berlin/Heidelberg, Germany, 2020; pp. 677–694.
146. Yang, X.; Hou, L.; Zhou, Y.; Wang, W.; Yan, J. Dense Label Encoding for Boundary Discontinuity Free Rotation Detection. In Proceedings of the 2021 IEEE/CVF Conference on Computer Vision and Pattern Recognition (CVPR), Nashville, TN, USA, 20–25 June 2021; pp. 15814–15824. [CrossRef]
147. Qian, W.; Yang, X.; Peng, S.; Yan, J.; Guo, Y. Learning modulated loss for rotated object detection. In Proceedings of the AAAI conference on artificial intelligence, Washington DC, USA, 7–14 February 2021; Volume 35, pp. 2458–2466.

148. Li, W.; Chen, Y.; Hu, K.; Zhu, J. Oriented RepPoints for Aerial Object Detection. In Proceedings of the 2022 IEEE/CVF Conference on Computer Vision and Pattern Recognition (CVPR), New Orleans, LA, USA, 18–24 June 2022; pp. 1819–1828. [CrossRef]
149. Wang, J.; Li, F.; Bi, H. Gaussian Focal Loss: Learning Distribution Polarized Angle Prediction for Rotated Object Detection in Aerial Images. *IEEE Trans. Geosci. Remote Sens.* **2022**, *60*, 4707013. [CrossRef]
150. Zhang, F.; Wang, X.; Zhou, S.; Wang, Y. DARDet: A Dense Anchor-Free Rotated Object Detector in Aerial Images. *IEEE Geosci. Remote Sens. Lett.* **2022**, *19*, 3024305. [CrossRef]
151. Yu, D.; Xu, Q.; Liu, X.; Guo, H.; Lu, J.; Lin, Y.; Lv, L. Dual-Resolution and Deformable Multihead Network for Oriented Object Detection in Remote Sensing Images. *IEEE J. Sel. Topics Appl. Earth Observ. Remote Sens.* **2023**, *16*, 930–945. [CrossRef]
152. Hua, Z.; Pan, G.; Gao, K.; Li, H.; Chen, S. AF-OSD: An Anchor-Free Oriented Ship Detector Based on Multi-Scale Dense-Point Rotation Gaussian Heatmap. *Remote Sens.* **2023**, *15*, 1120. [CrossRef]
153. He, K.; Zhang, X.; Ren, S.; Sun, J. Deep Residual Learning for Image Recognition. In Proceedings of the 2016 IEEE Conference on Computer Vision and Pattern Recognition (CVPR), Las Vegas, NV, USA, 27–30 June 2016; pp. 770–778. [CrossRef]
154. Xie, S.; Girshick, R.; Dollár, P.; Tu, Z.; He, K. Aggregated Residual Transformations for Deep Neural Networks. In Proceedings of the 2017 IEEE Conference on Computer Vision and Pattern Recognition (CVPR), Honolulu, HI, USA, 21–26 July 2017; pp. 5987–5995. [CrossRef]

remote sensing

Article

An Identification Method of Corner Reflector Array Based on Mismatched Filter through Changing the Frequency Modulation Slope

Le Xia †, Fulai Wang *,†, Chen Pang, Nanjun Li, Runlong Peng, Zhiyong Song and Yongzhen Li

The College of Electronic Science and Technology, National University of Defense Technology, Changsha 410073, China; xroyzhile@outlook.com (L.X.); pangchen@nudt.edu.cn (C.P.); lnj999214@163.com (N.L.); 18075652227@163.com (R.P.); songzhiyong08@nudt.edu.cn (Z.S.); liyongzhen@nudt.edu.cn (Y.L.)

* Correspondence: wflmadman@outlook.com; Tel.: +86-18673121317

† These authors contributed equally to this work.

Abstract: The corner reflector is an effective means of interference for radar seekers due to its high jamming intensity, wide frequency band, and combat effectiveness ratio. Properly arranging multiple corner reflectors in an array can form dilution jamming that resembles ships, substantially enhancing the interference effect. This results in a significant decline in the precision attack efficiency of radar seekers. Hence, it is critical to accurately identify corner reflector array. The common recognition methods involve extracting features on the high-resolution range profile (HRRP) and polarization domain. However, the former is constrained by the number of corner reflectors, while the latter is affected by the accuracy of polarization measurement, both of which have limited performance on the identification of corner reflector array. In terms of the evident variations in physical structures, there must be differences in their scattering characteristics. To highlight the differences, this paper proposes a new method based on the concept of mismatched filtering, which involves changing the frequency modulation slope of the chirp signal in the filter. Then, the variance of width and intervals within a specific scope are extracted as features to characterize these differences, and an identification process is designed in combination with the support vector machine. The simulation experiments demonstrate that the proposed method exhibits stable discriminative performance and can effectively combat dilution jamming. Its accuracy rate exceeds 0.86 when the signal-to-noise ratio is greater than 0 dB. Compared to the HRRP methods, the recognition accuracy of the proposed algorithm improves 15% in relation to variations in the quantity of corner reflectors.

Keywords: corner reflector array; combat dilution jamming; change the frequency modulation slope; mismatched filter; support vector machine

Citation: Xia, L.; Wang, F.; Pang, C.; Li, N.; Peng, R.; Song, Z.; Li, Y. An Identification Method of Corner Reflector Array Based on Mismatched Filter through Changing the Frequency Modulation Slope. *Remote Sens.* **2024**, *16*, 2114. https://doi.org/10.3390/rs16122114

Academic Editor: Paolo Tripicchio

Received: 22 April 2024
Revised: 6 June 2024
Accepted: 7 June 2024
Published: 11 June 2024

1. Introduction

In the field of radar electronic countermeasures, chaff jamming and corner reflector are the primary methods of passive jamming [1]. In contrast to the chaff jamming, corner reflector has the advantages of long−lasting interference duration and stable interference effects. In addition, corner reflector exhibits a number of advantageous properties with regard to scattering characteristics, spectral characteristics, polarization properties, and resistance to the technique of coherent accumulation. In recent years, the continuous advancement of structure and surface reflection materials has led to significant improvements in the performance metrics of corner reflectors, including coverage frequency bands, omnidirectionality, and the cost−effectiveness ratio of interference [2]. Consequently, many countries have devoted greater attention to the development of corner reflectors and deployed them in a variety of scenarios where they are used to counter the radar detection.

The corner reflector is composed of several orthogonal metallic planes that enable the incident wave to reflect multiple times. This property results in the corner reflector exhibiting a pronounced backward radar cross−section and the generation of a robust jamming signal [3]. Proper assignment of multiple corner reflectors can simulate false targets similar to ships, which will substantially enhance the interference effect [4–6]. Consequently, the array of corner reflectors has attracted greater interest [7,8]. According to the jamming performance, corner reflector array can be classified into two types: dilution jamming and centroid jamming [9,10]. The former takes effect in the tracking stage, in which the corner reflectors released by the ship are in the same resolution unit with ship. The radar echoes of the corner reflector are typically more pronounced than those of the ship, which results in a bias towards the corner reflector in the tracking direction. The dilution jamming is aimed at the seeking stage, generating multiple false targets covering the detection area of radar. The distance between the corner reflector array and the ship is sufficiently large to permit the clear division of two targets. In contrast to centroid jamming, the implementation of dilution jamming is less constrained, presenting a more significant and challenging issue. Consequently, it is of paramount importance to devise an identification method for the dilution jamming.

It is important to note that the distinction between corner reflectors and ships is not sufficiently obvious, which has made the identification of corner reflectors a challenging problem. A number of studies have been conducted to extract features from different domains with the objective of characterizing the differences between dilution jamming and ship. These domains can be mainly divided into the time domain, frequency domain, and polarization domain.

In the time domain, the majority of studies focus on the extraction of features in the high−resolution range profile (HRRP) [11,12]. The HRRP can be employed to reflect the construction information of the target, including its geometric shape, size, and material composition [13]. Ref. [14] extracted the features in HRRP such as radial size, scattering symmetry, and number of scattering points. Nevertheless, the efficacy of the HRRP method is constrained by the observation angle and spatial geometry, which may not yield expected precision in practical applications. In the frequency domain, the focus is typically on the variances in the motion characteristics. Due to the differing velocity and fluctuation of the corner reflector compared to the ship, the identification of the reflector is typically achieved through doppler frequency shift [15] or micro−doppler frequency shift [16,17]. However, the former cannot easily achieve the expected performance in the context of trailing the corner reflectors, given that the velocity of the corner reflector array closely matches that of the ship. And the latter cannot perform well in complex sea conditions. In the polarization domain, the polarization decomposition theory is widely used [18,19]. This method is one of the important parts of radar polarization technology, extracting the characteristics of the object by decomposing polarization data into various components [20,21]. Nevertheless, real targets exhibit pronounced angular sensitivity in their scattering responses, which makes it difficult to accurately measure the polarization scattering matrix based on polarization decomposition [22]. In order to reduce the impact of azimuth sensitivity of target polarization scattering response, some studies have employed polarimetric roll-invariant features for the identification of corner reflectors [23–25].

In light of the limitations of single−domain approaches, recent studies have sought to enhance performance by focusing on multi−dimensional joint features. The authors of [7], based on the theory of polarization modulation, extracted the correlation characteristic parameters on the polarization range 2D image. Simulation results indicate that this method has stable recognition performance. Ref. [26] proposed a novel method of discrimination for corner reflector arrays based on the time−spatial−polarization joint domains. Ref. [8] optimized features and proposed new characteristics in the polarization domain and HRRP. Then, based on the measured data, this paper provided performance analyses of different features and their combinations. However, this method has certain reference value but lacks universality.

Considering the aforementioned constraints, this paper utilizes a novel method based on mismatched filter, which involves modifying the frequency modulation slope of the linear frequency modulated (LFM) signal in the filter. The LFM signal is the most prevalent waveform employed in radar systems. Nevertheless, in comparison to the sophisticated waveforms proposed in recent years [27,28], it is relatively ineffective in mitigating interference and spoofing in radar detection. It is of great value to optimize the signal processing in order to enhance the radar performance [29]. In some previous studies, this technology was nearly applied at the transmitter by modifying the frequency modulation slope of the transmitting LFM signal to enhance the complexity of the waveform. This method resulted in the interference signal being mismatched with the transmitted signal, preventing it from acquiring the corresponding gain of pulse compression. It has since developed many applications, such as anti−interference [30] and defect detection [31]. On the contrary, we utilize the side effect of this technology to broaden the main lobe of the signal output, thereby reducing the degree of compression compared to matched filter. This will amplify the potential differences in scattering characteristics between ships and corner reflector arrays, thus improving the identification performance. Subsequently, this paper extracts the pertinent characteristics and develops an identification method in conjunction with the support vector machine (SVM). The advantages of this approach are as follows.

1. Stable performance. The recognition process aims to utilise the structural dissimilarities between the two targets in order to achieve recognition, rather than relying on some intuitive features, such as length or the number of scattering points, which is applicable to complex environments. Compared to the methods applied in HRRP, the proposed method is not limited to some environmental factors, such as the number of corner reflectors or the observation angle. In contrast to the aforementioned frequency domain features, this approach is not limited to scenarios where there are differences in the speed of targets.
2. Strong applicability. The primary objective of this method is to enhance the performance of the LFM radar, which is a common waveform in radar systems. Nevertheless, the efficacy of polarization decomposition is contingent upon the availability of a fully polarimetric radar and a signal possessing a high degree of polarization isolation, both of which are essential for the accurate measurement to guarantee its performance. The methods employed in the frequency domain similarly necessitate the capacity for coherent integration.

The remaining sections of this article are organized as follows. In Section 2, we establish the signal model and introduce the principle of mismatched filter by changing the frequency modulation slope. In Section 3, we mainly simulate the output of the ship and corner reflector array based on the proposed mismatch filter, identify the differences, and extract corresponding characteristics. Subsequently, based on the extracted features, we propose an identification method combine with SVM. In Section 4, based on the electromagnetic simulation data, we use the proposed method to evaluate the identification performance under different parameters, and compare with other methods in different conditions. In Section 5, some conclusions are drawn.

Notations: We use bold lowercase letters for vectors and bold uppercase letters for matrices. $(\cdot)^*$ represents the conjugate operation. $\otimes$ denotes the convolution operation. $|\cdot|$ denotes the modulus. The letter j denotes the imaginary unit (i.e., $j = \sqrt{-1}$). The letter c is the velocity of light.

2. Mismatched Filter by Changing Frequency Modulation Slope

2.1. Signal Model

The signal used in this paper is the common LFM signal, and its base band format can be expressed as below.

$$s(t) = \text{rect}\left(\frac{t}{T_p}\right)\exp\left(j\pi K t^2\right) \tag{1}$$

In Equation (1), $K = B/T_p$ is the frequency modulation slope of the LFM signal, where T_p and B, respectively, denote the pulse width and bandwidth. When $K > 0$, it is up−chirp; when $K < 0$, it is down−chirp. In addition, $\text{rect}(t)$ is the function of rectangular pulse, which can be expressed as below.

$$\text{rect}(t) = \begin{cases} 1, |t| \leq \dfrac{1}{2} \\ 0, \text{else} \end{cases} \tag{2}$$

When there exist targets within the detection range, the echoes can be represented as $s_r(t) = \sum_{i=1}^{I} a_i s(t - \tau_i)$, where I is the number of equivalent scattering points, a_i represents the intensity of the ith scattering point, $\tau_i = r_i/c$ denotes the time delay of the ith scattering point.

According to the usual process [32], the impulse response of the matched filter is an LFM signal, and the slopes of instantaneous frequency are opposite to K. For the signal in Equation (1), its matched filter impulse response is $h(t) = s^*(-t)$. To simplify the process of analysis, we set $I = 1$, $a = 1$, and $\tau = 0$. When the echo passes through the matched filter, the filter output can be expressed as

$$\begin{aligned} S_{\text{match}}(t) &= s(t) \otimes s^*(-t) \\ &= T_P \frac{\sin\left(\pi B \left(1 - \frac{|t|}{T_P}\right) t\right)}{\pi B t} \text{rect}\left(\frac{t}{2T_P}\right) \end{aligned} \tag{3}$$

It is evident that pulse compression can enhance the signal−to−noise ratio (SNR) and further highlight targets. However, the scattering characteristics of targets and interference can also be compressed, which can make it challenging to clearly distinguish them in the range profile. Therefore, in this paper, we propose using mismatched filtering to enhance and highlight these differences.

2.2. Mismatched Filter by Modifying Frequency Modulation Slope

In this paper, the way to change the frequency modulation slope is to modify the bandwidth but maintain the time width, as shown in Figure 1a. And the modified slope can be denoted as $K_1 = \beta B/T_p = \beta K$, where β characterizes the degree of change in bandwidth. The format of the modified LFM signal can be expressed as follows.

$$s_r(t) = \text{rect}\left(\frac{t}{T_p}\right) \exp\left(j\pi K_1 t^2\right) \tag{4}$$

To further simplify the analysis process, we use the same settings as Equation (3). The output of the radar echo passing through this mismatched filter can be expressed as

$$\begin{aligned} y(t) &= s(t) \otimes s_r^*(-t) \\ &= \int_{-\infty}^{\infty} \text{rect}\left(\frac{\tau}{T_p}\right) \exp\left(j\pi K \tau^2\right) \text{rect}\left(\frac{t-\tau}{T_p}\right) \exp\left[-j\pi K_1 \left(t - \tau\right)^2\right] d\tau \\ &= \exp\left(-j\pi K_1 t^2\right) \int_{-\infty}^{\infty} \text{rect}\left(\frac{\tau}{T_p}\right) \text{rect}\left(\frac{t-\tau}{T_p}\right) \exp\left[j\pi (K - K_1)\tau^2 + 2j\pi K_1 t\tau\right] d\tau \\ &= \exp\left(-j\pi K_1 t^2\right) \int_{-\infty}^{\infty} A(\tau) \exp\left[j\left(\pi (K - K_1)\tau^2 + 2\pi K_1 t\tau\right)\right] d\tau \end{aligned} \tag{5}$$

In Equation (5), $A(\tau)$ is the rectangular envelope, and its value is 1 in the range $[-T_p/2 + t, T_p/2]$. By using the method of stationary phase to analyze Equation (5), we can obtain its approximate analytical expression as follows.

$$y(t) \approx \sqrt{\frac{1}{|1-\beta|K}} \exp\left[j\pi\frac{\beta}{\beta-1}Kt^2 + \mathrm{sgn}(1-\beta)\frac{\pi}{4}\right], t \in \left[-\frac{|1-\beta|}{2\beta}T_p, \frac{|1-\beta|}{2\beta}T_p\right] \quad (6)$$

where $\mathrm{sgn}(\cdot)$ is the sign function.

From Equation (6), it can be observed that the width of main lobe of the output after mismatched filtering is approximately $|1-\beta|T_p/\beta$, while the one of pulse compression output is $1/B$, resulting in a ratio of $|1-\beta|BT_p/\beta$ between them. Since the time−bandwidth product of the LFM signal is much greater than 1, it can be inferred that the width of main lobe is significantly broadened after reception by changing the frequency modulation slope. Likewise, the output amplitude of pulse compression is $1/T_p$, so the decrease ratio of amplitude can be expressed as $\sqrt{(|1-\beta|BT_p)}$.

Then, we take an example of LFM signal with a bandwidth of 150 MHz and a time duration of 10 μs to validate the derivation above. The frequency modulation factors β chosen for the LFM signal in mismatched filter are 0.8, 0.9, 1.1, and 1.2. The filtered outputs of those factors are as graphed in Figure 1.

Now use the formula $|1-\beta|$ to express the deviation level of frequency modulation slope. Roughly speaking, Figure 1b shows that the larger the deviation level, the smaller the output amplitude. And it can be observed that at a given deviation level, the main lobe widens to a lesser extent when the modulation factor β is greater than 1. When the deviation level equals 0.1, the decrease amplitude is near 22 dB, which is close to the theoretical value.

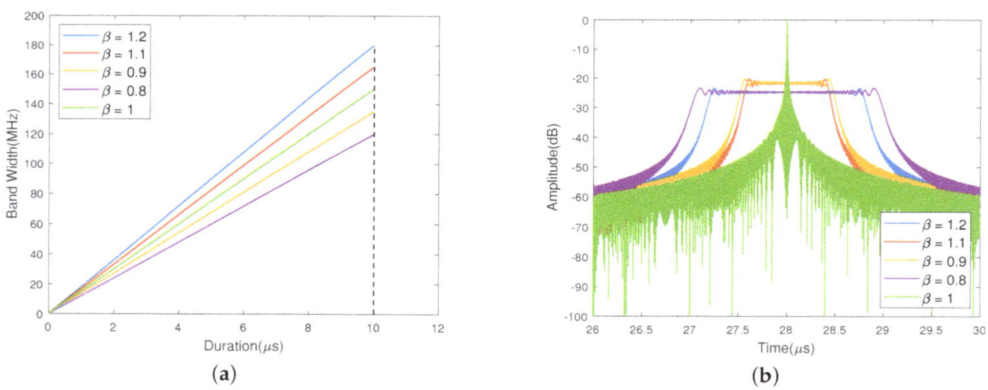

Figure 1. The mismatched filter by modifying frequency modulation slope. (**a**) Time−frequency scheme of LFM signal in mismatched filter. (**b**) The outputs of different modulation factor.

2.3. *Analysis on Echoes Simulated from Simple Scattering Points*

The mismatched filter illustrated in Section 2.2 can largely broaden the width of the main lobe, compared with pulse compression. Next, we simulate the mismatched filtering output of multiple scattering points to investigate the effects of this method on reflecting distribution, types, or other information of those scattering points.

Based on geometrical theory of diffraction (GTD) [33], we can establish backward scattering characteristics of the target and reconstruct the echo signal according to the transmitted signal. The backward scattering characteristics can be expressed as

$$E(f) = \sum_{i=1}^{I} A_i (\mathrm{j}\frac{f}{f_0})^{\alpha_i} \exp(-\frac{\mathrm{j}4\pi f r_i}{\mathrm{c}}) \tag{7}$$

where I is the number of scattering points, A_i represents the intensity of the ith scattering point, f_0 is the initial frequency of the transmitted signal, α_i denotes the type of the ith scattering point, r_i is the position of the ith scattering point.

Subsequently, four distinct scenarios are designed, with the requisite details presented in Table 1. The modulation factor–range two–dimensional images of these above scenarios are shown in Figure 2. The bandwidth of the LFM signal used in this simulation is 300 MHz, with a pulse duration of 20 µs.

Table 1. Four types of scenarios.

Scenario	Distribution of Position	Types of Scattering Points
1	Uniform	Consistent
2	Uniform	Inconsistent
3	Cluster [1]	Consistent
4	Cluster	Inconsistent

[1] clustered on two centers but the coverage range and the number of scattering points are the same.

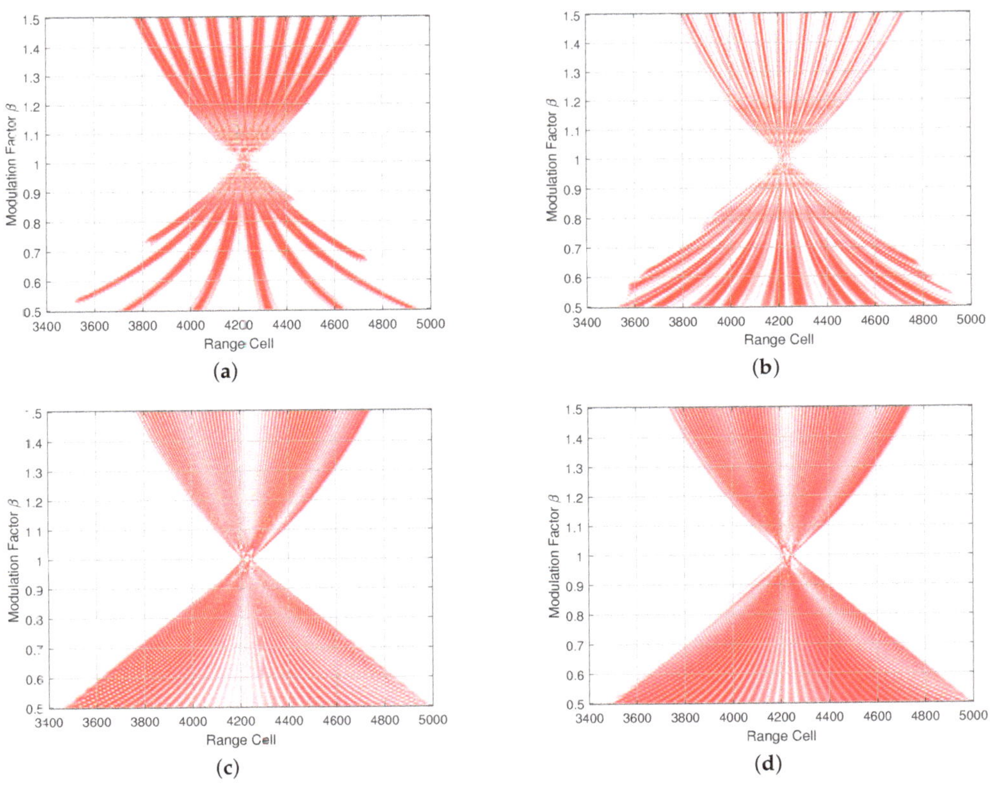

Figure 2. The modulation factor–range two–dimensional images of different scenarios. (**a**) Scenario 1. (**b**) Scenario 2. (**c**) Scenario 3. (**d**) Scenario 4.

Figure 2 illustrates the outputs of multiple mismatched filters, which are based on the received signal from scattering points with varying distributions and types. In these images, we perform a process of normalization at each value of modulation factor β in order to concentrate on the precise acquisition of alterations in the amplitude distribution. Comparing Figure 2a with Figure 2c, it can be observed that the former output amplitude is more concentrated and evenly distributed, forming a striped pattern. Conversely, the diagram of the latter exhibits a wrinkled pattern. In the same way, analyzing Figure 2a,b, we can find that the type of scatter points also contributes to a more intricate amplitude variation. However, the most significant factor influencing the output of the matched filter is the spatial distribution of the scattering points, which is closely related to the physical structure of the target. Consequently, based on this mismatched filter, we can capture the scattering characteristics of targets, including their structure and type, to a certain extent.

3. Character Extraction and Identification

The objective of this section is to employ this mismatched filter to distinguish between corner reflector arrays and ships and to succinctly summarize the discernible features. Then based on these features, an identification method is proposed.

3.1. Target Echo Acquisition

Due to the paucity of measured data concerning the scenarios of corner reflector arrays or ships at sea, we have employed electromagnetic simulation software (CST Studio Suite 2021) to acquire the backward scattering characteristics of the target. Figure 3 presents a pair of range profiles obtained through electromagnetic simulation software, which separately denote ship and corner reflector array. The blue solid lines in Figure 3 represent the range profiles acquired by electromagnetic simulation software. It is observed that the range profile of the ship is manifested as a few pronounced peaks, interspersed with a relatively weak region. In contrast, the range profile of the corner reflector array appears as a combination of similarly strong peaks. This is due to the complex and large structure of the ship, which can be considered as the superposition of echoes from multiple scattering centers. The scattering characteristics of these centers are usually different. The corner reflector is typically composed of multiple trihedral angles, with a simple structure and strong symmetry. This can be considered as strong scattering points with similar scattering characteristics.

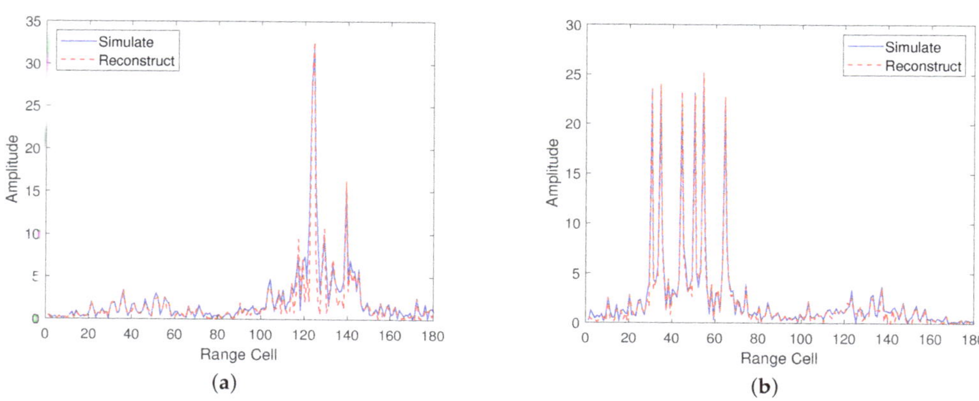

Figure 3. Simulated range profile and reconstructed range profile. (**a**) Ship. (**b**) Corner reflector array.

Due to the limitations imposed by the computational speed and time constraints, the number of sampling frequency points employed in the simulation of the one−dimensional range profile did not match that of the transmitted LFM signal, which is unable to generate the radar echoes through convolution in the frequency domain. Nevertheless, there are currently many methods for inverting the target echoes based on range profile [34,35]. In this paper, we utilize the total least squares–estimating signal parameter via rotational invariance techniques (TLS–ESPRIT) to reconstruct the range profile and to acquire the radar echoes [36,37]. The following outlines the brief operational processes.

The initial step is to utilize the electromagnetic simulation software to obtain the frequency response of the target. Based on this frequency response, the TLS–ESPRIT algorithm is employed to extract the parameters of the equivalent scattering centers, including amplitude, type, and relative position. Subsequently, the frequency response of the target is reconstructed based on the frequency sampling vector, according to the GTD listed in Equation (7). Finally, multiply the reconstructed frequency response with the transmitted signal in the frequency domain, and the target echoes can be obtained by Fourier transform.

The red dashed lines in Figure 3 are the reconstructed range profiles. A comparison of the reconstructed results with the electromagnetic calculations reveals that they only differ in regions heavily affected by clutter. Furthermore, the amplitudes at the peak positions are essentially consistent, which demonstrates the effectiveness of the aforementioned reconstruction method.

3.2. Character Extraction

Based on the reconstructed echoes of the target, we conduct the proposed mismatched filter through changing the frequency modulation slope. The bandwidth of the LFM signal used in this Section is 150 MHz, with a pulse duration of 10 μs.

From Figure 4, it can be observed that the image of the ship is concentrated on one side, while the image of the corner reflector array is more evenly distributed, consisting of multiple bright bands. The differences displayed in Figure 3 illustrate that the mismatched filter has the capacity to amplify the discrepancies between ship and corner reflector arrays to a considerable extent, which renders it more conducive to the process of identification.

In order to facilitate the process of identification, this section employs a process of feature extraction, whereby the intuitive distribution differences are translated into mathematical expressions. It can be observed from Figure 4 that the differences are concentrated on the distribution of bright bands, where the points with larger amplitude are located. Therefore, in order to accurately characterize the distinction, only those points whose normalized amplitude falls within a specific range are retained.

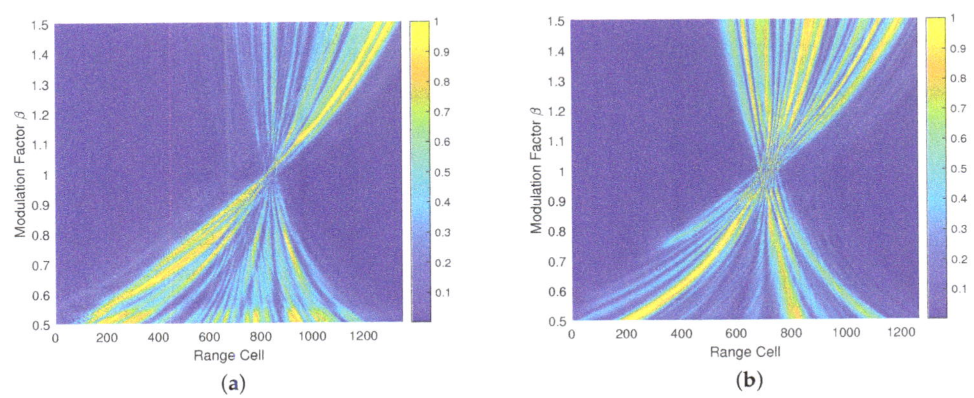

Figure 4. Modulation factor–range two–dimensional image. (**a**) Ship. (**b**) Corner reflector array.

Figure 5 illustrates the points within the range of −5 dB. As the preceding analysis, the points of the ship are concentrated on one side, whereas those of the corner reflector array are evenly and densely distributed. It can be observed that when β is less than 1, the widening of the image becomes more pronounced compared to when β is greater than 1. This is consistent with the analysis presented in Section 2.2. However, if the widening is too large, it will lead to excessive superposition in the outputs of the mismatched filter, thereby affecting the effectiveness of feature extraction. Consequently, in the following sections, this paper will only focus on the cases in which β is greater than 1.

Subsequently, this paper identifies two features that can be used to distinguish between ships and corner reflector arrays, based on the observed distribution differences.

(1) Variance of width.

A comparison of Figure 5a and Figure 5b reveals that the widths of the −5 dB regions in the ship are largely disparate, while those of the corner reflector array are relatively similar. Consequently, this paper calculates the variance of width to characterize this difference, which can be expressed as

$$\sigma^2_{\text{width},\alpha} = \frac{1}{N-1}\sum_{i=1}^{N}\left(\frac{x_i - \overline{x}}{N\overline{x}}\right)^2 \quad (8)$$

where N refers to the number of regions under the modulation factor β, x_i is the width of the ith region, $\overline{x}$ expresses the average width of the regions. In this feature, the summary of the width is used to normalize the variance.

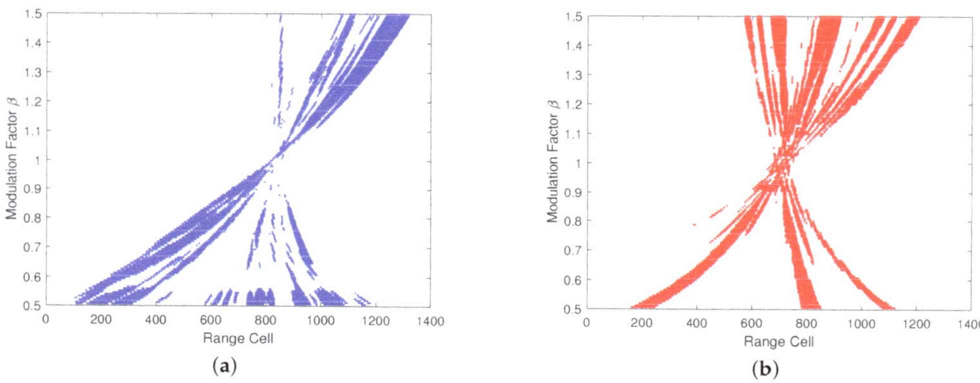

Figure 5. The distribution of points within the range of −5 dB. (**a**) Ship. (**b**) Corner reflector array.

(2) Variance of intervals.

Similarly, it can be demonstrated that the regions of the ship are relatively concentrated, with a few bright bands that are far apart. In contrast, the regions of the corner reflector array are more evenly distributed. Therefore, the variance of intervals is used to characterize the distribution characteristics, with the formula depicted below.

$$\sigma^2_{\text{gap},\alpha} = \frac{1}{N-2}\sum_{i=1}^{N-1}\left(\frac{g_i - \overline{g}}{L}\right)^2 \quad (9)$$

where g_i represents the interval between the ith region and the $i+1$th region, $\overline{g}$ denotes the average interval, L represents the total width under the modulation factor β. Figure 6 takes the condition of $\beta = 1.3$ in Figure 5a as an example, where the specific meanings of interval and width are explained.

3.3. Identification Method Based on SVM

SVM is a classifier used for solving binary classification problems [38]. It achieves non−linear classification through kernel functions that map data into higher dimensions, aiming to find a separation hyperplane that correctly divides the training data with the maximum geometric margin. SVM exhibits many unique advantages in addressing small sample sizes, non−linearity, and high−dimensional pattern recognition tasks. Compared to the SVM classifiers with linear kernel function, the SVM classifier using Gaussian radial basis kernel function has advantages such as diverse boundaries and higher classification accuracy. Therefore, this paper will use SVM based on the Gaussian radial basis kernel function to distinguish corner reflector arrays. The SVM identification process of corner reflector arrays is illustrated in Figure 7.

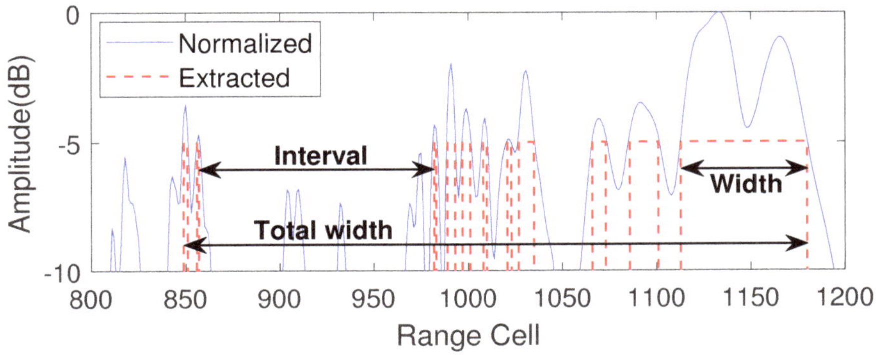

Figure 6. The intuitive illustration of the two extracted features.

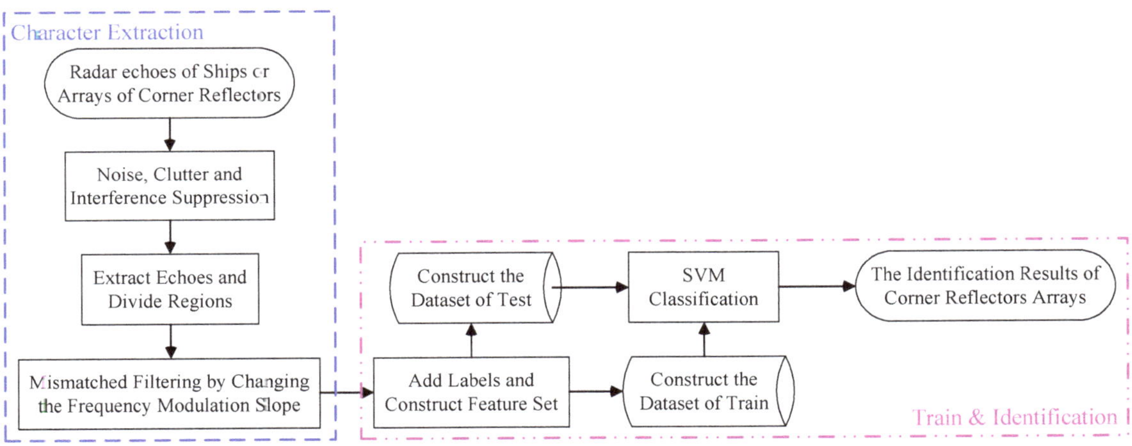

Figure 7. The identification process of corner reflector array.

The identification process can be divided into two parts. The initial stage of the process involves the extraction of characteristics. Firstly, the methods for interference suppression should be employed in order to moderate the influence of strong noise and clutter. Secondly, the search radius should be set to half of the maximum target size, after which the number of targets in the range profile can be determined. In the event that multiple targets exist, it is necessary to split their regions separately. Finally, the frequency modulation slope is modified in order to construct the mismatched filter, which is then employed to generate the modulation factor−range two−dimensional image and calculate the proposed features.

The second part is to train and to identify the corner reflector array. At first, add the appropriate labels in order to construct the training dataset of ships and corner reflector arrays. The format of the training dataset for SVM classifier can be expressed as $D = [x_1 \quad x_2 \quad y]$, where x_1 and x_2 separately represent the variance of width and intervals. And $y \in \{+1 \quad -1\}$ is the label set, where $+1$ represents ships and -1 represents corner reflector arrays. The SVM classifier can be used to train an optimal SVM classification model through simulated data of corner reflector arrays or ships on the sea surface. Subsequently, this model will be employed to identify corner reflector arrays within the testing dataset.

4. Simulation Experiment Analysis

4.1. Data Acquisition

In the experiments, we still utilize the electromagnetic simulation software (CST Studio Suite 2021) to obtain the backward scattering characteristics of different targets. Multiple models are employed to ensure the validity of the results, as illustrated in Figure 8. The shape parameters of four ship models are listed in Table 2.

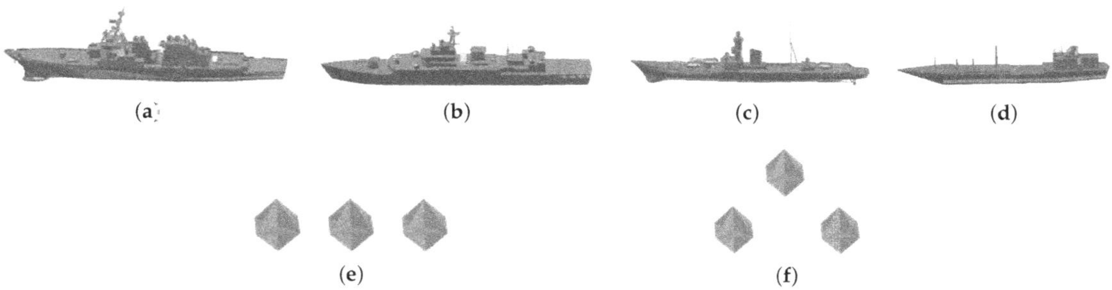

Figure 8. The models of ships and corner reflector arrays. (**a**) Ship 1. (**b**) Ship 2. (**c**) Ship 3. (**d**) Ship 4. (**e**) Corner reflector array 1. (**f**) Corner reflector array 2.

Table 2. Shape parameters of four ship models.

Type	Length (m)	Width (m)	Height (m)
Ship 1	169.39	22.90	56.25
Ship 2	145.35	17.74	34.27
Ship 3	107.74	11.39	29.14
Ship 4	130.80	10.12	23.32

Considering the scenario of dilution jamming, the corner reflector array is positioned in alignment with the ship's navigation direction on the sea surface, with a sufficient distance between them to ensure that both can be divided. Set the direction along the bow of the ship as 0° in azimuth angle, and downward from the deck as 0° in pitch angle. The bandwidth of the LFM signal used in these simulations is 150 MHz, with a pulse duration of 10 μs, and the center frequency is 10 GHz. The electromagnetic scattering data utilized in this paper are confined to the pitch angle range of 20° to 90° and azimuth angle range of 0° to 70°. In this method, the polarization information is not utilized. Consequently, the data employed in this method comprise only those of the same polarization type, both for the transmission and reception. To better approximate the ship's output, we also place one or two groups of identical corner reflector arrays to acquire similar length in range profile. The details of simulation experiments are listed in Table 3. It is important to note that the training dataset is derived from data generated by a single model with corresponding numbers, as indicated in Table 3. And the testing dataset comprises data simulated from each sub−row in Table 3.

Table 3. Groups of simulation data.

Type of Ship	Type of Array [1]	Number of Arrays
Ship 1	Array 1 Array 2	2 2
Ship 2	Array 1 Array 2	1 2
Ship 3	Array 1 Array 2	1 2
Ship 4	Array 1 Array 2	2 2

[1] Array denotes corner reflector array.

4.2. Identification Based on a Single Modification Factor

To further investigate the application conditions of this method and seek better discrimination performance, we first perform the identification process under a single modification factor β. In this section, we choose seven modification factors and calculate the proposed characters within the range of -5 dB. Meanwhile, we also test the condition of $\beta = 1.1$ within different ranges of selection area. Considering the impact of noise with varying amplitudes on classification accuracy, this paper introduces white Gaussian noise with varying SNR and employs 20 Monte Carlo simulations to compute the accuracy rate. The accuracy rates of these tests are graphed in Figure 9.

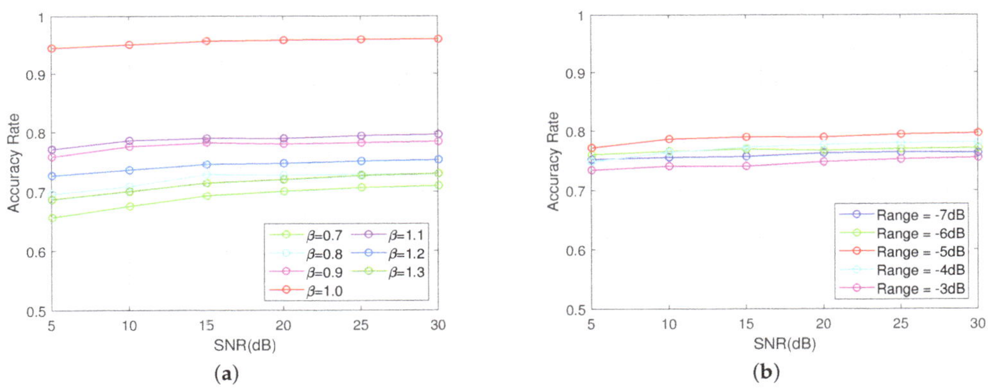

Figure 9. The identification performance of single modulation factor. (**a**) Different modulation factor but within the same range of -5 dB. (**b**) Same modulation factor within different ranges.

In Figure 9a, the red line represents the recognition accuracy rate under the condition of $\beta = 1$, which can also be regarded as the matched filter scenario. It is evident that the proposed features also demonstrate excellent discriminative performance on HRRP. But this performance is also constrained by some factors such as the quantity of corner reflectors, which will be discussed in Section 4.5. In addition, the bright−colored lines represent the conditions of factor β greater than 1, and the dark−colored lines represent the conditions of factor β less than 1. It can be observed that at the same deviation level $|1 - \beta|$, there is a deterioration in identification performance when the modulation factor β is less than 1.

Meanwhile, we can observe that the recognition performance under other single modulation factors is not satisfactory, with an accuracy rate that falls below 0.8 in each case. Figure 9b demonstrates that there is a better identification performance within the range of -5 dB. Nevertheless, the accuracy rate within each range remains not ideal.

The suboptimal discriminative performance under the condition of a single modulation factor is primarily attributable to the widening of the main lobe. The proposed mismatched filter does indeed amplify the differences of scattering characteristics to a certain extent. However, due to the effects of attenuation and superposition, it is more susceptible to the influence of incidental factors, thus making it difficult to accurately guarantee its performance under individual modulation factors. From Figure 9a, we can also find that the greater the extent of main lobe widening, the more unstable the robustness of its performance under this modulation factor.

4.3. Identification Based on a Range of Modulation Factors

Due to the unsatisfactory and unstable identification performance under a single modulation factor, this part will perform the identification process on a range of modulation factors. Consequently, the proposed features will be acquired by calculating the mean value of features in different modulation factors, thereby characterizing the average fluctuation of distribution differences.

As demonstrated in Section 4.2, the identification process exhibits a better performance when β is set to 1 and the range is selected to be -5 dB. Accordingly, this section sets the range of modulation factor as 1 to 1.1, with a step size of 100 (exclusive of $\beta = 1$). Based on the training dataset, the distributions of the two features for ships and corner reflectors arrays are, respectively, depicted below.

Figure 10 reveals a clear disparity in the distribution of the two features. In both characteristics, the values of the corner reflector arrays are relatively low, concentrated near the $X-$axis. In contrast, the ship's values exhibit a higher concentration, with a higher interquartile range compared to the corner reflector arrays. From the joint distribution in Figure 11, it can be observed that the feature points of corner reflector arrays are relatively concentrated, clustered near the origin. On the contrary, the ones of ships are relatively dispersed, forming an arc$-$shaped distribution with only a few overlapping regions. Consequently, subsequent validations of the method's performance will be conducted under these parameters.

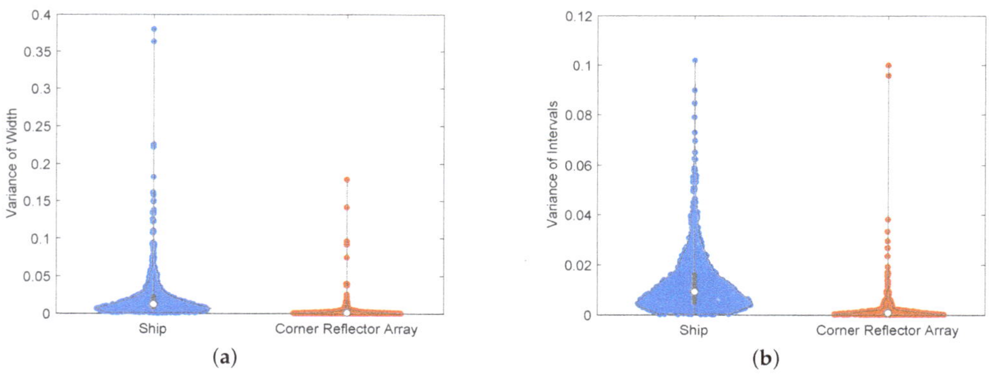

Figure 10. The distribution of characteristics. (**a**) The variance of width. (**b**) The variance of intervals.

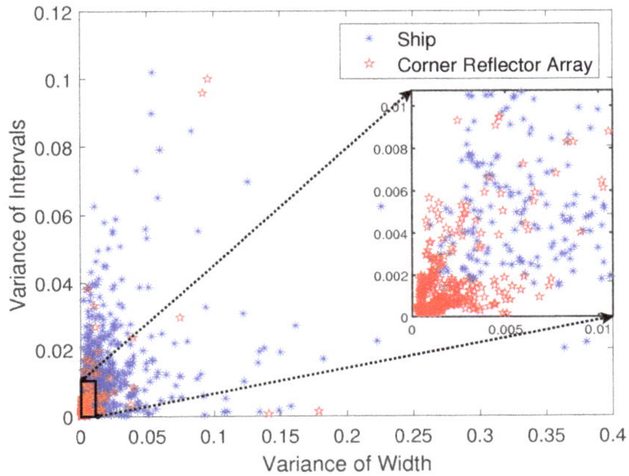

Figure 11. The joint distribution of the two features.

To further demonstrate the advantages of this method, several comparative methods were employed in this experiment. It is very regrettable to note that since electromagnetic simulation software is unable to accurately simulate the scattering characteristics of targets in motion, identification methods based on frequency domain features cannot be introduced for comparison, such as doppler frequency and micro−motion period. These methods for comparison in this section include the HRRP methods in [8], polarization decomposition method including Cloude decomposition [39] and Krogager decomposition [40], polarization−invariant method [41], and the method utilizing correlation characteristic parameters from the polarization−range 2D image [7].

Krogager decomposition divides the polarimetric scattering matrix of scattering points into three components: odd scattering, second−order scattering with rotation angle, and helix scattering. Thus, we choose the first two as discriminative features for the SVM classifier. In the method of Cloude decomposition, we utilize the scattering entropy and average scattering angle as the features for identification. Regarding the feature selection of polarization invariants, we chose three more important features based on the results in [24], which are the shape factor, depolarization coefficient, and target aspect ratio.

In the context of simulation experiment scenarios, the angular size of the corner reflector array and the ship are nearly similar on the HRRP. Consequently, the other comparative experiment primarily utilizes two HRRP features in [8]. The two features are total half−peak breadth (THPB) and mean differential amplitude (MDA), and their expressions are as follows.

$$THPB = \sum_{i=1}^{k} HPB_i \tag{10}$$

$$MDA = \frac{1}{p_e - p_s} \sum_{n=p_s}^{p_e-1} \frac{|x_{n+1} - x_n|}{\max_{p_s \le i \le p_e} x_i}, p_s \le n \le p_e - 1 \tag{11}$$

In Equations (10) and (11), $[p_s \quad p_e]$ represents the region of the target location, k represents the number of peaks above the threshold within the region, HPB_i refers to the half−peak width of the ith peak, x_i is the amplitude at the ith range unit.

Figure 12 shows the discriminative accuracy rate of various methods under different SNR conditions, and Table 4 is the numerical comparison table. When SNR is greater than −5 dB, it is evident that the identification accuracy of the proposed method is higher than other methods, almost exceeding 0.86 when the SNR is between −5 and 30. When the influence of noise is minimal, the accuracy of the proposed method can be approximated to 0.9. Compared to the methods in polarization, the proposed method and the HRRP method are significantly impacted by SNR, especially at a low SNR. This is mainly because the features selected in this method are the distribution characteristics within the area above the −5 dB region, which is relatively easily affected by strong noise or clutter. However, the polarization methods usually analyze the scattering matrix of peaks on polarization HRRP, which is minimally affected by noise. Nevertheless, it is noteworthy that the proposed method has demonstrated enhanced robustness with regard to noise in comparison to the HRRP method in [8].

Table 4. The comparison table of different methods.

Methods	−15	−10	−5	0	5	10	15	20	25	30
The Proposed Method	0.665	0.744	**0.858**	**0.877**	**0.884**	**0.887**	**0.889**	**0.891**	**0.893**	**0.896**
HRRP Method in [8]	0.548	0.741	0.827	0.857	0.856	0.862	0.864	0.866	0.869	0.871
Polarization Modulation [7]	**0.727**	**0.804**	0.838	0.855	0.863	0.876	0.878	0.878	0.878	0.879
Polarization Invariant [41]	0.654	0.742	0.801	0.824	0.829	0.831	0.832	0.836	0.837	0.840
Cloude Decomposition [39]	0.683	0.751	0.790	0.794	0.806	0.804	0.806	0.809	0.814	0.818
Krogager decomposition [40]	0.651	0.725	0.762	0.792	0.816	0.832	0.847	0.855	0.855	0.856

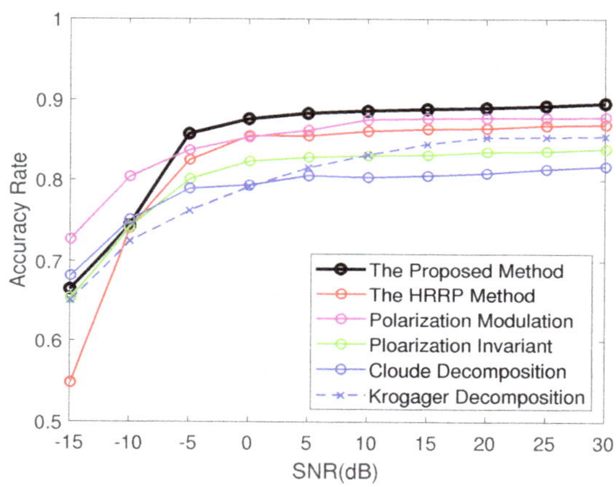

Figure 12. The identification performance of different methods. Where the HRRP method is in Ref. [8], polarization modulation is in Ref. [7], polarization invariant is in Ref. [41], Cloude decomposition is in Ref. [39], krogager decomposition is in Ref. [40].

However, it should be pointed out that the method in the polarization domain has extremely high requirements for measurement accuracy and utilizes information from multiple channels. Consequently, the polarization method has extremely high requirements on equipment, such as polarization measurement error and polarization isolation. On the contrary, the proposed method only requires multiple mismatched filters of the transmitted signal. Meanwhile, this method has relatively low requirements on target echoes, which can be combined with the common suppression method for noise, clutter, and interference. These methods include time−domain cancellation [42], blind source separation [43], cyclic cancellation [44], and so on. Therefore, in a low−SNR environment, it is first necessary to focus on improving the radar's noise resistance and target detection effectiveness, and then find solutions to suppress noise.

Considering that Gaussian noise may not easily simulate actual radar environments, we simulate the noise under other distribution functions to further assess the identification performance of the proposed method. These distribution functions include Rayleigh distribution, K−distribution, lognormal distribution, and Weibull distribution. In order to get closer to the actual situation, we simulate these distribution functions under different sea conditions. The data are simulated under these distribution functions with reference to the method in [45] and the parameters given in [46], which is most similar to the amplitude distribution of the IPIX dataset [47]. The identification performance under different distribution functions is plotted in Figure 13.

The black solid line in Figure 13 demonstrates the identification performance under the Gaussian complex noise. In other colors of lines, the solid lines represent the accuracy rate under high sea condition, and the dashed lines are the identification performance under low sea condition. The proposed method exhibits a superior performance under lognormal distribution in high sea conditions, while the other methods exhibit comparable performance. In comparison to the identification performance under other distributions, the maximum discrepancy in accuracy rate under Gaussian complex noise does not exceed 0.012 when SNR is greater than 0 dB. Therefore, in the remainder of this paper, we will still use Gaussian complex noise to study the identification performance under different SNR.

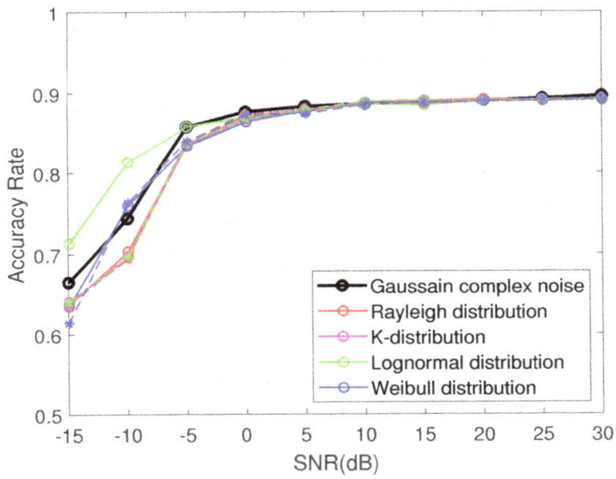

Figure 13. The identification performance under noise of different distributions.

4.4. Stability Tests under Different Conditions

The efficacy of the proposed method is contingent upon two parameters: the range of modulation factor and the range of point extraction. The impact of varying these two factors on the identification performance will be investigated in the following context. In this section, we denote ΔR as the degree of range variation, which is the ratio of changed amount to the original range. In the circumstances of different ΔR, the step size is maintained at 100.

Figure 14a illustrates the accuracy rate of different ranges of modulation factors. It can be observed that the identification performance remains relatively consistent across different ranges, with differences within 0.01. Similarly, Figure 14b demonstrates that the accuracy rate of identification varies slightly when the range of point extraction is above -3 dB, with differences within 0.015. Nevertheless, when the range of point extraction is equal to -3 dB, there is a slight decline in identification performance. This is attributed to the excessively high threshold setting, which results in a limited extraction area for the feature points. Consequently, this inadequate coverage fails to accurately represent the scattering characteristics of targets. However, in general, the proposed method demonstrates good robustness with respect to the variation of the two parameters.

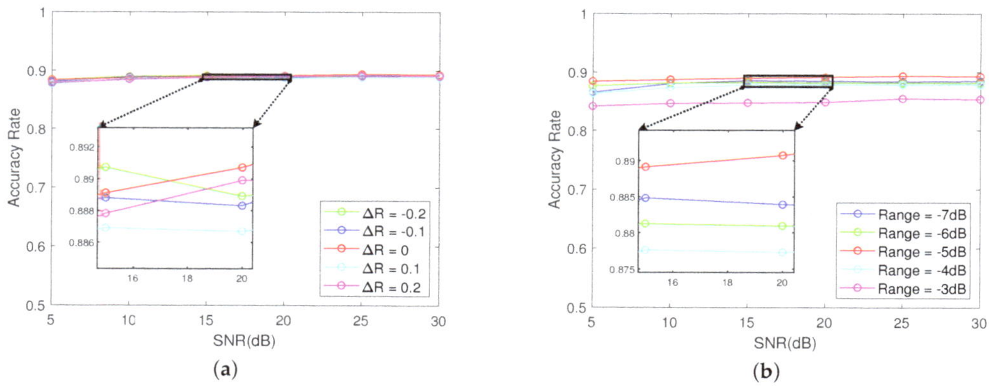

Figure 14. The identification performance under different parameters. (**a**) Different range of modulation factors but within the same range of −5 dB. (**b**) Different range of point extraction but within the same range of modulation factors.

The HRRP−based method is significantly affected by the observation angle. Therefore, it is imperative to investigate the recognition performance of the proposed method when subjected to varying observation angles or the pitch and yaw angles of the ship.

As indicated in Section 4.1, the electromagnetic scattering data utilized in this paper are confined to the pitch angle range of 20° to 90° and azimuth angle range of 0° to 70°, with steps of 10°. To further validate the performance of the method under different pitch and yaw angles, we designed relevant experiments based on the simulation data. By extracting a portion of the data from specific angles to serve as the test set, while using the remaining portion as the training set, we can evaluate the identification performance at new angles.

Figure 15 illustrates the identification performance under two sorts of conditions, which only change pitch angles or yaw angles. The designation "Train X & Test X" indicates that the condition employs X groups of angles for training and utilizes other X groups of angles for testing, resulting in a total of eight. To mitigate the impact of strong noise, SNR was set to a range of 5 to 30. Each condition was randomly sampled eight times, and the mean accuracy rate was calculated.

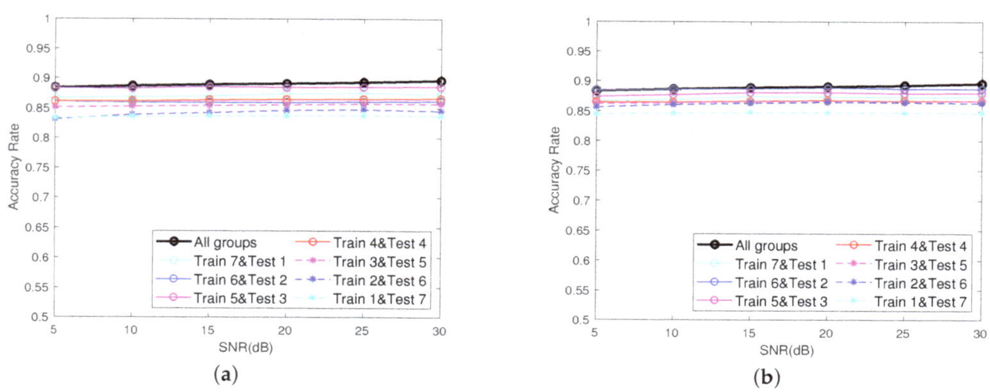

Figure 15. The identification performance of different conditions. (**a**) Training and testing using different pitch angles. (**b**) Training and testing using different yaw angles.

As can be seen from Figure 15, the recognition accuracy rate in each condition is higher than 0.83. When there are more training data, containing data from more angles, the recognition performance is better. However, the difference in recognition accuracy under different conditions is not large, and the maximum discrepancy does not exceed 0.05. This phenomenon indicates that when the training data do not contain the current angle of the ship, this method can still maintain good recognition performance. It is further explained that this method mainly relies on the structural differences between targets for identification, and its performance is less affected by factors such as yaw and pitch angles.

4.5. Stability Tests under Different Quantities of Corner Reflectors

This section compares the proposed method with the HRRP method in [8] to examine the stability of the discriminative performance when different quantities of corner reflectors are employed in the testing set. Furthermore, the method that utilizes the proposed features under the matched filter scenario is included as a comparison method. Based on the training dataset simulated from the single models presented in Table 3, the two−dimensional distributions of features for each method are plotted in Figure 16. To avoid the influence of strong noise, this section set the SNR as 5 to 30.

Comparing Figure 11 with Figure 16a,b, it can be observed that in the proposed method, the feature points of the corner reflector array are more concentrated in the area near the origin. However, in the other methods that extract features based on HRRP, they are centralized into several regions. This difference may be related to the number of corner reflectors. To validate this hypothesis, several sets of experiments were designed to assess the discriminative performance of each method when the training and testing datasets have different quantities of corner reflectors.

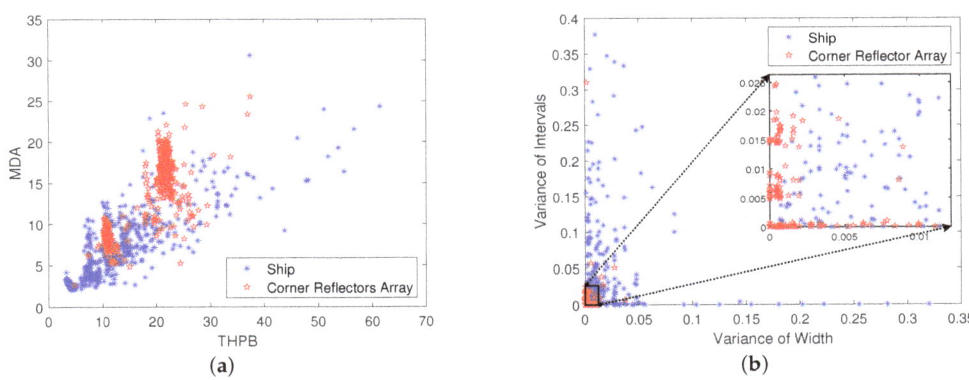

Figure 16. The two−dimensional distribution of different methods. (**a**) The HRRP method in [8]. (**b**) The proposed features in HRRP.

Table 3 presents the simulated dataset, comprising two sets of corner reflector arrays with different quantities, prompting us to conduct two sets of experiments. In each experimental set, the SVM classifier will be trained with simulated data using a specific quantity of corner reflectors, after which it will be tested with simulated data from a different quantity. From this process, we can conduct the assessment of discriminative performance across varying numbers of corner reflectors, based on different methods. The following diagrams illustrate the distribution of the testing dataset across varying quantities of corner reflector arrays, all under a SNR of 25 dB.

From Table 3, we can find that the dataset generated from a single group of corner reflector arrays is considerably smaller than that derived from two groups. Consequently, the number of feature points in the second row is markedly less than that in the first row in Figure 17. Due to the compressive property induced by pulse compression and the stretching variation of the range profile across different observation angles, the features proposed in this paper for matched filter scenarios may yield numerous meaningless values, necessitating filtration. As depicted in Figure 17c,f, it is evident that the number of feature points in this condition is comparatively fewer than in the other two methods.

Analyzing the vertical subplots in Figure 17, we can find that in the proposed method, the distribution areas of different quantities of corner reflectors are largely similar. However, under the other two methods, there are significant differences in the feature distribution among different quantities of corner reflectors. These figures suggest that the method proposed in this paper is less affected by variations in the quantity of corner reflectors. Subsequently, we will plot the identification accuracy rate under two sets of experiments in order to further compare the robustness of each method on this condition.

The dash lines in Figure 18 represent the identification performance under the condition where the quantity of corner reflector is held constant, whereas the solid line illustrates the alternative case. Correspondingly, the solid line denotes the experiment group. Comparing Figure 18a with Figure 18b, it can be observed that there exists an overall decrease in accuracy in the right graph. This is because the training dataset used in Figure 18b only consists of data from a single corner reflector array, resulting in a relatively small sample size, which may have led to a less effective identification performance.

We can observe that when the testing dataset contains data from different quantities of corner reflectors, the proposed method in this paper exhibits the smallest variation in accuracy. Specifically in Figure 18b, the variation is within the range of 0.03. In contrast, extracting features on the HRRP is significantly influenced by the quantity of corner reflectors, as evidenced by the sharp decrease in Figure 18. It is illustrated that the proposed method demonstrates good robustness with respect to variations in the quantity of corner reflectors.

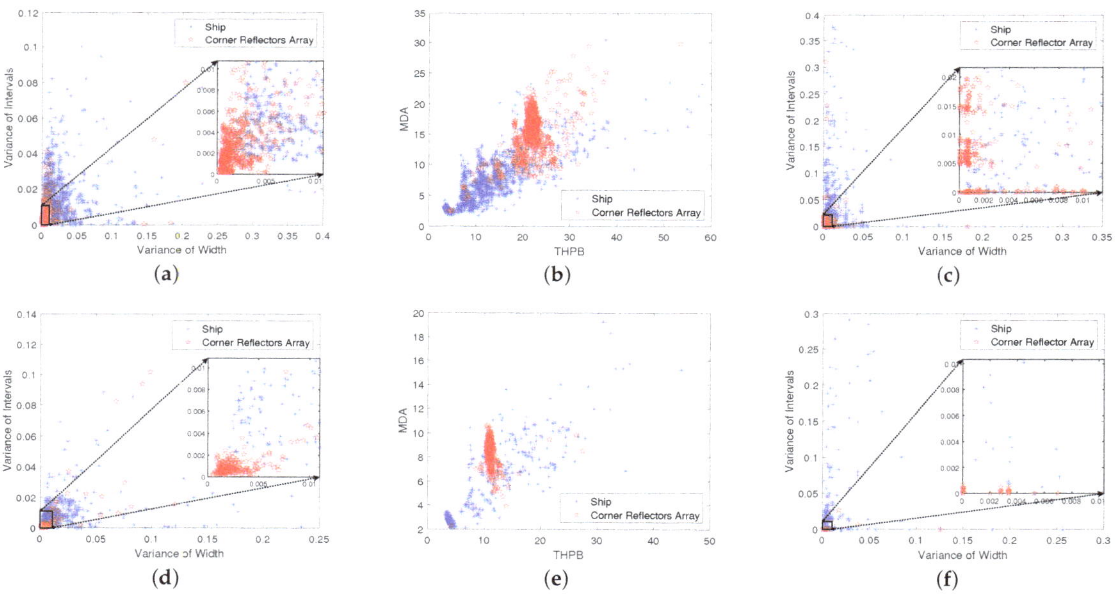

Figure 17. The two−dimensional distribution of different methods under different quantities of corner reflectors. (**a**,**d**) Based on the proposed method. (**b**,**e**) Based on the HRRP method in [8]. (**c**,**f**) Based on the proposed features in HRRP.

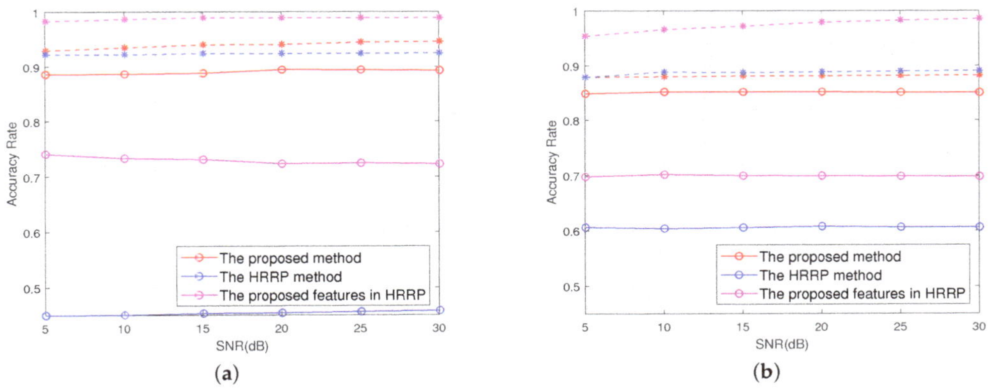

Figure 18. The identification accuracy rate of different methods under two sets of experiments. (**a**) A single array of corner reflectors. (**b**) Two arrays of corner reflectors.

5. Conclusions

To address the challenge of passive interference from corner reflector arrays in the anti−ship scenarios, this paper proposes a mismatched filtering method based on changing frequency modulation slope. Through analysis of the mismatched filtering output of simple scattering points in different distributions or types, it can be seen that the proposed method is capable of reflecting the scattering characteristics of the target to some extent. Then, based on the simulation data, we separately construct modulation factor−range two−dimensional images of ships and corner reflector arrays. Focusing on the differences in these images, this paper extracts the variance of width and intervals in a certain region as characteristics and designs an identification process based on the SVM. The results of numerical experiments conducted under different SNR conditions demonstrate that the proposed method exhibits excellent identification performance, consistently exceeding 0.86 when the SNR is greater than 0 dB. In terms of comparative experimental results among different methods, the proposed method is observed to exhibit superior discriminative performance when SNR exceeds 0 dB. In contrast to the method that extracts features in HRRP, this method demonstrates good robustness with respect to variations in the quantity of corner reflectors and is less susceptible to noise.

In the future, measured data will be collected to investigate the performance of the method in actual scenarios. Additionally, the data will be augmented by both simulation and measured data, seeking to enhance the performance of the method by developing a more effective classifier. In the meantime, additional research will focus on optimizing methods for area selection to enhance accuracy rate and exploring interference suppression techniques for application in environments with high levels of noise or clutter. The deployment strategy of corner reflector arrays will be optimized in order to make them more similar to ships in scattering characteristics or range profile to further validate the proposed method. Finally, we will be make efforts to combine other domain information, such as polarization, with the intention of enhancing the identification performance.

Author Contributions: Conceptualization, L.X. and F.W.; methodology, L.X. and F.W.; software, L.X., N.L. and R.P.; data curation, F.W. and L.X.; supervision, F.W., C.P. and Z.S.; writing–original draft preparation, L.X.; writing–review and editing, F.W., N.L. and C.P.; resources, Z.S., C.P. and Y.L. All authors have read and agreed to the published version of the manuscript.

Funding: This research was funded by the National Natural Science Foundation of China (No. 61921001 and No. 62301580).

Data Availability Statement: All data in this paper are generated by simulation and the details have been presented in Section 4.1.

Conflicts of Interest: The authors declare no conflicts of interest.

References

1. Li, H.; Chen, S. Electromagnetic scattering characteristics and radar identification of sea corner reflectors: Advances and prospects. *J. Radars* **2023**, *12*, 738–761.
2. Wu, L.; Xu, J.; Hu, S.; Liu, Z. High-frequency backscattering properties of quasi-omnidirectional corner reflector: The great-icosahedral-like reflector. *AIP Adv.* **2022**, *12*, 105225. [CrossRef]
3. Zhang, Z.; Zhang, J.; Qu, S.; Du, H. Research on radar corner reflector: Advances and perspectives. *Aerodyn. Missile J.* **2014**, *4*, 64–70.
4. Luo, Y.; Guo, L.X.; Zuo, Y.; Liu, W. Time-domain scattering characteristics and jamming effectiveness in corner reflectors. *IEEE Access* **2021**, *9*, 15696–15707. [CrossRef]
5. Jiang, T.; Luo, J.; Yu, Z. Research on corner reflector array fitting method for ship scattering characteristics. In Proceedings of the 2023 IEEE 2nd International Conference on Electrical Engineering, Big Data and Algorithms (EEBDA), Changchun, China, 24–26 February 2023.
6. Zhang, J.; Hu, S.; Wu, L.; Fan, X.; Yang, Q. Air-floating corner reflectors dilution jamming placement position. In Proceedings of the 2019 IEEE 8th Data Driven Control and Learning Systems Conference (DDCLS), Dali, China, 24–27 May 2019.
7. Wu, G.; Wang, L.; Pang, C.; Li, Y.; Wang, X. Radar polarization modulation countermeasures for combined corner reflector: Anti diluted jamming. *Acta Electron. Sin.* **2022**, *50*, 2969–2983.
8. Han, J.; Yang, Y.; Lian, J.; Wu, G.; Wang, X. Identification method of corner reflector based on polarization and HRRP feature fusion for radar seeker. *J. Syst. Eng. Electron.* 2023, *in press*.
9. Tang, G.; Li, H.; Gan, R.; Yuan, R. Analysis of corner reflector under naval battlefield. *Electron. Inf. Warf. Technol.* **2015**, *30*, 39–45+84.
10. Wang, L.; Jiang, N.; Sun, Y. The mechanism analyzing and use of corner reflector against anti-ship missiles. In Proceedings of the 2017 5th International Conference on Mechatronics, Materials, Chemistry and Computer Engineering (ICMMCCE 2017), Chongqing, China, 24–25 July 2017.
11. Zhou, X.; Zhu, J.; Yu, W.; Cui, T. Time-domain shooting and bouncing rays method based on beam tracing technique. *IEEE Trans. Antennas Propag.* **2015**, *63*, 4037–4048. [CrossRef]
12. Yuan, H.; Fu, X.; Zhao, C.; Xie, M.; Gao, X. Ship and Corner Reflector Identification Based on Extreme Learning Machine. In Proceedings of the 2019 IEEE International Conference on Signal, Information and Data Processing (ICSIDP), Chongqing, China, 11–13 December 2019.
13. Zeng, Z.; Sun, J.; Han, Z.; Hong, W. Radar HRRP target recognition method based on multi-input convolutional gated recurrent unit with cascaded feature fusion. *IEEE Geosci. Remote Sens. Lett.* **2022**, *19*, 4026005. [CrossRef]
14. Lv, F. Dynamic Echo Simulation and Characteristic Analysis of Sea Surface Targets. Master Dissertation, Xidian University, Xi'an, China, 2019.
15. Cui, K.; Wang, W.; Chen, X.; Yuan, N. A kind of method of anti-corner reflector interference for millimeter wave high resolution radar system. In Proceedings of the 2016 Progress in Electromagnetic Research Symposium (PIERS), Shanghai, China, 8–11 August 2016.
16. Shi, F.; Li, Z.; Zhang, M.; Li, J. Analysis and Simulation of the Micro-Doppler Signature of a Ship with a Rotating Shipborne Radar at Different Observation Angles. *IEEE Geosci. Remote Sens. Lett.* **2022**, *19*, 1504405. [CrossRef]
17. Hanif, A.; Muaz, M.; Hasan, A.; Adeel, M. Micro-Doppler Based Target Recognition With Radars: A Review. *IEEE Sens. J.* **2022**, *22*, 2948–2961. [CrossRef]
18. Fang, M.; Zhu, Y.; Huang, M.; Fu, Q. Sea surface target polarization feature extraction based on modified odd-time and even-time scattering models. In Proceedings of the 2013 2nd International Conference on Measurement, Information and Control, Harbin, China, 16–18 August 2013.
19. Liang, Z.; Yu, Y.; Zhang, B. Anti-corner reflector array method based on pauli polarization decomposition and BP neural network. In Proceedings of the 2021 IEEE 2nd International Conference on Pattern Recognition and Machine Learning (PRML), Chengdu, China, 16–18 July 2021.
20. Cloude, S.R.; Pottier, E. A review of target decomposition theorems in radar polarimetry. *IEEE Trans. Geosci. Remote Sens.* **1996**, *34*, 498–518. [CrossRef]
21. Krogager, E. New decomposition of the radar target scattering matrix. *Electron. Lett.* **1990**, *26*, 1525–1532. [CrossRef]
22. Chen, S.; Li, Y.; Wang, X.; Xiao, S.; Sato, M. Modeling and Interpretation of Scattering Mechanisms in Polarimetric Synthetic Aperture Radar: Advances and perspectives. *IEEE Signal Proc. Mag.* **2014**, *31*, 79–89. [CrossRef]
23. Li, H.; Li, M.; Cui, X.; Chen, S. Man-made target structure recognition with polarimetric correlation pattern and roll-invariant feature coding. *IEEE Geosci. Remote Sens. Lett.* **2022**, *19*, 8024105. [CrossRef]
24. Liang, Z.; Wang, Y.; Zhao, X.; Xie, M.; Fu, X. Identification of ship and corner reflector in sea clutter environment. In Proceedings of the 2020 15th IEEE International Conference on Signal Processing (ICSP), Beijing, China, 6–9 December 2020.
25. Chen, S.; Wu, G.; Dai, D.; Wang, X.; Xiao, S. Roll-Invariant Features in Radar Polarimetry: A Survey. In Proceedings of the IGARSS 2019–2019 IEEE International Geoscience and Remote Sensing Symposium, Yokohama, Japan, 28 July–2 August 2019.

26. He, Y.; Yang, H.; He, H.; Yin, J.; Yang, J. A Ship Discrimination Method Based on High-Frequency Electromagnetic Theory. *Remote Sens.* **2022**, *14*, 3893. [CrossRef]
27. Yu, K.; Zhu, S.; Lan, L.; Zhu, J.; Li, X. Mainbeam Deceptive Jammer Suppression With Joint Element-Pulse Phase Coding. *IEEE Trans. Veh. Technol.* **2024**, *73*, 2332–2344. [CrossRef]
28. Wang, F.; Li, N.; Pang, C.; Li, Y.; Wang, X. Algorithm for Designing PCFM Waveforms for Simultaneously Polarimetric Radars. *IEEE Trans. Geosci. Remote Sens.* **2024**, *62*, 5100716. [CrossRef]
29. Jin, L.; Wang, J.; Zhong, Y.; Wang, D. Optimal Mismatched Filter Design by Combining Convex Optimization with Circular Algorithm. *IEEE Access* **2022**, *10*, 56763–56772. [CrossRef]
30. Mattingly, R.G.; Martone, A.F.; Metcalf, J.G. Techniques for Mitigating the Impact of Intra-CPI Waveform Agility. *IEEE Trans. Radar Syst.* **2024**, *2*, 24–40. [CrossRef]
31. Liu, G.; Gao, W.; Liu, W.; Xu, J; Li, R.; Bai, W. LFM-Chirp-Square pulse-compression thermography for debonding defects detection in honeycomb sandwich composites based on THD-processing technique. *Nondestruct. Test. Eval.* **2024**, *39*, 832–845. [CrossRef]
32. Dai, H.; Zhao, Y.; Su, H.; Wang, Z.; Bao, Q.; Pan, J. Research on an intra-pulse orthogonal waveform and methods resisting interrupted-sampling repeater jamming within the same frequency band. *Remote Sens.* **2023**, *15*, 3673. [CrossRef]
33. Kouyoumjian, R.G.; Pathak, P.H. A uniform geometrical theory of diffraction for an edge in a perfectly conducting surface. *IEEE Trans. Signal Process.* **1974**, *62*, 1448–1461. [CrossRef]
34. Ghasemi, M.; Sheikhi, A. Joint Scattering Center Enumeration and Parameter Estimation in GTD Model. *IEEE Trans. Antenn. Propag.* **2020**, *68*, 4786–4798. [CrossRef]
35. Hu, P.; Xu, S.; Zou, J.; Chen, Z. Parameter estimation of GTD model using iterative adaptive approach. In Proceedings of the 2017 IEEE SENSORS, Glasgow, UK, 29 October–1 November 2017.
36. Chen, S.; Pan, M. Analytical Model and Real-Time Calculation of Target Echo Signals on Wideband LFM Radar. *IEEE Sens. J.* **2021**, *21*, 10726–10734. [CrossRef]
37. Li, S.; Wang, X.; Fu, Z.; Zhang, J. Extraction of scattering center parameter and RCS reconstruction based on the improved TLS-ESPRIT algorithm of Hankel matrix. *J. Syst. Eng. Electron.* **2021**, *43*, 62–73.
38. Zhou, Z. *Machine Learning*, 1st ed.; Tsinghua University Press: Beijing, China, 2016; pp. 121–139.
39. Lu, Z.; Wang, Z.; Dan, B. Ship target identification method based on the characteristic of target polarimetric HRRP of radars. In Proceedings of the 2022 Global Reliability and Prognostics and Health Management (PHM-Yantai), Yantai, China, 13–16 October 2022.
40. Zhu, Z.; Tang, G.; Cheng, Z; Huang, P. Discrimination method of ship and corner reflector based on polarization decomposition. *Shipboard Electron. Countermeas.* **2010**, *33*, 15–21.
41. Wang, M.; Xie, M.; Su, Q.; Fu, X. Identification of ship and corner reflector based on invariant features of the polarization. In Proceedings of the 2019 IEEE 4th International Conference on Signal and Image Processing (ICSIP), Wuxi, China, 19–21 July 2019.
42. Shi, Y.; Shui, P. Target detection in high-resolution sea clutter via block-adaptive clutter suppression. *IET Radar Sonar Navigat.* **2011**, *5*, 48–57. [CrossRef]
43. Chen, M.; Li, L.; Geng, Z.; Xie, X. Single-channel Blind Source Separation Algorithm Based on Water Area Noise Characteristics. In Proceedings of the 2022 IEEE International Conference on Mechatronics and Automation (ICMA), Guilin, China, 7–10 August 2022.
44. Lv, M.; Zhou, C. Study on Sea Clutter Suppression Methods Based on a Realistic Radar Dataset. *Remote Sens.* **2019**, *11*, 2721. [CrossRef]
45. Zhu, J.; Tang, J. K-distribution Clutter Simulation Methods Based on Improved ZMNL and SIRP. *J. Radars* **2014**, *3*, 533–540. [CrossRef]
46. Gao, Z.; Zhang, A. Simulation Analysis of Typical Amplitude Distribution Model of Sea Clutter. *Ship Electron. Eng.* **2018**, *38*, 75–79.
47. The McMaster IPIX Radar Sea Clutter Database. Available online: http://soma.mcmaster.ca/ipix/ (accessed on 29 May 2024).

remote sensing

MDPI

Article

Scattering-Point-Guided Oriented RepPoints for Ship Detection

Weishan Zhao [1,2,3], Lijia Huang [1,2,3,*], Haitian Liu [4] and Chaobao Yan [1,2,3]

1 Key Laboratory of Technology in Geo-Spatial Information Processing and Application System, Chinese Academy of Sciences, Beijing 100190, China; zhaoweishan22@mails.ucas.ac.cn (W.Z.); yanchaobao22@mails.ucas.ac.cn (C.Y.)
2 Aerospace Information Research Institute, Chinese Academy of Sciences, Beijing 100094, China
3 School of Electronic, Electrical and Communication Engineering, University of Chinese Academy of Sciences, Beijing 100049, China
4 School of Software Engineering, Xi'an Jiaotong University, Xi'an 710049, China; liuhaitian@stu.xjtu.edu.cn
* Correspondence: iecas8huanglijia@163.com

Abstract: Ship detection finds extensive applications in fisheries management, maritime rescue, and surveillance. However, detecting nearshore targets in SAR images is challenging due to land scattering interference and non-axisymmetric ship shapes. Existing SAR ship detection models struggle to adapt to oriented ship detection in complex nearshore environments. To address this, we propose an oriented-reppoints target detection scheme guided by scattering points in SAR images. Our method deeply integrates SAR image target scattering characteristics and designs an adaptive sample selection scheme guided by target scattering points. This incorporates scattering position features into the sample quality measurement scheme, providing the network with a higher-quality set of proposed reppoints. We also introduce a novel supervised guidance paradigm that uses target scattering points to guide the initialization of reppoints, mitigating the influence of land scattering interference on the initial reppoints quality. This achieves adaptive feature learning, enhancing the quality of the initial reppoints set and the performance of object detection. Our method has been extensively tested on the SSDD and HRSID datasets, where we achieved mAP scores of 89.8% and 80.8%, respectively. These scores represent significant improvements over the baseline methods, demonstrating the effectiveness and robustness of our approach. Additionally, our method exhibits strong anti-interference capabilities in nearshore detection and has achieved state-of-the-art performance.

Keywords: ship detection; reppoints; adaptive sample selection; guided learning; synthetic aperture radar (SAR); scattering point

Citation: Zhao, W.; Huang, L.; Liu, H.; Yan, C. Scattering-Point-Guided Oriented RepPoints for Ship Detection. *Remote Sens.* **2024**, *16*, 933. https://doi.org/10.3390/rs16060933

Academic Editor: Paolo Tripicchio

Received: 29 January 2024
Revised: 28 February 2024
Accepted: 4 March 2024
Published: 7 March 2024

1. Introduction

Currently, radar application scenarios are continuously expanding, with various radar systems emerging and advancing rapidly [1]. Synthetic aperture radar (SAR) functions as an active microwave imaging system, impervious to natural conditions such as illumination, clouds, and weather. Consequently, it boasts the capability for all-weather, day-and-night observation of the Earth's surface, establishing itself as a primary tool for current maritime applications [2]. Ship detection, a main direction in the maritime domain [3], holds a critical role in monitoring maritime transportation, managing ports, and overseeing maritime zones [4]. In recent years, more and more SAR satellites have been successfully launched [5–7], with continuous advancements in collaborative observation techniques [8]. The improvement of data quality [9], the further diversity of imaging scenarios, and the continuous establishment and upgrading of SAR datasets [10,11] have greatly promoted the development of intelligent technology for the interpretation of SAR images [12].

Among traditional algorithms, CFAR [13] stands out as one of the most widely used detection methods, relying on manually crafted features. This method entails modeling cluttered backgrounds by setting the background threshold to a predetermined level, thus identifying abnormal pixel points that deviate from the background distribution. Various CFAR-based detection algorithms have emerged by employing different background-modeling models [14–17]. However, it is still susceptible to interference from complex environments, and its nearshore environment detection performance is low.

Due to advancements in deep learning algorithms, numerous CNN-based object detection techniques tailored to natural scenes have been applied in SAR target detection [18]. These algorithms leverage the robust feature extraction and representation capabilities inherent in CNNs [19], exhibiting superior performance in detection when compared to traditional methods such as CFAR [20]. However, a significant limitation arises from the fact that most of these detection networks are designed based on horizontal bounding boxes, commonly used in natural scenes. Traditional horizontal bounding boxes exhibit overlap in nearshore ship detection, introducing interference from land areas beyond the target region and hindering the extraction of detailed target features. Consequently, this challenge limits the network's ability to effectively capture a target's fine-grained structural texture features [21]. In contrast, oriented bounding boxes avoid these issues.

In reference to the previously mentioned concern, numerous algorithms centered around oriented object detection have been introduced. These algorithms are predominantly categorized into two main types: two-stage algorithms employing anchor boxes and single-stage algorithms without anchor boxes.

The first type of anchor box detection algorithm based on the rotated box (RBOX) often oversamples anchor boxes with a specified aspect ratio and generates a large number of anchor boxes. On the one hand, it greatly increases the number of parameters and the computational complexity. On the other hand, the anchor frame with artificially fixed proportions is difficult to adapt to multi-scale and multi-directional ship targets. At the same time, when distinguishing between positive and negative samples through intersection over union (IoU), it will further aggravate the problem of the imbalance between positive and negative samples and the insufficiency of positive samples. Accordingly, the recall rate of the model will be reduced, and problems such as category skew and network degradation will occur, which make it difficult to achieve the network's generalization ability. The second type of detection algorithm improves the representation of bounding boxes by implementing a combined strategy of initializing anchor points and fine-tuning anchor points. The model scale and computational complexity are reduced. However, due to the lack of a target position prior, there is a lack of effective guidance when performing feature learning and initializing anchor points, and strong land scattering points in complex nearshore environments further interfere with the generation of initialized anchor points, resulting in lower learning sample quality and poor network detection performance.

In response to the aforementioned challenges, we propose a directional anchor-free detection network guided by significant scattering features in SAR images. Firstly, we adopt a lightweight single-stage reppoints detection architecture, which generates target boxes through reppoints and exhibits stronger adaptability and higher detection granularity for nearshore directional targets. Secondly, by comprehensively considering the imaging mechanism of SAR and the physical characteristics of strong scatterers such as ships, we integrate SAR image scattering properties for the first time. We extract strong scattering points from SAR images and design an adaptive sample selection scheme guided by these scattering points to select high-quality samples for network training. Additionally, we design a supervision guidance mechanism that utilizes target scattering points to guide the initialization of reppoints, thus achieving adaptive feature learning. The main contributions of our work can be summarized as follows:

1. A reppoints-based object detection network deeply fused with SAR scattering characteristics is proposed, which leverages the profound integration of SAR image scattering properties to guide the network for high-quality learning, enabling fine-grained nearshore detection.

2. To address the issue of low sample quality, this study introduces an innovative adaptive sample selection scheme known as SPG-ASS (Scattering-Point-Guided Adaptive Sample Selection). The method integrates the positional features of strong scattering points on ships to enhance the overall quality of samples. By extracting scattering points and clustering their positions into the optimal number of clusters, the method measures the similarity between the scattering point clusters and the set of sample points using the cosine similarity metric to achieve the best match. This, in turn, determines the quality score of the sample points. Finally, the adaptive selection of high-quality samples is achieved using the TOP K algorithm. This method further improves the quality of reppoints.
3. To reduce land scattering interference and further improve the quality of initialized reppoints, a novel reppoints supervision guidance paradigm is proposed. This paradigm aligns target scattering points with initialized reppoints at the point level by employing an intermediary framework. Using the KLD (Kullback–Leibler Divergence) loss, it integrates the structural and positional attributes of scattering point clusters into the supervised learning process of initialized reppoints. During the training phase, this paradigm effectively guides the reppoints to extract the semantic features of key regions in targets.

2. Related Works

2.1. Deep Learning-Based Object Detection

Object detection, as one of the fundamental visual tasks in deep learning, has seen the emergence of numerous algorithms with the advancements in deep learning. These algorithms can mainly be categorized into two classes: two-stage methods using anchor boxes and single-stage detectors without anchor boxes.

Two-stage methods using anchor boxes: Candidate regions are generated in the first stage, followed by the mapping of these regions to a fixed size in the second stage for classification and bounding box regression. For instance, R-CNN [22] utilizes selective search algorithms to produce candidate regions and then employs convolutional operations to obtain bounding boxes and their respective classes. SPP-Net [23] addresses the drawbacks of repeated convolutions and fixed output sizes. Fast R-CNN [24], building upon the aforementioned methods, utilizes ROI (Region of Interest) pooling to extract target features, sharing the tasks of bounding box regression and classification, thereby achieving end-to-end training. Faster R-CNN [25] replaces the selective search algorithm with an RPN (Region Proposal Network) to generate candidate boxes, significantly reducing algorithmic complexity. FPN [26] introduces a pyramid structure to leverage information from various scales, considerably enhancing the performance of object detection tasks.

Single-stage detectors without anchor boxes: These detectors do not rely on complex anchor box designs and accomplish object detection in a single stage. For example, YOLO-V1 [27] divides the image into a grid and predicts bounding boxes and confidence for all objects within each grid cell in one go. SSD [28] efficiently detects objects of various scales and aspect ratios by introducing multi-scale feature extraction and the Default Boxes mechanism. RetinaNet [29] addresses the issue of imbalanced positive and negative samples through the design of focal loss. CenterNet [30] models objects as the center points of bounding boxes, where the detector finds the center point through keypoint estimation and regresses other attributes of the target. Reppoints [31], considering the limited granularity in existing feature learning, utilizes a set of representative points to adaptively learn key semantic positions in the image, thereby achieving classification and regression.

2.2. Oriented Object Detection

Traditional horizontal bounding boxes often lack the capability to capture target orientation information and are prone to background interference, especially in intricate environments. Consequently, oriented object detection has emerged as a pivotal research area. For instance, ROI Transformer [32] employs spatial transformations of Regions of Interest, learning transformation parameters supervised by annotated directional bounding boxes. Oriented-RCNN [33] adjusts the regression parameters of the Region Proposal Network (RPN) to six, directly generating oriented proposals for corresponding targets. Utilizing KLD [34] to construct the Gaussian representation of oriented bounding boxes, it redesigns rotation regression losses, dynamically adjusting parameter gradients for object alignment. G-Rep [35] devises a unified Gaussian representation to construct Gaussian distributions for both OBBs and PointSets, accompanied by a Gaussian regression loss to further enhance object detection performance. Oriented Reppoints [36] utilizes an adaptive point learning methodology to capture the geometric information of arbitrary orientation instances and formulate schemes for adaptive point quality assessment and sample allocation.

Various oriented bounding box (OBB) detection algorithms have found applications in SAR ship detection. For instance, Zhang et al. [37] proposed a Rotated Region Proposal Network to generate multi-directional proposals with ship azimuth information, thereby enhancing the performance of multi-angle target detection. Yang et al. [38] introduced R-RetinaNet, which utilizes scale calibration methods to contrast scale distributions. They leveraged a task-level Feature Pyramid Network to fuse features, alleviating conflicts between different targets. Additionally, an adaptive IOU threshold training method was introduced to address imbalance issues. Yue et al. [39] proposed a method for detecting oriented ships in synthetic aperture radar (SAR) images, which improved the accuracy of detecting small oriented ships by fusing high-resolution feature maps and dynamically mining rotated positive samples (DRPSM). Sun et al. [40] proposed the SPAN, which integrates scattering characteristics for ship detection and classification. This addresses the weak detection performance caused by the lack of SAR features in conventional detectors. Zhang et al. [2] proposed an object detection network based on scatter-point-guided region proposal, combining SAR image scattering characteristics to guide an RPN in generating crucial proposals. They incorporated supervised contrastive learning to mitigate category differences, thereby enhancing the target detection performance.

2.3. Sample Assignment for Object Detection

Sample allocation plays a crucial role in the performance of object detection. Various sample allocation methods have been proposed, such as Faster R-CNN [25], SSD [28], and RetinaNet [29], which employ IOU for positive sample selection, relying on manually designed thresholds. ATSS dynamically adjusts thresholds based on the statistical features of sample groups. OTA [41] extends the consideration to ambiguous sample allocation (one-to-many) by transforming sample allocation into a dynamic programming problem. Furthermore, PAA [42] adapts sample allocation in a probabilistic manner. APAA [36] addresses the limitations of IOU in directional scenes by proposing an adaptive sample point set allocation scheme based on a comprehensive evaluation of orientation, classification, localization, and pixel-wise correlation. Although the above algorithms have proven their effectiveness in the field of optical images, in the field of SAR, we still need to further explore methods tailored to the characteristics of SAR images.

3. Materials and Methods

3.1. Overview of Model Structure

Figure 1 illustrates an overview of our proposed method. Our method can be mainly divided into four parts. The first part is the FPN backbone network, the second part is scattering point extraction and matching evaluation, the third part is adaptive sample selection, and the fourth part is reppoints generation of shared headers. This method starts with inputting the SAR image, which enters two feature extraction channels at the same time. One channel is the scattering point extraction branch based on corner points, and a strong scattering point set is obtained and adaptive clustering processing is performed to obtain several point clusters. The other channel is the deep semantic feature extraction part based on FPN, which extracts high-level features and then sends them to the shared headers. Through this two-stage operation, initialized reppoints and finely corrected reppoints are obtained. In the training phase, the initialized reppoints are sent to the adaptive sample selection part to evaluate the point set quality using matched and aligned scattering point clusters so as to select high-quality samples for learning. In addition, in order to improve the quality of the initialized reppoints, guided learning is performed through the SPG learning part. In the testing phase, the oriented detection results are directly generated by the finely corrected reppoints through the conversion function.

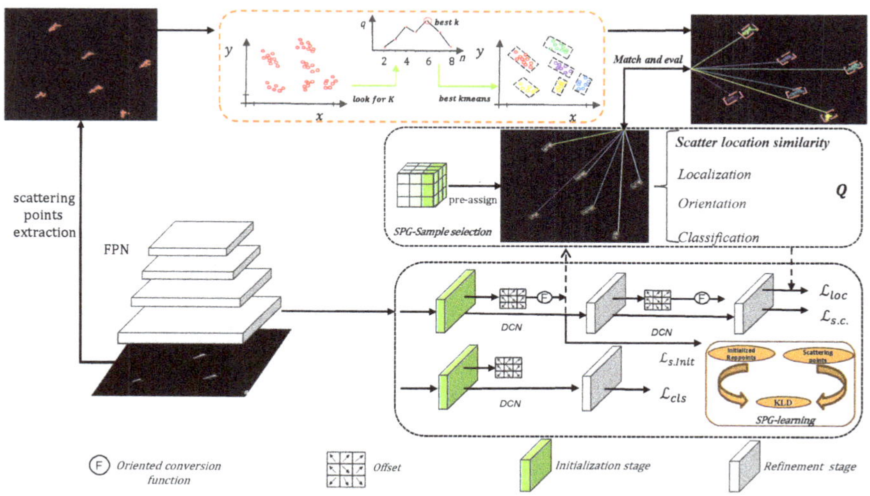

Figure 1. The model structure It consists of the FPN backbone network, scattering point extraction and matching evaluation, adaptive sample selection, and shared header for reppoints generation. Additionally, $L_{s.c}$ indicates the spatial constraint loss, L_{loc} indicates the localization loss, and $L_{s.Init}$ indicates the SPG learning loss.

3.2. *Scattering-Point-Guided Adaptive Sample Selection (SPG-ASS)*

3.2.1. Extraction of Scattering Points and Clustering

Ships usually have metal shells composed of a large number of strong scattering structures, such as dihedral angles, trihedral angles, etc., which, in turn, result in the strong scattering phenomena of ships in SAR images when combined with the unique imaging mechanism of the SAR system. These strong scattering points often contain the structure and location characteristics of the ship itself. For this reason, we use the Harris corner detector to extract corner points. In order to reduce the interference of land scattering, the corner point threshold is set to 0.2, and a part of the low-quality corner point responses are filtered. In order to better capture the global ship scattering characteristics, the maximum number of corner points is set to 100. In order to better capture the local characteristics of the ship, the minimum Euclidean distance is set to 10. The obtained scattering point set is $R_{sp}\{(x_i, y_i)\}_{i=1}^{N}$. In order to better realize the guiding role of the SAR scattering point set, cluster processing [43] is performed on the point set, with $K \in [2, 8]$, and the silhouette coefficient is used as the evaluation metric for the cluster quality. By iteratively looping through this process, the optimal cluster number K is determined, and its clustering effect is shown in Figure 2.

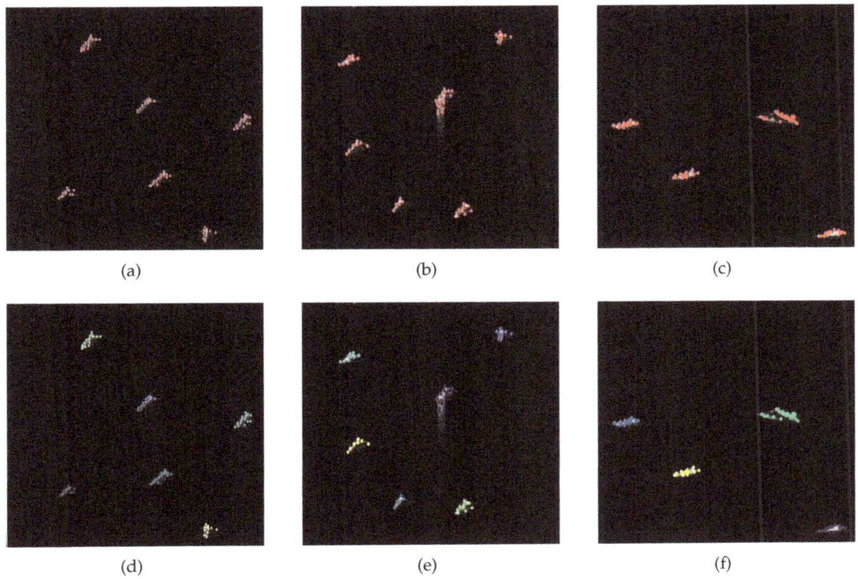

Figure 2. The extracted scattering points(red color) and their clustered results(different colors). (**a–c**) The extracted scattering points from the ships. (**d–f**) The scattering points after clustering.

Afterward, based on the allocation strategy, they are assigned to the corresponding initialized reppoints for quality evaluation.

3.2.2. Feature Matching and Adaptive Sample Selection

We improve the quality metric for adaptive sample reppoints, which is different from APAA [36], and we introduce a scattering position metric in the quality assessment to more comprehensively measure the quality of the sample reppoints and provide higher-quality samples for model training.

The extracted set of scattering points is $R_{sp}\{(x_i, y_i)\}_{i=1}^{N}$, which is clustered to obtain a number of clusters. The number of clusters is obtained by measuring the optimal silhouette coefficient of the clusters; the scattering cluster center point set is $C\{(x_i, y_i)\}_{i=1}^{K}$, where K denotes the optimal number of clusters. The point set of sample reppoints is $R = \{S_1, S_2, S_3, \ldots, S_m\}$, where $S_i = \{(x_j, y_j)\}_{j=1}^{9}$, and m denotes the generation of m samples. We use cosine similarity as the similarity measure between point set clusters and initialized reppoints, thereby achieving matching and assessment between scattering cluster center points and initialized reppoints. The average cosine similarity between the scattering cluster center points and the sample reppoints can be expressed as E_{ik}:

$$E_{ik} = \frac{1}{N}\sum_{j=1}^{9}(cos < e_{ij}, e_k^* >) \tag{1}$$

where $N = 9$ indicates that each sample's reppoints consists of nine points. e_{ij} represents the vector denoting each point within every sample's reppoints, while e_k^* denotes the vector representing the scattering cluster center points. By traversing all scattering cluster center points, we obtain the cosine similarity matrix $\mathbf{E}$ between sample reppoints and the scattering cluster center points:

$$\mathbf{E} = \begin{bmatrix} (E_{1,1}) & (E_{1,2}) & \dots & (E_{1,n}) \\ (E_{2,1}) & (E_{2,2}) & \dots & (E_{2,n}) \\ \vdots & \vdots & \ddots & \vdots \\ (E_{m,1}) & (E_{m,2}) & \dots & (E_{m,n}) \end{bmatrix} \quad (2)$$

By computing the average cosine similarity between the extracted scattering cluster center points and the sample reppoints, we obtain the similarity matrix $\mathbf{E} \in \mathbb{R}^{m \times n}$. Taking the maximum along the dimension of n, we derive the optimal match between the samples and the scattering cluster center points, along with their corresponding cosine similarity, denoted as follows:

$$\max_n(\mathbf{E}) = \begin{bmatrix} \max(E_{1,1}, E_{1,2}, \dots, E_{1,n}) \\ \max(E_{2,1}, E_{2,2}, \dots, E_{2,n}) \\ \vdots \\ \max(E_{m,1}, E_{m,2}, \dots, E_{m,n}) \end{bmatrix} \quad (3)$$

In order to facilitate integration with other quality scores, the corresponding cosine distance is obtained based on the cosine similarity, thereby generating the quality score $Q_{sp} \in \mathbb{R}^m$.

$$Q_{sp} = 1 - \max_n(\mathbf{E}) \quad (4)$$

In summary, the score Q_{sp}, measured by the scattering position, is obtained, and then our quality evaluation part can be divided into the following:

$$Q_{union} = Q_{cls} + \mu_1 Q_{sp} + \mu_2 Q_{loc} + \mu_2 Q_{ori} \quad (5)$$

Among these measures, Q_{sp} denotes the similarity measure for scattering positions, while Q_{loc} represents the assessment of the spatial positioning quality, computed through the positioning loss converted by GIoU [44]. Q_{ori} employs the Chamfer distance [45] to gauge directional disparities, whereas Q_{cls} utilizes FocalLoss [29] to evaluate the correlation in category quality. A dynamic TOP K sample selection scheme is devised based on their quality assessment scores. Quality score lists are generated during different iterations, and these scores are arranged in ascending order. Additionally, a random sampling rate σ is set to adaptively adjust the number of positive samples, denoted by k .

$$k = \sigma \times N_t \quad (6)$$

where N_t represents the total number of generated samples, and the initial default setting for σ is 0.2. Subsequently, during the training phase, the top k samples with the highest quality assessment scores are selected as positive samples for training. Considering the practical application scenarios of SAR target detection, we restrict the utilization of this approach solely to the training phase, aiming to reduce the computational load during the detection phase.

3.3. Scattering-Point-Guided Reppoints Dynamic Learning (SPG Learning)

Ship misdetection tends to happen in nearshore scenarios due to the presence of land scattering interference, which results in the poor generation of initial reppoints. Additionally, some outliers appear in the adaptive learning stage of key semantic features for reppoints, which further reduces the performance of ship detection.

We add supervisory information based on scattering point location priors during the initialization point generation stage, thereby guiding reppoints to learn features from key semantic parts of the target. This reduces the land scattering interference and makes the extracted features more accurate and complete.

Specifically, scattering points play a critical role in representing ship features within SAR images. Once we perform positive sample selection using SPG-ASS, we acquire the corresponding positive sample set of reppoints and assign ground truth (GT) boxes to them. Subsequently, we utilize these GT boxes to identify matching clusters of scattering points positioned accordingly. This alignment process ensures a cohesive match between scattering points and sample reppoints, thereby consolidating more of the inherent structural and positional features of targets into the supervised information. Consequently, this offers valuable guidance for initializing reppoints, facilitating a more effective learning process regarding the key semantic features of the target. Ultimately, this procedure significantly elevates the quality of the initialized point set. The learning process is depicted in Figure 3.

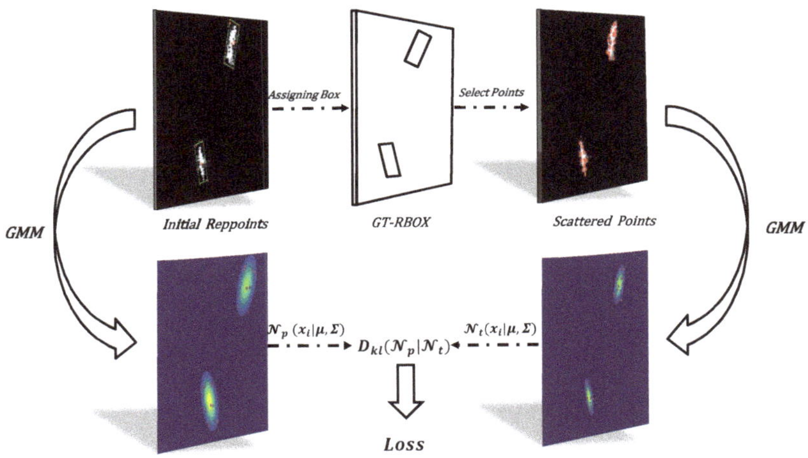

Figure 3. SPG learning. After alignment via GT bounding boxes, Gaussian distribution is employed to fit the scattering points and initialize reppoints, followed by computing the KLD loss, achieving supervised-guided learning.

After initialization, the reppoints are generated as $R = \{S_1, S_2, S_3, \ldots, S_m\}$, where $S_i = \{(x_j, y_j)\}_{j=1}^{9}$. To better achieve the adaptive learning of the target's crucial semantic parts by initializing reppoints, we utilize Kullback–Leibler Divergence (KLD) [34] as the regression loss for supervised optimization. Specifically, Gaussian distributions are employed to individually model the scattering point clusters and the reppoints generated during initialization. Subsequently, the KLD loss is computed based on the Gaussian distributions between these two sets of points. This enables the dynamic adaptive adjustment of gradients for each parameter based on the loss between the two point sets. Such an approach is advantageous for facilitating the adaptive collaborative optimization of the initialized reppoints and, consequently, enabling the learning of key semantic features of the target. The computation of the Gaussian distribution of the point set is as follows:

$$\mathcal{N}(\mathbf{x}_i|\mu,\Sigma) = \frac{1}{2\pi\sqrt{\det(\Sigma)}}\exp\left(-\frac{1}{2}(\mathbf{x}_i-\mu)^T\Sigma^{-1}(\mathbf{x}_i-\mu)\right) \quad (7)$$

where μ represents the mean value, and Σ represents the covariance matrix. The calculation of KLD for the Gaussian distribution of point sets is as follows:

$$D_{KL}(\mathcal{N}_p\|\mathcal{N}_t) = \frac{1}{2}\left((\mu_p-\mu_t)^T\Sigma_t^{-1}(\mu_p-\mu_t) + \mathrm{Tr}(\Sigma_t^{-1}\Sigma_p) + \ln\left(\frac{|\Sigma_t|}{|\Sigma_p|}\right)\right) - 1 \quad (8)$$

where $\mathcal{N}_p$ and $\mathcal{N}_t$ represent the Gaussian distributions of the initialized reppoints and the corresponding position scattering point cluster, respectively. Consequently, the loss between the two sets of points is obtained as follows:

$$L_{\text{s.Init}} = 1 - \frac{1}{2 + f(D)} \tag{9}$$

where $f(\cdot)$ denotes a non-linear function applied to distances, in this case using $sqrt(D)$. The overall loss function for the entire training process is as follows:

$$L_{\text{total}} = L_{\text{cls}} + \alpha L_{\text{s1}} + \beta L_{\text{s2}} + \gamma L_{\text{s.Init}} \tag{10}$$

where α, β, and γ represent balancing weighting coefficients, and L_{s1} and L_{s2}, respectively, represent the spatial localization losses during the initialization stage and the fine-tuning stage. The spatial localization loss comprises two components: positioning loss [44] and spatial constraint loss [36]. Additionally, $L_{\text{s.Init}}$ indicates the guidance loss of SPG learning.

4. Results and Discussion

4.1. Dataset and Its Evaluation Metrics

We conducted experiments on the SSDD [46] and HRSID datasets [10]. The SSDD dataset consists of 1160 images, with 928 images used for training and 232 images (including 46 nearshore images and 186 offshore images) used for testing. These images are sourced from RADARSAT-2, TerraSAR-X, and Sentinel-1, with resolutions ranging from 1 m to 15 m and with the C and X bands. The HRSID dataset comprises 5604 images, with 65% used for training and 35% for testing. The image slice resolutions in this dataset range from 0.5 m to 3 m. All images were resized to 800 × 800 pixels, and the data augmentation approach exclusively employs random flipping to enhance the sample set.

In our experiments, we used mAP (mean Average Precision) to evaluate the performance of the network. Its expression is as follows:

$$\text{mAP} = \frac{1}{K} \sum_{j=1}^{K} \int_0^1 \text{precision}(\text{recall})\, d(\text{recall}) \tag{11}$$

Besides mAP, we also utilized Recall as another important metric to reflect the performance of our method. Its expression is as follows:

$$\text{Recall} = \frac{N_{TP}}{N_{TP} + N_{FN}} \tag{12}$$

where N_{TP} represents the number of true positives, and N_{FN} represents the number of false negatives.

4.2. The Details of the Experiment

The entire experiment was implemented within the mmrotate codebase. The total number of training epochs was 50, and the SGD optimizer was used with a learning rate of 0.0025, a momentum parameter of 0.9, and a weight decay of 0.0001. Learning rate adjustments were made using a stepwise strategy with adjustment nodes at (38, 40, 42, 44, 46, 48). All training and testing experiments in this paper were conducted on the Ubuntu 18.04 operating system. As for hardware specifications, the experiments were performed using an Intel i5-13490F CPU (Intel, Santa Clara, CA, USA) and an NVIDIA RTX 4080 GPU (NVIDIA, Santa Clara, CA, USA).

4.3. Comparison with State-of-the-Art Methods

In order to validate the effectiveness of our SPG oriented-reppoints method, we compared our method with ten other state-of-the-art directional target detection algorithms on a unified SSDD dataset; their mAP and Recall values were computed for nearshore scenarios, offshore scenarios, and hybrid scenarios, as shown in Table 1.

Table 1. Comparison with state-of-the-art methods (SSDD).

Method	Backbone	mAP			Recall			Params (M)	Flops (GFLOPs)
		Nearshore	Offshore	Total	Nearshore	Offshore	Total		
Oriented-rcnn	R-50-FPN	0.794	0.906	0.902	0.847	0.963	0.934	41.13	198.53
Rotated-faster-rcnn	R-50-FPN	0.776	0.904	0.896	0.824	0.963	0.929	41.12	198.40
Roi-trans	R-50-FPN	0.703	0.903	0.893	0.779	0.943	0.902	55.03	200.41
Rotated-retinanet	R-50-FPN	0.629	0.904	0.864	0.763	0.945	0.900	36.13	209.58
Gliding-vertex	R-50-FPN	0.703	0.903	0.892	0.763	0.935	0.893	41.13	198.40
Fcos	R-50-FPN	0.668	0.904	0.810	0.740	0.938	0.889	31.89	206.20
S2anet	R-50-FPN	0.680	0.897	0.804	0.786	0.928	0.893	38.54	196.21
Kld	R-50-FPN	0.693	0.904	0.892	0.771	0.955	0.910	36.13	229.95
R3det	R-50-FPN	0.691	0.906	0.890	0.740	0.965	0.910	41.58	328.70
Oriented reppoints (baseline)	R-50-FPN	0.747	0.904	0.895	0.863	0.965	0.940	36.60	171.70
Our method	R-50-FPN	0.780	0.904	0.898	0.885	0.963	0.944	36.60	171.70

The compared methods include (1) two-stage object detectors based on anchor boxes: oriented-rcnn [33], Roi-Transformer [32], and Rotated-faster-rcnn [25], Gliding-vertex [47]; (2) anchor-free object detectors: Fcos [48] and oriented reppoints (baseline) [36]; (3) single-stage detectors based on anchor points: Rotated-retinanet [29], S2anet [49], Kld [34], and R3det [50]. Ultimately, our method achieved 78% mAP and 88.5% Recall in nearshore scenes and 89.8% mAP and 94.4% Recall in hybrid scenes, surpassing all methods except oriented-rcnn. Particularly in nearshore scenes, in terms of mAP, our method significantly outperforms other anchor-free algorithms and some anchor-based methods. Furthermore, compared to our baseline method (oriented reppoints), our method demonstrated a 3.3% mAP and 2.2% Recall improvement in nearshore scenes and a 0.26% mAP and 0.4% Recall improvement in mixed scenes. This validates the effectiveness of our method over the baseline.

The ship detection capabilities of these methods for the nearshore scenario are directly exhibited in Figure 4. As shown in the figure, our algorithm can detect ships in complex nearshore environments, while other methods exhibit varying degrees of false positives and misses. In Figure 4b–g, the targets on land were mistakenly detected as ships. In Figure 4c,e, the clutter on the sea surface was mistakenly detected as a ship. In Figure 4e,g, the nearshore ship was missed.

Simultaneously, we considered the practical application scenarios of the models and compared their parameter sizes and computational complexities. As illustrated in Figure 5, our method's model parameter size constitutes 88.98% of that of oriented-rcnn, while its computational complexity represents 86% of oriented-rcnn's. However, the difference in detection accuracy between our method and oriented-rcnn is merely 1.4%. To some extent, our method achieves comparable precision to oriented-rcnn while possessing a smaller parameter size and reduced computational overhead. Moreover, in contrast to the baseline method, our approach significantly enhances model performance without increasing the model's parameters or computational complexity. This reinforces the practical superiority of our method in detection scenarios.

Furthermore, to further analyze the performance of our algorithm, we conducted tests on the HRSID dataset. The results are shown in Table 2. Our method achieved an improvement of 3.6% in nearshore environments and 4.5% in mixed scenarios compared to the baseline. Multiple metrics reached the state-of-the-art (SOTA) level, further demonstrating the effectiveness and robustness of our method.

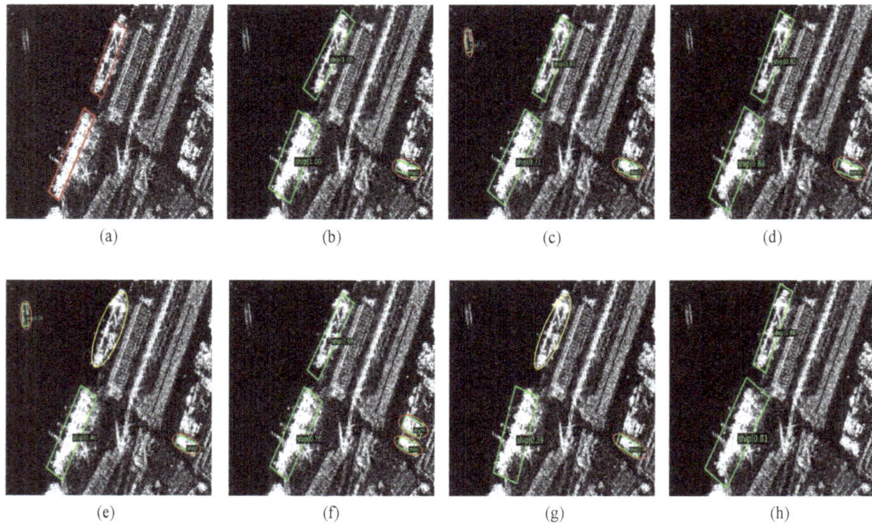

Figure 4. A comparison of methods for nearshore detection. The red, yellow, and green circles represent false positives, misses, and correct detections, respectively. (**a**) Ground truth. (**b**) Fcos. (**c**) R3det. (**d**) Oriented-rcnn. (**e**) Rotated-retinanet. (**f**) Roi-trans. (**g**) S2anet. (**h**) Our method.

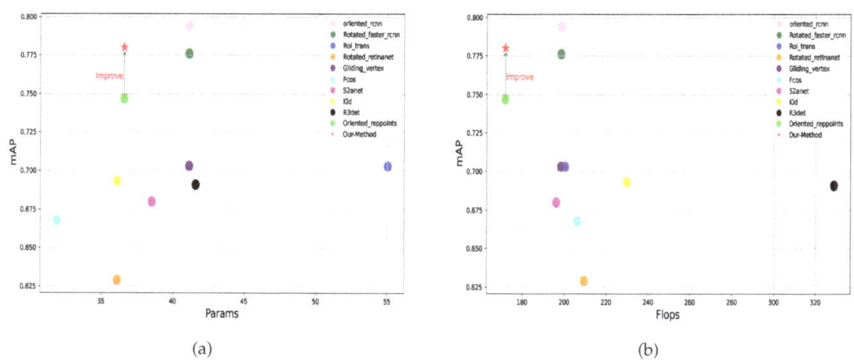

Figure 5. Model parameters and their flops against mAP,the red star represents the performance of our method. (**a**) Model parameters against mAP. (**b**) Model flops against mAP.

Table 2. Comparison with state-of-the-art methods (Hrsid).

Method	Backbone	mAP			Recall			Params (M)	Flops (GFLOPs)
		Nearshore	Offshore	Total	Nearshore	Offshore	Total		
Oriented-rcnn	R-50-FPN	0.543	0.905	0.781	0.670	0.957	0.830	41.13	198.53
Rotated-faster-rcnn	R-50-FPN	0.523	0.901	0.774	0.639	0.955	0.815	41.12	198.40
Roi-trans	R-50-FPN	0.521	0.906	0.774	0.668	0.953	0.827	55.03	200.41
Rotated-retinanet	R-50-FPN	0.481	0.903	0.689	0.632	0.945	0.812	36.13	209.58
Gliding-vertex	R-50-FPN	0.509	0.903	0.709	0.628	0.934	0.799	41.13	198.40
Fcos	R-50-FPN	0.383	0.903	0.696	0.544	0.954	0.772	31.89	206.20
S2anet	R-50-FPN	0.493	0.905	0.759	0.645	0.949	0.814	38.54	196.21
Kld	R-50-FPN	0.506	0.904	0.776	0.771	0.969	0.844	36.13	229.95
R3det	R-50-FPN	0.463	0.904	0.708	0.605	0.935	0.789	41.58	328.70
Oriented reppoints (baseline)	R-50-FPN	0.532	0.905	0.763	0.669	0.955	0.840	36.60	171.70
Our method	R-50-FPN	0.568	0.906	0.808	0.810	0.963	0.869	36.60	171.70

4.4. Ablation Experiments

In this section, to analyze the effectiveness of various proposed components within our method, we employed the original oriented-reppoints method as a baseline and evaluated its performance first. Subsequently, we conducted a series of ablation experiments and compared their results. To ensure the reliability of these experimental outcomes, all experiments were conducted under identical conditions and with identical settings.

We incorporated two parts into the baseline method to study their impacts separately: adaptive sample selection guided by scattering points (SPG-ASS) and adaptive reppoints learning guided by scattering points (SPG learning). The experimental results are presented in Table 3. When solely incorporating SPG-ASS, the mAP is increased by 1.3%, and the network's detection Recall is increased by 1.5%, benefiting from the exclusivity of high-quality samples in the network training and learning processes. When solely incorporating the SPG learning part, the mAP and Recall for nearshore detection are improved by 1.6% and 0.2%, respectively. As indicated in row IV of Table 3, when both components were integrated into our network, it exhibited greater performance improvements. The mAP and Recall for nearshore detection are increased by 3.3% and 2.2%, respectively. Additionally, these components were applied during the training phase of our network, without increasing the computational load during the testing phase.

Table 3. Ablation experiments.

	SPG-ASS	SPG Learning	Map (Nearshore) ↑	Recall (Nearshore) ↑
I			0.747	0.863
II	✓		0.760	0.878
III		✓	0.763	0.865
IV	✓	✓	0.780	0.885

4.4.1. SPG-ASS

Given the utilization of an anchor-free mechanism within our network architecture, the acquisition of high-quality samples stands as a pivotal factor in effectively detecting intricate nearshore targets. We introduced SPG-ASS into the baseline model. By incorporating scattering point positional information, during the training phase, we can select higher-quality samples for learning, thereby avoiding issues of model degradation caused by low-quality samples. In Figure 6a, due to the lack of scattering point position information of the target with the adaptive sampling scheme, the correlation between the sample's classification confidence and localization score (IoU) is low. Moreover, a considerable number of samples are concentrated in regions with both lower classification confidence and lower localization scores, indicating low sample quality overall.

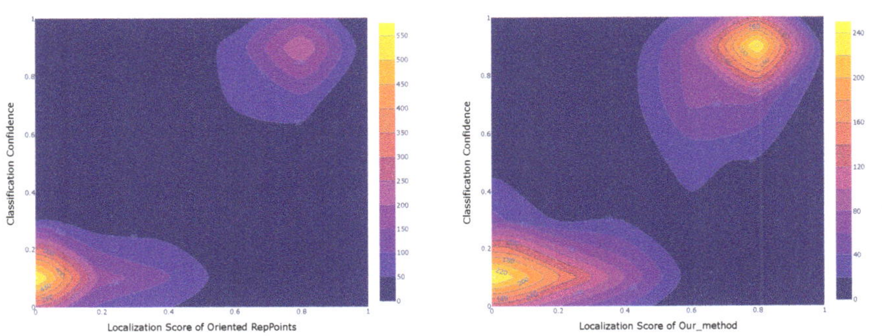

(a) (b)

Figure 6. The impact of SPG-ASS on the correlation between the classification confidence and localization score of oriented reppoints. (**a**) Without SPG-ASS. (**b**) With SPG-ASS.

In contrast, as depicted in Figure 6b, by incorporating the scattering position information, the localization quality scores and classification confidence of the samples are significantly increased compared to Figure 6a without this integration. This approach has led to the selection of numerous high-quality samples exhibiting higher classification confidence and localization scores. This substantiates the effectiveness of our method in selecting high-quality samples. In addition, we conducted a comparative analysis in nearshore environments using both the baseline method and the improved approach with SPG-ASS.

As depicted in Figure 7b,e, it is evident that the baseline method is prone to false positives and false negatives in nearshore detection. Conversely, the detection outcomes of the improved approach, as illustrated in Figure 7c,f, exhibit a significant decrease in false negatives and the absence of false positives. This further corroborates the effectiveness of our SPG-ASS component.

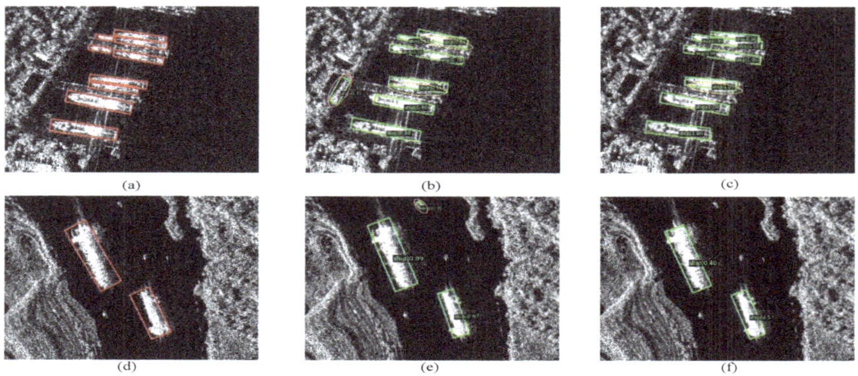

Figure 7. Nearshore detection comparison. (**a**,**d**) Ground truth. (**b**,**e**) Without SPG-ASS. (**c**,**f**) With SPG-ASS. The red, yellow, and green represent false positives, misses, and correct detections, respectively.

4.4.2. SPG Learning

To further explore the impact of SPG learning, we independently incorporated it into the baseline task. As indicated in Table 3, by employing SPG learning techniques, the adaptive feature learning capability of the initial reppoints is focused on the semantic features at critical target locations so as to mitigate the impact of land-based scattering interference. Furthermore, to further illustrate the effectiveness of our approach, we visualized the features of the backbone network.

As shown in Figure 8b,e, in the baseline task, the network is more sensitive to land scattering, making it difficult for adaptive points to learn the key semantic features of the target itself in nearshore detection. However, after applying adaptive point learning guided by scattering points, the results in Figure 8c,f show that land scattering interference is suppressed. The scattering-guided initialization points move toward the key semantic areas of the ship, enabling the network to highlight the significance of the target itself while reducing attention to land regions. This improves the robustness and accuracy of nearshore ship detection. Furthermore, we conducted tests on both the baseline and improved methods with SPG learning in nearshore environments. The results are shown in Figure 9. False detections and missed detections occur with the baseline. Meanwhile, the detection results generated by the improved method are consistent with the ground-truth bounding boxes. This further validates the effectiveness of the SPG learning component.

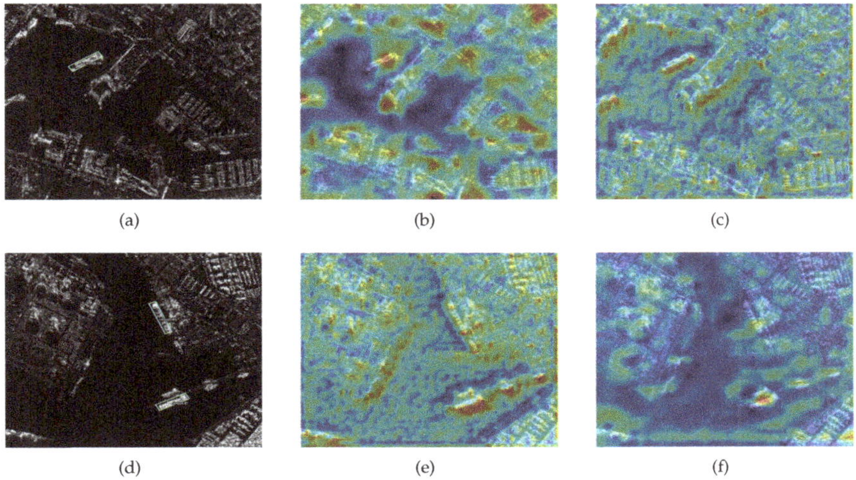

Figure 8. Visualization of the confidence heatmaps,the gradient from blue to red represents the increasing level of attention focus. (**a**,**d**) Ground truth. (**b**,**e**) Without SPG learning. (**c**,**f**) With SPG learning.

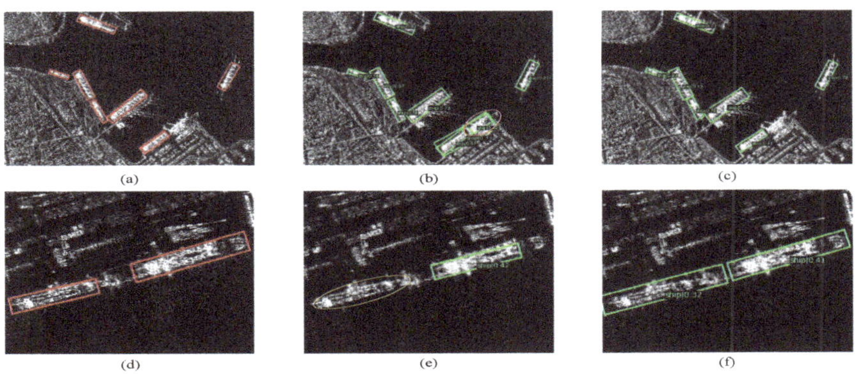

Figure 9. Nearshore detection comparison. (**a**,**d**) Ground truth. (**b**,**e**) Without SPG learning. (**c**,**f**) With SPG learning. The red circles indicate false detections, green circles indicate correct detections,while the yellow circles indicate missed detections.

4.4.3. SPG Oriented Reppoints Detection

In addition, we simultaneously incorporated the two proposed modules into the baseline network. The detection results are shown in Figure 10. Figure 10a–c represent the ground truth, while Figure 10d–f illustrate the detection results of our proposed method. As depicted in Figure 10, our approach achieves the precise detection of objects in various scenes, such as offshore and nearshore, by adapting reppoints transformations. This demonstrates the effectiveness of our approach.

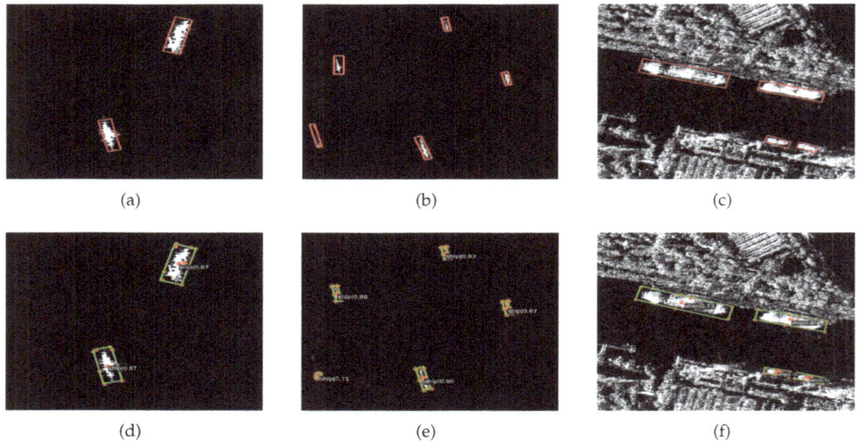

Figure 10. The detection results of our proposed method. (**a–c**) Ground truth in the SAR image. (**d–f**) The detection results from our proposed method.

4.5. Qualitative Evaluation

Additionally, to assess the generalization performance of our method, a SAR image captured in the vicinity of the Zhoushan port area was chosen for ship detection, with the specific details outlined in Table 4.

Table 4. Details of SAR images.

Satellite	Center (Longitude/Latitude)	Scene Date (UTC)	Imaging Mode	Resolution	Band	Polarization
Sentinel-1	122.147°/29.369°	12 May 2019 09:53:10.579856	IW	13.957 m	C	VH

We selected three representative areas within the image for analysis, as illustrated in Figure 11. Area 1 comprises a mixed scene of nearshore and offshore areas, while area 2 depicts an offshore scene, and area 3 portrays a nearshore scene. The oriented_rcnn method, having the best mAP value on the SSDD dataset, was chosen for comparison with the proposed method. The left three subplots Figure 11A–C showcase the detection results obtained using the oriented_rcnn method. In contrast, the right three subplots Figure 11D–F display the detection outcomes achieved by our proposed method. In subplot A, there are likely to be false and missed detections in the nearshore area when using oriented_rcnn. However, our method, as depicted in subplot D, not only effectively detects nearshore ships but also avoids false positives in the strong scattering areas on land. This also shows that our method has better ability to resist land scattering interference.

From subplot B in Figure 11, it is evident that there were some missed detections during offshore detection. However, our proposed method, as depicted in subplot E, presents more comprehensive detection results, with a significantly reduced rate of missed detections. Additionally, in nearshore scenarios, such as the area illustrated in Figure 11, there exists prominent strong scattering areas on land, closely adjacent to the ships, significantly increasing the difficulty of ship detection. The detection results of oriented_rcnn, as shown in subplot C, exhibit both missed detections in nearshore areas and false positives on land. In contrast, our method's detection results, displayed in subplot F, identify all ship targets in that area without producing false detections on land targets. Overall, our method demonstrates superior detection and generalization performance in practical scenarios.

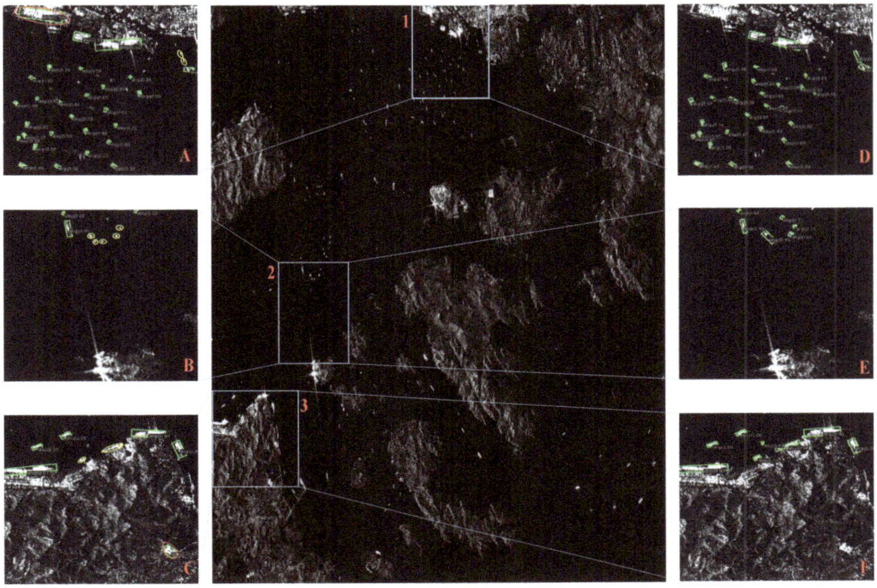

Figure 11. Ship detection results of Sentinel-1 SAR image. Areas 1, 2, and 3 represent mixed scenes, offshore scenes, and nearshore scenes, respectively. Subplots (**A–C**) are the detection results of oriented_rcnn for the three areas, and subplots (**D–F**) are the detection results of our method. Red circles indicate false detections, and yellow circles indicate missed detections.

4.6. Discussion

The experimental results on the SSDD and HRSID datasets validate the effectiveness of our proposed method. On the SSDD dataset, our method outperformed the baseline by 3.2% and performed comparably to oriented-rcnn in nearshore environments, achieving a suboptimal level. To further verify the method's generalization and reliability, we conducted a comparative study on the HRSID dataset, which is larger in scale, richer in imaging modes, and more complex in nearshore environments. The results show that our proposed method outperformed the baseline by 3.6% and achieved the state-of-the-art level on this dataset. Additionally, we observed performance fluctuations on different datasets, mainly due to differences in dataset characteristics. The HRSID dataset has a more complex nearshore environment with diverse slice characteristics, and the detection results in these complex scenarios also reflect the robustness and generalization of our method. Our method benefits from the anchor-free detection framework guided by scattering points, which provides higher granularity for recognizing ships in complex nearshore environments and has higher perceptual adaptability for detecting directional ships. Moreover, the SPG learning mechanism can better learn the features of nearshore ships, reduce false alarms on land, achieve feature focusing, and thus achieve higher detection accuracy. We also conducted ablation experiments to explore the roles of various parts of the proposed method. However, this method currently has some shortcomings. For example, both the adaptive sample selection scheme and the adaptive learning part rely on the extraction of scattering points from the target. If the area occupied by ships is limited or the scattering from ships is weak, resulting in fewer or no corner points being extracted, the method may fail. In the future, we plan to redesign the scattering point extraction part and introduce more efficient and advanced network structures for scattering feature extraction and fusion.

5. Conclusions

In summary, we propose an anchor-free detection scheme based on oriented reppoints guided by the scattering characteristics of SAR images. This scheme addresses the challenges of detecting oriented ships in complex nearshore environments. Initially, considering the scattering mechanism of metal-made ships, the strong points, such as corner points, are extracted as positional prior information. Then, we use the positional information of scattering points for adaptive sample selection, enabling the superior selection of high-quality sample points during the training phase and thus avoiding model degradation caused by low-quality samples. Furthermore, we enhance the reppoints quality in the initializing phase by a novel supervised guidance paradigm, allowing the network to learn more refined representations of the electromagnetic features of ships, consequently reducing land scattering interference in complex nearshore environments. Our method offers new insights into the integration of scattering features and demonstrates effectiveness in various environments, especially in nearshore scenes with significant land interference. On the SSDD dataset, our method achieves an mAP of 78% for nearshore detection, which is a 3.3% improvement over the baseline. To further validate the robustness of our method, we tested it on the HRSID dataset, where it achieves an mAP of 56.8% for nearshore detection, a 3.6% improvement over the baseline, reaching the state-of-the-art (SOTA) level compared to other methods. In the future, we will try to extend this methodology to other application scenarios so as to improve other object detection tasks with SAR images.

Author Contributions: Conceptualization, W.Z. and L.H.; methodology, W.Z.; software, W.Z.; validation, W.Z., H.L. and L.H.; formal analysis, W.Z.; investigation, W.Z.; resources, L.H.; data curation, W.Z.; writing—original draft preparation, W.Z.; writing—review and editing, W.Z., L.H. and C.Y.; visualization, W.Z.; supervision, L.H.; project administration, L.H.; funding acquisition, L.H. All authors have read and agreed to the published version of the manuscript.

Funding: This research was funded by the Youth Innovation Promotion Association No. 2019127,Chinese Academy of Sciences.

Data Availability Statement: The majority of the dataset is available at https://github.com/TianwenZhang0825/Official-SSDD, (accessed on 15 September 2023).

Acknowledgments: We sincerely appreciate the constructive comments and suggestions of the anonymous reviewers, which have greatly helped to improve this paper.

Conflicts of Interest: The authors declare no conflicts of interest.

References

1. Zhou, G.; Lin, G.; Liu, Z.; Zhou, X.; Li, W.; Li, X.; Deng, R. An optical system for suppression of laser echo energy from the water surface on single-band bathymetric LiDAR. *Opt. Lasers Eng.* **2023**, *163*, 107468. [CrossRef]
2. Zhang, Y.; Lu, D.; Qiu, X.; Li, F. Scattering-Point-Guided RPN for Oriented Ship Detection in SAR Images. *Remote Sens.* **2023**, *15*, 1411. [CrossRef]
3. Zheng, Y.; Liu, P.; Qian, L.; Qin, S.; Liu, X.; Ma, Y.; Cheng, G. Recognition and depth estimation of ships based on binocular stereo vision. *J. Mar. Sci. Eng.* **2022**, *10*, 1153. [CrossRef]
4. Reigber, A.; Scheiber, R.; Jager, M.; Prats-Iraola, P.; Hajnsek, I.; Jagdhuber, T.; Papathanassiou, K.P.; Nannini, M.; Aguilera, E.; Baumgartner, S.; et al. Very-High-Resolution Airborne Synthetic Aperture Radar Imaging: Signal Processing and Applications. *Proc. IEEE* **2013**, *101*, 759–783. [CrossRef]
5. Castelletti, D.; Farquharson, G.; Stringham, C.; Duersch, M.; Eddy, D. Capella space first operational SAR satellite. In Proceedings of the 2021 IEEE International Geoscience and Remote Sensing Symposium IGARSS, Brussels, Belgium, 11–16 July 2021; pp. 1483–1486.
6. Jordan, R.L.; Huneycutt, B.L.; Werner, M. The SIR-C/X-SAR synthetic aperture radar system. *IEEE Trans. Geosci. Remote Sens.* **1995**, *33*, 829–839. [CrossRef]
7. Orzel, K.; Fujimaru, S.; Obata, T.; Imaizumi, T.; Arai, M. The on-orbit demonstration of the small SAR satellite. Initial calibration and observations. In Proceedings of the 2022 IEEE Radar Conference (RadarConf22), New York City, NY, USA, 21–25 March 2022; pp. 1–5.
8. Mao, Y.; Zhu, Y.; Tang, Z.; Chen, Z. A novel airspace planning algorithm for cooperative target localization. *Electronics* **2022**, *11*, 2950. [CrossRef]

9. Zhang, F.; Yao, X.; Tang, H.; Yin, Q.; Hu, Y.; Lei, B. Multiple mode SAR raw data simulation and parallel acceleration for Gaofen-3 mission. *IEEE J. Sel. Top. Appl. Earth Obs. Remote Sens.* **2018**, *11*, 2115–2126. [CrossRef]
10. Wei, S.; Zeng, X.; Qu, Q.; Wang, M.; Su, H.; Shi, J. HRSID: A high-resolution SAR images dataset for ship detection and instance segmentation. *IEEE Access* **2020**, *8*, 120234–120254. [CrossRef]
11. Bastani, F.; Wolters, P.; Gupta, R.; Ferdinando, J.; Kembhavi, A. SatlasPretrain: A Large-Scale Dataset for Remote Sensing Image Understanding. In Proceedings of the IEEE/CVF International Conference on Computer Vision, Paris, France, 2–6 October 2023; pp. 16772–16782.
12. Yasir, M.; Niang, A.J.; Hossain, M.S.; Islam, Q.U.; Yang, Q.; Yin, Y. Ranking Ship Detection Methods Using SAR Images Based on Machine Learning and Artificial Intelligence. *J. Mar. Sci. Eng.* **2023**, *11*, 1916. [CrossRef]
13. Kuttikkad, S.; Chellappa, R. Non-Gaussian CFAR techniques for target detection in high resolution SAR images. In Proceedings of the 1st International Conference on Image Processing, Austin, TX, USA, 3–16 November 1994; Volume 1, pp. 910–914.
14. El-Darymli, K.; McGuire, P.; Power, D.; Moloney, C. Target detection in synthetic aperture radar imagery: A state-of-the-art survey. *J. Appl. Remote Sens.* **2013**, *7*, 071598. [CrossRef]
15. Leng, X.; Ji, K.; Yang, K.; Zou, H. A bilateral CFAR algorithm for ship detection in SAR images. *IEEE Geosci. Remote Sens. Lett.* **2015**, *12*, 1536–1540. [CrossRef]
16. Dai, H.; Du, L.; Wang, Y.; Wang, Z. A modified CFAR algorithm based on object proposals for ship target detection in SAR images. *IEEE Geosci. Remote Sens. Lett.* **2016**, *13*, 1925–1929. [CrossRef]
17. Liao, M.; Wang, C.; Wang, Y.; Jiang, L. Using SAR Images to Detect Ships From Sea Clutter. *IEEE Geosci. Remote Sens. Lett.* **2008**, *5*, 194–198. [CrossRef]
18. Ai, J.; Tian, R.; Luo, Q.; Jin, J.; Tang, B. Multi-Scale Rotation-Invariant Haar-Like Feature Integrated CNN-Based Ship Detection Algorithm of Multiple-Target Environment in SAR Imagery. *IEEE Trans. Geosci. Remote Sens.* **2019**, *57*, 10070–10087. [CrossRef]
19. Yasir, M.; Liu, S.; Mingming, X.; Wan, J.; Pirasteh, S.; Dang, K.B. ShipGeoNet: SAR Image-Based Geometric Feature Extraction of Ships Using Convolutional Neural Networks. *IEEE Trans. Geosci. Remote Sens.* **2024**, *62*, 1–13. [CrossRef]
20. Chang, Y.L.; Anagaw, A.; Chang, L.; Wang, Y.C.; Hsiao, C.Y.; Lee, W.H. Ship detection based on YOLOv2 for SAR imagery. *Remote Sens.* **2019**, *11*, 786. [CrossRef]
21. Chen, S.; Zhang, J.; Zhan, R. R2FA-Det: Delving into high-quality rotatable boxes for ship detection in SAR images. *Remote Sens.* **2020**, *12*, 2031. [CrossRef]
22. Girshick, R.; Donahue, J.; Darrell, T.; Malik, J. Rich feature hierarchies for accurate object detection and semantic segmentation. In Proceedings of the IEEE Conference on Computer Vision and Pattern Recognition, Columbus, OH, USA, 23–28 June 2014; pp. 580–587.
23. He, K.; Zhang, X.; Ren, S.; Sun, J. Spatial Pyramid Pooling in Deep Convolutional Networks for Visual Recognition. In *Computer Vision—ECCV 2014*; Springer International Publishing: Berlin/Heidelberg, Germany, 2014; pp. 346–361. [CrossRef]
24. Girshick, R.B. Fast R-CNN. *arXiv* **2015**, arXiv:1504.08083.
25. Ren, S.; He, K.; Girshick, R.B.; Sun, J. Faster R-CNN: Towards Real-Time Object Detection with Region Proposal Networks. *arXiv* **2015**, arXiv:1506.01497.
26. Lin, T.; Dollár, P.; Girshick, R.B.; He, K.; Hariharan, B.; Belongie, S.J. Feature Pyramid Networks for Object Detection. *arXiv* **2016**, arXiv:1612.03144.
27. Redmon, J.; Divvala, S.; Girshick, R.; Farhadi, A. You Only Look Once: Unified, Real-Time Object Detection. In Proceedings of the IEEE Conference on Computer Vision and Pattern Recognition (CVPR), Las Vegas, NV, USA, 27–30 June 2016.
28. Liu, W.; Anguelov, D.; Erhan, D.; Szegedy, C.; Reed, S.; Fu, C.Y.; Berg, A.C. Ssd: Single shot multibox detector. In Proceedings of the Computer Vision—ECCV 2016: 14th European Conference, Amsterdam, The Netherlands, 11–14 October 2016; Part I; Springer: Berlin/Heidelberg, Germany, 2016; Volume 14, pp. 21–37.
29. Lin, T.Y.; Goyal, P.; Girshick, R.; He, K.; Dollár, P. Focal loss for dense object detection. In Proceedings of the IEEE International Conference on Computer Vision, Venice, Italy, 22–29 October 2017; pp. 2980–2988.
30. Zhou, X.; Wang, D.; Krähenbühl, P. Objects as Points. *arXiv* **2019**, arXiv:1904.07850.
31. Yang, Z.; Liu, S.; Hu, H.; Wang, L.; Lin, S. RepPoints: Point Set Representation for Object Detection. *arXiv* **2019**, arXiv:1904.11490.
32. Ding, J.; Xue, N.; Long, Y.; Xia, G.S.; Lu, Q. Learning RoI transformer for oriented object detection in aerial images. In Proceedings of the IEEE/CVF Conference on Computer Vision and Pattern Recognition, Long Beach, CA, USA, 16–17 June 2019; pp. 2849–2858.
33. Xie, X.; Cheng, G.; Wang, J.; Yao, X.; Han, J. Oriented R-CNN for object detection. In Proceedings the IEEE/CVF International Conference on Computer Vision, Montreal, BC, Canada, 11–17 October 2021; pp. 3520–3529.
34. Yang, X.; Yang, X.; Yang, J.; Ming, Q.; Wang, W.; Tian, Q.; Yan, J. Learning high-precision bounding box for rotated object detection via kullback-leibler divergence. *Adv. Neural Inf. Process. Syst.* **2021**, *34*, 18381–18394.
35. Hou, L.; Lu, K.; Yang, X.; Li, Y.; Xue, J. G-Rep: Gaussian Representation for Arbitrary-Oriented Object Detection. *Remote Sens.* **2023**, *15*, 757. [CrossRef]
36. Li, W.; Zhu, J. Oriented RepPoints for Aerial Object Detection. *arXiv* **2021**, arXiv:2105.11111.
37. Zhang, Z.; Guo, W.; Zhu, S.; Yu, W. Toward Arbitrary-Oriented Ship Detection With Rotated Region Proposal and Discrimination Networks. *IEEE Geosci. Remote Sens. Lett.* **2018**, *15*, 1745–1749. [CrossRef]
38. Yang, R.; Pan, Z.; Jia, X.; Zhang, L.; Deng, Y. A Novel CNN-Based Detector for Ship Detection Based on Rotatable Bounding Box in SAR Images. *IEEE J. Sel. Top. Appl. Earth Obs. Remote Sens.* **2021**, *14*, 1938–1958. [CrossRef]

39. Yue, T.; Zhang, Y.; Wang, J.; Xu, Y.; Liu, P.; Yu, C. A Precise Oriented Ship Detector in SAR Images Based on Dynamic Rotated Positive Sample Mining. *IEEE J. Sel. Top. Appl. Earth Obs. Remote Sens.* **2023**, *16*, 10022–10035. [CrossRef]
40. Sun, Y.; Wang, Z.; Sun, X.; Fu, K. SPAN: Strong Scattering Point Aware Network for Ship Detection and Classification in Large-Scale SAR Imagery. *IEEE J. Sel. Top. Appl. Earth Obs. Remote Sens.* **2022**, *15*, 1188–1204. [CrossRef]
41. Ge, Z.; Liu, S.; Li, Z.; Yoshie, O.; Sun, J. OTA: Optimal Transport Assignment for Object Detection. *arXiv* **2021**, arXiv:2103.14259
42. Kim, K.; Lee, H.S. Probabilistic Anchor Assignment with IoU Prediction for Object Detection. *arXiv* **2020**, arXiv:2007.08103.
43. Zhou, G.; Wu, G.; Zhou, X.; Xu, C.; Zhao, D.; Lin, J.; Liu, Z.; Zhang, H.; Wang, Q.; Xu, J.; et al. Adaptive model for the water depth bias correction of bathymetric LiDAR point cloud data. *Int. J. Appl. Earth Obs. Geoinf.* **2023**, *118*, 103253. [CrossRef]
44. Rezatofighi, H.; Tsoi, N.; Gwak, J.; Sadeghian, A.; Reid, I.; Savarese, S. Generalized intersection over union: A metric and a loss for bounding box regression. In Proceedings of the IEEE/CVF Conference on Computer Vision and Pattern Recognition, Long Beach, CA, USA, 16–17 June 2019; pp. 658–666.
45. Butt, M.A.; Maragos, P. Optimum design of chamfer distance transforms. *IEEE Trans. Image Process.* **1998**, *7*, 1477–1484. [CrossRef] [PubMed]
46. Zhang, T.; Zhang, X.; Li, J.; Xu, X.; Wang, B.; Zhan, X.; Xu, Y.; Ke, X.; Zeng, T.; Su, H.; et al. SAR ship detection dataset (SSDD): Official release and comprehensive data analysis. *Remote Sens.* **2021**, *13*, 3690. [CrossRef]
47. Xu, Y.; Fu, M.; Wang, Q.; Wang, Y.; Chen, K.; Xia, G.S.; Bai, X. Gliding Vertex on the Horizontal Bounding Box for Multi-Oriented Object Detection. *IEEE Trans. Pattern Anal. Mach. Intell.* **2021**, *43*, 1452–1459. [CrossRef] [PubMed]
48. Tian, Z.; Shen, C.; Chen, H.; He, T. FCOS: Fully Convolutional One-Stage Object Detection. *arXiv* **2019**, arXiv:1904.01355.
49. Han, J.; Ding, J.; Li, J.; Xia, G.S. Align deep features for oriented object detection. *IEEE Trans. Geosci. Remote Sens.* **2022**, *60*, 1–11. [CrossRef]
50. Yang, X.; Liu, Q.; Yan, J.; Li, A. R3Det: Refined Single-Stage Detector with Feature Refinement for Rotating Object. *arXiv* **2019**, arXiv:1908.05612.

remote sensing

Article

LMSD-Net: A Lightweight and High-Performance Ship Detection Network for Optical Remote Sensing Images

Yang Tian [1,2], Xuan Wang [1,*], Shengjie Zhu [1,2], Fang Xu [1] and Jinghong Liu [1]

1 Changchun Institute of Optics, Fine Mechanics and Physics, Chinese Academy of Sciences, Changchun 130033, China; tianyang19@mails.ucas.ac.cn (Y.T.); shengjie_zhu@foxmail.com (S.Z.); xufang59@126.com (F.X.); liujinghong@ciomp.ac.cn (J.L.)
2 University of Chinese Academy of Sciences, Beijing 100049, China
* Correspondence: ally637@163.com

Abstract: Ship detection technology has achieved significant progress recently. However, for practical applications, lightweight ship detection still remains a very challenging problem since small ships have small relative scales in wide images and are easily missed in the background. To promote the research and application of small-ship detection, we propose a new remote sensing image dataset (VRS-SD v2) and provide a fog simulation method that reflects the actual background in remote sensing ship detection. The experiment results show that the proposed fog simulation is beneficial in improving the robustness of the model for extreme weather. Further, we propose a lightweight detector (LMSD-Net) for ship detection. Ablation experiments indicate the improved ELA-C3 module can efficiently extract features and improve the detection accuracy, and the proposed WGC-PANet can reduce the model parameters and computation complexity to ensure a lightweight nature. In addition, we add a Contextual Transformer (CoT) block to improve the localization accuracy and propose an improved localization loss specialized for tiny-ship prediction. Finally, the overall performance experiments demonstrate that LMSD-Net is competitive in lightweight ship detection among the SOTA models. The overall performance achieves 81.3% in AP@50 and could meet the lightweight and real-time detection requirements.

Keywords: optical remote sensing; small-ship detection; lightweight detection; convolutional neural network

Citation: Tian, Y.; Wang, X.; Zhu, S.; Xu, F.; Liu, J. LMSD-Net: A Lightweight and High-Performance Ship Detection Network for Optical Remote Sensing Images. *Remote Sens.* **2023**, *15*, 4358. https://doi.org/10.3390/rs15174358

Academic Editor: Paolo Tripicchio

Received: 12 June 2023
Revised: 21 August 2023
Accepted: 25 August 2023
Published: 4 September 2023

1. Introduction

Ship detection has gained much attention in the field of marine remote sensing. It has been widely used in sea area management, maritime intelligent traffic, and military target reconnaissance [1–4]. In sea area management, ship detection can improve sea area security, such as assisting in combating illegal smuggling, illegal oil dumping, and illegal fishing [5,6]. Both maritime intelligent traffic and military target reconnaissance rely on Automatic Identification System (AIS) and Vessel Traffic System (VTS) to determine the current position of a ship. Although AIS and VTS integrate multiple technologies such as Very High Frequency (VHF), Global Positioning System (GPS), and Electronic Chart Display and Information System (ECDIS) technologies, an essential prerequisite is that the ship must be equipped with the corresponding transponder. However, ships below the standard tonnage specified by the International Maritime Organization (IMO) can be unnecessarily equipped with AIS or VTS, which means the Electronic Charts and GPS will not work. In addition to tonnage restrictions, some other special-purpose ships often deliberately turn off their transceivers to avoid radar detection. Therefore, optical image-based remote sensing detection techniques can provide an effective means in these cases. In addition, lightweight research for detection is essential to improve efficiency further.

In recent years, a large number of high-resolution optical remote sensing images (ORSI) have been collected for ship detection since the optimization of optical sensors and accurate geometric correction. However, the following challenges remain in ORSI for ship detection:

Large field of view: Due to different parameter settings of imaging sensors and changes in the flight altitude of the acquisition platform, the target scale changes sharply, which increases the model burden. In addition, the objects of interest in nearshore remote sensing images are usually tiny and densely clustered. Rapid low-altitude flight causes motion blur in dense target areas, posing challenges for detection.

Background interference: In high-resolution images, some environmental conditions, such as fog and low light, will indirectly amplify the interference of sea clutter, wake waves, islands, and other false alarms in the detection. Therefore, it is necessary to consider the impact of complex weather conditions on the image.

Application limitations: Some embedded processors have limited computational performance and storage space. Reducing the computation and spatial complexity of the model with guaranteed performance is crucial for lightweight deployment.

To solve the above problems, traditional methods based on supervised learning are highly dependent on feature descriptors, such as HOG [7], DPM [8], and FourierHOG [9]. For the sparse distribution of small ships on the sea, if feature extraction and calculation are directly implemented within the global sea area, it will greatly increase memory and time consumption. Subsequently, some studies [10–13] have added a candidate region extraction stage, which could significantly improve the detection speed. However, nearshore dense ships often cause candidate regions to overlap, which is not conducive to feature discrimination. Therefore, these traditional methods are not very robust for unified marine–nearshore ship detection.

With the tremendous success of Convolutional Neural Networks (CNNs) in image classification, CNN has been migrated to object detection frameworks and has played a significant role. Furthermore, the construction of datasets, such as PASCAL VOC challenges [14,15] (VOC2007 and VOC2012), ImageNet large-scale visual recognition challenges [16,17] (ILSVRC2014), and MS-COCO detection challenges [18], has laid a data-driven foundation for the broad application of CNN in object detection.

In the past few decades, two-stage detectors based on CNN have inherited the traditional detection approach, which involves extracting candidate regions first and then discriminating targets, such as SPP-Net [19], R-FCN [20], and Faster R-CNN [21]. Progressively, instead of traditional candidate region extraction methods, related research attempts to use learnable regional proposal networks (RPNs) and achieve state-of-the-art (SOTA) performance in terms of accuracy. For instance, Hu [22] proposed a two-stage detector to improve the accuracy of multi-scale ship targets in complex backgrounds. However, the higher accuracy comes at the cost of detection speed loss. In contrast, single-stage detectors have faster detection velocities, such as RetinaNet [23], Centernet [24], and YOLO series v3–v8 [25–30]. For instance, Wang [31] used Yolov4 for ship inspections. Despite a large increase in speed, multi-scale detection performance was poor. For this reason, Ye [32] proposed an adaptive attention fusion mechanism (AAFM) to cope with multi-scale target detection in remote sensing scenes and achieved a better performance. Xu [33] proposed a specific model named LMO-YOLO for ship detection. However, for the detection of small and tiny ship targets, the current accuracy is still low. The low accuracy of these single-stage detectors is the result of sample imbalance. Subsequently, Zhang [34] proposed a balanced learning method to solve the problem of imbalance in the target, scene, and feature pyramid network and classification regression network and achieved better results. In addition, since being inspired by Visual Transformer in Natural Language Processing (NLP), some single-stage detectors have shown great potential, such as Swin Transformer [35,36], Detr [37], and MobileViT [38]. Transformer-based detectors usually use attention matrices to establish the dependencies of sequence elements, which focuses more on contextual information. Remote feature interactions in the transformer can compensate for CNN's shortcomings. However, high computation complexity and large numbers of parameters are not favorable for deployment. In a word, designing a model should take into account multiple properties such as detection speed, accuracy at multiple scales, and lightweight nature. Therefore, there is still room for improvement to perfect these aspects mentioned above.

With the increasing demand for deployment, lightweight detection has become a necessary evolutionary process. Since the breakthrough of network depth, the vast majority of existing advanced models are pursuing real-time performance and accuracy and have indeed reached a high level. However, to deploy to edge platforms, the detection model must occupy a small amount of memory and participate in less computation. Therefore, some studies have designed model scaling to address different device parameter limitations. For example, Yolov6 [28] has three models with different widths and depths. Two of the three models are used for lightweight deployment. However, one drawback of model scaling is that lightweight models reduce network size while significantly reducing performance. EfficientDet [39] demonstrated in ablation experiments that mixed scaling can reduce the loss of accuracy. In addition, some studies focus on model compression, which minimizes model size as much as possible while ensuring performance. Specifically, SqueezeNext [40] and CondenseNet [41] improved inference speed with parameter pruning and network optimization. The IGC series [42–44] pointed out that group convolution could help to reduce the number of parameters. Based on group convolution, ShuffleNetV2 [45] adopted a channel split for feature reusing. While group convolution shares parameters, it still retains redundant features, and parameter sharing affects the accuracy of the prediction box, leading to the missed detection of small targets. It seems to have reached the bottleneck regarding lightweight and performance improvement. Based on the defects mentioned above, there is still room for improvement in designing the detection backbone and shared parameter modes suitable for remote sensing images.

On account of the significant differences in ship scales, it is necessary to design a multi-layer detection model. Most existing layered detection models are based on Feature Pyramid Networks [46] (FPNs). Forming the feature pyramid requires multiple downsamplings and pooling, which may lead to the loss of tiny targets. For example, a small ship with a 12×12 dimension has only about one pixel after three layers of pooling, which makes it difficult to distinguish due to its low dimensionality. SSD [47] applied FPN by multiple downsamplings. The receptive field of the underlying feature map is small, which makes it difficult for the network to learn the features of the small targets. Yolov3-spp [25] proposed a spatial pooling pyramid to increase the receptive field of the network, which has a certain improvement in small-target detection. In fact, according to the detection ranking of MS-COCO Challenge1, the detection accuracy of small objects is still far lower than that of large objects. At present, due to differences in resolution, insufficient appearance information, and limited prior knowledge of ORSI, the current technology is still not ideal for detecting tiny ships.

We notice that the expansion of network depth facilitates the mining of higher-level semantic features. High-level semantic features and low-level localization features can reflect the differences of observers well, which brings more potential room to fuse the layered features. For efficient fusion, the layered detection models usually employ bidirectional mapping, including top-down paths and bottom-up paths, such as PANet [48], NAS-FPN [49], BiFPN [39], ASFF [50], and SFAM [51]. Moreover, after feature aggregation, the number of channels of fused features mostly remains consistent with the original features to ensure the width of the network. However, the larger the width of the network, the better it may not necessarily be. Numerous studies have demonstrated an upper limit to network width. When the width reaches a certain scale, the performance will not improve or may even decrease.

We also notice that the design of the detection head is crucial for prediction. The coupled head that is widely used obtains a unified output for localization and classification by sharing convolutional layers between two branches. In contrast, decoupled head designs separate convolutional layers for the localization and classification vectors to obtain more accurate outputs. FCOS [52] pointed out that the decoupled head can speed up model convergence and improve detection accuracy but also brings additional parameters and computational costs. Therefore, the coupled head that shares convolutional layers may be more in line with the lightweight requirement. But how to compensate for the lost performance? With the entry of the transformer into the object detection field, THP-yolov5 [53] treated the transformer as the convolution and utilized the Swin Transformer encoding block [35] to capture the global feature. However, the fully connected layer and residual connections are not optimized enough for the parameters. We urgently need to design a lightweight detection head that combines the advantage of CNN's inductive bias and the global receptive field capability of ViT, which would improve the detection performance of tiny targets.

As mentioned, although the performance of the above models is impressive, existing frameworks cannot meet the requirements of lightweight and practical remote sensing images. This paper provides an advanced detection model for marine remote sensing applications. The main contributions of this article can be summarized as follows:

- We propose a method to generate fog images in remote sensing datasets to simulate actual background disturbances and compensate for the lack of images with extreme weather. From the perspective of data augmentation and data driven, fog simulation indirectly improves the model's robustness and detection performance.
- Based on the analysis of the difficulties in optical remote sensing, we have designed a lightweight and layered detection framework (LMSD-Net). Inspired by the detection paradigm of "backbone–neck–head", in LMSD-Net, an improved module (ELA-C3) is proposed for efficient feature extraction. In the neck, we design a weighted fusion connection (WFC-PANet) to compress the network neck and enhance the representation ability of channel features. In the prediction, we introduce a Contextual Transformer (CoT) to improve the accuracy of dense targets in complex offshore scenes. During the training process, we discovered the degradation problem of CIoU in dealing with small ships and proposed V-CIoU to improve the detection performance of vessels marked by small boxes.
- Based on the VRS ship dataset [54], we added more nearshore images to construct a new ship dataset (VRS-SD v2). The dataset covers different nearshore and offshore scenes, multiple potential disturbances, different target scales, and more dense distributions of tiny ships. Then, we used the proposed fog simulation to process the dataset and obtained the dataset for the actual scenes.

The rest of the paper is organized as follows: Section 2 provides a detailed introduction to the fog simulation and detection framework. In Section 3, we conduct extensive ablation experiments to demonstrate the innovative and efficient framework, and then, we demonstrate the detection results of our model on typical datasets. According to the experiments, Section 4 emphatically discusses the problems solved by the corresponding methods and the experiment results. The final section summarizes the entire paper and briefly discusses future research directions.

2. Methods

An advanced and lightweight ship detection framework consists of three main components: effective data augmentation, efficient feature extraction and fusion, and accurate target prediction. Given the detection difficulties and lightweight requirements mentioned above, these three parts need to be reconsidered. In this section, we have provided a detailed introduction to the methods proposed, including the data augmentation combination and the lightweight detection framework.

2.1. Data Augmentation–Fog Simulation on Actual Remote Sensing Scenes

Whether at sea or near shore, ships are arbitrary in direction and random in distribution. Therefore, we selected several common data augmentation methods, such as cropping, translation, rotation, and random scaling. Then, we adjusted the images' hue, brightness, and saturation values to address photometric distortion and intensity differences. In addition, we adopted Mosaic [26], which concatenates four images and computes the activation statistics of multiple images together. It has been proven that Mosaic can enrich the detection of backgrounds and improve training efficiency. Essentially, the above data augmentation methods are aimed at achieving more complex representations of the data. Enriched data reduces the gap between the validation, training, and final test sets, so that the network can learn the data distribution better.

In optical remote sensing images, the background of ship targets is often complex and has significant interference with detection. The difficulty of detecting nearshore ships is related to the complex scene of the shore, while the interference of ship detection at sea is mainly caused by islands, wake waves, and sea clutter. Considering more actual scenes, detection work will be carried out under different lighting and weather conditions, especially extreme weather. However, there are few images of existing extreme weather. Due to the absence of cloud and fog scenes in the training and validation sets, the detection performance of the network would be poor. Therefore, simulating the dataset close to the actual scene is necessary to improve the robustness of the model. Thus, we proposed an image degradation method to simulate foggy scenes.

According to the optical model and the imaging mechanism in Figure 1, the influence of fog is modeled as a radiation attenuation function that maps the radiance of a clear scene to the camera sensor. According to the standard optical model, the degradation formula is expressed as follows:

$$D(x) = I(x)t(x) + L_{atmo}(1 - t(x)) \tag{1}$$

where $I(x)$ and $D(x)$ represent the original image intensity and observed fog-simulated image intensity at pixel x, respectively, L_{atmo} is global atmospheric light, and $t(x)$ is the transmission transmittance, which depends on the distance from the lens to the scene and the noise particles in the air. Therefore, the key to simulating fog lies in the estimation of atmospheric light noise and transmission transmittance.

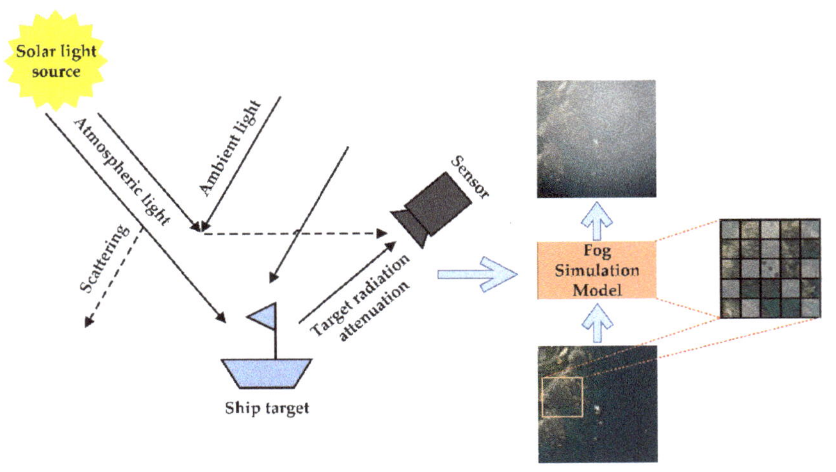

Figure 1. Fog simulation based on the optical model.

Considering the impact of noise on transmission, fog consistently exhibits spatial randomness and density nonuniformity. Therefore, we established the random diffusion of regional noise brightness. The input image was divided into different regions $R_i^{n \times n}$, and parts of the regions were randomly selected to participate in the diffusion processing. Based on the principle of center point diffusion, the diffusion degree at pixels (j, k) is defined as follows:

$$C(x) = -0.04\sqrt{(j-m)^2 + \left(k-n\right)^2} + 17 \tag{2}$$

where (m, n) is the central point of the region $R_i^{n\times n}$. It can be inferred that the closer to the center point, the higher the diffusion degree value.

Considering the impact of the distance from the camera to the scene on transmission, unlike common scenes, the top-view angle of remote sensing results in minimal spatial distance differences between the foreground and background. Strictly speaking, the difference is preserved and regarded as a weak distance attenuation. In the case of random diffusion noise, transmission transmittance is defined as distance attenuation:

$$t(x) = e^{-\beta C(x)} \tag{3}$$

where β represents the attenuation factor, which effectively controls the thickness of the fog: the smaller the attenuation factor, the thicker the fog is. According to the theory of semantic foggy scene understanding [55], the attenuation factor always obeys $\beta \geq 2.996 \times 10^{-3}\ \mathrm{m}^{-1}$. In this experiment, for convenience, β was limited in set S: {0.01 0.02 0.04 0.06 0.08 0.12 0.16}.

Global atmospheric light is related to lighting and is often set as a relative value. In this experiment, considering different lighting conditions, global atmospheric light was randomly selected in set T: {0.8 0.85 0.9 0.95 1}. Finally, the fog simulation was added to part of the data to improve the generalization performance of the model.

2.2. *The Proposed LMSD-Net*

Most lightweight frameworks mainly consider factors such as parameter size and computation complexity. Some models [45,56] achieve less computation complexity but sacrifice accuracy. Therefore, it is important to design a framework focusing on both lightweight and high performance. In this section, we proposed a lightweight multi-scale ship detector network (LMSD-Net) that can simultaneously locate and classify ship targets in ORSI, especially small-target ships.

2.2.1. Overall Architecture

Based on the classic detection paradigm, the overall architecture consists of three parts shown in Figure 2. The first part is a CNN backbone, which extracts feature maps of different layers. The second part is a bidirectional fusion process based on feature pyramids, and the third part includes a detection head used to predict the categories and bounding boxes of ships.

In terms of the architecture backbone, we continued the idea of the YOLO series models, which have proven their strong feature extraction capabilities in detection and other issues. It is worth noting that, unlike the C3 module (Yolov5), Repvgg Block (Yolov6), and E-LHAN (Yolov7), we designed a new functional module (ELA-C3 Block). Rethinking C3 and bottleneck-CSP, we added a branch containing Bottleneck structural units. After branch expansion, ELA-C3 Block has a more efficient feature extraction ability than C3.

Regarding the architecture neck, we proposed an improved fusion structure with a weighted-channel network (WFC-PANet). In WFC-PANet, the features of different channels are given weighted specificity. In addition, we abandoned the principle of equal channels for feature aggregation but designed half of the convolutional kernels to control the number of channels. Therefore, the number of channels for fused features was reduced to half of the original number, greatly reducing the parameters and Floating Point Operations (FLOPs).

In the detection head, a Contextual Transformer encoder (CoT) was added to effectively locate targets, further improving the detection performance of small ships. Thus, a more detailed network structure is shown in Table 1.

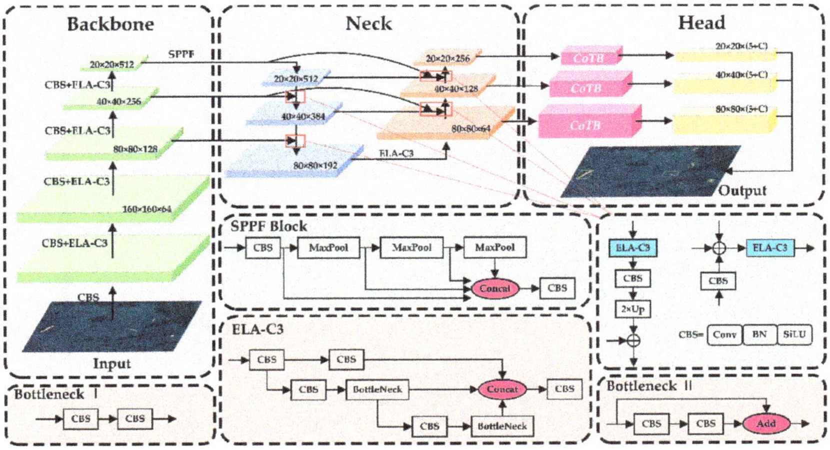

Figure 2. Overall architecture of the LMSD-Net framework.

Table 1. Information about each layer of the LMSD-Net structure.

The *n*th Layer	From	Module	Num	Output Shape	Params
	/	Input	/	[640,640,3]	/
0	−1	Convolution	1	[320,320,32]	3520
1	−1	Convolution	1	[160,160,64]	18,560
2	−1	ELA-C3 Block	1	[160,160,64]	18,816
3	−1	Convolution	1	[80,80,128]	73,984
4	−1	ELA-C3 Block	2	[80,80,128]	115,712
5	−1	Convolution	1	[40,40,256]	295,424
6	−1	ELA-C3 Block	3	[40,40,256]	625,152
7	−1	Convolution	1	[20,20,512]	1,180,672
8	−1	ELA-C3 Block	1	[20,20,512]	1,182,720
9	−1	SPPF	1	[20,20,512]	656,896
10	−1	Convolution	1	[20,20,128]	65,792
11	−1	Nearest Upsample	1	[40,40,128]	-
12	−1,6	WFC_Concat_2	1	[40,40,384]	2
13	−1	ELA-C3 Block	1	[40,40,128]	107,264
14	−1	Convolution	1	[40,40,64]	8320
15	−1	Nearest Upsample	1	[80,80,64]	-
16	−1,4	WFC_Concat_2	1	[80,80,192]	2
17	−1	ELA-C3 Block	1	[80,80,64]	27,008
18	−1	Convolution	1	[40,40,64]	36,992
19	−1,14,6	WFC_Concat_3	1	[40,40,384]	3
20	−1	ELA-C3 Block	1	[40,40,128]	107,264
21	−1	Convolution	1	[20,20,128]	147,712
22	−1,10,8	WFC_Concat_3	1	[20,20,768]	3
23	−1	ELA-C3 Block	1	[20,20,256]	427,520
24	17	CoTB	3	[80,80,64]	18,944
25	20	CoTB	3	[40,40,128]	74,240
26	23	CoTB	3	[20,20,256]	293,888
27	24,25,26	Detect	1	/	8118
366 Conv layers		12.8 GFLOPs		5.5×10^6 parameters	

Each row in Table 1 represents the forward propagation of the corresponding feature layer. By executing the corresponding number of modules, the shape of the feature output is marked in the "Output Shape" and the parameters are recorded in the "Params". "Num" represents the number of repetitions. For example, in the sixth row of the table, the features of the fourth layer of the network will be used as the input of the ELA-C3 module to further extract the features, the extracted feature scale is $80 \times 80 \times 128$, and the number of process parameters is 115,712. From the output shape of the 24th–26th rows, the model provides three scales of feature output, which would serve for multi-scale ship detection. From the output shape and "Params" of the 17th, 20th, and 23rd rows, the improved feature fusion part preserves small parameters and channels. The last line summarizes the model's convolution layers, total parameters, and computational complexity values.

2.2.2. Efficient Layer Aggregation Block

The backbone and neck focus more on obtaining efficient features, especially in lightweight models. As shown in Figure 3a,b, C3, as a variant of CSP-ResNeXt, still retains the CSP architecture and adopts CSP-Bottleneck as the modified unit with fewer parameters. In lightweight models, sharing current layer weights often achieves efficient layer aggregation. Based on this idea, we proposed a variant named Efficient Layer Aggregation of C3 (ELA-C3) in Figure 3d. In addition to reducing repetitive gradient learning, we also analyzed the gradient path. Compared to the Efficient Layer Aggregation Network (ELAN) [29], ELA-C3 removes the base layer paths with less contribution and assigns different channel numbers to different layers. For example, in Figure 3d, the number of channels in the three paths from left to right is c, c/2, and c/2, respectively. In this way, different layers can learn more various features without damaging the original gradient path, which is beneficial in enhancing learning ability.

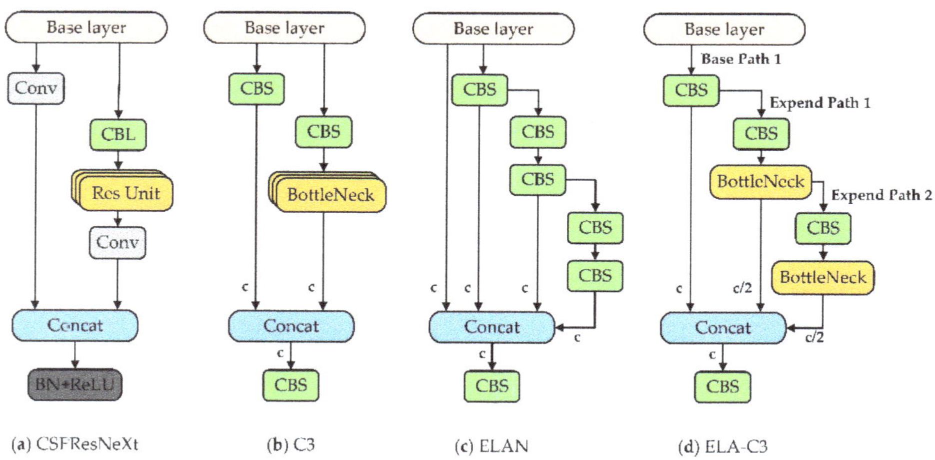

Figure 3. Evolution and exploration of the ELA-C3 module.

From the perspective of gradient diversion, the base path only performs ordinary transformations, while the two extended paths use efficient transformations to obtain extended features. Based on group convolution, ELA-C3 forms a local "extend–transform–merge" structure. Assume that feature x is obtained from the base path by the CBS operation. On the one hand, x is exported to participate in the final merger. On the other hand, x serves as the input for extended path features. In Extension Path 1, x performs an efficient transformation to obtain $\psi(x)$. Then, $\psi(x)$, as the input of Extension Path 2, participates in an efficient transformation of c/2 convolution kernels. Finally, the output results are merged by concatenating operations. The "split–transform–merge" structure can be expressed as follows:

$$F_c = \Theta(x_c, \psi(x)_{\frac{c}{2}}, \psi(\psi(x))_{\frac{c}{2}}) \quad (4)$$

where Θ represents the merge operation, and ψ represents the efficient transformation. Output F_c of the structure has c channels.

In the implementation, we adopted group convolution (group = g) to expand the channel and cardinality of the computational block. First, we applied the same parameters and channel multipliers to the two extended paths. Then, we concatenated the tensors of the three paths together. The number of channels in each group of feature maps will be the same as that in the base layer. Finally, we added g sets of feature maps to obtain the complete features. Therefore, ELA-C3 could construct efficient layer aggregation blocks by group convolution to learn more diverse features.

2.2.3. Lightweight Fusion with Weighted-Channel Concatenation

For the single-stage detector, multi-layer detection is an important method to address scale differences. As we all know, FPN has inconsistency of features among the different scales of the target. Specifically, large targets are typically associated with higher-feature maps, while small targets are typically associated with lower-feature maps. After sampling and fusion, the high-level feature responsible for large targets has rich semantic information but fuzzy spatial information. In contrast, the low-level feature responsible for small targets has an accurate location but less semantic information. This may result in a low classification accuracy for small targets and an inaccurate positioning for large targets. In Figure 4b, PANet adds a bottom-up fusion path, which is a "soft fusion" to ensure that spatial features are mapped to global features. However, not only does it bring more parameters and computational complexity, but also the loss from sampling is irreparable. For these issues, we proposed a lightweight fusion with the weighted channel based on PANet (WFC-PANet).

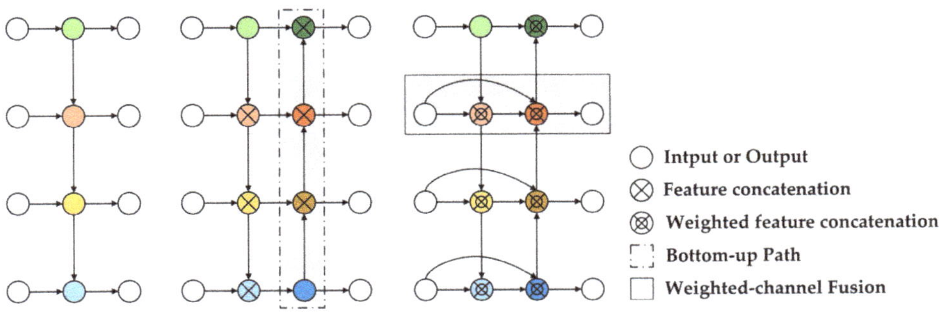

Figure 4. Fusion structure improvement of WFC-PANet.

Specifically, WFC-PANet adds learnable weights to all the channels in bidirectional fusion. Since different feature maps have different resolutions before stacking or adding, their contributions to the fusion are also different. Therefore, we established a feature competition mechanism based on the contribution to the fused feature map. Once a channel becomes more important in the fusion of features, it will occupy a greater weight. Then the weight is expressed by a fast normalization fusion formula:

$$W = \sum_i \frac{w_i}{\varepsilon + \sum_j w_j} \quad (5)$$

where $w_i > 0$ and $\varepsilon = 0.0001$ for stabilizing the value. Then, the number of channels of the output features is reduced to half of the original features, which avoids the reuse of similar features and reduces training parameters. Although it sacrifices some of the compelling features, the cross-layer weighted concatenation basically guarantees the expressiveness of the fusion.

To illustrate the fusion in Figure 4c, we used the concept of set to describe the features. As shown in Figure 5a, the entire detection neck is divided into three layers horizontally and three columns vertically. The available feature sets X, Y, and Z contain three scales of feature maps with different receptive fields. Then, based on the number of branches, the fusion includes two specific forms: two-node fusion and multi-node fusion. In Figure 5b,c, external mapping expands the fusion scales, while internal mapping only increases the diversity of features. Multi-node fusion adds cross-layer weighted fusion compared to two-node fusion. Because of more available feature map choices, multi-node fusion will be more inclined to select efficient features. Therefore, it seems this part of the features is screened and participates in feature refactoring. Moreover, both of them adopt Formula 5, and the values of each normalized weight are limited to [0, 1]. As for the layers corresponding to set Y, two-node weighted fusion is used. For example, the M_y layer is generated by the weighted fusion of corresponding M_x and S_y in the X set. As for the feature layers corresponding to set Z, multi-node weighted fusion is used because of the addition of cross-layer channels. For example, M_z is generated by weighted splicing of M_x, M_y, and L_z.

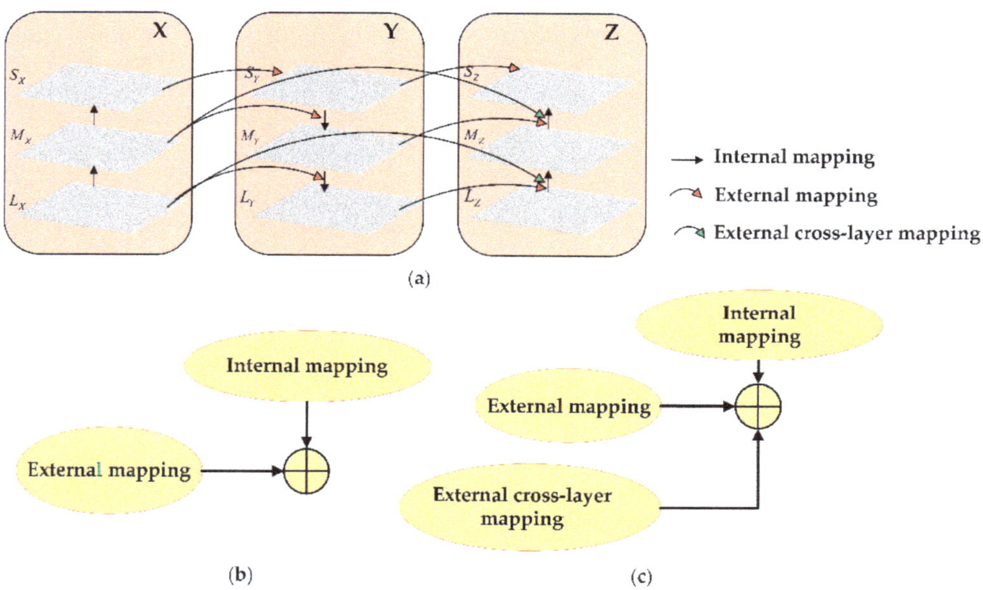

Figure 5. Abstract representation of fusion mapping. (**a**) Schematic diagram of a bidirectional fusion set. (**b**,**c**) Specific integration forms. The available features include the native feature set X, the top-down feature set Y, and the bottom-up feature set Z.

2.2.4. Contextual Transformer Block for the Detection Head

Discrete convolution operators impose spatial locality variance, which is beneficial for reflecting local differences. However, the limited acceptance field affects the modeling of global relationships and makes it less apparent to the remote feature interactions. Inspired by visual transformers, interactions in pairs of queries and keys can measure the global attention matrix, which reflects contextual self-attention expression well. Based on CNN, we added a lightweight Contextual Transformer (CoT) block before the shared decoupled head for more accurate classification and localization.

Specifically, as shown in Figure 6, given a ship feature map $X \in R^{H \times W \times C}$, it can be transformed into queries, keys, and values, which are defined as follows:

$$Q = XM_q \tag{6}$$

$$K = XM_k^E \tag{7}$$

$$V = XM_v \tag{8}$$

where M_q, M_k^E, and M_v are the embedding matrices, which transform the sparse image into a dense matrix. Assuming the central key of the context area is X_{cen}, the surrounding key is the region with $k \times k$ ($k = 3$ in Figure 6). Centered around each key in the surrounding area, the $k \times k$ convolution can calculate the contextual information of each key. Similar to sliding window convolution in CNN, the learned contextual key $K_{Static} \in R^{H \times W \times C}$ reflects the static information of the center and surrounding.

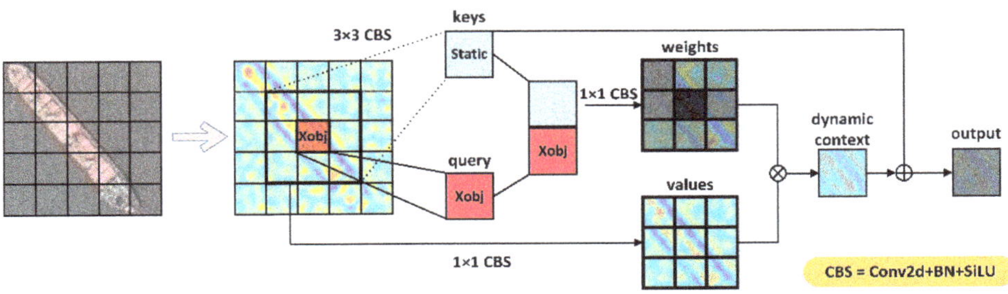

Figure 6. Measurement of the attention matrix in the CoT block.

Then, the learned contextual keys and queries are concatenated to synthesize new keys $[K_{Static}, Q]$. By using two consecutive 1×1 convolutions to perform self-attention:

$$W_{att} = [K_{Static}, Q] \times M_{att}^{SiLu} \times M_{att} \tag{9}$$

where M_{att}^{SiLu} represents the convolution with SiLU while M_{att} represents the convolution without activation. Obviously, the learned attention weight matrix considers the context keys and queries. In other words, the purpose of mining contextual information is to improve the self-attention of local regions. Next, Softmax is used to form the attention weight matrix $W_{att}^{Softmax}$. Aggregating the value matrix, a dynamic contextual self-attention weight matrix is calculated and represented as follows:

$$K_{dynamic} = V \bigotimes W_{att}^{Softmax} \tag{10}$$

During the forward transmission process, static context K_{Static} and dynamic context $K_{dynamic}$ integrate through the overlay fusion mechanism [57]. The hardware algorithm implementation is shown in Figure 7.

Essentially, CoT is a self-attention block that combines transformers. Therefore, treating CoT as a convolution module is feasible. In the ablation experiment, we increased the number of CoT blocks to obtain the best response.

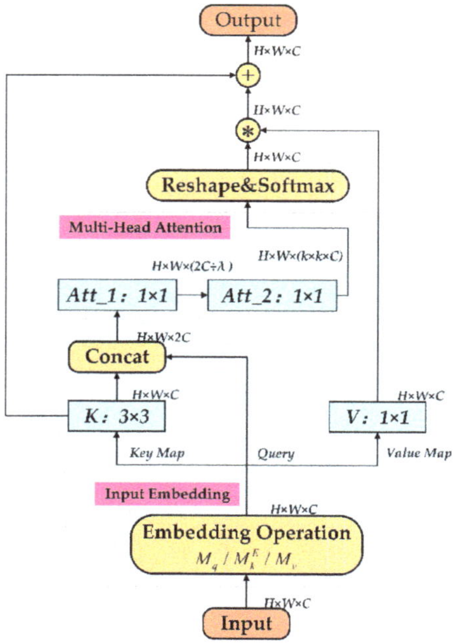

Figure 7. The detailed structures of the Contextual Transformer (CoT) block. ⊛ denotes local matrix multiplication, and ⊕ denotes the fusion of dynamic and static keys. For two consecutive 1 × 1 convolutions, channel scaling factor λ is set as 4 in the experiment.

2.2.5. Prediction

As mentioned above, three prediction branches are elicited to accurately detect multi-scale ships. In the output of each branch, the positive sample grids, which are used to predict the real target, need to be filtered and serve for location prediction. Since the ship targets are mostly distinctly elongated, the aspect ratio of the label has a positive effect on the prediction. In addition, we expanded the prediction location to three cell grids to filter positive samples with a multi-sample label matching strategy [27]. In this way, the labels are assigned to all the anchors simultaneously during training, thus alleviating the problem of unbalanced positive and negative samples during training to some extent. Once the positive samples are identified, the positive sample loss is calculated as the sum of grid confidence loss, target classification loss, and target bounding box regression loss. The negative samples only need to calculate the confidence loss.

In the training process, we inherited the Binary Cross-Entropy as the class loss and confidence loss of the positive and negative samples of the grid. Considering the prediction output grid ($S \times S$), each cell in the grid generates N bounding boxes, whose center coordinate is (x, y), prediction confidence is c, and the prediction vector points to the kth class with prediction value p_k. Class loss and confidence loss are defined as follows:

$$L_{class} = \sum_{i=0}^{S^2} \sum_{j=0}^{B} \sum_{k=0}^{N} 3_{ij}^{obj} [\hat{p}_k \ln(p_k) + (1 - \hat{p}_k)\ln(1 - p_k)] \tag{11}$$

$$L_{obj} = \sum_{i=0}^{S^2} \sum_{j=1}^{B} 3_{ij}^{obj} [\hat{c}\ln(c) + (1 - \hat{c})\ln(1 - c)] \tag{12}$$

where $\hat{p}$, $\hat{c}$ are the truth of p, c. 3_{ij}^{obj} denotes whether the object appears in the bounding box j predictor in cell i. It is worth noting that the positive sample only contains three grids, while the negative sample contains other grids as well as grids from other detection layers. Due to the labels of the negative samples $\hat{c} = 0$, the confidence loss calculation for negative samples can be optimized approximately as follows:

$$L_{obj} = \sum_{i=0}^{num(neg)} \sum_{j=1}^{B} 3_{ij}^{obj} \lim_{\hat{c} \to 0} [-\hat{c}\ln(c) - (1-\hat{c})\ln(1-c)] = \sum_{i=0}^{num(neg)} \sum_{j=1}^{B} 3_{ij}^{obj} \ln(1-c) \quad (13)$$

For the bounding box regression loss of positive samples, we proposed an improved version named V-CIoU based on CIoU [58]. First, consider the formula of CIoU:

$$L_{Bbox}^{CIoU} = IoU - \left(\frac{(x-\hat{x})^2 + (y-\hat{y})^2}{c^2} + \alpha v \right) \quad (14)$$

$$IoU = \left| \frac{Area(B) \bigcap Area(\hat{B})}{Area(B) \bigcup Area(B)} \right| \quad (15)$$

$$v = \frac{4}{\pi^2} \left(arctan \frac{\hat{w}}{\hat{h}} - arctan \frac{w}{h} \right)^2 \quad (16)$$

$$\alpha = \frac{v}{(1 - IoU) + v} \quad (17)$$

where B and $\hat{B}$ represent the areas of the prediction box and the ground-truth box, respectively, $(\hat{x}, \hat{y}, \hat{w}, \hat{h})$ is the matched truth value of (x, y, w, h), c is the diagonal length of the smallest closed box covering both boxes, α is the weight parameter, and v is the penalty representing the aspect ratio's consistency.

CIoU loss adds the distance offset and aspect ratio of the prediction box to the IoU, and both of them are beneficial for improving the regression accuracy of the ship. However, a problem that needs to be considered is that the penalty term v in Formula (16) will fail when the aspect ratio of the truth and prediction is equal or approximately equal. Especially for some small-ship targets, the similar aspect ratio results in incomplete convergence. In this case, we proposed a penalty function based on the variance of the ground truth and the prediction for each corresponding aspect ratio. This penalty term u is defined as follows:

$$u = \begin{cases} \frac{4}{\pi^2} (arctan \frac{\hat{w}}{\hat{h}} - arctan \frac{w}{h})^2, & \left| \hat{w}h - w\hat{h} \right| \geq 0.001 \\ \frac{8}{\pi^2} \left[\left(arctan \frac{\hat{w}}{w} - \frac{\pi}{4} \right)^2 + \left(arctan \frac{\hat{h}}{h} - \frac{\pi}{4} \right)^2 \right], & \left| \hat{w}h - w\hat{h} \right| < 0.001 \end{cases} \quad (18)$$

The penalty term v is preserved as a part of the new penalty function. Normally, the penalty term v can solve the problem of offset. The variance penalty term is activated when the ratio between the prediction and the ground truth is consistent. Therefore, V-CIoU not only embodies the advantages of CIoU but also solves the degradation problem, in that the aspect ratio of the ground truth equals that of the prediction. Once the aspect ratio of the prediction and ground truth are maintained within a small range, the convergence behavior reaches its limit, and then the penalty loses efficacy. Finally, the bounding box regression loss is defined as follows:

$$L_{Bbox}^{VCIoU} = IoU - \left(\frac{(x-\hat{x})^2 + (y-\hat{y})^2}{c^2} + \alpha u \right) \quad (19)$$

Furthermore, the implementation process is summarized in Algorithm 1.

Algorithm 1. V-CIoU computation

Input: Bounding box of ground truth $B^{gt} = \left(w^{gt}, h^{gt}, x^{gt}, y^{gt}\right)$
Input: Bounding box of prediction $B^{p} = (w^{p}, h^{p}, x^{p}, y^{p})$
Output: VCIoU between the ground-truth box and the prediction boxes
If $\left(B^{gt} \neq \varnothing\right) \cup \left(B^{p} \neq \varnothing\right)$ do
  For A and B, find the smallest enclosing convex object C.
  within C, calculate $IoU = \left|\frac{Area(B^{p}) \cap Area(B^{gt})}{Area(B^{p}) \cup Area(B^{gt})}\right|$.
  If $\left|w^{gt}h^{p} - w^{p}h^{gt}\right| \leq 0.001$:
    then $u = \frac{8}{\pi^2}\left[\left(arctan\frac{\hat{w}}{w} - \frac{\pi}{4}\right)^2 + \left(arctan\frac{\hat{h}}{h} - \frac{\pi}{4}\right)^2\right]$,
      $\alpha = \frac{u}{(1-IoU)+u}$,
      $L_{Bbox} = IoU - \left(\frac{(x-\hat{x})^2+(y-\hat{y})^2}{c^2} + \alpha u\right)$.
  else
    then $v = \frac{4}{\pi^2}\left(arctan\frac{\hat{w}}{\hat{h}} - arctan\frac{w}{h}\right)^2$.
else
  $L_{Bbox} = 0$.

3. Results and Experiments

This section provides a detailed introduction to the dataset and a description of the evaluation metric. Then, we conduct a large number of experiments to demonstrate the effectiveness of the framework. On the one hand, we perform ablation experiments for the proposed data argument and self-designed modules with relevant advanced methods. On the other hand, we perform a detailed comparison with the current excellent lightweight detection frameworks. Finally, the detection results using the most advanced methods are presented, leading to a profound discussion in the next section.

3.1. Dataset

The increase in high-resolution optical images has greatly contributed to the advancement of target detection. Improving the detection performance of small ships relies on collecting small-target ship datasets. However, existing open data sources still need to be extended in the diversity of scenes and targets. For example, in HRSC2016 [59], there are only two or three targets in an image, most of which are large-scale targets. The scenes of NWPU VHR-10 [60] and the Airbus ship dataset [61] are more singular with the coastal background. Subsequently, we have proposed the VRS ship dataset [54] (VRS-SD) in our previous study, which contains various maritime disturbances, such as thin clouds, islands, sea waves, and wake waves. Therefore, the application of VRS-SD is oriented toward detection tasks in maritime scenes. In order to meet the unified detection requirements for nearshore and maritime scenes, we furthermore construct VRS-SD v2, which covers different nearshore scenes, marine environments, maritime disturbances, target scales, and dense small-target distributions. The detailed differences among the current datasets are summarized in Table 2.

Table 2. Comparison of ship datasets.

Dataset	Images	Class	Ship Instances	Image Size	Source	Fog
NWPU VHR-10	800	10	302	/	Google Earth	×
HRSC2016	1061	3	2976	300 × 300~1500 × 1900	Google Earth	×
Airbus ship dataset	192,570	2	/	768 × 768	Google Earth	×
MASATI [62]	6212	7	7389	512 × 512	Aircraft	×
FGSD2021 [63]	636	20	5274	157 × 224~6506 × 7789	Google Earth	×
AI-TOD [64]	28,036	8	700,621	/	Google Earth	√
VRS-SD	893	6	1162	512 × 512	Google Earth	√
VRS-SD v2	2368	8	4054	512 × 512	Google Earth and Aircraft	√

According to the statistics in Table 2, most of the existing ship datasets are from Google Earth and are mostly taken under sunny conditions. Both VRS-SD and VRS-SD v2 are collected under a variety of weather conditions. Compared with VRS-SD, VRS-SD v2 has significantly expanded the amounts of images, and the two additional classes are near-shore ships and river-distribution ships. In addition, to address the problem of insufficient fog interference background in VRS-SD, we provided more images of such scenes through fog simulation. Since AI-TOD focuses more on the differences in nearshore target scale, it usually better reflects the complexity of the scenes. Therefore, in the final validation, we implemented our method on the AI-TOD dataset.

3.1.1. The Analysis of VRS-SD v2

VRS-SD v2 increases the number of ship targets at different scales. To compare the targets at different scales, we first refer to the definition of the small target. The small-target scale has different absolute definitions in different remote sensing datasets. For example, the MS COCO dataset defines small targets within 32 × 32 pixels. TinyPerson [65] defines small targets as those with pixel values in the interval [20, 50]. Furthermore, the aerial image dataset DOTA [66] defines a small target with pixel values in the range of 10–50. It is difficult to unify the definition of small targets for different datasets, so we introduced a relative definition of small-target scale. Ref. [67] states that the relative areas of small-target instances in the same class, the median ratio of the area of the ground truth to the image, should be limited to between 0.08% and 0.58%. In addition, the ratio of the target bounding box area to the image area is open-squared to less than a certain value, the more general value being 0.03. Based on the above considerations, we compared the two datasets at a finer scale as shown in Table 3. It can be seen that there is a significant increase in tiny ships, and the number of small targets has increased to varying degrees at the subdivision scales. Figure 8 counts the relative areas of all ship instances and the number of targets in different intervals. In addition, Figure 9 shows the distribution of ship positions at different scales, and VRS-SD v2 has more targets and a denser distribution.

Table 3. Quantitative statistics of multi-scale ships.

Relative Scales	Relative Area Rates	VRS-SD/pcs	VRS-SD v2/pcs
Tiny ship	(0, 0.0008)	312	2284
	(0.0008, 0.0016)	761	943
Small ship	(0.0016–0.0025)	244	381
	(0.0025–0.0058)	300	335
Medium ship	(0.0058–0.04)	46	111

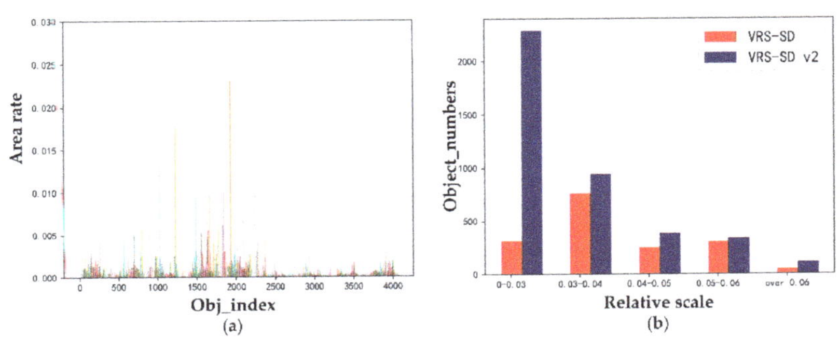

Figure 8. Target statistics of VRS-SD v2 and comparison with VRS-SD. (**a**) Relative scale statistics in VRS-SD v2. (**b**) Comparison of target-relative scale distribution between VRS-SD and VRS-SD v2.

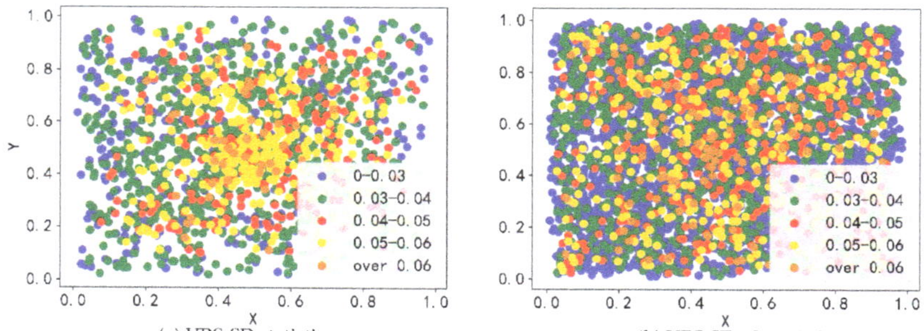

(a) VRS-SD statistics. (b) VRS-SD v2 statistics.

Figure 9. Distribution of target positions at different scales in VRS-SD and VRS-SD v2. The X and Y axes indicate the relative positions of the ships, and the image scale is normalized to a relative scale of 1.0 × 1.0. Different colors indicate the targets at different scales.

3.1.2. Fog Simulation

VRS-SD v2 includes a few cloud images and fog images. We performed the fog simulation on a certain proportion of images to simulate the real-world detection background. These images have been fogged at random spatial locations with varying degrees. In Figure 10, we present some simulation examples of some typical scenes. The fog simulation in the coastal area represents the real situation. Once the model is trained to resist the disturbances caused by fog, it can be deployed to industrial equipment, especially those devices under severe weather conditions.

(a) Open areas at sea (b) Coastal areas

Figure 10. Examples of fog simulation. (**a**) The open areas contain lakes, island shores, and sea clutter. (**b**) The coast scene with dense ship targets.

3.2. Evaluation Metrics

Similar to the general target detection task, we used precision rate, recall rate, and average precision to evaluate the performance of the proposed network. By setting a threshold for the intersection over union (IoU), the prediction results can be filtered and divided as true positive (TP), true negative (TN), false positive (FP), and false negative (FN). The formulas for precision, recall, and F1 score are as follows:

$$Precision = \frac{TP}{TP + FP} \tag{20}$$

$$Recall = \frac{TP}{TP + FN} \tag{21}$$

$$F1 = 2 \times \frac{Precsion \times Recall}{Precsion + Recall} \tag{22}$$

Furthermore, average precision (AP) calculates the total precision of the recall value from 0 to 1, that is, AP is the area enclosed by the *P-R* curve and the coordinate axis. Let r be the recall rate and $P(r)$ be the accuracy corresponding to the curve. By interpolation, AP as the line integral is calculated as follows:

$$AP = \int_0^1 P(r)dr = \sum_{k=1}^{N} P(k)\Delta r(k) \tag{23}$$

For the lightweight comparison, we use the GFLOPs and parameters, which could reflect the network complexity and memory usage. Additionally, frames per second (FPS) is calculated to quantify the detection speed. In consideration of the limitation of the device, FPS is tested with *batch size* = 1 or 16 in the experiments.

3.3. Ablation Study

All the experiments were tested and evaluated on a computer with an Intel Core i7-10900 2.90 GHz CPU, 24 GB memory, and GeForce GTX 3060Ti GPU with 8 GB. In the preparation phase, the dataset was divided into a training set, a validation set, and a test set in a ratio of 8:1:1. By k-means clustering, the criteria for the three classes of anchors were automatically generated based on the ship scale in the specific dataset. During the training process, we applied the AdamW optimizer and trained 200 epochs to ensure convergence. For all experiments, the IoU was set to 0.6.

3.3.1. Effect of Fog Simulation

To verify the importance of fog simulation for practical detection work, as shown in Table 4, we tested the fog simulation on MASATI and VRS-SD v2, which are both the small-ship dataset. It is worth noting that we set three rates, 0, 50%, and 100%, to test the effect of fog simulation on the results. The best results of the three rates are highlighted in red.

Table 4. Fog simulation for data enhancement.

Dataset	Train/Val Set With Fog	Test Set with Fog	Recall	Precision	F1	AP@0.5	AP@0.5:0.95
MASATI	×	×	0.813	0.825	0.82	0.813	0.407
	×	√ (100%)	0.609	0.679	0.64	0.587	0.264
	√ (100%)	√ (100%)	0.738	0.766	0.75	0.758	0.345
	√ (50%)	√ (50%)	0.731	0.833	0.78	0.783	0.358
VRS-SOD v2	×	×	0.771	0.832	0.80	0.817	0.395
	×	√ (100%)	0.612	0.718	0.66	0.615	0.283
	√ (100%)	√ (100%)	0.650	0.744	0.69	0.718	0.32
	√ (50%)	√ (50%)	0.662	0.848	0.74	0.741	0.342

Taking MASATI as an example, the model can give the best results at AP@0.5 of 0.813 and AP@0.5:0.95 of 0.407 without fog interference. However, when the training set lacks fog images, the testing achieves the worst results, with AP@0.5 of 0.587 and AP@0.5:0.95 of 0.264. Adding a certain percentage of fog images in the dataset can match the real remote sensing detection and improve the robustness of the model to weather conditions. On VRS-SD v2, when the training and test sets are mixed with fog images simultaneously, the detection results are better than the in case of all fog images, and AP@0.5 and AP@0.5:0.95 reach 0.741 and 0.342. It also provides an experimental basis for obtaining the best ratio of fog images.

3.3.2. Effect of ELA-C3

ELA-C3 is an improved version of the C3 module. To verify the validity of ELA-C3, we used C3 as a baseline in LMSD-Net. Additionally, we applied all remaining components of LMSD-Net. As shown in Table 5, the model obtains results by replacing the C3 module in the backbone and neck.

Table 5. Ablation of ELA-C3.

Input Size	Backbone + ELA-C3	Neck + ELA-C3	AP@0.5	AP@0.5:0.95	FPS bs@16	Params (M)	GFLOPs (G)
640 × 640	×	×	0.782	0.363	126	6.97	17.3
640 × 640	√	×	0.821 (+3.9%)	0.381 (+1.8%)	204	5.09	12.1
640 × 640	×	√	0.797 (+1.5%)	0.382 (+1.9%)	161	6.51	16.5
640 × 640	√	√	0.837 (+5.5%)	0.396 (+3.3%)	181	5.5	12.8

When ELA-C3 is added to the backbone or neck, the AP@50 values are 3.9% or 1.5% higher than the baseline model. In addition, the AP value with ELA-C3 exclusively is 5.5% higher than that using C3. As a lightweight feature extraction module, ELA-C3 has less increase of parameters. Therefore, the ELA-C3 module facilitates the efficient acquisition of rich contextual spatial features to improve the detection performance of ship targets.

3.3.3. Effect of WFC-PANet

In the detection neck, we designed the cross-layer and weighted-channel concatenation based on PANet. To avoid the influence of ELA-C3, all the following networks uniformly used the Yolov5s-backbone. Then, we quantified the experimental results of the current advanced feature fusion methods in Table 6.

Table 6. Comparison of different feature fusion methods in the neck.

Neck	Recall	Precision	AP@0.5	AP@0.5:0.95	FPS bs@16	Params (M)	GFLOPs (G)
PANet	0.811	0.823	0.831	0.41	181	7.02	15.8
BiFPN_Add	0.783	0.789	0.809	0.38	169	9.32	22.9
BiFPN_Concat	0.771	0.844	0.823	0.404	181	7.08	16.0
WFC-PANet(ours)	0.790	0.832	0.817	0.39	**208**	**5.10**	**12.1**

The experiment results show that using WGC-PANet leads to an increase in speed and a more lightweight model. In addition, there is a small sacrifice in average accuracy compared with PANet. Nevertheless, the model still maintains good performance and enough to finish the detection task. Similar to BiFPN, WGC-PANet also mentions a cross-layer connection. However, the use of adding BiFPN increases the computation complexity significantly. On the contrary, using Concat guarantees the model's performance and reduces the computation complexity. Taken together, the cross-channel and weighted-channel concatenation adopted by WGC-PANet can maintain the model's expressiveness and provide the possibility of lightweight implementation.

3.3.4. Structure Exploration of the Detection Head

The prediction head is crucial for the decoupling of the feature map. Based on the general structure of LMSD-Net, the comparison results of applying different mainstream detection heads are presented in Table 7. Further, to explore the effect of the number of CoT blocks, we embedded different numbers of CoT blocks and obtained the optimal choice according to the comparison. Note that CoT_x denotes the use of x CoT blocks.

Table 7. Exploration and comparison of detection heads.

Detection Head	Recall	Precision	AP@0.5	AP@0.5:0.95	FPS (bs@16)	Params (M)	GFLOPs (G)
YOLO head	0.743	0.784	0.793	0.368	208	5.10	12.1
Decoupled head [28]	0.792	0.821	0.800	0.386	188	6.09	13.9
Swin+ YOLO head [53]	0.773	0.804	0.796	0.392	181	5.54	25.7
CoT_1+ YOLO head	0.756	0.837	0.817	0.375	208	4.97	12.1
CoT_2+ YOLO head	0.787	0.821	0.831	0.384	185	5.18	12.5
CoT_3+ YOLO head	0.781	**0.847**	**0.837**	**0.396**	171	5.49	12.8
CoT_4+ YOLO head	0.784	0.850	0.839	0.398	162	5.90	13.2

In the YOLO head, the classification and localization branches are fused to share the convolutional layers. In the decoupled head, the two branches are convolved separately to obtain higher accuracy. Therefore, applying the YOLO head has fewer parameters and computation complexity than the decoupled head but poorer performance. With the addition of CoT blocks, the detection performs more powerfully. Compared with Swin Transformer block, CoT_3 obtains less computation complexity as well as higher precision. In addition, the number of CoT blocks affect the performance. More CoT blocks will bring a slight increase in parameters and GFLOPs but a decrease in speed. Considering the performance and hardware consumption, we finally chose CoT_3 in the network.

3.3.5. Validation of Regression Loss Function

According to the analysis of VRS-SD v2 in Table 3, the relative area ratios of tiny and small targets are primarily of [0,0.0016]. Therefore, the observation will have a similar aspect ratio between the ground truth and the predicted bounding box, which leads to the failure of the aspect ratio penalty term of CIoU. To verify the validity of the proposed variance penalty term for V-CIoU, we designed experiments of regression loss, as shown in Table 8. We set three different thresholds for the following loss functions in the valid. On the whole, V-CIoU has the best effect. Compared with CIoU, V-CIoU improves by 2.9% at AP@75 and 2.2% at AP@50:95. The experiments demonstrated that adding the variance penalty term makes V-CIoU more adaptable to tiny- and small-ship detection.

Table 8. Validation of the improved V-CIoU.

Regression Loss	AP_{50}^{val}	AP_{75}^{val}	$AP_{50:95}^{val}$
CIoU	0.821	0.309	0.382
DIoU [68]	0.817	0.293	0.371
EIoU [69]	0.796	0.294	0.375
SIoU [70]	0.787	0.318	0.379
Wise-IoU [71]	0.817	0.326	0.378
V-CIoU	**0.823**	**0.338**	**0.404**

3.3.6. Multi-Scale Performance of the Model

Based on the statistics of the dataset, the proposed VRS-SD v2.0 contains ship targets that are mostly small- and medium-sized, whereas VRS-SD proposed in previous work contains more large targets. Therefore, we combined the two datasets to explore the model's detection performance for different-sized ship targets. Table 9 lists the comparison results of the lightweight SOTA detectors.

Table 9. Comparison of detection performance at different scales.

Model	Params (M)	GFLOPs(G)	Size	AP^{val}	AP_S^{val}	AP_M^{val}	AP_L^{val}
Yolov7-tiny	6.01	13.2	640	0.208	0.211	0.149	0.342
Yolov5s-6.1	7.03	15.9	640	0.376	0.369	0.549	0.581
Yolov6n-3.0	4.63	11.34	640	0.323	0.316	0.513	0.604
Yolov8s	11.1	28.6	640	0.380	0.360	0.595	0.683
LMSD-Net	**5.50**	**12.8**	640	0.392	**0.372**	**0.591**	0.644

From the results, we see that LMSD-Net is comparable to the latest Yolov6-3.0n in terms of being lightweight, while LMSD-Net performs better on small targets and medium-sized targets, with an improvement of 5.6% and 7.8%, respectively. Considering this enhancement, on the one hand, the small and medium targets are well trained due to the large number of small and medium samples in the dataset. On the other hand, V-CIOU specifically solves the problem of the inconsistent aspect ratio of small targets, thus improving detection accuracy. In addition, the AP for large-ship targets reaches 0.644, which is lower than Yolov8s by about 3.9%. Nevertheless, the parameters of LMSD-Net are only half of those of Yolov8s, and the computation is reduced by 45%.

3.4. Overall Detection Performance

To validate the overall detection performance, we first compared the proposed models with the current lightweight state-of-the-art on the VRS-SD v2. These comparison methods include lightweight versions of the universal detectors, such as EfficientDet (D0-D3), Yolov7-tiny, and Yolov8n, and specialized lightweight detectors, such as the Nanodet family. In addition, we added a variant of Yolov5s called Yolov5-Ghost, which introduces the lightweight backbone GhostNet into the CSP architecture. For this part of the experiments, we used the training and validation setup of the ablation study. To ensure great and fast convergence, we increased the pre-training weights and performed 200 epochs of training. In addition, we set the *batch size* = 1 to test the general real-time performance. The comparison experiments were fair and extensive. We directly trained and tested all the comparison methods using official open-source codes.

Generally, as shown in Table 10, the proposed method performs best on this small-ship dataset. In terms of AP@50, LMSD-Net achieves the highest value with 81.3%. Compared with Yolov8s and Yolov6-3.0-s, which have high average accuracy, LMSD-Net has more advantages in terms of parameters and computation complexity. Therefore, it can meet the needs of ship-target detection tasks better. In addition, we observed that parts of the anchor-free detectors in Table 10, like Yolov6s-3.0 and Yolov8s, performed better than the Yolov5 series, Yolov4-tiny and Yolov7-tiny, which are anchor-based detectors. Since tiny targets are more sensitive to IoU than large targets, the anchor-based detectors, such as Yolov7-tiny and Yolov5n, cannot accurately predict the bounding box. Especially in AP@50:95, which has a stricter limitation than AP@50, common IoU loss will lead to less improvement. With the proposed V-CIOU, we could improve the average accuracy and cope with the tiny-target detection.

In terms of lightweight, the Nanodet series perform the best. However, they are mainly applied to mobile target detection and are not well adapted to small-ship target detection in the remote sensing field. Due to the small model input scale, such as 320 × 320 or 416 × 416, the feature description capability is limited, which leads to low detection accuracy. Differently, the model input scale of the EfficientDet series increases with the expansion of the backbone. Based on DWConv, the scaled model gradually adapts to lightweight but sacrifices more accuracy and improves a little in speed. In contrast, the accuracy advantage of LMSD-Net is very obvious and ensures efficient detection performance.

Table 10. Comparison of the lightweight SOTA performance on VRS-SD v2 (30% foggy images).

Method	Backbone	Input Size	Recall	Precision	F1	AP@0.5	AP@0.5:0.95	FPS (bs@1)	Params (M)	GFLOPs (G)
EfficientDet-D0 [39]	Efficient-B0	512	0.233	0.766	0.36	0.291	0.125	23	3.83	4.7
EfficientDet-D1 [39]	Efficient-B1	640	0.404	0.833	0.54	0.444	0.213	19	6.56	11.5
EfficientDet-D2 [39]	Efficient-B2	768	0.458	0.842	0.59	0.561	0.266	16	8.01	20.5
EfficientDet-D3 [39]	Efficient-B3	896	0.671	0.780	0.72	0.638	0.300	13	11.90	46.9
Nanodet-m [72]	ShuffleNetV2 1.0x	320	0.355	0.879	0.51	0.420	0.162	78	0.94	0.72
Nanodet-plus-m [72]	ShuffleNetV2 1.5x	416	0.556	0.656	0.60	0.585	0.278	67	2.44	2.97
Nanodet-EfficientLite [72]	EfficientNet-Lite1	416	0.586	0.677	0.63	0.578	0.288	59	4.00	4.06
Nanodet-EfficientLite [72]	EfficientNet-Lite2	512	0.635	0.691	0.66	0.596	0.284	48	4.70	7.12
Yolov4-tiny [26]	CSPDarknet53-tiny	640	0.576	0.751	0.65	0.683	0.235	130	5.87	16.2
Yolov7-tiny [29]	CSP-ELAN	640	0.699	0.891	0.78	0.731	0.282	80	6.01	13.2
Yolox-nano [73]	CSPDarknet-C3	640	0.689	0.661	0.67	0.705	0.283	57	0.90	2.5
Yolox-tiny [73]	CSPDarknet-C3	640	0.763	0.827	0.79	0.782	0.324	53	5.06	15.4
Yolov5n6 [27]	CSPDarknet-C3	640	0.665	0.842	0.74	0.756	0.329	91	1.77	4.2
Yolov5s6 [27]	CSPDarknet-C3	640	0.724	0.856	0.78	0.787	0.370	79	7.03	15.9
Yolov5-Ghost [27]	CSPDarknet-C3Ghost	640	0.725	0.781	0.75	0.771	0.347	84	4.90	10.6
Yolov6-3.0-nano [74]	EfficientRep	640	0.726	0.829	0.77	0.744	0.380	81	4.63	11.34
Yolov6-3.0-s [74]	EfficientRep	640	0.743	0.884	0.81	0.789	0.392	73	18.50	45.17
Yolov8n [30]	CSPDarknet-C2f	640	0.716	0.877	0.79	0.772	0.345	82	3.1	8.2
Yolov8s [30]	CSPDarknet-C2f	640	0.760	0.886	0.82	0.809	0.358	79	11.1	28.6
LMSD-Net(ours)	CSPDarknet-ELA-C3 (ours)	640	0.790	0.824	0.81	**0.813**	**0.384**	**68**	**5.50**	**12.8**

Although the speed of LMSD-Net is not the fastest, it is acceptable compared to most of the advanced detectors mentioned earlier. Its detection speed reaches 68 FPS, which could meet the real-time requirement (FPS > 30).

Further, in Figure 11, we show the detection results using LMSD-Net on AI-TOD, MASATI, and VRS-SD v2. It can be observed that our model performs well on all three datasets with no missed and false detections essentially, which indicates that the model has a high generalization ability. Despite the large interference caused by clouds and fog to the ship target, the detection still performs well.

Figure 11. *Cont.*

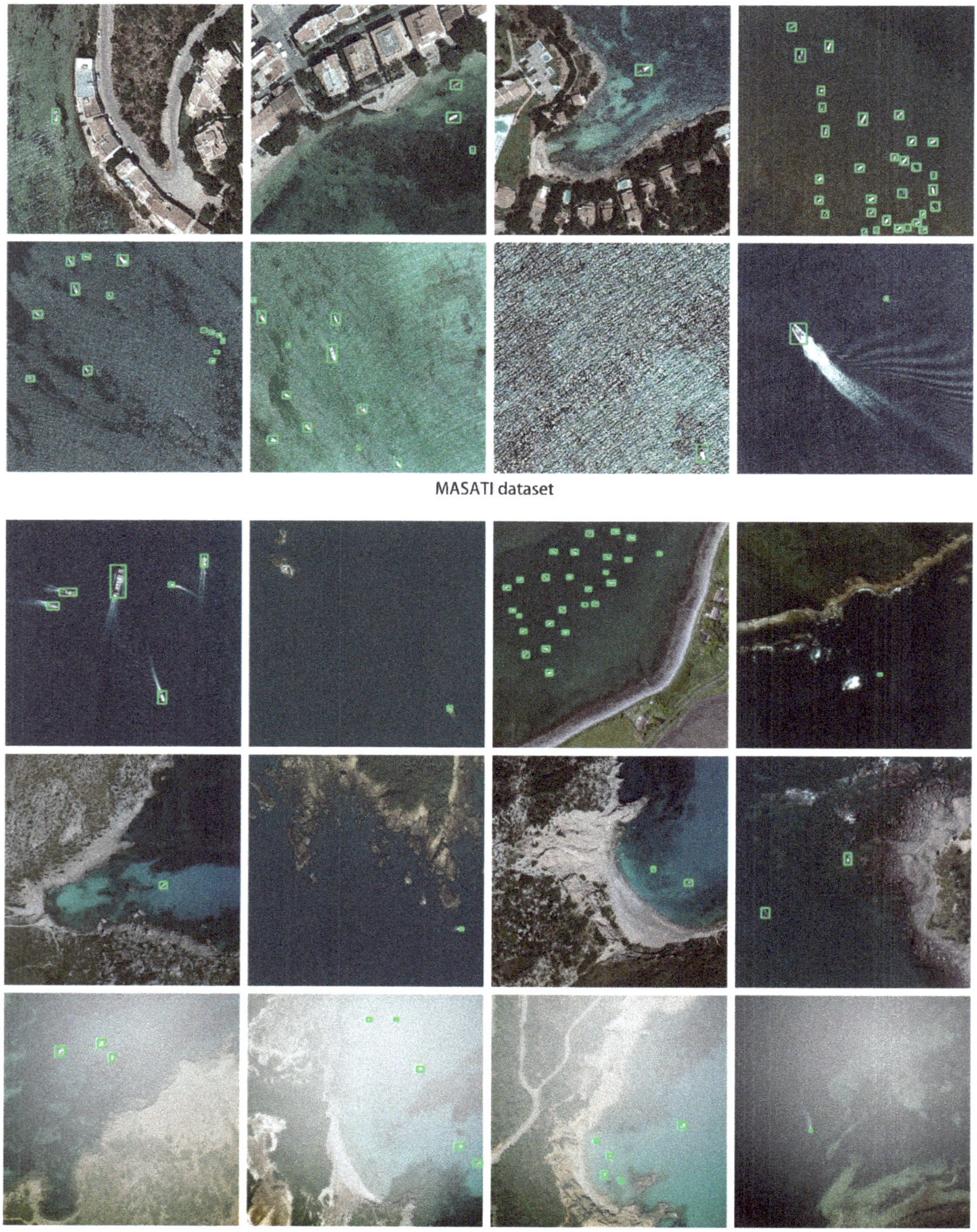

Figure 11. Detection results of the proposed LMSD-Net on different datasets.

4. Discussion

In this study, we propose a new ship dataset VRS-SD v2, which adds more small- and tiny-ship targets located nearshore and in rivers. The dataset covers different open coast scenes, marine environments, maritime disturbances, target scales, and more dense distributions. In addition, we propose a new fog simulation method for increasing the proportion of fog images in the dataset. This method can improve the robustness of the model in severe weather conditions. We have demonstrated the importance of fog simulation for actual detection by implementing different proportions of fog simulation on the dataset in the ablation experiment.

Then, we propose a new lightweight model (LMSD-Net) specifically for ship detection. In the network, we design the ELA-C3 module for efficient feature extraction. In the feature-fusion process, we propose a fusion method with compressed channels and weighted connections to ensure lightweight and low computational complexity. In the detection head, we introduce a contextual transformer (CoT) block to improve the detection accuracy. In the prediction process, the variance penalty term is added, and the prediction performance is improved for the relative scale consistency of the targets.

Furthermore, we validate the effectiveness of each module and the overall detection performance on two small-ship datasets (VRS-SD v2 and MASATI). The ablation experiments indicate that the ELA-C3 module, CoT block, and V-CIoU are beneficial in improving accuracy. Meanwhile, WGC-PANet mainly enhances lightweight performance while ensuring the expressiveness of the model. The overall comparison demonstrates that the proposed model can reach 81.3% at AP@50 and 38.4% at AP@50:95 in VRS-SD v2, while with only 5.5M parameters and 12.8 GFLOPs. Among the existing lightweight detection models, LMSD-Net has better detection capability for small and tiny ships and achieves SOTA performance. In addition, the detection speed reaches 68 FPS, which could meet the real-time requirement.

5. Conclusions

The proposed lightweight model presents a feasible solution for remote sensing ship detection and project deployment. The model performs well in dealing with complex background disturbances near shore and at sea. Fog simulation has positive implications for ship detection in bad weather conditions. In the future, reducing the computation complexity will remain a challenging research task. In addition, we will further improve our research in weighted-feature fusion and more comprehensive weather simulations. Inspired by the Transformer, we believe that remote feature interaction will be the key to improving detection performance in lightweight ship detection.

Author Contributions: Conceptualization, Y.T. and S.Z.; methodology, Y.T.; validation, Y.T. and S.Z.; investigation, Y.T., F.X. and X.W.; resources, Y.T. and J.L. writing—original draft preparation, Y.T.; writing—review and editing, F.X. All authors have read and agreed to the published version of the manuscript.

Funding: This work was supported by the National Natural Science Foundation of China under Grant 61905240.

Institutional Review Board Statement: Not applicable.

Informed Consent Statement: Not applicable.

Data Availability Statement: Not applicable.

Acknowledgments: The authors are grateful for anonymous reviewers' critical comments and constructive suggestions.

Conflicts of Interest: The authors declare no conflict of interest.

References

1. Zou, H.; He, S.; Wang, Y.; Li, R.; Cheng, F.; Cao, X. Ship detection based on medium-low resolution remote sensing data and super-resolved feature representation. *Remote Sens. Lett.* **2022**, *13*, 323–333. [CrossRef]
2. Cui, D.; Guo, L.; Zhang, Y. Research on the development of ship target detection based on deep learning technology. In Proceedings of the ACM International Conference on Frontier Computing (FC), Turin, Italy, 17–19 May 2022.
3. Wu, J.; Li, J.; Li, R.; Xi, X. A fast maritime target identification algorithm for offshore ship detection. *Appl. Sci.* **2022**, *12*, 4938 [CrossRef]
4. Yue, T.; Yang, Y.; Niu, J. A Light-weight Ship Detection and Recognition Method Based on YOLOv4. In Proceedings of the 2021 International Conference on Advanced Electronic Materials, Computers and Software Engineering (AEMCSE), Changsha, China, 26–28 March 2021.
5. Joseph, S.I.T.; Karunakaran, V.; Sujatha, T.; Rai, S.B.E.; Velliangiri, S. Investigation of deep learning methodologies in satellite image based ship detection. In Proceedings of the International Conference on Sustainable Computing and Data Communication Systems (ICSCDS), Erode, India, 7–9 April 2022.
6. Ding, J.; Xue, N.; Long, Y.; Xia, G.; Lu, Q. Learning RoI transformer for detecting oriented objects in aerial images. In Proceedings of the IEEE/CVF Conference on Computer Vision and Pattern Recognition (CVPR), Long Beach, CA, USA, 15–20 June 2019.
7. Dalal, N.; Triggs, B. Histograms of Oriented Gradients for Human Detection. In Proceedings of the IEEE Conference on Computer Vision and Pattern Recognition (CVPR), San Diego, CA, USA, 20–25 June 2005.
8. Felzenszwalb, P.; McAllester, D.; Ramanan, D. A discriminatively trained, multiscale, deformable part model. In Proceedings of the IEEE Conference on Computer Vision and Pattern Recognition (CVPR), Anchorage, AK, USA, 23–28 June 2008.
9. Wu, X.; Hong, D.; Tian, J.; Chanussot, J.; Li, W.; Tao, R. ORSIm Detector: A Novel Object Detection Framework in Optical Remote Sensing Imagery Using Spatial-Frequency Channel Features. *IEEE Trans. Geosci. Remote Sens.* **2019**, *57*, 5146–5158. [CrossRef]
10. Shi, Z.; Yu, X.; Jiang, Z.; Li, B. Ship Detection in High-Resolution Optical Imagery Based on Anomaly Detector and Local Shape Feature. *IEEE Trans. Geosci. Remote Sens.* **2014**, *52*, 4511–4523.
11. Tang, J.; Deng, C.; Huang, G.; Zhao, B. Compressed-Domain Ship Detection on Spaceborne Optical Image Using Deep Neural Network and Extreme Learning Machine. *IEEE Trans. Geosci. Remote Sens.* **2015**, *53*, 1174–1185. [CrossRef]
12. Zou, Z.; Shi, Z. Ship Detection in Spaceborne Optical Image with SVD Networks. *IEEE Trans. Geosci. Remote Sens.* **2016**, *54*, 5832–5845. [CrossRef]
13. Nie, T.; Han, X.; He, B.; Li, X.; Liu, H.; Bi, G. Ship Detection in Panchromatic Optical Remote Sensing Images Based on Visual Saliency and Multi-Dimensional Feature Description. *Remote Sens.* **2020**, *12*, 152. [CrossRef]
14. Everingham, M.; Van Gool, L.; Williams, C.K.I.; Winn, J.; Zisserman, A. The Pascal Visual Object Classes (VOC) Challenge. *Int. J. Comput. Vis.* **2010**, *88*, 303–338. [CrossRef]
15. Everingham, M.; Eslami, S.M.A.; Van Gool, L.; Williams, C.K.I.; Winn, J.; Zisserman, A. The PASCAL Visual Object Classes Challenge: A Retrospective. *Int. J. Comput. Vis.* **2015**, *111*, 98–136. [CrossRef]
16. Russakovsky, O.; Deng, J.; Su, H.; Krause, J.; Satheesh, S.; Ma, S.; Huang, Z.; Karpathy, A.; Khosla, A.; Bernstein, M.; et al. ImageNet large scale visual recognition challenge. *Int. J. Comput. Vis.* **2015**, *115*, 211–252. [CrossRef]
17. Deng, J.; Dong, W.; Socher, R.; Li, L.; Li, K.; Li, F. ImageNet: A large-scale hierarchical image database. In Proceedings of the IEEE Conference on Computer Vision and Pattern Recognition (CVPR), Miami Beach, FL, USA, 20–25 June 2009.
18. Lin, T.Y.; Maire, M.; Belongie, S.; Hays, J.; Perona, P.; Ramanan, D.; Dollar, P.; Zitnick, C.L. Microsoft COCO: Common Objects in Context. In Proceedings of the European Conference on Computer Vision (ECCV), Zurich, Switzerland, 6–12 September 2014.
19. He, K.; Zhang, X.; Ren, S.; Sun, J. Spatial pyramid pooling in deep convolutional networks for visual recognition. *IEEE Trans. Pattern Anal. Mach. Intell.* **2015**, *37*, 1904–1916. [CrossRef] [PubMed]
20. Dai, J.; Li, Y.; He, K.; Sun, J. R-FCN: Object Detection via Region-based Fully Convolutional Networks. In Proceedings of the Conference on Neural Information Processing Systems (NIPS), Barcelona, Spain, 5–10 December 2016.
21. Ren, S.; He, K.; Girshick, R.; Sun, J. Faster R-CNN: Towards Real-Time Object Detection with Region Proposal Networks. *IEEE Trans. Pattern Anal. Mach. Intell.* **2017**, *39*, 1137–1149. [CrossRef] [PubMed]
22. Hu, J.; Zhi, X.; Jiang, S.; Tang, H. Supervised Multi-Scale Attention-Guided Ship Detection in Optical Remote Sensing Images. *IEEE Trans Geosci Remote Sens.* **2022**, *60*, 1–14. [CrossRef]
23. Lin, T.-Y.; Goyal, P.; Girshick, R.; He, K.; Dollar, P. Focal Loss for Dense Object Detection. *IEEE Trans. Pattern Anal. Mach. Intell.* **2020**, *42*, 318–327. [CrossRef]
24. Zhou, X.; Wang, D.; Krhenbühl, P. Objects as Points. *arXiv* **2019**, arXiv:1904.07850.
25. Redmon, J.; Farhadi, A. YOLOv3: An Incremental Improvement. *arXiv* **2020**, arXiv:1804.02767.
26. Bochkovskiy, A.; Wang, C.; Liao, H. YOLOv4: Optimal Speed and Accuracy of Object Detection. *arXiv* **2020**, arXiv:2004.10934v1.
27. GitHub: Ultralytics. YOLOv5-v 6.1. 2022. Available online: https://github.com/ultralytics/yolov5 (accessed on 23 December 2022).
28. Li, C.; Li, L.; Jiang, H.; Weng, K.; Geng, Y.; Li, L.; Ke, Z.; Li, Q.; Cheng, M. YOLOv6: A Single-Stage Object Detection Framework for Industrial Applications. *arXiv* **2022**, arXiv:2209.02976.
29. Wang, C.-Y.; Bochkovskiy, A.; Liao, H.-Y.M. YOLOv7: Trainable bag-of-freebies sets new state-of-the-art for real-time object detectors. *arXiv* **2022**, arXiv:2207.02696.

30. GitHub: Airockchip. YOLOv8. 2023. Available online: https://github.com/airockchip/ultralytics_yolov8. (accessed on 10 February 2023).
31. Wang, B.; Han, B.; Yang, L. Accurate Real-time Ship Target detection Using Yolov4. In Proceedings of the International Conference on Transportation Information and Safety (ICTIS), Wuhan, China, 27 June 2022.
32. Ye, Y.; Ren, X.; Zhu, B.; Tang, T.; Tan, X.; Gui, Y.; Yao, Q. An Adaptive Attention Fusion Mechanism Convolutional Network for Object Detection in Remote Sensing Images. *Remote Sens.* **2022**, *14*, 516. [CrossRef]
33. Xu, Q.; Li, Y.; Shi, Z. LMO-YOLO: A Ship Detection Model for Low-Resolution Optical Satellite Imagery. *IEEE J-STARS* **2022**, *15*, 4117–4131. [CrossRef]
34. Zhang, T.; Zhang, X.; Liu, C.; Shi, J.; Wei, S. Balance learning for ship detection from synthetic aperture radar remote sensing imagery. *ISPRS J. Photogramm. Remote Sens.* **2021**, *182*, 190–207. [CrossRef]
35. Liu, Z.; Lin, Y.; Cao, Y.; Hu, H.; Wei, Y.; Zhang, Z.; Lin, S.; Guo, B. Swin transformer: Hierarchical vision transformer using shifted windows. In Proceedings of the IEEE/CVF International Conference on Computer Vision (2021), Montreal, QC, Canada, 10–17 October 2021.
36. Liu, Z.; Hu, H.; Lin, Y.; Yao, Z.; Xie, Z.; Wei, Y.; Ning, J.; Cao, Y.; Zhang, Z.; Dong, L.; et al. Swin Transformer V2: Scaling Up Capacity and Resolution. In Proceedings of the IEEE/CVF Conference on Computer Vision and Pattern Recognition (CVPR), New Orleans, LA, USA, 18–24 June 2022.
37. Carion, N.; Massa, F.; Synnaeve, G.; Usunier, N.; Kirillov, A.; Zagoruyko, S. End-to-End Object Detection with Transformers. *arXiv* **2020**, arXiv:2005.12872.
38. Mehta, S.; Rastegari, M. MobileViT: Light-weight, General-purpose, and Mobile-friendly Vision Transformer. *arXiv* **2021**, arXiv:2110.02178.
39. Tan, M.; Pang, R.; Le, Q.V. EfficientDet: Scalable and Efficient Object Detection. *arXiv* **2020**, arXiv:1911.09070.
40. Gholami, A.; Kwon, K.; Wu, B.; Tai, Z.; Yue, X.; Jin, P.; Zhao, S.; Keutzer, K. SqueezeNext: Hardware-Aware Neural Network Design. *arXiv* **2018**, arXiv:1803.10615.
41. Huang, G.; Liu, S.; Maaten, L.; Weinberger, K.Q. CondenseNet: An Efficient DenseNet using Learned Group Convolutions. *arXiv* **2017**, arXiv:1711.09224.
42. Zhang, T.; Qi, G.; Xiao, B.; Wang, J. Interleaved Group Convolutions for Deep Neural Networks. *arXiv* **2017**, arXiv:1707.02725.
43. Xie, G.; Wang, J.; Zhang, T.; Lai, J.; Hong, R.; Qi, G. IGCV2: Interleaved Structured Sparse Convolutional Neural Networks. *arXiv* **2018**, arXiv:1804.06202.
44. Sun, K.; Li, M.; Liu, D.; Wang, J. IGCV3: Interleaved Low-Rank Group Convolutions for Efficient Deep Neural Networks. *arXiv* **2018**, arXiv:1806.00178.
45. Ma, N.; Zhang, X.; Zheng, H.; Sun, J. ShuffleNet V2: Practical Guidelines for Efficient CNN Architecture Design. *arXiv* **2018**, arXiv:1807.11164.
46. Lin, T.Y.; Dollar, P.; Girshick, R.; He, K.M.; Hariharan, B.; Belongie, S. Feature Pyramid Networks for Object Detection. In Proceedings of the IEEE /CVF Conference on Computer Vision and Pattern Recognition (CVPR), Honolulu, GA, USA, 21–26 July 2017.
47. Liu, W.; Anguelov, D.; Erhan, D.; Szegedy, C.; Reed, S.; Fu, C.Y.; Berg, A.C. SSD: Single Shot MultiBox Detector. In Proceedings of the 14th European Conference on Computer Vision (ECCV), Amsterdam, The Netherlands, 8–16 October 2016.
48. Liu, S.; Qi, L.; Qin, H.; Shi, J.; Jia, J. Path Aggregation Network for Instance Segmentation. *arXiv* **2018**, arXiv:1803.01534.
49. Ghiasi, G.; Lin, T.-Y.; Pang, R.; Le, Q.V. NAS-FPN: Learning Scalable Feature Pyramid Architecture for Object Detection. *arXiv* **2019**, arXiv:1904.07392.
50. Liu, S.; Huang, D.; Wang, Y. Learning Spatial Fusion for Single-Shot Object Detection. *arXiv* **2019**, arXiv:1911.09516.
51. Zhao, Q.; Sheng, T.; Wang, Y.; Tang, Z.; Chen, Y.; Cai, L.; Ling, H. M2Det: A Single-Shot Object Detector based on Multi-Level Feature Pyramid Network. *arXiv* **2018**, arXiv:1811.04533. [CrossRef]
52. Tian, Z.; Shen, C.; Chen, H.; He, T. FCOS: Fully Convolutional One-Stage Object Detection. *arXiv* **2019**, arXiv:1904.01355.
53. Zhu, X.; Lyu, S.; Wang, X.; Zhao, Q. TPH-YOLOv5: Improved YOLOv5 Based on Transformer Prediction Head for Object Detection on Drone-captured Scenarios. *arXiv* **2021**, arXiv:2108.11539.
54. Tian, Y.; Liu, J.; Zhu, S.; Xu, F.; Bai, G.; Liu, C. Ship Detection in Visible Remote Sensing Image Based on Saliency Extraction and Modified Channel Features. *Remote Sens.* **2022**, *14*, 3347. [CrossRef]
55. Sakaridis, C.; Dai, D.; Gool, L.V. Semantic Foggy Scene Understanding with Synthetic Data. *arXiv* **2019**, arXiv:1708.07819. [CrossRef]
56. Sandler, M.; Howard, A.; Zhu, M.; Zhmoginov, A.; Chen, L.-C. MobileNetV2: Inverted Residuals and Linear Bottlenecks. *arXiv* **2019**, arXiv:1801.04381.
57. Li, Y.; Yao, T.; Pan, Y.; Mei, T. Cot Contextual Transformer Networks for Visual Recognition. *arXiv* **2021**, arXiv:2107.12292.
58. Zheng, Z.; Wang, P.; Ren, D.; Liu, W.; Ye, R.; Hu, Q.; Zuo, W. Enhancing Geometric Factors in Model Learning and Inference for Object Detection and Instance Segmentation. *arXiv* **2021**, arXiv:2005.03572. [CrossRef] [PubMed]
59. Liu, Z.; Yuan, L.; Weng, L.; Yang, Y. A High Resolution Optical Satellite Image Dataset for Ship Recognition and Some New Baselines. In Proceedings of the International Conference on Pattern Recognition Applications and Methods (ICPRAM), Porto, Portugal, 24–26 February 2017.

60. Cheng, G.; Han, J.; Zhou, P.; Guo, L. Multi-Class Geospatial Object Detection and Geographic Image Classification Based on Collection of Part Detectors. *ISPRS J. Photogramm. Remote Sens.* **2014**, *98*, 119–132. [CrossRef]
61. Al-Saad, M.; Aburaed, N.; Panthakkan, A.; Al Mansoori, S.; Al Ahmad, H.; Marshall, S. Airbus Ship Detection from Satellite Imagery using Frequency Domain Learning. In Proceedings of the Conference on Image and Signal Processing for Remote Sensing XXVII, online, Spain, 13–17 September 2021.
62. Gallego, A.J.; Pertusa, A.; Gil, P. Automatic Ship Classification from Optical Aerial Images with Convolutional Neural Networks. *Remote Sens.* **2018**, *10*, 511. [CrossRef]
63. Chen, K.; Wu, M.; Liu, J.; Zhang, C. FGSD: A Dataset for Fine-grained Ship Detection in High Resolution Satellite Images. *arXiv* **2021**, arXiv:2003.06832.
64. Wang, J.; Xu, C.; Yang, W.; Yu, L. A Normalized Gaussian Wasserstein Distance for Tiny Object Detection. *arXiv* **2021**, arXiv:2110.13389.
65. Yu, X.; Gong, Y.; Jiang, N.; Ye, Q.; Han, Z. Scale Match for Tiny Person Detection. *arXiv* **2020**, arXiv:1912.10664.
66. Xia, G.; Bai, X.; Ding, J.; Zhu, Z.; Belongie, S.; Luo, J.; Datcu, M.; Pelillo, M.; Zhang, L. DOTA: A Large-scale Dataset for Object Detection in Aerial Images. *arXiv* **2019**, arXiv:1711.10398.
67. Chen, C.; Liu, M.; Tuzel, O.; Xiao, J. R-CNN for Small Object Detection. In Proceedings of the 13th Asian Conference on Computer Vision (ACCV), Taipei, Taiwan, 20–24 November 2016.
68. Zheng, Z.; Wang, P.; Liu, W.; Li, J.; Ye, R.; Ren, D. Distance-IoU Loss: Faster and Better Learning for Bounding Box Regression. *arXiv* **2019**, arXiv:1911.08287. [CrossRef]
69. Zhang, Y.; Ren, W.; Zhang, Z.; Jia, Z.; Wang, L.; Tan, T. Focal and Efficient IOU Loss for Accurate Bounding Box Regression. *arXiv* **2021**, arXiv:2101.08158. [CrossRef]
70. Gevorgyan, Z. SIoU Loss: More Powerful Learning for Bounding Box Regression. *arXiv* **2022**, arXiv:2205.12740.
71. Tong, Z.; Chen, Y.; Xu, Z.; Yu, R. Wise-IoU: Bounding Box Regression Loss with Dynamic Focusing Mechanism. *arXiv* **2023**, arXiv:2301.10051.
72. GitHub: RangiLyu. NonoDet-Plus. 2021. Available online: https://github.com/RangiLyu/nanodet (accessed on 12 February 2023).
73. Ge, Z.; Liu, S.; Wang, F.; Li, Z.; Sun, J. YOLOX: Exceeding YOLO Series in 2021. *arXiv* **2021**, arXiv:2107.08430.
74. Li, C.; Li, L.; Geng, Y.; Jiang, H.; Cheng, M.; Zhang, B.; Ke, Z.; Xu, X.; Chu, X. YOLOv6 v3.0: A Full-Scale Reloading. *arXiv* **2023**, arXiv:2301.05586.

remote sensing

MDPI

Article

MVT: Multi-Vision Transformer for Event-Based Small Target Detection

Shilong Jing [1,2], Hengyi Lv [1,*], Yuchen Zhao [1], Hailong Liu [1] and Ming Sun [1]

1 Changchun Institute of Optics, Fine Mechanics and Physics, Chinese Academy of Sciences, Changchun 130033, China; jingshilong22@mails.ucas.ac.cn (S.J.); zhaoyuchen@ciomp.ac.cn (Y.Z.); liuhailong@ciomp.ac.cn (H.L.); sunming@ciomp.ac.cn (M.S.);
2 University of Chinese Academy of Sciences, Beijing 100049, China
* Correspondence: lvhengyi@ciomp.ac.cn

Abstract: Object detection in remote sensing plays a crucial role in various ground identification tasks. However, due to the limited feature information contained within small targets, which are more susceptible to being buried by complex backgrounds, especially in extreme environments (e.g., low-light, motion-blur scenes). Meanwhile, event cameras offer a unique paradigm with high temporal resolution and wide dynamic range for object detection. These advantages enable event cameras without being limited by the intensity of light, to perform better in challenging conditions compared to traditional cameras. In this work, we introduce the Multi-Vision Transformer (MVT), which comprises three efficiently designed components: the downsampling module, the Channel Spatial Attention (CSA) module, and the Global Spatial Attention (GSA) module. This architecture simultaneously considers short-term and long-term dependencies in semantic information, resulting in improved performance for small object detection. Additionally, we propose Cross Deformable Attention (CDA), which progressively fuses high-level and low-level features instead of considering all scales at each layer, thereby reducing the computational complexity of multi-scale features. Nevertheless, due to the scarcity of event camera remote sensing datasets, we provide the Event Object Detection (EOD) dataset, which is the first dataset that includes various extreme scenarios specifically introduced for remote sensing using event cameras. Moreover, we conducted experiments on the EOD dataset and two typical unmanned aerial vehicle remote sensing datasets (VisDrone2019 and UAVDT Dataset). The comprehensive results demonstrate that the proposed MVT-Net achieves a promising and competitive performance.

Keywords: event cameras; multi-scale fusion; remote sensing; small target detection

Citation: Jing, S.; Lv, H.; Zhao, Y.; Liu, H.; Sun, M. MVT: Multi-Vision Transformer for Event-Based Small Target Detection. *Remote Sens.* **2024**, *16*, 1641. https://doi.org/10.3390/rs16091641

Academic Editor: Paolo Tripicchio

Received: 11 March 2024
Revised: 17 April 2024
Accepted: 24 April 2024
Published: 4 May 2024

1. Introduction

The event camera is a novel vision sensor inspired by biology, also known as Dynamic Vision Sensor (DVS) [1] or Dynamic and Active-Pixel Vision Sensor (DAVIS) [2]. Compared to conventional cameras that capture images at a fixed frame rate, event cameras independently measure and output the logarithmic intensity changes of each pixel instead of capturing images. When it comes to capturing fast-moving objects, traditional cameras require a significant cost to achieve satisfactory performance. In contrast, event cameras can effectively circumvent the limitations, providing asynchronous information with sub-millisecond latency. As a result, event cameras possess characteristics such as low latency, low power consumption, high dynamic range, and high temporal resolution. In addition, due to the fact that event cameras only capture changes in light intensity at different pixel locations, they can capture objects even in low-light conditions or extremely bright lighting. Thanks to these advantages, event cameras have demonstrated significant applications in both the military and civilian sectors. Figure 1 illustrates the theory of event generation in DVS.

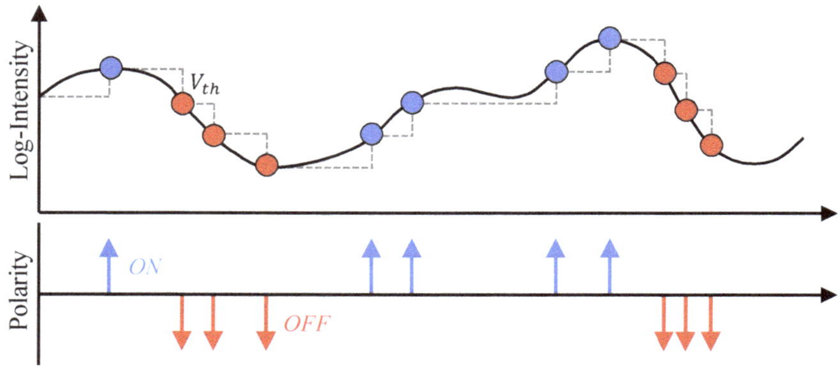

Figure 1. The process of DVS generates events. Each pixel serves as an independent detection unit for changes in brightness. An event is generated when the logarithmic intensity change at a pixel exceeds a specified threshold V_th. The continuous generation of events forms an event stream, which consists of two types of polarity: when the light intensity changes from strong to weak and reaches the threshold, DVS outputs a negative event (red arrow); when the light intensity changes from weak to strong and reaches the threshold, DVS outputs a positive event (blue arrow).

Utilizing drones equipped with event cameras for object detection or tracking is an innovative approach that holds great potential for a wide range of applications including satellite imaging, transportation, and early warning systems. However, due to the scarcity of remote sensing datasets based on the event cameras, we present the first event-based remote sensing dataset named Event-based Object Detection Dataset (EOD Dataset), which utilizes a DAVIS346 event camera mounted on an unmanned aerial vehicle (UAV) to capture various scenes. Furthermore, in practical processing, a high flying altitude results in ground targets occupying only a small portion of the image output, which poses challenges for object detection. Recently, advanced approaches for enhancing the detection performance of small targets often apply Feature Pyramid Networks (FPN) to concatenate multi-scale features. However, these methods have significant limitations as they are unable to differentiate between distinct feature layers. So how can we address this problem? Deformable DETR [3] provides an answer by introducing Scale-Level Embedding to differentiate the positional encoding of different features at the same location. Therefore, we draw inspiration from this embedding operation to concatenate multi-scale features, with the aim of enhancing the detection performance of small targets. Moreover, solely considering multi-scale features undoubtedly incurs significant computational and memory overhead, making convergence more challenging. For instance, in the Transformer Encoder of Deformable DETR, the model needs to extract features for all scales, even though deformable attention is used to reduce computational complexity, which still remains redundant.

In this work, we propose Cross-Deformable-Attention (CDA) to further enhance the performance of the model while significantly reducing its computational complexity. Specifically, by applying CDA between low-level and high-level features, we continuously propagate the fused information from lower layers to higher layers. In addition to reducing computational complexity, CDA can also reduce model training time and improve inference speed. What is more, we propose an efficient feature extraction model called Multi-Vision Transformer (MVT), which consists of three modules: Downsampling Module, Channel Spatial Attention Module (CSA), and Global Spatial Attention Module (GSA). Firstly, the downsampling module employs a simple overlapped convolution for scale reduction, resulting in better performance compared to non-overlapped convolution and patch merging operations. Then, we apply CSA for attention querying between spatial and channel dimensions. Compared to the original SE Block, CSA applies adaptive max pooling operations to preserve more high-frequency information. Finally, we employ GSA including Window-Attention and Grid-Attention for local and global search. Compared to Swin-Attention, which requires more computational resources and complex offset vectors, Grid-Attention and Window-Attention are similar but only require local grid attention to extend them to the entire domain, achieving higher performance and fewer parameters. Additionally, we also provide three model variants (MVT-B, MVT-S, MVT-T) by setting different embedding dimensions and output scales. Employing MVT-B trained for 36 epochs, we achieve 28.7% mAP@0.5:0.95, outperforming all current state-of-the-art methods on the EOD dataset. With the application of multiple efficient attention modules that consider multi-scale features, the detection performance is improved especially for small objects, achieving 16.6% AP_S. While due to the scarcity of remote sensing datasets based on event cameras, we select the VisDrone2019 dataset [4] and UAVDT dataset [5], which are similar to our own dataset and consist of images captured by drones equipped with cameras. In this case, we employ MVT-B, which is trained for 36 epochs and achieve 31.7% mAP@0.5:0.95 and 24.3% AP_S on the VisDrone2019 Dataset, as well as 28.2% mAP@0.5:0.95 and 23.7% AP_S on the UAVDT Dataset.

Our contributions can be summarized as follows:

1. The first remote sensing dataset based on event cameras has been proposed, called the Event Object Detection Dataset (EOD Dataset), which consists of over 5000 event streams and includes six categories of objects like car, bus, pedestrian, two-wheel, boat, and ship.
2. We propose a novel multi-scale extraction network named Multi-Vision Transformer (MVT), which consists of three efficient modules proposed by us. The downsampling module, the Channel Spatial Attention (CSA) module, and the Global Spatial Attention (GSA) module. Overall, The MVT incorporates efficient modules, achieving a substantial reduction in computational complexity with high performance.
3. Considering that extracting information at all scales consumes massive computing resources, we propose a novel cross-scale attention mechanism that progressively fuses high-level features with low-level features, enabling the incorporation of low-level information. The Cross-Deformable-Attention (CDA) reduces the computational complexity of the Transformer Encoder and entire network by approximately 82% and 45% while preserving the original performance.
4. As a multi-scale object detection network, MVT achieves state-of-the-art performance trained from scratch without fine-tuning, which trained for 36 epochs, achieving 28.7% mAP@0.5:0.95 and 16.6% AP_S on the EOD Dataset, 31.7% mAP@0.5:0.95 and 24.3% AP_S on the VisDrone2019 Dataset, 28.2% mAP@0.5:0.95 and 23.7% AP_S on the UAVDT Dataset.

2. Related Work

2.1. Multi-Scale Feature Learning

Convolutional neural networks extract features of objects through hierarchical abstractions, and an important concept in this process is the receptive field. Higher-level feature maps have larger receptive fields, which make them strong in representing semantic information, while they have lower spatial resolution and lack detailed spatial geometric features. On the other hand, lower-level feature maps have smaller receptive fields, which makes them strong in representing geometric details with higher resolution, but they exhibit weaker semantic information representation. For remote sensing object detection, the accuracy of small target recognition greatly affects the performance of the network. Therefore, multi-scale feature representation is a commonly used approach in small target detection [6,7].

The concept of the Feature Pyramid Networks (FPN) [8] is initially introduced for multi-scale object detection. However, the computation-intensive nature of the FPN significantly influences the detection speed. For this reason, various improvement methods have been developed. Centralized Feature Pyramid (CFP) [9] focuses on optimizing the representation of features within the same level, particularly in the corners of the image. Path Aggregation Network (PANet) [10] extends the FPN with a bottom-up path to capture deeper-level features using shallow-level features. Additionally, the U-Net, originally designed for segmentation tasks, has also demonstrated outstanding performance in object detection [11–13].

In addition, there are methods that specifically utilize low-scale features for small target detection. Unlike approaches that recover high-resolution representation from low-resolution ones, the High-Resolution Network (HRNet) [6] maintains high-resolution representation during forward propagation. Lite-High-Resolution Network (Lite-HRNet) [14] can rapidly estimate feature points, thereby reducing the computational complexity of the model. Feature-Selection High-Resolution network (FSHRNet) [15] adopts HRNet as the backbone and introduces a Feature Selection Convolution (FSConv) layer to fuse multi-resolution features, enabling adaptive feature selection based on object characteristics. The Improved U-Net (IU-Net) [16] enhances the HRNetv2 [17] by incorporating the csAG module, composed of spatial attention and channel attention, to improve model performance. However, solely relying on low-scale features often leads to inferior performance, and the FPN operation fails to distinguish between different feature levels.

Scale-Level Embedding [3] was proposed for multi-scale fusion, which has the significant advantage of encoding different feature levels to enable the model to differentiate the same position information across different feature levels, and is widely applied in various types of models.

2.2. Attention Mechanism

The Attention Mechanism (AM) originated from studies on human vision. Due to the limitations in information processing, humans selectively focus on important information while disregarding less significant details [18]. In deep learning, AM is employed to mimic the human cognitive system by adding weights to different regions of feature maps, ensuring a prioritized processing order for neural networks [19,20]. Currently, AM can be broadly categorized into two branches: (1) applying pooling operations to extract salient information in channel or spatial dimensions [21]; (2) employing self-attention mechanisms to model global information and capture long-range dependencies [19].

There are several representative approaches in the first branch. Squeeze-and-Excitation Networks (SENet) [22] operate in the channel dimension, applying global pooling and fully connected layers to downsample feature maps to a single point and employ a multilayer perceptron (MLP) to generate weights for different regions. Then, the Hadamard product is computed between the weights applied sigmoid activation function and the original input to obtain channel-weighted feature maps, establishing relationships between channels. Efficient Channel Attention Networks (ECA-Net) [20] is an improved version of SENet that uses 1D convolution instead of fully connected layers to achieve channel-wise information interaction, which avoids the degradation of a part of feature representations during the scale variation process. The Convolutional Block Attention Module (CBAM) [23] further introduces the Spatial Attention Module (SAM), which calculates weights for both the channel and spatial domains, selectively assigning importance to different features. SAM first generates distinct global information feature maps through pooling operations. Subsequently, the Hadamard product is computed between the result applied sigmoid and the original input to enhance the target region. Due to the lightweight and plug-and-play advantages, these methods have been widely applied. However, their drawback lies in the loss of features for small objects due to their limitations in long-range regions.

Transformer [19] is the representative approach in the second branch, capable of effectively extracting features from long-range regions. The Vision Transformer (VIT) [24] provides a novel approach to extract features by treating images as tokens, similar to sentences, to capture global information. Due to its simplicity and strong scalability, sparking subsequent research. However, VIT still faces the challenge of high computational complexity with excessively long tokens. Therefore, VIT only extracts features from images with an input resolution of 224. To solve these problems, Swin Transformer [25] introduces a window shift strategy to overcome the limitation of input resolution and utilizes a window sliding mechanism with convolutional operations to enable interaction between different windows, thus achieving global attention. Despite achieving remarkable results in various tasks, the Swin Transformer still faces the redundancy of using offset vectors. Furthermore, Multi-Axis Vision Transformer (MAXVIT) [26] proposes Multi-axis Self-Attention (MaxSA), which decomposes the conventional self-attention mechanism into two sparse forms: Window-Attention and Grid-Attention. This approach reduces the quadratic complexity of traditional computation methods to linear complexity. Importantly, it discards redundant window offset operations and instead employs a simpler form of window attention and grid attention to consider both local and global information. Additionally, Deformable DETR [3] introduces Deformable-Attention, which can be summarized as each feature pixel does not need to interact with all other feature pixels for computation. Instead, it only needs to interact with a subset of other pixels obtained through sampling. This mechanism significantly accelerates model convergence while reducing computational complexity. The aforementioned studies discuss the capability of Transformer Attention to model global information for accurate target localization. While these methods have made improvements in terms of computational resources, they still encounter challenges regarding the excessive computational complexity caused by remote sensing images. Therefore, we propose a novel Cross-Deformable-Attention (CDA) structure to achieve a balance between performance and computational cost.

2.3. Remote Sensing Images Object Detection

Currently, the mainstream frameworks for event camera object detection include CNN-based [27–29] and Transformer-based [3,30,31] approaches. Specifically, the event streams are encoded into spatiotemporal tensors, which are then fed into deep neural networks for some complex downstream tasks. While this process is similar to conventional image detection frameworks, the representation of the event tensor is significantly different from the image. Therefore, the performance of the network is directly influenced by the extracted information. Meanwhile, RNN-based [32] models have also shown great potential in event camera detection.

The small targets in remote sensing are often susceptible to interference from complex backgrounds. There are several studies have shown that enhancing multi-scale features can significantly improve small target detection. Compared to R-CNN [33] and Faster R-CNN [34], which generate redundant bounding boxes during small object detection, Events-SSD [35] introduces Single-Shot MultiBox Detector to improve detection efficiency. However, due to its relatively weak representation capability in shallow feature maps, it is not robust for small targets. Events-YOLO [36] improves upon Events-SSD by introducing a multi-scale detection mechanism that combines visible frames to supplement event representations with finer details, enabling the detection of objects at different scales. RVT [32] introduces a novel recurrent neural network that incorporates the temporal dimension of event tensors, achieving excellent performance on ground vehicle datasets. EMS-YOLO [37] directly trains a deep Spiking Neural Network, aiming for better applicability to neurocomputing hardware by binary data communication. While RVT and EMS-YOLO both take into account the temporal sequence of the event stream, they are frameworks in the field of ground object detection, utilizing FPN for multi-scale fusion rather than Scale-Level Embedding that can differentiate information of different scales.

In general, event cameras have seen emerging developments in ground-based detection, while research in the field of remote sensing remains notably scarce. Moreover, due to the extreme challenges (e.g., smaller targets, more severe motion blur, more complex backgrounds) associated with event camera remote sensing detection, designing an efficient backbone becomes particularly crucial. However, existing research lacks the capability of global modeling, and is unable to extract long-dependence information, especially in high-resolution remote sensing images. In this work, our objective is to propose a novel end-to-end object detection framework that better inherits the advantages of multi-scale features and attention mechanisms to address small target detection in complex backgrounds of remote sensing.

3. Method

3.1. Overall Architecture

The proposed MVT Network is illustrated in Figure 2, which is composed of four main components, namely Data Processing, MVT Backbone, Feature Fusion Module, and Prediction Head.

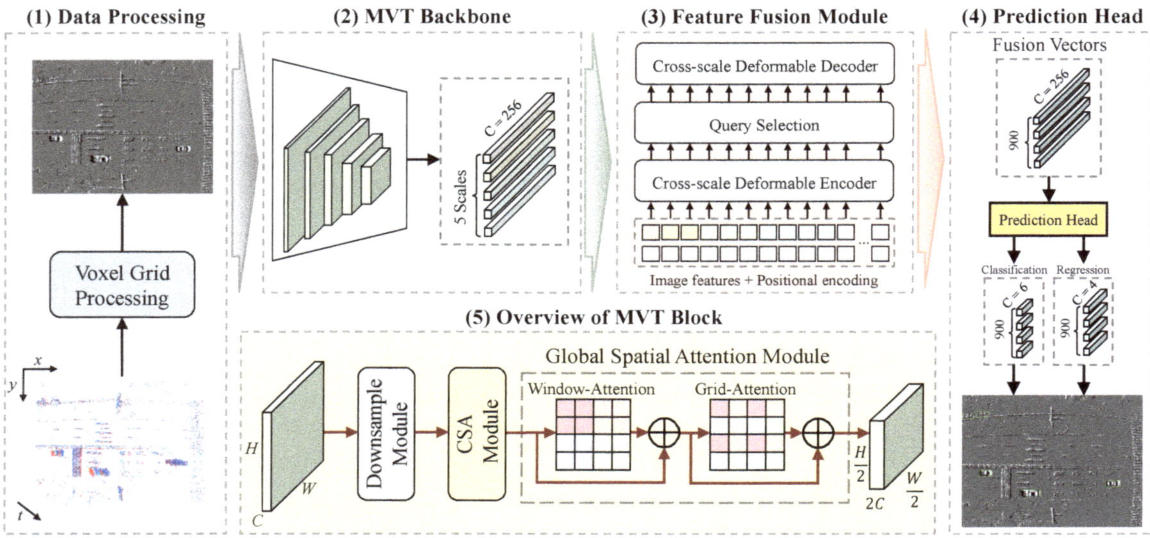

Figure 2. Overview of the MVT framework, which contains five main components, including: (1) the data preprocessing method of converting event streams into dense tensors; (2) the proposed MVT Backbone used to extract multi-scale features; (3) the designed feature fusion module for encoding and aggregating features at different scales; (4) the detection head that applies bipartite matching strategy; (5) Each MVT Block, composed of three designed components.

The original chaotic event sequence cannot be directly used as an input tensor for deep neural networks. Therefore, we encode the event stream in the form of voxel grid representation [38], which has a channel regarding the temporal dimension generated by a time partitioning function, and described in detail in Section 3.2. In this work, we do not consider the correlation of the temporal order. Thus, the processed event tensor has a shape of $X = \mathbb{R}^{H\times W\times 1}$. Different scale features of the event tensor are extracted by the backbone, which utilizes CSA to attend to short-range dependencies and GSA to attend to long-range dependencies, which is specifically described in Section 3.3. Subsequently, the multi-scale features with rich semantic information are fed into the Transformer Encoder, where CDA is applied to fuse tokens at different levels, which is described in detail in Section 3.4. Finally, regression calculations are performed on the 900 vectors generated by the Feature Fusion Module to obtain the detection results.

3.2. Event Representation

The event camera captures the brightness changes of individual pixels, generating an asynchronous event stream. An event with polarity is generated at time t when the logarithmic change of light intensity $I^t(u)$ exceeds the threshold V_{th} within a small time interval Δt, which satisfies

$$p[I^t(u) - I^{t-\Delta t}(u)] \geqslant V_{th} \quad (1)$$

where $p \in \{0, 1\}$ is the event polarity, V_{th} is the threshold. The event camera will generate an ordered set of events $\varepsilon = \{e_k\}^{E_x, E_y, E_p, E_t} \in \mathbb{R}^4$ according to Equation (1). Afterwards, the polarity of each pixel in the same time window is aggregated by performing bilinear voting, which requires the standardization of event timestamps as

$$E_{t\_norm} = T \frac{E_t - E_t(0)}{E_t(N) - E_t(0)} \tag{2}$$

where T represents the number of non-overlapping time windows, which we set to 1 in this work. $E_t(N)$ indicates the timestamp corresponding to the last event. Equation (2) demonstrates the normalization of temporal dimensions for the event stream. In addition, the two encoded polarities are represented as

$$E_{p\_left} = E_p(E_{t\_norm} - [E_{t\_norm}]) \tag{3}$$

$$E_{p\_right} = E_p(1.0 - (E_{t\_norm} - [E_{t\_norm}])) \tag{4}$$

where $[\cdot]$ represents the floor function. Equations (3) and (4), respectively, denote the product of the time distance from the current event to the start and end points of the time window and the polarity. Finally, by accumulating the encoded polarities at the corresponding pixel position (E_x, E_y), we obtain the event tensor in the form of voxel grid representation. The Algorithm 1 for event representation is as follows:

Algorithm 1 Voxel grid encoding from event stream

Input: Event stream containing N number of events $\varepsilon = \{e_k\}^{E_x, E_y, E_p, E_t} \in \mathbb{R}^4$.
Output: Voxel grid tensor $X = \mathbb{R}^{H \times W}$.

```
X = R^{H×W}; // Create a tensor with all values set to 0;
Compute the normalized event stream time E_{t_norm} according to Equation (2);
TI = [E_{t_norm}]; // Perform time windowing based on the setting values;
Compute the encoded polarity fused event time E_{p_left} and E_{p_right} according to
  Equations (3) and (4);
if (TI < T) then
  for (i = 0, i < len(T), i++) do
    X(E_x[i], E_y[i]) += E_{p_left}[i]; // Accumulate the left polarity at the corresponding
      pixel positions where events occur.
  end for
end if
if (TI + 1 < T) then
  for (i = 0, i < len(T), i++) do
    X(E_x[i], E_y[i]) += E_{p_right}[i]; // Accumulate the right polarity at the correspond-
      ing pixel positions where events occur.
  end for
end if
return X
```

3.3. Multi-Vision Transformer (MVT)

The MVT Backbone consists of four layers, with each layer stacked with a varying number of MVT Blocks to extract features at different scales, which is specifically demonstrated in illustration (5) of Figure 2. The MVT Block consists of three components: the downsampling module applying overlapping convolutions, the CSA module utilizing spatial and channel attention to consider short-term attention, and the GSA module employing Window-Attention and Grid-Attention to consider long-term attention.

3.3.1. Downsample Module

We design an extremely simple downsampling module, which consists of an overlapping convolution. Specifically, in the first layer, we use a 7×7 convolution kernel with a stride of 4 to achieve fourfold downsampling, while the remaining layers apply a 3×3 convolution kernel with a stride of 2 for two-fold downsampling. Furthermore, we demonstrate that the overlapping convolution outperforms non-overlapping convolutions and patch merging operations in Section 4.3.

3.3.2. Channel Spatial Attention Module (CSA)

In this section, we introduce CSA for extracting short-term dependency attention, which assigns more weight to focal channels and spatial locations in the feature map, thereby enhancing the capability of feature representation, as illustrated in Figure 3.

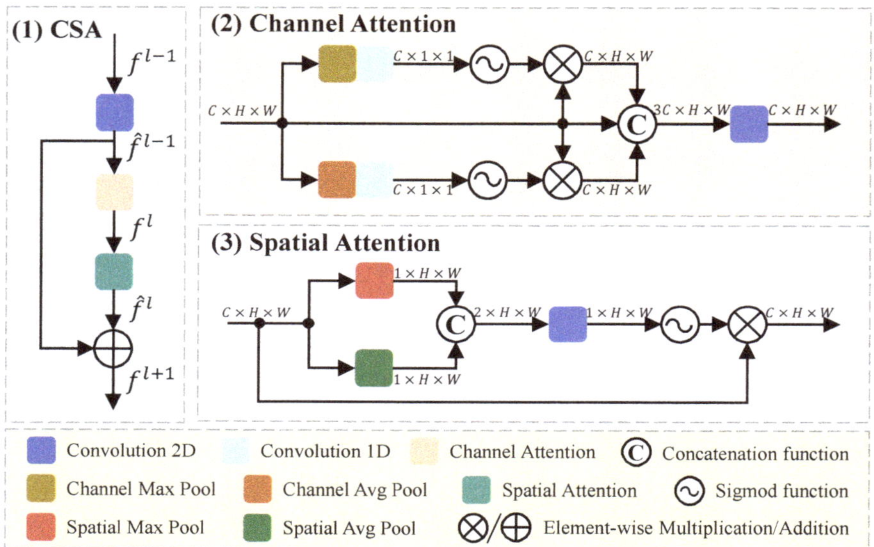

Figure 3. Architecture of CSA module, which consists of channel attention and spatial attention module to extract short-term dependent attention.

The input feature map f^{l-1} is fed into CSA for feature extraction. Firstly, f^{l-1} undergoes a 2D convolution ($Conv2D$) operation with a 1×1 kernel, resulting in $\widehat{f}^{l-1}$, which has the same dimensions as f^{l-1}. Then, $\widehat{f}^{l-1}$ is separately fed into channel attention ($\mathcal{L}^{C}_{Attn}$) and spatial attention ($\mathcal{L}^{S}_{Attn}$) modules, producing the intermediate feature map $\widehat{f}^{l}$, which is added to $\widehat{f}^{l-1}$ to obtain the output feature map f^{l+1}. The entire CSA computation process can be represented by Equation (5).

$$
\begin{aligned}
\widehat{f}^{l-1} &= Conv2D(f^{l-1}) \\
\widehat{f}^{l} &= \mathcal{L}^{S}_{Attn}(\mathcal{L}^{C}_{Attn}(\widehat{f}^{l-1})) \\
f^{l+1} &= \widehat{f}^{l-1} \oplus \widehat{f}^{l}
\end{aligned} \tag{5}
$$

The main components of CSA can be divided into Channel Attention and Spatial Attention. Within the Channel Attention module, there are three branches: in the first branch, the input (F) is fed into Channel Max Pooling ($\mathcal{P}^{C}_{Max}$) and a 1D convolution ($Conv1D$), resulting in a tensor (F_{max}) with $C \times 1 \times 1$ dimensions, then, applying a sigmoid function (σ) to obtain attention weights, which are multiplied with the shortcut layer to produce a feature map ($\widehat{F}_{max}$) with $C \times H \times W$ dimensions; the second branch utilizes a residual connection to preserve the original features, enhancing the representational and generalization capabilities of the network; the third branch differs from the first branch only in applying Channel Average Pooling ($\mathcal{P}^C_{Avg}$) to extract features ($\widehat{F}_{avg}$). In addition, a concatenation function (*Concat*) is employed to transform the three feature maps with $C \times H \times W$ dimensions into a single feature map with $3C \times H \times W$ dimensions. Finally, utilizing a 2D convolution to map the channels back to $C \times H \times W$ and obtain the feature map (F^C_{Attn}). The entire Channel Attention computation process can be represented by Equation (6).

$$\begin{aligned} F_{max} &= Conv1D(\mathcal{P}^C_{Max}(F)) \\ \widehat{F}_{max} &= \sigma(F_{max}) \otimes F \\ F_{avg} &= Conv1D(\mathcal{P}^C_{Avg}(F)) \\ \widehat{F}_{avg} &= \sigma(F_{avg}) \otimes F \\ F^C_{Attn} &= Conv2D(Concat[F, \widehat{F}_{max}, \widehat{F}_{avg}]) \end{aligned} \tag{6}$$

Within the Spatial Attention module, there are two branches: in the first branch, the input (F) is fed into both Spatial Max Pooling ($\mathcal{P}^S_{Max}$) and Spatial Average Pooling ($\mathcal{P}^S_{Avg}$) to obtain features ($\widehat{F}_{max}$) and ($\widehat{F}_{avg}$), which are then concatenated to form a tensor ($\widehat{F}$) with $2 \times H \times W$ dimensions. Next, the feature map ($\widehat{F}$) undergoes a 2D convolution (*Conv2D*) followed by a sigmoid function (σ), resulting in spatial attention weights ($\widehat{F}^S_{Attn}$), which are multiplied element-wise with the original input (F) to achieve the final feature map (F^S_{Attn}) in the second branch. The entire Spatial Attention computation process can be represented by Equation (7).

$$\begin{aligned} \widehat{F} &= Concat[\mathcal{P}^S_{Max}(F), \mathcal{P}^S_{Avg}(F)] \\ \widehat{F}^S_{Attn} &= \sigma(Conv2D(\widehat{F})) \\ F^S_{Attn} &= \widehat{F}^S_{Attn} \otimes F \end{aligned} \tag{7}$$

The CSA module improves the feature extraction performance for short-range regions by incorporating attention mechanisms for both channels and spatial dimensions. However, convolutional attention modules suffer from a loss of features for small objects due to their limitations in long-range regions. Therefore, we propose GSA, considering global attention to enhance the detection performance of small targets.

3.3.3. Global Spatial Attention Module (GSA)

In this section, we introduce GSA for extracting long-term dependency attention, which is able to obtain global information and long-distance connections in one single operation, as illustrated in Figure 4.

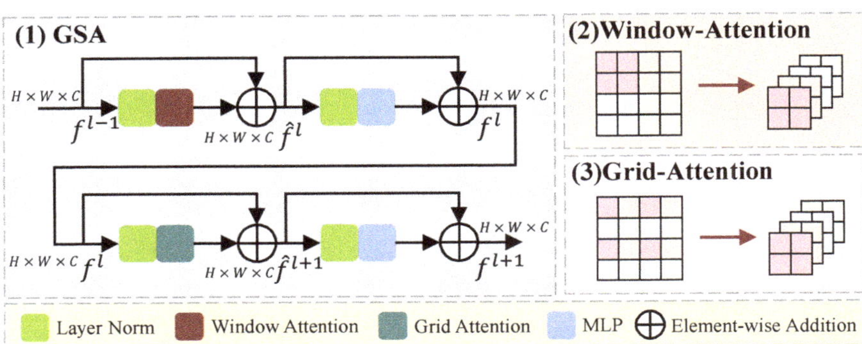

Figure 4. Architecture of GSA module, which consists of window attention and grid attention to extract long-term-dependent attention.

We propose an efficient modeling solution with two window configurations: Window-based Multi-head Self-Attention (W-MSA) and Grid-based Multi-head Self-Attention (G-MSA). Firstly, the input (f^{l-1}) is normalized and fed into the W-MSA to obtain the local attention feature map, which is then added to the original input (f^{l-1}) through a residual path, resulting in the hidden feature map ($\widehat{f^l}$). Subsequently, ($\widehat{f^l}$) is separately processed through Layer Normalization (LN) + Multilayer Perceptron (MLP) and the shortcut path to obtain the feature map (f^l). In addition, (f^l) undergoes LN and G-MSA to obtain the global attention feature map ($\widehat{f^{l+1}}$), which is further processed through LN and MLP to obtain the global spatial feature map (f^{l+1}). The entire Global Spatial Attention computation process can be represented by Equation (8).

$$
\begin{aligned}
\widehat{f^l} &= W\text{-}MSA(LN(f^{l-1})) + f^{l-1} \\
f^l &= MLP(LN(\widehat{f^l})) + \widehat{f^l} \\
\widehat{f^{l+1}} &= G\text{-}MSA(LN(f^l)) + f^l \\
f^{l+1} &= MLP(LN(\widehat{f^{l+1}})) + \widehat{f^{l+1}}
\end{aligned}
\tag{8}
$$

3.4. Cross Deformable Attention (CDA)

The framework of the Cross-scale Deformable Attention (CDA) is shown in Figure 5. Different from the repeated iterative feature extraction operation of multi-scale cross fusion, we propose CDA to achieve layer-by-layer feature fusion to better fuse feature maps of different scales and reduce computational complexity. Accordingly, enhances the representation of high-level features with both high-level semantics and high-resolution details.

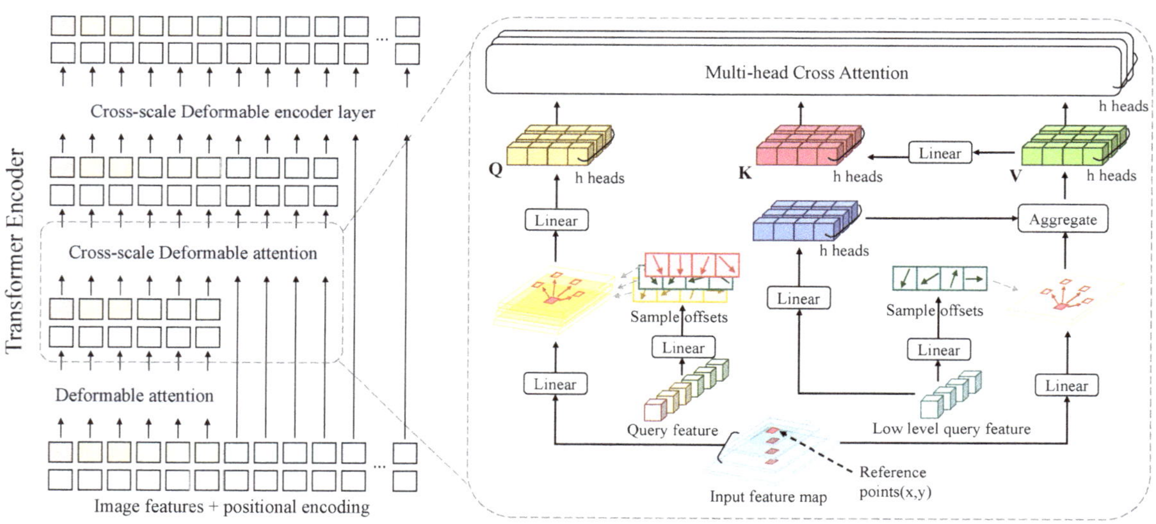

Figure 5. Overview of the Cross-scale Deformable Encoder layer. The three high-level features are used as the basic tokens to fuse low-level features layer by layer using Cross-scale Deformable Attention, finally building the architecture of the transformer encoder.

The encoder layer contains deformable self-attention and cross-scale attention. Considering that the feature map size of the high level is much smaller than the low level. Thus only the middle and final encoder layers are needed for cross-scale attention to the low and high scale instead of extracting all tokens, as shown in Figure 5. In this module, high-level features $\mathbf{F_H} \in \mathbb{R}^{N_H \times d_{model}}$ will serve as queries to extract features from the low-level features $\mathbf{F_L} \in \mathbb{R}^{N_L \times d_{model}}$, each query feature will be split into M heads, and each head will sample K points from each of the L feature scales as query $\mathbf{Q}$. Therefore, the total number of queries sampled for a query feature is $N_p = 2 \times M \times L \times K$, Δp is sampling offsets, and their corresponding attention weights are directly predicted from query features using two linear projections denoted as $W_p \in \mathbb{R}^{d_{model} \times N_p}$ and $W_A \in \mathbb{R}^{d_{model} \times d_{model}}$. Formally, we have

$$\mathbf{Q} = \sum_{m=1}^{M} W_m [\sum_{l=1}^{L} \sum_{k=1}^{K} W_m' S\left(x^l, \phi(p^l) + \Delta p_{mlk}\right)] \tag{9}$$

$$\mathbf{K} = \sum_{m=1}^{M} W_m [\sum_{k=1}^{K} W_A \cdot W_m' S(x, p + \Delta p_{mk})] \tag{10}$$

where m is the attention head, p are the reference points of the query features, x indexes the different scale feature, $W_m \in \mathbb{R}^{d_{model} \times N_m}$ and $W_m' \in \mathbb{R}^{N_m \times d_{model}}$ are of learnable weights ($N_m = d_{model} / M$ by default). With the sampled offsets ($\Delta p = \mathbf{F}W_p$), bilinear interpolation is applied in computing the features with the function $S(x, p + \Delta p)$ in the sampled locations $(p + \Delta p)$ of the corresponding feature x. As all the high-level features will sample locations to query the key consisting of low-level features, the original model can quickly learn which sampled location given the queries is important. Finally, we can obtain the value ($\mathbf{V} = \mathbf{K}W_V$) with a parameter matrices $W_V \in \mathbb{R}^{d_{model} \times d_{model}}$, and the cross-scale deformable attention can be formulated as

$$CDA(\mathbf{Q}, \mathbf{K}, \mathbf{V}) = Cat(\mathbf{F_L}, Softmax(\frac{\mathbf{QK}^T}{\sqrt{d_K}})\mathbf{V}) \tag{11}$$

In words, the cat function is to concatenate low-level features and other multi-scale features, d_k is the key dimension of a head. Equation (11) indicates more reliable attention weights predicted by stacking CDA when updating layer-by-layer features from different scales.

4. Experiments

In this section, we test the proposed method on the EOD, VisDrone [4], and UAVDT [5] datasets, and the mean average precision (mAP) [39] is the main metric that we consider. In addition, we performe ablation experiments to verify the effectiveness of each module. Finally, the experimental results demonstrate the superiority of the proposed method.

4.1. Datasets

4.1.1. EOD Dataset

The EOD dataset consists of 5317 event streams captured in various scenes, where each event stream is a collection of events within 33 ms. The dataset includes 3722 event streams for training, 530 event streams for validation, and 1065 event streams for testing, and contains six categories: car, bus, pedestrian, two-wheel, boat, and ship.

4.1.2. VisDrone Dataset

The VisDrone-DET2019 dataset [4] consists of 8599 images, including 6471 images for training, and 1580 images for testing. The dataset contains ten categories: person, pedestrian, car, bus, truck, bicycle, tricycle, awning-tricycle, van, and motor.

4.1.3. UAVDT Dataset

The UAVDT dataset [5] consists of 40,409 images, selected from 10 h long videos that cover various scene variations (e.g., weather, viewpoint, and illumination), including 23,829 images for training and 16,580 images for testing. The images in this dataset have a resolution of 540 × 1024 pixels and include three categories: car, bus, and truck.

4.2. Implementation Details

4.2.1. Evaluation Metrics

We quantitatively evaluate the performance of our method through the mAP, which is used to comprehensively evaluate the precision and recall of a model across different categories, commonly used in object detection. Specifically, the mAP can be defined as the area under the precision–recall (P-R) curve when plotted with the recall (R) on the horizontal axis and precision (P) on the vertical axis. mAP@0.5 refers to the IOU (Intersection of Union) is greater than 0.5. mAP@0.5:0.95 refers to the average of IOU values from 0.5 to 0.95 with an interval of 0.05. The P and R are defined as

$$\begin{aligned} Precision &= \frac{TP}{TP+FP} \\ Recall &= \frac{TP}{TP+FN} \end{aligned} \tag{12}$$

where TP (True Positive) indicates the number of positive samples correctly classified as positive by the model, FP (False Positive) represents the number of negative samples incorrectly classified as positive by the model, and FN (False Negative) represents the number of positive samples incorrectly classified as negative by the model. By calculating the area under the P-R curve, the mAP is defined as

$$mAP = \int_0^1 P(R)dR \tag{13}$$

where $P(R)$ is a function of P and R. In addition, we also evaluate the model size and computational complexity through Params and GFLOPs (giga floating point of operations).

4.2.2. Training Settings

In this work, we use a 6-layer Transformer encoder and a 6-layer Transformer decoder, utilize 8 as multi-heads and 4 as sampling offset, adopt 2048 as Transformer feed-forward, and 256 as the hidden feature dimension, apply 1×10^{-4} as the initial learning rate and 1×10^{-5} as the backbone learning rate. In addition, we use the AdamW optimizer with a weight decay of 1×10^{-4} and train our model by using the PyTorch framework with 8 Nvidia GeForce RTX3090 GPUs on Ubuntu22.04 with batch size 32 for all datasets. Particularly, our model is trained from scratch without pre-training and fine-tuning.

4.2.3. Model Variants

By setting different dimensions and final output scales for each layer, we constructed three variants of MVT-B/S/T. Where MVT-B is the base form with five stages, MVT-S and MVT-T are small and tiny forms with four and three stages, respectively. After the backbone feature extraction, the feature maps are missing the final stage that is obtained by applying a convolutional block to the last-second feature map. Furthermore, the first layer utilizes a 7×7 kernel with a stride of 4 for overlapping convolution to reduce the input feature resolution and computational cost. The second to fourth layers adopt a 3×3 kernel with a stride of 2 for overlapping convolution to extract higher-level feature information. "✓" indicates that the stage serves as a multi-scale feature output. Table 1 shows the specific parameters for different variants.

Table 1. MVT parameters and variations. Except for the channel numbers at each stage, all model variants share the same parameter set.

Stage	Size	Kernel	Stride	Channels		
				MVT-B	MVT-S	MVT-T
S1	1/4	7	4	96 ✓	64	32
S2	1/8	3	2	192 ✓	128 ✓	64
S3	1/16	3	2	384 ✓	256 ✓	128 ✓
S4	1/32	3	2	768 ✓	512 ✓	256 ✓

We utilize three variants, MVT-B/S/T, for detection on the EOD dataset. Figure 6 presents the detection results in different scenarios. Since the event camera outputs asynchronous data, and generates corresponding events even in low light and overexposure without limitation by the intensity of light.

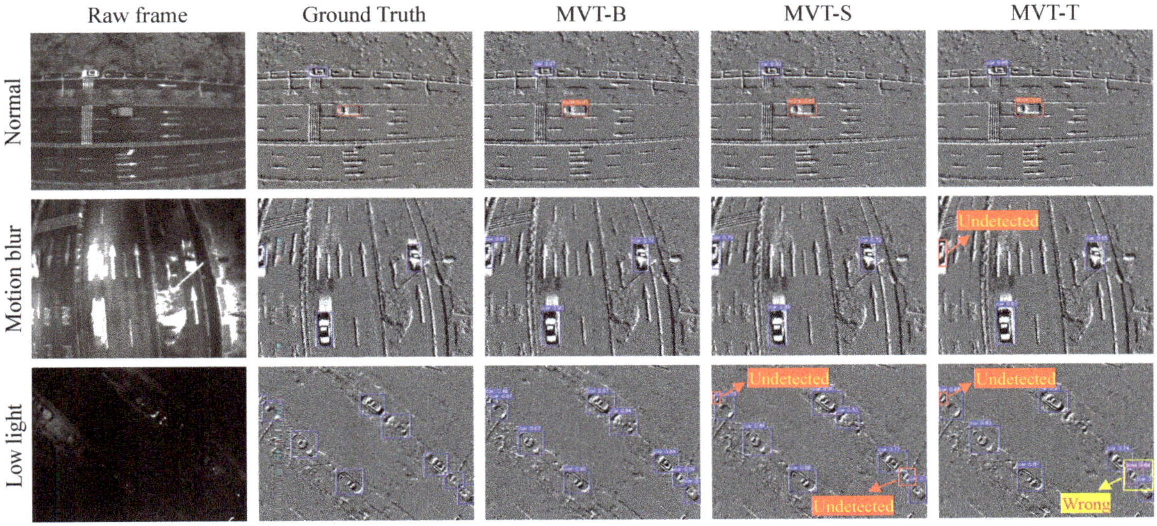

Figure 6. Prediction examples on the EOD dataset. The MVT-B/S/T variants are applied to detect in normal, motion blur, and low-light scenarios, respectively.

MVT-B outperforms the other variants due to its higher-resolution feature scale information, resulting in superior performance in detecting small objects. Specifically, in scenarios with motion blur and low light, MVT-S and MVT-T occasionally fail to detect small targets located in the top-left corner. However, despite the inherent advantages of event cameras over traditional cameras in terms of efficiency, they suffer from the loss of high-frequency information in the images, leading to the degradation of image details. Consequently, under low-light conditions, MVT-T erroneously misclassifies a car as a boat.

4.3. Ablation Experiments

In this section, we conducted ablation experiments on the EOD dataset to assess the contribution of each proposed module to the results. The contribution to the final results by evaluating the detection performance before and after applying each module. Table 2 represents the performance of ablation experiments.

Table 2. Ablation experiment on the EOD dataset. "✓" indicates that the module is used in the MVT network, while "-" indicates that it is not used, best results in **bold**, underlined denotes the second best performance, and the same colors indicate the same benchmarks except for CDA.

Model	Structure			mAP	mAP	Entire	Encoder	Params
	CSA	GSA	CDA	@0.5:0.95	@0.5	GFLOPs	GFLOPs	
Baseline	-	-	-	0.214	0.403	67.6	47.6	25.6 M
MVT-B	✓	-	-	0.238	0.474	69.7	47.6	34.5 M
	-	✓	-	0.265	0.527	84.7	47.6	97.3 M
	-	-	✓	0.212	0.401	28.4	**8.4**	25.7 M
	✓	✓	-	**0.288**	**0.569**	86.8	47.6	106.3 M
	✓	✓	✓	0.287	0.565	47.6	**8.4**	106.4 M

By progressively incorporating the designed modules, we achieve higher mAP. Compared to the baseline that only considers downsampling, the inclusion of CSA for extracting channel and spatial information improves mAP@0.5:0.95 by 2.4%, the incorporation of GSA for extracting global spatial information enhances mAP@0.5:0.95 by 5.1%, the introduction of CDA reduces model computational complexity by approximately 58% in terms of GFLOPs while maintaining the original performance. Combining CSA and GSA results in a 7.4% increase in mAP@0.5:0.95. Finally, by considering CSA, GSA, and CDA, we achieve 28.7% mAP@0.5:0.95, reducing Entire GFLOPs and Encoder GFLOPs by approximately 45% and 82% compared to the model without CDA. Figure 7 shows the attention visualization both without and with CDA.

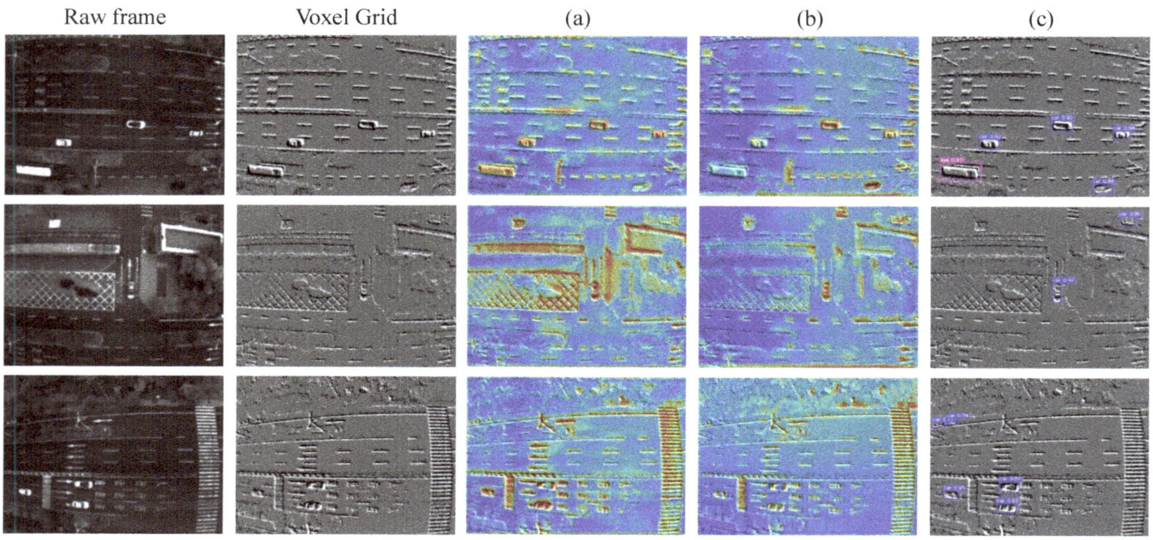

Figure 7. Visualization of attention maps. (**a**) Visualization of feature maps generated by the model without CDA. (**b**) Visualization of feature maps generated by the model with CDA. It can be observed that the attention applied by CDA is more focused on small targets. (**c**) Detection results applied CDA.

4.3.1. Ablation of Downsample Module

In the original Vision Transformer [24], downsampling operations primarily employ patch merging, which involves fewer parameters compared to pooling layers while fully preserving the input feature maps. Specifically, elements are divided and concatenated with two strides in both rows and columns, and a linear layer is utilized to scale the input feature maps from 4× to 2×. Thus, we have separately compared patch merging, overlapping, and non-overlapping convolutional downsampling blocks. Table 3 demonstrates that the use of convolutional downsampling outperforms patch merging.

Table 3. Ablation of the downsampling module. Best results in **bold**. The usage of Conv. overlapping outperforms other downsample approaches.

Downsampling Type	mAP @0.5:0.95	mAP @0.5	mAP @0.75	mAP_S	mAP_M	mAP_L	Params
Patch Merging	0.281	0.557	0.254	0.159	0.336	0.566	6.21 M
Conv. non-overlapping	0.283	0.559	0.257	0.160	0.337	0.565	**6.20** M
Conv. overlapping	**0.290**	**0.573**	**0.264**	**0.166**	**0.358**	**0.582**	13.94 M

4.3.2. Ablation of GSA Module

We apply Swin-Attention [25] and Grid-Attention [26] as global attention modules, respectively, to consider the all tokens. Table 4 shows that using Swin-Attention consumes more computational resources and has lower performance than Grid-Attention.

Table 4. Ablation of the global spatial attention module. Best results in **bold**. The usage of Grid-Attention outperforms Swin-Attention.

Attention Type	mAP @0.5:0.95	mAP @0.5	mAP @0.75	mAP_S	mAP_M	mAP_L	Params
Swin-Attn	0.280	0.558	0.251	0.162	0.347	0.545	99.5 M
Grid-Attn	**0.287**	**0.565**	**0.263**	**0.166**	**0.353**	**0.580**	**66.9** M

4.3.3. Effect of Multi-Vision Transformer Network

The introduction of CSA and GSA aims to efficiently extract features at different scales while considering dependencies between short-range and long-range features. The second and third rows in Table 1 demonstrate the effects of incorporating CSA and GSA into the baseline respectively. CSA improves mAP@0.5:0.95 by 2.4% and mAP@0.5 by 7.1% with a slight increase in the number of GFLOPs and parameters. GSA improves mAP@0.5:0.95 by 5.1% and mAP@0.5 by 12.4% while at the cost of higher GFLOPs and parameters. Furthermore, the joint utilization of CSA and GSA results in an improvement of 7.4% mAP@0.5:0.95 and 16.6% mAP@0.5. By introducing CDA, the computational complexity of the model was reduced by approximately 58% and 45% compared to the baseline and the model that combines CSA and GSA. Figure 8 provides a visual comparison of the detection results obtained by MVT when using CSA alone, GSA alone, and both CSA and GSA.

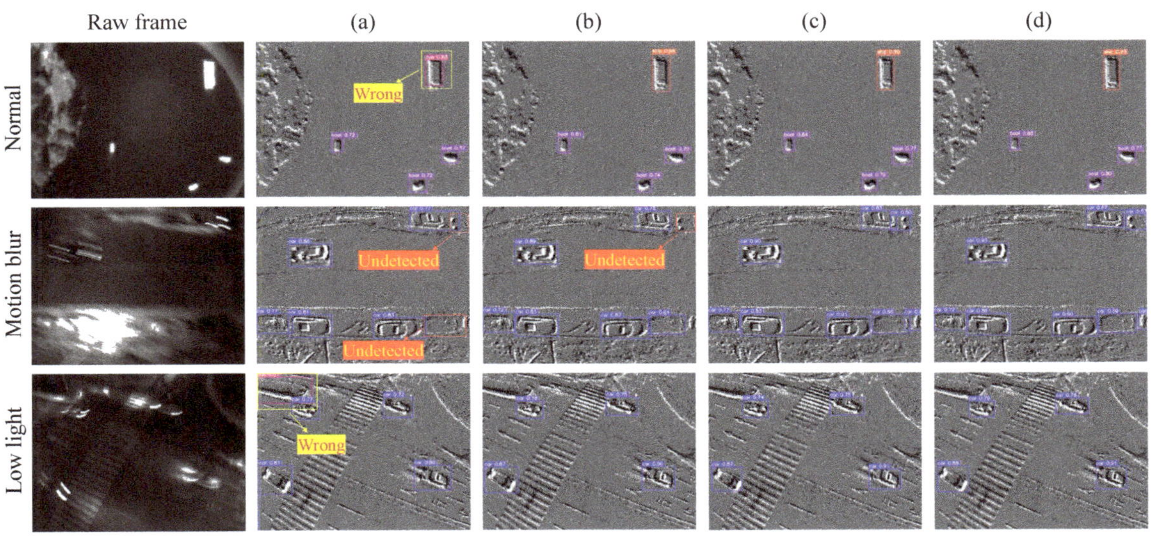

Figure 8. Comparison of the detection results before and after using CSA alone, GSA alone, and both CSA and GSA in the MVT network. (**a**) Baseline. (**b**) Baseline + CSA. (**c**) Baseline + GSA. (**d**) Baseline + CSA + GSA.

There are significant visual improvements as the baseline progressively incorporates CSA and GSA. Specifically, in Figure 8, the first and third rows exhibit cases of false detections, which arise from the lack of effective feature extraction operations. In the second row, illustrations (a) and (b) show cases of missed detections, attributed to dense object interference that hampers feature distinction between foreground and background or feature overlap. It is worth noting that in illustrations (c) and (d) of the second row, a small target in the bottom-right corner is detected, even though it is not annotated in the ground truth (GT), which demonstrates that the model incorporating global attention can achieve better detection performance for small targets.

4.4. Benchmark Comparisons

We conduct comparative experiments using three variants of MVT on the EOD, VisDrone, and UAVDT datasets, benchmarking against state-of-the-art methods with mAP.

4.4.1. Results on the EOD Dataset

We compare our method with several state-of-the-art detectors as shown in Table 5. MVT-B achieves 28.7% mAP@0.5:0.95, 56.5% mAP@0.5, and 26.3% mAP@0.75 on the EOD dataset, outperforming all other state-of-the-art methods. Figure 9 presents the results of our method for detecting objects in various scenes within the EOD dataset. As we expected, MVT has demonstrated superior detection performance for small objects, achieving 16.6% mAP_S.

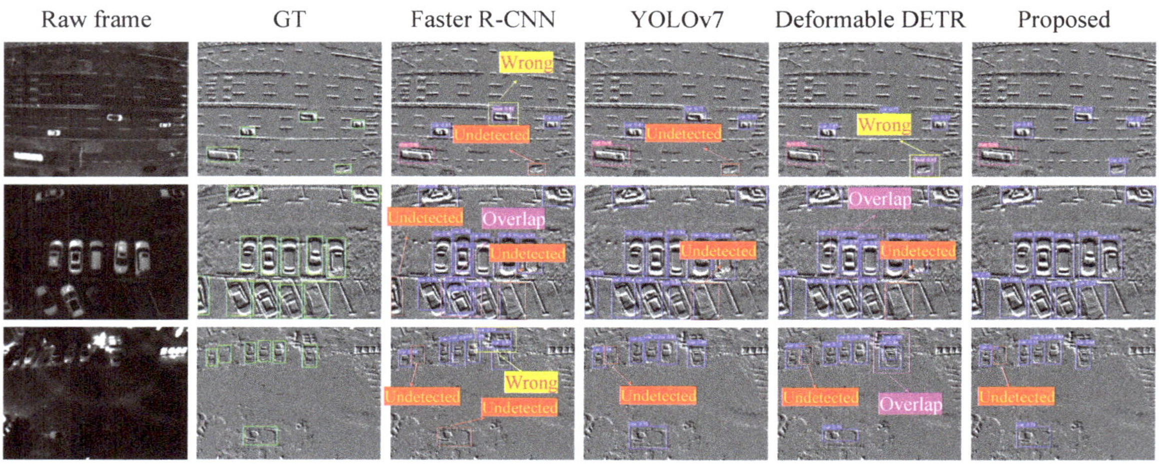

Figure 9. Prediction examples on the EOD dataset using different approaches involving Faster R-CNN, YOLOv7, Deformable DETR, and proposed method.

Table 5. Comparison of detection performance on the EOD dataset. The best result is highlighted with **bold**.

Model	Backbone	mAP @0.5:0.95	mAP @0.5	mAP @0.75	mAP_S	mAP_M	mAP_L	Params
Faster R-CNN [34]	ResNet 50	0.183	0.392	0.122	0.089	0.202	0.371	42.0 M
DetectoRS [40]	ResNet 101	0.194	0.433	0.154	0.103	0.235	0.389	540.1 M
YOLOv5 [27]	CSPDarkNet 53	0.232	0.469	0.190	0.113	0.263	0.466	93.0 M
Cascade R-CNN [28]	Transformer	0.234	0.445	0.208	0.122	0.276	0.485	335.0 M
YOLOv7 [41]	CSPDarkNet 53	0.237	0.480	0.197	0.118	0.286	0.479	135.8 M
DMNet [42]	CSPDarkNet 53	0.255	0.503	0.228	0.142	0.311	0.529	96.7 M
Sparse R-CNN [43]	Transformer	0.259	0.510	0.215	0.133	0.312	0.521	352.0 M
Deformable DETR [3]	ResNet 50	0.262	0.521	0.238	0.145	0.317	0.536	41.0 M
CLusDet [44]	ResNeXt 101	0.266	0.543	0.244	0.127	0.332	0.547	-
MVT-B (ours)	Transformer	**0.287**	**0.565**	**0.263**	**0.166**	**0.353**	**0.580**	106.4 M
MVT-S (ours)	Transformer	0.273	0.557	0.255	0.159	0.341	0.551	56.6 M
MVT-T (ours)	Transformer	0.258	0.525	0.230	0.144	0.315	0.542	**26.1** M

4.4.2. Results on VisDrone2019 Dataset

We compare our method with several state-of-the-art detectors as shown in Table 6. MVT-B achieves 31.7% mAP@0.5:0.95, 52.2% mAP@0.5, and 34.2% mAP@0.75 on the VisDrone2019 dataset, outperforming all other state-of-the-art methods expect mAP@0.5.

Figure 10 presents the results of our method for detecting objects in various scenes within the VisDrone2019 dataset.

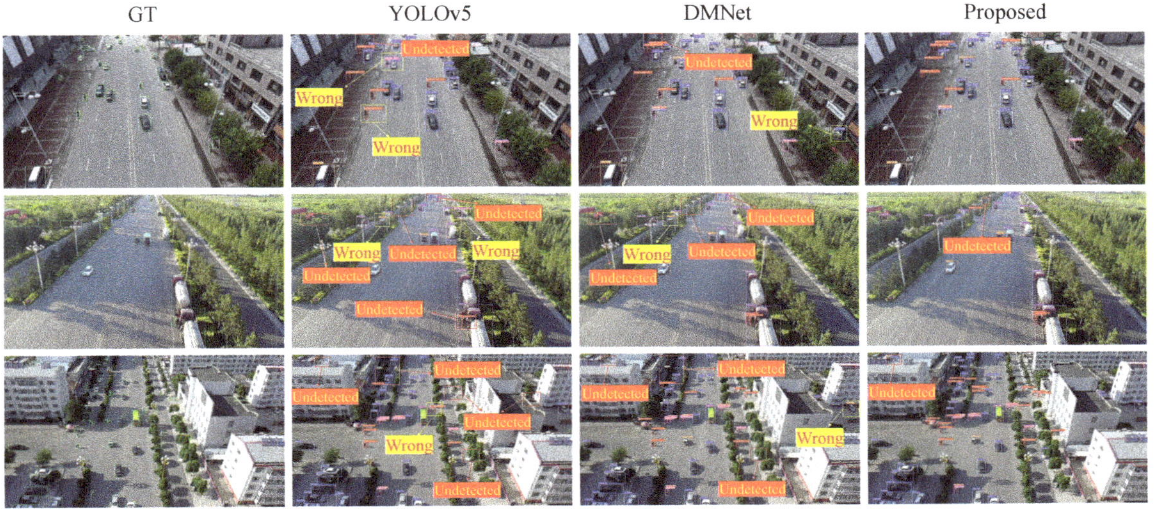

Figure 10. Prediction examples on the VisDrone2019 dataset using different approaches involving YOLOv5, DMNet, and proposed method.

Table 6. Comparison of detection performance on the VisDrone2019 dataset. The best result is highlighted with **bold**.

Model	Backbone	mAP @0.5:0.95	mAP @0.5	mAP @0.75	mAP$_S$	mAP$_M$	mAP$_L$	Params
Cascade R-CNN [28]	ResNet 50	0.232	0.399	0.234	0.165	0.368	0.394	273.2 M
YOLOv5 [27]	CSPDarknet 53	0.241	0.441	0.247	0.153	0.356	0.384	93.0 M
RetinaNet [45]	ResNet 101	0.243	0.443	0.187	0.187	0.352	0.378	251.7 M
Libra RCNN [29]	ResNet 50	0.243	0.412	0.249	0.168	0.340	0.368	185.4 M
Cascade R-CNN [28]	Transformer	0.247	0.424	0.265	0.177	0.372	0.403	335.0 M
HawkNet [46]	ResNet 50	0.256	0.443	0.258	0.199	0.360	0.391	130.9 M
VFNet [47]	ResNet 50	0.259	0.421	0.270	0.168	0.373	0.414	296.2 M
DetectoRS [40]	ResNet 101	0.268	0.432	0.280	0.175	0.382	0.417	540.1 M
Sparse R-CNN [43]	Transformer	0.276	0.463	0.282	0.188	0.392	0.433	352.0 M
DMNet [42]	CSPDarknet 53	0.282	0.476	0.289	0.199	0.396	**0.558**	96.7 M
ClusDet [44]	ResNeXt 101	0.284	**0.532**	0.264	0.191	0.408	0.544	-
SDMNet [48]	CSPDarknet 53	0.302	0.525	0.306	0.226	0.396	0.398	96.6 M
MVT-B (ours)	Transformer	**0.317**	0.522	**0.342**	**0.243**	**0.421**	0.552	106.4 M
MVT-S (ours)	Transformer	0.296	0.497	0.321	0.225	0.405	0.533	56.6 M
MVT-T (ours)	Transformer	0.277	0.465	0.303	0.202	0.388	0.502	**26.1** M

4.4.3. Results on UAVDT Dataset

We compare our method with several state-of-the-art detectors as shown in Table 7. MVT-B achieves 28.2% mAP@0.5:0.95, 42.1% mAP@0.5, and 32.2% mAP@0.75 on the UAVDT dataset, outperforming all other state-of-the-art methods expect mAP@0.5. Figure 11 presents the results of our method for detecting objects in various scenes within the UAVDT dataset.

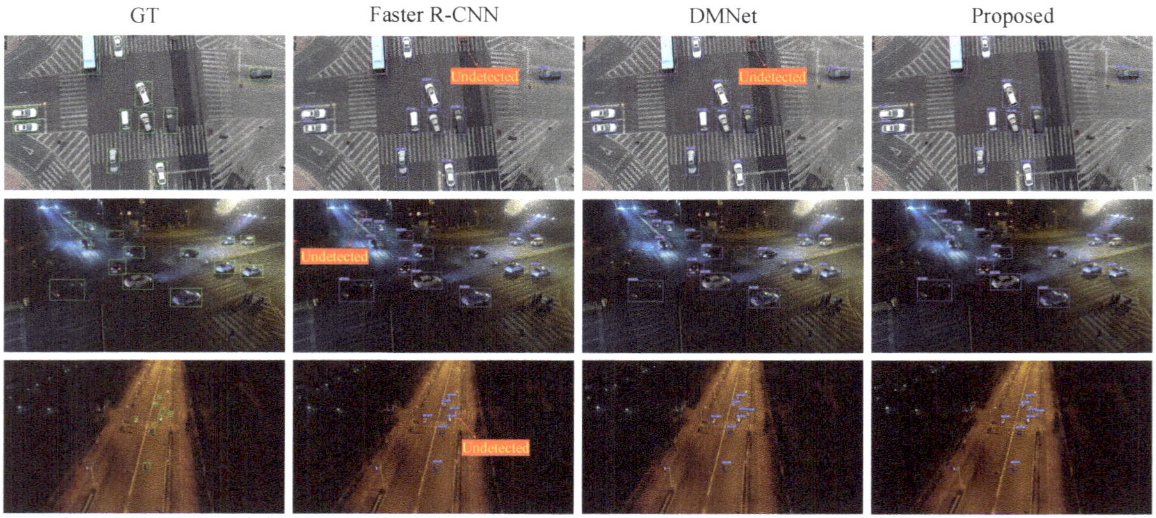

Figure 11. Prediction examples on the UAVDT dataset using different approaches involving Faster R-CNN, DMNet, and proposed method.

Table 7. Comparison of detection performance on the UAVDT dataset. The best result is highlighted with **bold**.

Model	Backbone	mAP @0.5:0.95	mAP @0.5	mAP @0.75	mAP_S	mAP_M	mAP_L	Params
Faster R-CNN [34]	ResNet 50	0.110	0.234	0.084	0.081	0.202	0.265	42.0 M
Cascade R-CNN [28]	ResNet 50	0.121	0.235	0.108	0.084	0.215	0.147	273.2 M
ClusDet [44]	ResNet 101	0.137	0.265	0.125	0.091	0.251	0.312	-
Cascade R-CNN [28]	Transformer	0.138	0.244	0.117	0.090	0.232	0.268	335.0 M
DMNet [42]	CSPDarkNet 53	0.147	0.246	0.163	0.093	0.262	0.352	96.7 M
Sparse R-CNN [43]	Transformer	0.153	0.266	0.171	0.118	0.253	0.288	352.0 M
GLSAN [49]	CSPDarkNet 53	0.170	0.281	0.188	-	-	-	-
AdaZoom [50]	CSPDarkNet 53	0.201	0.345	0.215	0.142	0.292	0.284	-
ReasDet [51]	CSPDarkNet 53	0.218	0.349	0.248	0.153	0.327	0.308	-
EVORL [52]	ResNet 50	0.280	**0.438**	0.315	0.218	**0.404**	0.359	-
MVT-B (ours)	Transformer	**0.282**	0.421	**0.322**	**0.237**	0.397	**0.368**	106.4 M
MVT-S (ours)	Transformer	0.267	0.405	0.297	0.206	0.373	0.350	56.6M
MVT-T (ours)	Transformer	0.238	0.367	0.271	0.162	0.356	0.322	**26.1M**

5. Discussion

The UAVDT dataset only annotates three categories of objects and has simpler scenes compared to the VisDrone2019 dataset. However, the UAVDT dataset exhibits lower detection performance due to its challenging scenes (e.g., low lighting, motion blur), as exemplified in Figure 12. Therefore, applying event cameras to improve visual effects in extreme environments will greatly improve the accuracy of object detection. Despite event cameras being capable of capturing moving objects in various challenging scenarios, they only retain intensity features while losing color information, resulting in the loss of object details. While traditional cameras are limited by a fixed frame rate, they preserve more high-frequency information. Therefore, it is a meaningful step to simultaneously consider the event and traditional cameras for detection, aiming to achieve improved performance in any challenging scenario.

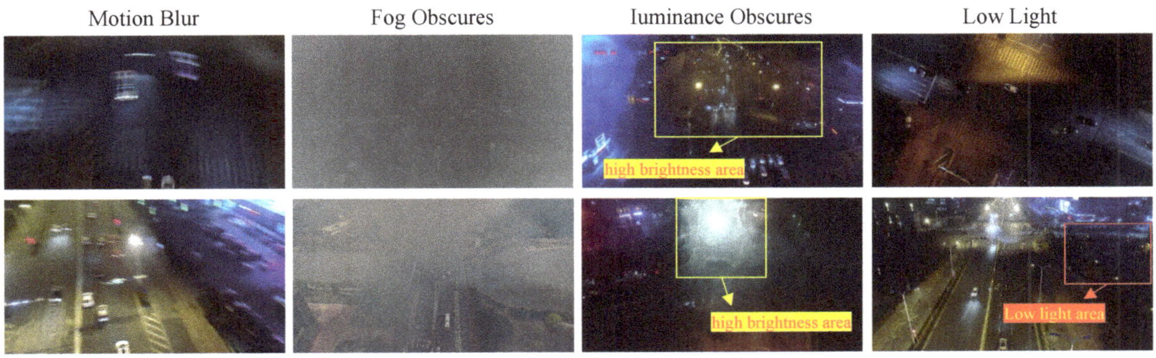

Figure 12. Extreme scenarios in UAVDT dataset. These scenes captured by traditional cameras pose challenges for object detection.

6. Conclusions

In this paper, we aim to capture details in challenging remote sensing images (e.g., low light, motion blur scenarios) to improve the detection performance of small targets. We propose a method called Multi-Vision Transformer (MVT), which employs Channel Spatial Attention (CSA) to enhance short-range dependencies and extract high-frequency information features, utilizing Global Spatial Attention (GSA) to strengthen long-range dependencies and retain more low-frequency information. Specifically, the proposed MVT backbone generates more accurate object locations with enhanced features by maintaining multi-scale high-resolution features with rich semantic information. Subsequently, we use Scale-Level Embedding to extract multiple scales features and apply Cross Deformable Attention (CDA) to progressively fuse information from different scales, significantly reducing the computational complexity of the network. Furthermore, we introduce a dataset called EOD, captured by a drone equipped with an event camera. Finally, all experiments are conducted on the EOD dataset and two widely used UAV remote sensing datasets. The results demonstrate that our method outperforms widely used methods in terms of detection performance on the EOD dataset, VisDrone2019 dataset, and UAVDT dataset.

Author Contributions: Conceptualization, methodology, software, S.J.; data curation, visualization, investigation, H.L. (Hengyi Lv); software, validation, Y.Z.; software, Writing—original draft preparation, H.L. (Hailong Liu); writing—reviewing and editing, M.S. All authors have read and agreed to the published version of the manuscript.

Funding: This work was supported by the National Natural Science Foundation of China (62005269) and the 2023 Jilin Province industrialization project for the specialized program under Grant 2023C031-6.

Data Availability Statement: The VisDrone2019 dataset and UAVDT dataset are available from the websites https://github.com/VisDrone/VisDrone-Dataset and https://sites.google.com/view/grli-uavdt.

Acknowledgments: The authors thank the editors and reviewers for their hard work and valuable advice.

Conflicts of Interest: The authors declare no conflicts of interest.

References

1. Brandli, C.; Berner, R.; Yang, M.; Liu, S.C.; Delbruck, T. A 240× 180 130 db 3 µs latency global shutter spatiotemporal vision sensor. *IEEE J.-Solid-State Circuits* **2014**, *49*, 2333–2341. [CrossRef]
2. Delbruck, T. Frame-free dynamic digital vision. In Proceedings of the International Symposium on Secure-Life Electronics, Advanced Electronics for Quality Life and Society, Tokyo, Japan, 6–7 March 2008 ; Volume 1, pp. 21–26.
3. Zhu, X.; Su, W.; Lu, L.; Li, B.; Wang, X.; Dai, J. Deformable detr: Deformable transformers for end-to-end object detection. *arXiv* **2020**, arXiv:2010.04159.

4. Du, D.; Zhu, P.; Wen, L.; Bian, X.; Lin, H.; Hu, Q.; Peng, T.; Zheng, J.; Wang, X.; Zhang, Y.; et al. VisDrone-DET2019: The vision meets drone object detection in image challenge results. In Proceedings of the IEEE/CVF International Conference on Computer Vision (ICCV), Seoul, Republic of Korea, 27 October–2 November 2019.
5. Du, D.; Qi, Y.; Yu, H.; Yang, Y.; Duan, K.; Li, G.; Zhang, W.; Huang, Q.; Tian, Q. The unmanned aerial vehicle benchmark: Object detection and tracking. In Proceedings of the European Conference on Computer Vision (ECCV), Munich, Germany, 8–14 September 2018; pp. 370–386.
6. Sun, K.; Xiao, B.; Liu, D.; Wang, J. Deep high-resolution representation learning for human pose estimation. In Proceedings of the IEEE/CVF Conference on Computer Vision and Pattern Recognition, Long Beach, CA, USA, 15–20 June 2019; pp. 5693–5703.
7. Newell, A.; Yang, K.; Deng, J. Stacked hourglass networks for human pose estimation. In Proceedings of the Computer Vision–ECCV 2016: 14th European Conference, Amsterdam, The Netherlands, 11–14 October 2016; Proceedings, Part VIII 14; Springer: Berlin/Heidelberg, Germany, 2016; pp. 483–499.
8. Lin, T.Y.; Dollár, P.; Girshick, R.; He, K.; Hariharan, B.; Belongie, S. Feature pyramid networks for object detection. In Proceedings of the IEEE Conference on Computer Vision and Pattern Recognition, Honolulu, HI, USA, 21–26 July 2017; pp. 2117–2125.
9. Quan, Y.; Zhang, D.; Zhang, L.; Tang, J. Centralized feature pyramid for object detection. *IEEE Trans. Image Process.* **2023**, *32*, 4341–4354. [CrossRef] [PubMed]
10. Wang, W.; Xie, E.; Song, X.; Zang, Y.; Wang, W.; Lu, T.; Yu, G.; Shen, C. Efficient and accurate arbitrary-shaped text detection with pixel aggregation network. In Proceedings of the IEEE/CVF International Conference on Computer Vision, Seoul, Republic of Korea, 27 October–2 November 2019; pp. 8440–8449.
11. Mboga, N.; Grippa, T.; Georganos, S.; Vanhuysse, S.; Smets, B.; Dewitte, O.; Wolff, E.; Lennert, M. Fully convolutional networks for land cover classification from historical panchromatic aerial photographs. *ISPRS J. Photogramm. Remote Sens.* **2020**, *167*, 385–395. [CrossRef]
12. Abriha, D.; Szabó, S. Strategies in training deep learning models to extract building from multisource images with small training sample sizes. *Int. J. Digit. Earth* **2023**, *16*, 1707–1724. [CrossRef]
13. Solórzano, J.V.; Mas, J.F.; Gallardo-Cruz, J.A.; Gao, Y.; de Oca, A.F.M. Deforestation detection using a spatio-temporal deep learning approach with synthetic aperture radar and multispectral images. *ISPRS J. Photogramm. Remote Sens.* **2023**, *199*, 87–101. [CrossRef]
14. Yu, C.; Xiao, B.; Gao, C.; Yuan, L.; Zhang, L.; Sang, N.; Wang, J. Lite-hrnet: A lightweight high-resolution network. In Proceedings of the IEEE/CVF Conference on Computer Vision and Pattern Recognition, Nashville, TN, USA, 19–25 June 2021; pp. 10440–10450.
15. Xu, H.; Tang, X.; Ai, B.; Yang, F.; Wen, Z.; Yang, X. Feature-selection high-resolution network with hypersphere embedding for semantic segmentation of VHR remote sensing images. *IEEE Trans. Geosci. Remote Sens.* **2022**, *60*, 4411915. [CrossRef]
16. Hao, X.; Yin, L.; Li, X.; Zhang, L.; Yang, R. A Multi-Objective Semantic Segmentation Algorithm Based on Improved U-Net Networks. *Remote Sens.* **2023**, *15*, 1838. [CrossRef]
17. Sun, K.; Zhao, Y.; Jiang, B.; Cheng, T.; Xiao, B.; Liu, D.; Mu, Y.; Wang, X.; Liu, W.; Wang, J. High-resolution representations for labeling pixels and regions. *arXiv* **2019**, arXiv:1904.04514.
18. Li, R.; Shen, Y. YOLOSR-IST: A deep learning method for small target detection in infrared remote sensing images based on super-resolution and YOLO. *Signal Process.* **2023**, *208*, 108962. [CrossRef]
19. Vaswani, A.; Shazeer, N.; Parmar, N.; Uszkoreit, J.; Jones, L.; Gomez, A.N.; Kaiser, Ł.; Polosukhin, I. Attention is all you need. *Adv. Neural Inf. Process. Syst.* **2017**, *30*, 6000–6010. .
20. Wang, Q.; Wu, B.; Zhu, P.; Li, P.; Zuo, W.; Hu, Q. ECA-Net: Efficient channel attention for deep convolutional neural networks. In Proceedings of the IEEE/CVF Conference on Computer Vision and Pattern Recognition, Seattle, WA, USA, 14–19 June 2020; pp. 11534–11542.
21. Zhang, M.; Zhang, R.; Zhang, J.; Guo, J.; Li, Y.; Gao, X. Dim2Clear network for infrared small target detection. *IEEE Trans. Geosci. Remote Sens.* **2023**, *61*, 5001714. [CrossRef]
22. Hu, J.; Shen, L.; Sun, G. Squeeze-and-excitation networks. In Proceedings of the IEEE Conference on Computer Vision and Pattern Recognition, Salt Lake City, UT, USA, 18–22 June 2018; pp. 7132–7141.
23. Woo, S.; Park, J.; Lee, J.Y.; Kweon, I.S. Cbam: Convolutional block attention module. In Proceedings of the European Conference on Computer Vision (ECCV), Munich, Germany, 8–14 September 2018; pp. 3–19.
24. Dosovitskiy, A.; Beyer, L.; Kolesnikov, A.; Weissenborn, D.; Zhai, X.; Unterthiner, T.; Dehghani, M.; Minderer, M.; Heigold, G.; Gelly, S.; et al. An image is worth 16x16 words: Transformers for image recognition at scale. *arXiv* **2020**, arXiv:2010.11929.
25. Liu, Z.; Lin, Y.; Cao, Y.; Hu, H.; Wei, Y.; Zhang, Z.; Lin, S.; Guo, B. Swin transformer: Hierarchical vision transformer using shifted windows. In Proceedings of the IEEE/CVF International Conference on Computer Vision, Montreal, BC, Canada, 11–17 October 2021; pp. 10012–10022.
26. Tu, Z.; Talebi, H.; Zhang, H.; Yang, F.; Milanfar, P.; Bovik, A.; Li, Y. Maxvit: Multi-axis vision transformer. In *European Conference on Computer Vision*; Springer: Berlin/Heidelberg, Germany, 2022, pp. 459–479.
27. Jocher, G.; Stoken, A.; Borovec, J.; Changyu, L.; Hogan, A.; Diaconu, L.; Poznanski, J.; Yu, L.; Rai, P.; Ferriday, R.; et al. ultralytics/yolov5: v3. 0. *Zenodo* **2020** . [CrossRef]
28. Cai, Z.; Vasconcelos, N. Cascade r-cnn: Delving into high quality object detection. In Proceedings of the IEEE Conference on Computer Vision and Pattern Recognition, Salt Lake City, UT, USA, 18–22 June 2018; pp. 6154–6162.

29. Pang, J.; Chen, K.; Shi, J.; Feng, H.; Ouyang, W.; Lin, D. Libra r-cnn: Towards balanced learning for object detection. In Proceedings of the IEEE/CVF Conference on Computer Vision and Pattern Recognition, Long Beach, CA, USA, 15–20 June 2019; pp. 821–830.
30. Carion, N.; Massa, F.; Synnaeve, G.; Usunier, N.; Kirillov, A.; Zagoruyko, S. End-to-end object detection with transformers. In *European Conference on Computer Vision*; Springer: Berlin/Heidelberg, Germany, 2020; pp. 213–229.
31. Zhang, H.; Li, F.; Liu, S.; Zhang, L.; Su, H.; Zhu, J.; Ni, L.M.; Shum, H.Y. Dino: Detr with improved denoising anchor boxes for end-to-end object detection. *arXiv* **2022**, arXiv:2203.03605.
32. Gehrig, M.; Scaramuzza, D. Recurrent vision transformers for object detection with event cameras. In Proceedings of the IEEE/CVF Conference on Computer Vision and Pattern Recognition, Vancouver, BC, Canada, 18–22 June 2023; pp. 13884–13893.
33. Girshick, R.; Donahue, J.; Darrell, T.; Malik, J. Rich feature hierarchies for accurate object detection and semantic segmentation. In Proceedings of the IEEE Conference on Computer Vision and Pattern Recognition, Columbus, OH, USA, 23–28 June 2014; pp. 580–587.
34. Girshick, R. Fast r-cnn. In Proceedings of the IEEE International Conference on Computer Vision, Santiago, Chile, 7–13 December 2015; pp. 1440–1448.
35. Iacono, M.; Weber, S.; Glover, A.; Bartolozzi, C. Towards event-driven object detection with off-the-shelf deep learning. In Proceedings of the 2018 IEEE/RSJ International Conference on Intelligent Robots and Systems (IROS), IEEE, Madrid, Spain, 1–5 October 2018; pp. 1–9.
36. Jiang, Z.; Xia, P.; Huang, K.; Stechele, W.; Chen, G.; Bing, Z.; Knoll, A. Mixed frame-/event-driven fast pedestrian detection. In Proceedings of the 2019 International Conference on Robotics and Automation (ICRA), IEEE, Montreal, QC, Canada, 20–24 May 2019; pp. 8332–8338.
37. Su, Q.; Chou, Y.; Hu, Y.; Li, J.; Mei, S.; Zhang, Z.; Li, G. Deep directly-trained spiking neural networks for object detection. In Proceedings of the IEEE/CVF International Conference on Computer Vision, Paris, France, 2–6 October 2023; pp. 6555–6565.
38. Zhu, A.Z.; Yuan, L.; Chaney, K.; Daniilidis, K. Unsupervised event-based learning of optical flow, depth, and egomotion. In Proceedings of the IEEE/CVF Conference on Computer Vision and Pattern Recognition, Long Beach, CA, USA, 15–20 June 2019; pp. 989–997.
39. Lin, T.Y.; Maire, M.; Belongie, S.; Hays, J.; Perona, P.; Ramanan, D.; Dollár, P.; Zitnick, C.L. Microsoft coco: Common objects in context. In Proceedings of the Computer Vision–ECCV 2014: 13th European Conference, Zurich, Switzerland, 6–12 September 2014; Proceedings, Part V 13; Springer: Berlin/Heidelberg, Germany, 2014; pp. 740–755.
40. Xu, C.; Wang, J.; Yang, W.; Yu, H.; Yu, L.; Xia, G.S. Detecting tiny objects in aerial images: A normalized Wasserstein distance and a new benchmark. *ISPRS J. Photogramm. Remote Sens.* **2022**, *190*, 79–93. [CrossRef]
41. Wang, C.Y.; Bochkovskiy, A.; Liao, H.Y.M. YOLOv7: Trainable bag-of-freebies sets new state-of-the-art for real-time object detectors. In Proceedings of the IEEE/CVF Conference on Computer Vision and Pattern Recognition, Vancouver, BC, Canada, 18–22 June 2023; pp. 7464–7475.
42. Li, C.; Yang, T.; Zhu, S.; Chen, C.; Guan, S. Density map guided object detection in aerial images. In Proceedings of the IEEE/CVF Conference on Computer Vision and Pattern Recognition Workshops, Seattle, WA, USA, 14–19 June 2020; pp. 190–191.
43. Sun, P.; Zhang, R.; Jiang, Y.; Kong, T.; Xu, C.; Zhan, W.; Tomizuka, M.; Li, L.; Yuan, Z.; Wang, C.; et al. Sparse r-cnn: End-to-end object detection with learnable proposals. In Proceedings of the IEEE/CVF Conference on Computer Vision and Pattern Recognition, Nashville, TN, USA, 19–25 June 2021; pp. 14454–14463.
44. Yang, F.; Fan, H.; Chu, P.; Blasch, E.; Ling, H. Clustered object detection in aerial images. In Proceedings of the IEEE/CVF International Conference on Computer Vision, Seoul, Republic of Korea, 27 October–2 November 2019; pp. 8311–8320.
45. Lin, T.Y.; Goyal, P.; Girshick, R.; He, K.; Dollár, P. Focal loss for dense object detection. In Proceedings of the IEEE International Conference on Computer Vision, Venice, Italy, 22–29 October 2017; pp. 2980–2988.
46. Lin, H.; Zhou, J.; Gan, Y.; Vong, C.M.; Liu, Q. Novel up-scale feature aggregation for object detection in aerial images. *Neurocomputing* **2020**, *411*, 364–374. [CrossRef]
47. Zhang, H.; Wang, Y.; Dayoub, F.; Sunderhauf, N. Varifocalnet: An iou-aware dense object detector. In Proceedings of the IEEE/CVF Conference on Computer Vision and Pattern Recognition, Nashville, TN, USA, 19–25 June 2021; pp. 8514–8523.
48. Ma, Y.; Chai, L.; Jin, L. Scale decoupled pyramid for object detection in aerial images. *IEEE Trans. Geosci. Remote Sens.* **2023**, *61*, 4704314. [CrossRef]
49. Deng, S.; Li, S.; Xie, K.; Song, W.; Liao, X.; Hao, A.; Qin, H. A global-local self-adaptive network for drone-view object detection. *IEEE Trans. Image Process.* **2020**, *30*, 1556–1569. [CrossRef]
50. Xu, J.; Li, Y.; Wang, S. Adazoom: Adaptive zoom network for multi-scale object detection in large scenes. *arXiv* **2021**, arXiv:2106.10409.
51. Ge, Z.; Qi, L.; Wang, Y.; Sun, Y. Zoom-and-reasoning: Joint foreground zoom and visual-semantic reasoning detection network for aerial images. *IEEE Signal Process. Lett.* **2022**, *29*, 2572–2576. [CrossRef]
52. Zhang, J.; Yang, X.; He, W.; Ren, J.; Zhang, Q.; Zhao, T.; Bai, R.; He, X.; Liu, J. Scale Optimization Using Evolutionary Reinforcement Learning for Object Detection on Drone Imagery. *arXiv* **2023**, arXiv:2312.15219.

remote sensing

Article

Multiscale Feature Extraction U-Net for Infrared Dim- and Small-Target Detection

Xiaozhen Wang [1,2], Chengshan Han [1], Jiaqi Li [1,2], Ting Nie [1], Mingxuan Li [1], Xiaofeng Wang [1,2] and Liang Huang [1,*]

1 Changchun Institute of Optics, Fine Mechanics and Physics, Chinese Academy of Sciences, Changchun 130033, China; wangxiaozhen22@mails.ucas.ac.cn (X.W.); hanchengshan@ciomp.ac.cn (C.H.); lijiaqi221@mails.ucas.ac.cn (J.L.); nieting@ciomp.ac.cn (T.N.); limingxuan17@mails.ucas.ac.cn (M.L.); wangxiaofeng201@mails.ucas.edu.cn (X.W.)

2 University of Chinese Academy of Sciences, Beijing 100049, China

* Correspondence: huangliang@ciomp.ac.cn or hezqxfk9@126.com

Abstract: The technology of infrared dim- and small-target detection is irreplaceable in many fields, such as those of missile early warning systems and forest fire prevention, among others. However, numerous components interfere with infrared imaging, presenting challenges for achieving successful detection of infrared dim and small targets with a low rate of false alarms. Hence, we propose a new infrared dim- and small-target detection network, Multiscale Feature Extraction U-Net for Infrared Dim- and Small-Target Detection (MFEU-Net), which can accurately detect targets in complex backgrounds. It uses the U-Net structure, and the encoders and decoders consist of ReSidual U-block and Inception, allowing rich multiscale feature information to be extracted. Thus, the effectiveness of algorithms in detecting very small-sized targets can be improved. In addition, through the multidimensional channel and spatial attention mechanism, the model can be adjusted to focus more on the target area in the image, improving its extraction of target information and detection performance in different scenarios. The experimental results show that our proposed algorithm outperforms other advanced algorithms in detection performance. On the MFIRST, SIRST, and IRSTD-1k datasets, we achieved detection rates of 0.864, 0.962, and 0.965; IoU values of 0.514, 0.671, and 0.630; and false alarm rates of 3.08×10^{-5}, 2.61×10^{-6}, and 1.81×10^{-5}, respectively.

Keywords: convolutional neural network; multiscale features; infrared image; small-target detection

Citation: Wang, X.; Han, C.; Li, J.; Nie, T.; Li, M.; Wang, X.; Huang, L. Multiscale Feature Extraction U-Net for Infrared Dim- and Small-Target Detection. *Remote Sens.* **2024**, *16*, 643. https://doi.org/10.3390/rs16040643

Academic Editor: Paolo Tripicchio

Received: 26 December 2023
Revised: 30 January 2024
Accepted: 6 February 2024
Published: 9 February 2024

1. Introduction

Infrared detection systems can distinguish between a target and its background by collecting the different radiation signatures and comparing between the two. They are a type of passive detection system able to work under all-weather conditions without being influenced by light and can realize long-distance detection with high detection accuracy. As they are not affected by the shortcoming of interference from other electromagnetic waves, in contrast to detection based on radar and visible light, they have become one of the important means of acquiring strategic perception data, experiencing very high application in both military and civil contexts [1]. However, in practical applications, such as involving guidance, early-warning, airborne, or satellite surveillance, the very long distance of targets from the detector results in them representing a very small percentage of the image output from the detector; at the same time, such targets are generally not the brightest in the image due to the effect of atmospheric scattering and absorption, and this kind of typical target is usually referred to as an infrared dim and small target (IDST) [2].

IDSTs usually present as a speckle in the image, thus lacking geometrical and textural feature information, and the target is often submerged in the background, which makes it impossible to extract the target through global grayscale characteristics [3]. Compared with sky and sea backgrounds, ground backgrounds are more complex, and there are often sources of interference, such as noise and small edges, close to the IDST in the background, which will lead to a more complex and variable grayscale distribution in the target neighborhood. All these factors lead to IDSTs being difficult to detect. Therefore, IDST detection represents both a difficulty and a hotspot in the field of target detection. The ability to function under real-time detection conditions is an important application requirement in the practical projects of detection algorithms, which have high research and application value in many fields [4].

Numerous traditional target detection algorithms have previously been proposed by researchers [5]. Filter-based methods use a specific filter that can eliminate the background of the infrared image to detect the IDST. Filter-based methods require less computation but have low efficacy. They can thus only be used in specific scenes to suppress the background of a gentle change and cannot solve the problem of complex background [6]. The LCM-based methods take advantage of the difference in gray values between the target and the background to boost the gray values of the target while reducing those of the background, but good detection results can mostly only be obtained when there is high image contrast, so the algorithm's generalization ability is poor, and it cannot be effectively applied to complex backgrounds [7]. Data structure-based methods mainly transform the IDST detection problem into a convex optimization problem with low-rank and sparse matrix recovery. This type of algorithm has good applicability to images with complex backgrounds. However, the algorithm is very computationally intensive, so it is not suitable for imaging in real-time applications where latency is significantly compromised [8].

Due to the many advantages of deep learning-based algorithms, numerous researchers have proposed their use in IDST detection [9]. Since the size of an IDST is very small, and they are very sensitive to bounding box perturbation, image segmentation methods are adopted in most approaches for IDST detection such that more fine target information can be obtained [10]. In order to detect very-small-size and general-size targets, some algorithms enhance the information fusion between different layers so they can extract the information for different sizes and improve the detection effect for differently sized targets [11]. Due to the sparse nature of IDSTs, some algorithms enhance the visibility of targets by suppressing the background [12]. There are also algorithms that use GAN networks to separately address the problem of missed and false alarms, using different generators to address the difficult balance between them [13].

The existing IDST detection algorithms still have some limitations. The traditional methods are overly dependent on a priori knowledge and have poor detection performance in real scenes [14]. Although the above deep learning algorithms have achieved good detection results, most cannot achieve a good balance between the detection rate and the false alarm rate. In addition, some algorithms with insufficient generalization ability can only be used with specific datasets and cannot meet the requirements for real-scene detection [15].

In this paper, we propose a new convolutional network-based system for IDST detection: the Multiscale Feature Extraction U-Net for Infrared Dim- and Small-Target Detection (MFEU-net). The network uses U-Net, and ReSidual U-block (RSU) and Inception modules are introduced in the encoders and decoders to extract multiscale feature information, making it possible to detect very small IDSTs. There are multidimensional channels and spatial attention mechanisms in each encoder and decoder, and the global information is extracted by the attention mechanism such that the model can give a greater weight to the target area, thereby improving the ability of the model to adapt to different scenarios as well as the detection performance with complex backgrounds. The algorithm proposed in this paper has the lowest leakage detection rates and false alarm rates.

Overall, the main contributions of this paper are as follows:

(1) We design a multiscale feature extraction network using a combination of ReSidual U-block (RSU) and Inception, which enables the network to have different receptive fields at one level, allowing the network to adapt to scenarios containing targets of different sizes;

(2) We design a multidimensional channel and spatial attention mechanism (MCSAM) that can make full use of the different information in the feature map and more effectively determine the region where the target is located;

(3) Compared to other state-of-the-art algorithms, our algorithm achieved better detection results on different datasets.

2. Materials and Methods

2.1. Related Work

2.1.1. IDST Detection

Traditional single-frame IDST detection methods can be divided into filter-based methods, local contrast measure (LCM)-based methods, and data structure-based methods.

Filter-based methods can be divided into spatial- and transform-domain filtering. In the spatial-domain filtering methods, a specific filtering kernel is used to remove the background in the infrared image [16]. For these methods, start by designing a filter kernel based on the characteristics of the background and the target to eliminate the background, then use the estimated background to perform a difference operation with the original image, and finally threshold the difference image to segment and detect IDSTs. With the frequency-domain filtering approach, the background is considered to be low frequency and the target to be high frequency, and by designing an appropriate high-pass filter, the low-frequency background and the high-frequency target can be separated [17]. Overall, filtering-based methods require less computation but have low efficacy, being only applicable to scenes with very little background change. Thus, they cannot be used to solve the problem of complex backgrounds and, moreover, have high false alarm rates and poor algorithm robustness [18].

The LCM-based algorithm uses the different gray values of the target images and other images to calculate different gain factors such that the difference between the two can be increased, making the target more prominent [19]. In the LCM approach, a kernel is used to traverse the entire image, multiplying the gray value at the center of the kernel by the ratio of the center gray value to the average gray value of the surrounding area, and when the center gray value of the kernel greatly exceeds the surrounding gray value, the center of the kernel is considered to be the target, and a saliency map can be obtained. Then, the small targets are segmented in the saliency map via thresholding. Finally, the position of the targets in the saliency map must correspond to the original image to achieve IDST detection [20]. The key to this algorithm is the way in which the saliency map is acquired, which will greatly affect the algorithm's performance. These LCM-based methods can be used to suppress background enhancement targets through certain means, but most them can only detect targets when there is high image contrast, and the generalization ability of the algorithm is poor, so it cannot be effectively applied to complex backgrounds [21].

The methods based on image data structure involve transforming the small-target detection problem into a convex optimization problem for low-rank and sparse matrix restoration based on the sparsity of the target and the low rank of the background [22]. These algorithms are based on the two prerequisites of having few targets and strong background correlation in infrared images, so when these two conditions are not met, these algorithms are much less effective in detection. The methods based on image data structure have good applicability for images with fewer targets and complex backgrounds, but these algorithms will have leakage detection in the case of more targets, and the computational weight is very high, so they are difficult to apply to remote sensing images [23].

Deep learning algorithms can realize complex nonlinear computations and surpass traditional algorithms in many areas, so they are increasingly being applied in IDST detection [24].

Wang et al. used two independent generators, each accomplishing the task of reducing false alarms and missed detections, and the two models were based on a contextual aggregation network that could utilize different feature information, thus achieving low rates of missed detections and false alarms in IDST detection [25]. In addition, they published a large synthetic IDST detection dataset that can be used in advancing the development of IDST detection algorithms.

Lee et al. incorporated fusion and augmentation modules at each level of the network, and through repeated augmentation and fusion, different levels of information could be fused to retain more information about the target [26]. However, it was necessary to retain many of the previous feature maps, thereby consuming high amounts of storage resources, which poses a problem for practical use.

Chen et al. designed a global attention mechanism that can be used to separately extract local and global features, eliminate most of the background pixels, and highlight the target location; by fusing global and local features, the target can be detected using multiscale information [27]. However, its post-processing is complex, blurring the target with loss of detailed information.

Hou et al. utilized ResNet to extract features in the form of groups, making it possible increase the weight of important groups; furthermore, the addition of a fully connected layer to the jump connections of U-Net allows the network to extract global information to improve target extraction [28]. However, the use of the mean square error (MSE) as a loss function results in the network being prone to predicting the target as background during training due to the imbalance in positive and negative samples.

Yu et al. proposed a multiscale local contrast learning mechanism, which can generate multiscale local contrast feature maps during the training process such that more detailed information about the target can be extracted, enabling the network to better localize the target position [29]. However, the use of normal convolutional layers and dilation convolution to extract local information introduces a grid effect when the dilation parameter is excessively large, which tends to result in the loss of target information.

2.1.2. Attention Mechanism

In deep learning, an attention mechanism (AM) can be used to ensure neural networks prioritize important regions when processing data by mimicking the human visual and cognitive systems and adding different weights to different regions in the feature map [30]. By introducing an attention mechanism, different regions of the input feature map can be multiplied by different weighting factors, and the neural network is able to focus on important local information from the global information and more important information can be extracted by the network such that the model can make more accurate predictions or classifications without consuming more computational and storage resources. Therefore, AMs have been widely used in deep learning networks, such as SE-Net, ECA-Net, CBAM, etc. [31].

Squeeze-and-Excitation Networks (SE-Nets) [32] are representative of work in the field of CV where the attention mechanism is applied to the channel dimension. They have a simple and effective structure and can adaptively adjust the feature responses between channels by means of feature recalibration. This network extracts global information using the global average pooling operation and downsamples all feature maps to a single point. After that, it utilizes a two-layer multilayer perceptron network to change the weights of different regions. The sigmoid activation function is then used to generate the channel weights, after which the Hadamard product is computed with the input to obtain the channel-weighted feature map.

Efficient Channel Attention (ECA-Net) [33] is an improvement of the feature transformation part of SE-Nets. The channel information interaction of SE-Nets is realized through the full connection, which damages a part of the feature expression in the process of downscaling and upscaling, while ECA-Net utilizes one-dimensional convolution to realize channel information interaction, which significantly reduces the computational complexity, basically with no loss of performance.

The Convolutional Block Attention Module (CBAM) [34] can be understood as adding a spatial attention module (SAM) to an SE-Net, which separately calculates weights in the channel and spatial domains, allowing it to more precisely localize the region where the target is located compared to a single-channel attention mechanism. A SAM generates two feature maps containing different global information through two pooling operations, which are concatenated together and then fused by a 7 × 7-sized convolutional layer. Finally, a sigmoid operation is performed to generate a weight map, after which the Hadamard product is computed from the original input feature map to enhance the target region.

2.2. Method

2.2.1. Overall Architecture

U-Net can fuse different information at different levels through skip connections such that detailed information at the low level can be directly passed to the high level, thus providing richer contextual and detailed information. This skip connection design helps the network to better capture the boundaries and details of the target and results in improved accuracy of detection. Another advantage of U-Net is its efficient architectural design, especially the skip connections and symmetric expansion paths, which contribute to the network's good performance even on small datasets. Thus, we use the U-Net structure in our deep learning network.

Structurally, the upsampling stage and the downsampling stage are basically symmetrical. The downsampling stage consists of an encoder module and global maximum pooling for extracting the multiscale information of the input feature maps and downsampling the feature maps. The upsampling phase consists of an upsampling and decoder module in which linear interpolation is used to upsample the low-resolution feature map, and the multiscale information from different layers is then fused. In stages one to four, the encoder and decoder are RSU and MCSAM, while in stages five to six, the encoder and decoder are Inception and MCSAM. The downsampling stage and the upsampling stage are connected by the Merge module. The structure of MFEU-Net is shown in Figure 1.

Inside the Merge module is a ResNet consisting of convolutional layers with a convolutional kernel size of 1 × 1. Through these 1 × 1 convolutional layers, the information of different channels can be fused, and the nonlinear ability of the model can be increased after convolution through the activation function.

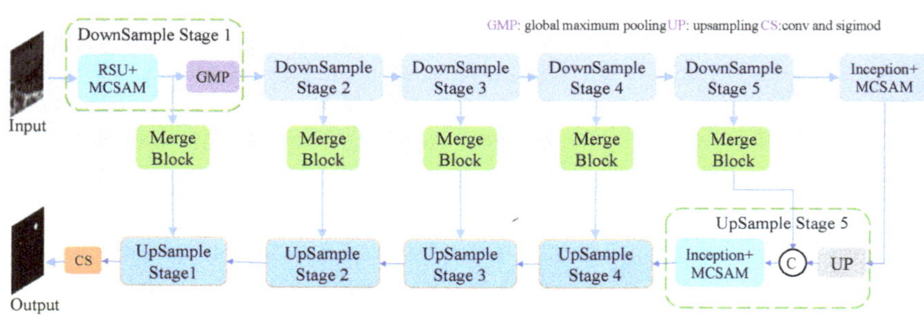

Figure 1. MFEU-Net structure.

2.2.2. Encoder and Decoder

Conventional convolutional layers have a fixed convolutional kernel size, which means they have a fixed sense field for the input image. Therefore, they cannot fully utilize the contextual information and have poor detection performance when encountering very small targets. Multiscale feature extraction methods can enable a network to have different receptive field sizes at different layers by adding parallel convolutional branches or using pooling operations at different scales. Thus, they enable the algorithm to better detect very small targets in the image.

ReSidual U-block (RSU) [35] uses small U-Net modules instead of single-stream convolution, so it can have a variety of different-sized receptive fields at different layers, which allows it to better capture contextual information at different scales. RSU uses pooling operations to increase the overall architecture depth as well as the network's ability to sense global and semantic information through multiple downsampling.

However, excessive downsampling will lead to a large reduction in detail information, and u-sampling will bring invalid information when concatenating with high-resolution feature maps, affecting the retention of detail information. In addition, U-Net's structure is dependent on retention of the feature maps before downsampling, and multiple rounds of downsampling will increase the number of feature maps to be retained, which will consume a large amount of storage resources. For this reason, we reduce the number of downsampling events in the RSU module and remove the feature maps that will not subsequently be used. As a result, more information in the feature map can be retained, and the consumption of storage resources is reduced.

Inception uses parallel convolution and pooling operations of different sizes or different depths to capture rich multiscale information, allowing the model to handle richer spatial features and increase feature diversity [36]. Inception modules can be repeatedly stacked to form larger networks, which can effectively extend the depth and width of the network, preventing overfitting phenomena while improving the accuracy of deep learning networks. However, for parallel multibranching, a large number of parameters are introduced to the model, increasing the requirement for computational resources and the time for training and inference. Therefore, we decrease the parameters by reducing the number of channels in each branch.

Therefore, a combination of RSU and Inception is used such that the U-Net has different multiscale features at different levels. In the initial stage, RSU is used and the amount of downsampling is limited. Its structure is shown in Figure 2. First, the number of channels of the input feature map is changed by a convolutional layer of size 1×1. The data are then fed into the RSU module. In the RSU there is a small U-Net, whose encoder and decoder employ ResNet and are connected by skip connections. The data are then fed into the AM to add different weights to different regions of the feature map. Finally, they data are added to the feature map after changing the number of channels and output to the next module.

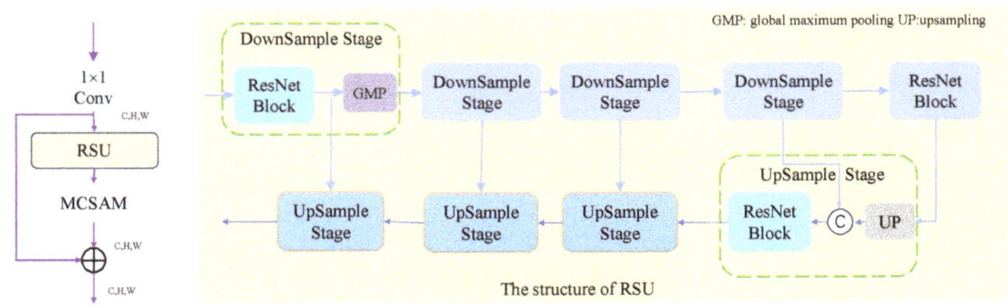

Figure 2. Architecture diagram of an encoder and decoder using the RSU module.

By changing the number of downsampling events, the depth of the RSU module can be changed to accommodate different-sized feature maps. Specifically, encoder and decoder block one uses four rounds of downsampling, encoder and decoder block two uses three rounds of downsampling, encoder and decoder block three uses two rounds of downsampling, and encoder and decoder block four uses one round of downsampling.

Following this, Inception is used. In order to avoid having excessive parameters, four different branches are used, and the number of channels in each branch is one-quarter of the number of output channels. Since the parameters of the convolutional layer are proportional to the square of the number of channels, the parameters and computation of the model can be drastically reduced by reducing the number of channels. Its structure is shown in Figure 3. First, the number of channels of the input feature map is changed to the number of output channels by a convolutional layer of size 1×1. It is then fed into four different branches.

Through different branches, different examples of feature information can be learned and synthesized to improve model performance. Afterwards, the outputs of these four different branches are concatenated together and fed into a 1×1-sized convolutional layer, exchanging information between the different channels. The output is then fed into the AM to add different weights to different regions of the feature map. Finally, it is added to the feature map after changing the number of channels and output to the next module.

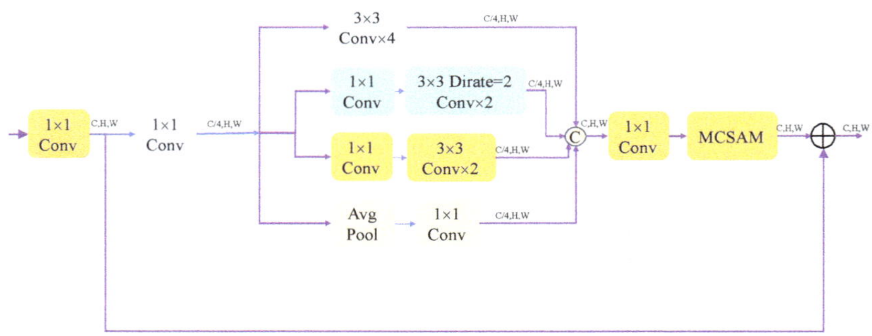

Figure 3. Architecture diagram of an encoder and decoder using Inception.

The backbone network and the number of downsampling events at different stages are shown in Table 1. This allows the model to have different receptive fields without significantly increasing the number of model parameters, resulting in improving the efficacy of IDST detection.

Table 1. Backbone network and number of rounds of downsampling at different stages.

Stage	Backbone	Downsampling Number
Stage one	RSU	4
Stage two	RSU	3
Stage three	RSU	2
Stage four	RSU	1
Stage five	Inception	0
Stage six	Inception	0

2.2.3. Attention Mechanism

In this section, we describe the design of a Multidimensional Channel Attention and Spatial Attention Mechanism (MCSAM) to extract global information. Through the attention mechanism, more weight can be given to the focus area in the feature map. The channel AM is first utilized to generate different weights for each channel in its channel domain for the input feature map. Then, the spatial AM is utilized to generate different weights for each region in the spatial domain for the channel-weighted feature maps. The structure diagram is shown in Figure 4.

In the channel attention mechanism, to extract more advanced information, we additionally add pooling operations. Two $1 \times 1 \times C$ feature maps (F^c_{max}, F^c_{avg}) are generated by performing global maximum pooling (GMP^c) and global average pooling (GAP^c) on the input feature maps (F). The two feature maps are concatenated on the channel domain to obtain a $2 \times 1 \times C$ feature map. After that, a $1 \times 1 \times C$ feature map is generated by one-dimensional convolution. A $C \times 1 \times 1$ channel weight feature map ($W_c(F)$) is then obtained by using the sigmoid function (σ) and transpose operation on it. Finally, the Hadamard product ($\otimes$) is computed using the input feature map to get a channel-weighted feature map. The process is illustrated in Equation 1. The structure diagram is shown in Figure 5.

$$\begin{aligned} F^c_{max} &= GAP^c(F) \\ F^c_{avg} &= GMP^c(F) \\ W_c(F) &= \sigma(Conv([F^c_{avg}, F^c_{max}])) \\ F_c &= W_c(F) \otimes F \end{aligned} \tag{1}$$

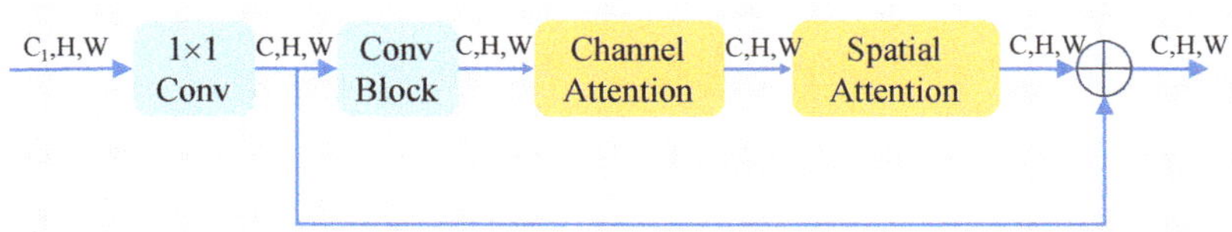

Figure 4. MCSAM structure.

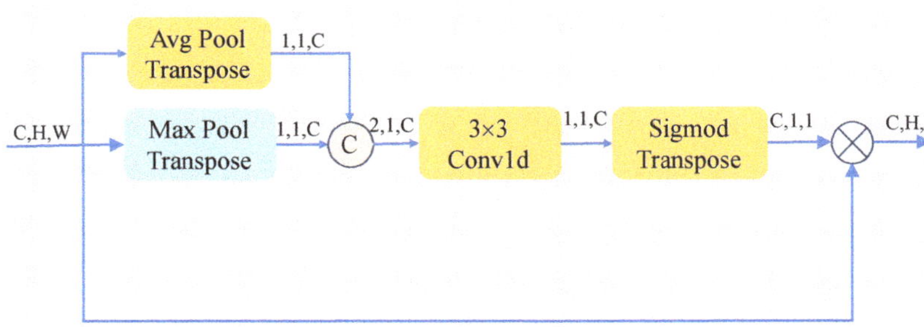

Figure 5. Channel attention structure.

SAM extracts information using only pooling operations, which can result in a significant loss of local information. In order to retain more information, we additionally add a convolution operation that can retain feature information differently from the pooling retention operation. This is beneficial in generating better spatial weights and enabling the model to better localize the target.

The feature maps (F) are fed into the spatial attention mechanism, which first generates feature maps ($F^s_{avg}, F^s_{max}, F^s_{conv}$) of sizes $1 \times H \times W$, $1 \times H \times W$, and $2 \times H \times W$ using global average pooling (GAP^s), global maximum pooling (GMP^s), and a convolutional layer ($Conv^s$) of size 1×1, respectively. Through the convolution and pooling operations, different features can be extracted and more information can be retained. These feature maps are then concatenated together and fed into a convolutional layer (Conv) of size 7×7 to fuse different types of feature information. After that, the spatial weights ($W_s(F)$) are generated using the sigmod function (σ), and then the Hadamard product ($\otimes$) is computed using the input feature map to generate a spatially weighted feature map (F_s) [34]. The process is illustrated in Equation (2). The structure diagram is shown in Figure 6.

$$\begin{aligned} F_{avg}^{s} &= GAP^{s}(F) \\ F_{max}^{s} &= GMP^{s}(F) \\ F_{conv}^{s} &= Conv^{s}(F) \\ W_s(F) &= \sigma(Conv([F_{avg}^{s}, F_{max}^{s}, F_{conv}^{s}])) \\ F_s &= W_s(F) \otimes F \end{aligned} \quad (2)$$

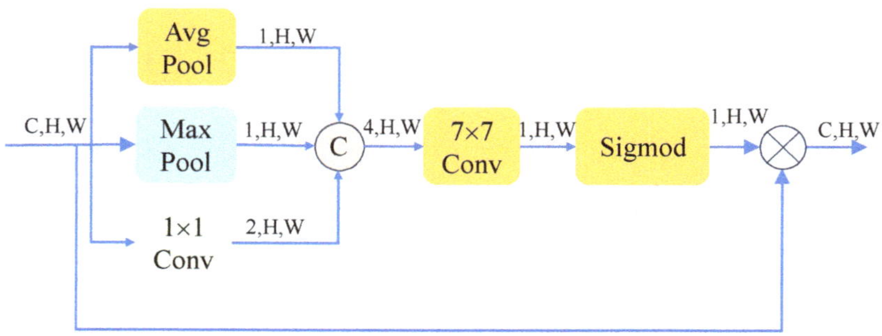

Figure 6. Spatial attention structure.

MCSAM uses channel and spatial attention in tandem, where the input feature maps (F) are first fed into the channel attention mechanism to generate channel-weighted feature maps (F_1) and later into the spatial attention mechanism to generate spatially weighted feature maps (F_2). The formula for the entire MCSAM is shown in Equation (3). By varying the parameters of the convolutional layer, the weights generated by the attention mechanism can be changed and therefore increase the visibility of the area containing the image, thereby improving the perception and discrimination abilities of the model.

$$\begin{aligned} F_1 &= W_c(F) \otimes F \\ F_2 &= W_s(F_1) \otimes F_1 \end{aligned} \quad (3)$$

2.2.4. Loss Function

Due to the small sizes and low numbers of IDSTs, they comprise only a small portion of an image, and the sum of target pixels as positive samples is much less than the sum of background pixels as negative samples. Therefore, when using infrared images to train the model, there is a very serious imbalance in positive and negative samples, which leads to a decrease in the model's ability to recognize the target category, and it can easily misclassify the target as background.

For this reason, we use the sum of focal loss and dice loss as the loss function of the algorithm. When calculating the value of the loss function, different weights are separately added for different samples such that each of these samples have a roughly equal share in the loss function during training. As a result, the model can learn the different features of different samples simultaneously in becoming fully trained, thus reducing the possibility of the algorithm predicting all samples as negative.

Dice loss (DL) [37] is a region-dependent loss function, where the value of the loss function is independent of the whole image and is related only to the intersection and concatenation of the actual and predicted target regions. The formula for DL is shown in Equation (4):

$$DiceLoss = 1 - \frac{2TP + s}{2TP + FP + FN + s} \tag{4}$$

Here, TP represents true positive, FP represents false positive, FN represents false negative, and s takes the value 1×10^{-5} to avoid having a denominator of 0.

Focal loss (FL) [38] is a loss function specialized in solving the problem of too many negative samples in the training data. The formula for FL is shown in Equation (5).

$$FocalLoss = -\alpha(1-p)^{\gamma} y \lg(p) - (1-\alpha)(1-y)p^{\gamma} \lg(1-p) \tag{5}$$

where α is an adjustable balancing parameter that regulates the proportion of different samples in the loss function. γ is a regulatory factor used to control the weight difference between samples that are easy to classify and those that are difficult to classify. p represents the prediction probability, wherein the closer p is to 0 or 1, the easier it is to categorize. y is the true labeling, where 1 indicates the target and 0 indicates the background.

DL focuses on the overall target, while FL focuses on individual pixels, so the final loss function is

$$Loss = DL + FL. \tag{6}$$

3. Results

3.1. Evaluation Metrics

The probability of detection (P_d) and false alarm rate (F_a) were used to assess whether the algorithm can accurately detect the target, and IoU was used to estimate whether the algorithm can retain the shape of the target. For these three metrics, we used a fixed threshold of 0.5. In addition, ROC curves were used to evaluate whether the algorithm can accurately detect the target under dynamic thresholds [39].

Probability of detection (P_d) reflects the ability to correctly detect targets and is the ratio of the sum of correctly detected targets $T_{correct}$ to the actual sum of targets T_{act}. Its formula is shown in Equation (7):

$$P_d = \frac{T_{correct}}{T_{act}} \tag{7}$$

The false alarm rate (F_a) reflects the accuracy of the algorithm in detecting the target and is the ratio of the sum of false predicted pixels P_{false} to the sum of pixels in the whole image P_{All}. It is defined by the formula shown in Equation (8):

$$F_a = \frac{P_{false}}{P_{All}} \tag{8}$$

IoU reflects the degree of shape resemblance between the predicted and actual targets and is the ratio of the intersection and union of the two (intersection/union of the two). It takes a value between 0 and 1, where 0 means there is no overlap at all, and 1 means there is perfect overlap. The calculation formula is

$$IoU = \frac{TP}{TP + FP + FN}. \tag{9}$$

where TP represents true positive, FP represents false positive, and FN represents false negative.

The ROC curve represents the classification effect of a classifier under different thresholds; specifically, the curve from left to right can be thought of as a change in threshold from 0 to 1. Its vertical axis is the true positive rate (TPR) and its horizontal axis is the false positive rate (FPR). The closer the curve is to the coordinates (0, 1), the better the performance of the algorithm. The TPR and FPR are calculated as follows:

$$FPR = \frac{FP}{N}$$
$$TPR = \frac{TP}{N} \tag{10}$$

where N is the sum of pixels in the whole image, TP represents true positive, and FP represents false positive.

3.2. Implementation Details

For the proposed network MFEU-Net, we performed ablation experiments and comparisons with other algorithms using three publicly available datasets: SIRST [40], MFIRST [25], IRSTD-1k [41]. We used an NVIDIA RTX A6000 (48 GB memory) for our graphics cards, and the algorithms were all based on a Pytorch neural network framework.

The training set image size of MFIRST is 128 × 128, and the batch size (BS) can be up to 128 on the A6000, but in order to avoid it being too large such that it would negatively impact the model, we set the BS to 32, the epoch to 100, and the learning rate (LR) to 1×10^{-5}. The test set image size of MFIRST is not fixed, so we set the BS as 1.

There are 427 images in the SIRST dataset, which is separated into a training set and a test set with 332 and 85 images, respectively. The image size is not fixed in the SIRST data, so we resized all the images in the training set to 320 × 320, and the size of the images in the test set was kept unchanged. For training, we set the BS, epoch, and LR to 8, 100, and 1×10^{-5}, respectively. For testing, the BS was 1.

There are 1001 images in the IRSTD-1k dataset, which is separated into a training set and a test set with 901 and 100 images, respectively. The image sizes in the SIRST data are all 512 × 512. For the training, we set the BS, epoch, and LR to 8, 100, and 1×10^{-5}, respectively. For testing, the BS was 8.

3.3. Ablation Study

To validate the effectiveness of our proposed algorithm, we performed an ablation experiment on the aforementioned dataset. Specifically, the performance of networks using different backbones was compared with the overall structure unmodified, and the performance of networks with and without the attention mechanism was compared with all other structures unchanged. For each comparison experiment, we ensured that the structure of the other parts remained the same.

3.3.1. Different Backbones

We compared the detection performance of networks using classical residual networks and networks using RSU without Reduced Downsampling Times (RSURD). A comparison of their specific performance metrics is shown in Table 2. It can be found that the P_d of MFEU-Net was higher than that of the network using RSURD, while the P_d of the network using RSURD was higher than that of the network using ResNet. Our proposed multiscale feature extraction network can extract rich multiscale information, and our algorithm can retain more detail for this information compared with RSURD, thereby outperforming RSURD in different quantitative metrics. Compared with the single-stream ResNet, the detection effect of the model can be substantially improved by multiscale feature extraction. MFEU-Net achieved the highest P_d and the lowest F_a, which demonstrates that our proposed backbone network of RSU combined with Inception is able to extract more information about different features, enabling the model to detect targets of different sizes in different scenarios.

Table 2. Comparison of quantitative metrics for the different backbone networks. The best of these metrics are shown in red bold font.

Backbone	MFIRST Dataset			SIRST Dataset		
	P_d	F_a	IoU	P_d	F_a	IoU
RSU+Inception	**0.864**	**3.08×10^{-5}**	**0.514**	**0.963**	**2.61×10^{-6}**	**0.671**
RSURD	0.8	7.22×10^{-5}	0.463	0.935	1.42×10^{-4}	0.585
ResNet	0.764	4.08×10^{-5}	0.444	0.915	6.89×10^{-5}	0.506

3.3.2. Attention Mechanism

We compared the detection performance of networks without and using MCSAM, and the specific indicators are shown in Table 3. It is obvious from the different evaluation metrics that the networks that used attention mechanisms outperformed those that did not. The above analysis clearly demonstrates that our proposed MCSAM can effectively determine the IDST location, which demonstrates the necessity of introducing MCSAM.

Table 3. Comparison of detection performance of the different backbone networks.The best of these metrics are shown in red bold font.

Attention	MFIRST Dataset			SIRST Dataset		
	P_d	F_a	IoU	P_d	F_a	IoU
With attention	**0.864**	**3.08×10^{-5}**	**0.514**	**0.963**	**2.61×10^{-6}**	**0.6714**
Without attention	0.714	6.32×10^{-5}	0.393	0.88	4.54×10^{-5}	0.487

3.4. Comparison to State-of-the-Art Methods

We selected different algorithms for comparison, including Infrared Patch Image (IPI) [42], MPCM [21], FKRW [43], MDvsFA cGAN (MDFA) [25], Dense Nested Attention Network (DNA) [26], Infrared Small Target Detection U-Net (ISTDU) [28], Local Patch Network with Global Attention (LPNet) [27], and Multiscale Local Contrast Learning (MLCL) networks [29].

3.4.1. Quantitative Comparison

The quantitative metrics for these algorithms are shown in Table 4. The best of these quantitative metrics are shown in red bolded font and the second best in blue font. Overall, thanks to the feature representation capability, the quantization metrics for the deep learning-based algorithms were significantly higher than the traditional algorithms.

The MPCM algorithm is very sensitive to edges and drastic grayscale changes, so it could detect most of the targets and had a high P_d; however, it also had a high F_a, one of the highest among the evaluated algorithms. The FKRW algorithm removes the edges and noise in an image but also part of the detail information, so the P_d of this algorithm was relatively low. The IPI algorithm achieved better F_a and P_d compared to the other two conventional algorithms. However, its detection efficacy depends on the sparsity of the targets, which is affected when there are multiple targets in the image. This is also illustrated by the fact that the IPI algorithm did not achieve as good a P_d in the IRSTD-1k dataset as in the other two datasets.

ISTDU groups feature maps and enhances the weights of IDST feature map groups to improve IDST characterization, but it uses the mean squared error (MSE) as the loss function, and due to the imbalance between positive and negative samples, it tends to predict the target as background, so its detection rate was not very high. DNA can make full use of contextual information through a large number of jump connections, but it does not have an attention mechanism, so its detection performance was not very good. MDFA uses two generators responsible for the P_d and F_a, respectively, and its P_d was very high. However, its network is relatively simple and cannot adapt to complex scenarios, and its F_a

was also high. MLCL uses a combination of convolutional-layer and dilated-convolutional-layer approaches to learn local contrast feature information, but the dilation is too large to lead to the grid effect, resulting in the target being easily lost, so its detection rate was very low. LPNet can extract global and local information at the same time, which can improve the detection effect of the algorithm, but the target becomes fuzzy in post-processing, so the IoU was not high.

The deep learning algorithms proposed in this paper outperformed the other methods The proposed algorithm achieved the lowest F_a, highest IoU, and high P_d on the MFIRST dataset. It also achieved the best P_d, F_a, and IoU on the SIRST and IRSTD-1k datasets. Our algorithm also outperformed others in terms of ROC curves on different datasets, as shown in Figure 7. Taken together, our algorithm outperformed the other algorithms.

Table 4. Comparison of quantitative metrics for the different algorithms on different datasets. The best of these metrics are shown in red bold font, and the second-best metrics are shown in blue font.

Method	MFIRST Dataset			SIRST Dataset			IRSTD-1k Dataset		
	P_d	F_a	IoU	P_d	F_a	IoU	P_d	F_a	IoU
IPI	0.861	3.86×10^{-4}	0.411	0.923	2.22×10^{-3}	0.532	0.75	3.15×10^{-5}	0.469
MPCM	0.828	9.58×10^{-3}	0.402	0.945	1.30×10^{-2}	0.120	0.956	6.09×10^{-3}	0.483
FKRW	0.607	4.82×10^{-4}	0.233	0.814	3.43×10^{-4}	0.229	0.709	1.31×10^{-4}	0.235
ISTDU	0.828	3.67×10^{-4}	0.439	0.954	1.07×10^{-4}	0.470	0.780	2.41×10^{-4}	0.563
DNA	0.692	2.35×10^{-4}	0.351	0.889	2.63×10^{-4}	0.46436	0.815	1.84×10^{-5}	0.611
MDFA	**0.928**	5.94×10^{-3}	0.445	0.917	2.82×10^{-4}	0.579	0.962	1.86×10^{-4}	0.610
MLCL	0.478	9.46×10^{-5}	0.251	0.565	1.65×10^{-5}	0.350	0.808	2.81×10^{-5}	0.616
LPNet	0.785	9.39×10^{-4}	0.247	0.929	8.89×10^{-5}	0.577	0.621	1.64×10^{-4}	0.320
Ours	0.864	$\mathbf{3.08 \times 10^{-5}}$	**0.514**	**0.962**	$\mathbf{2.61 \times 10^{-6}}$	**0.671**	**0.965**	$\mathbf{1.81 \times 10^{-5}}$	**0.630**

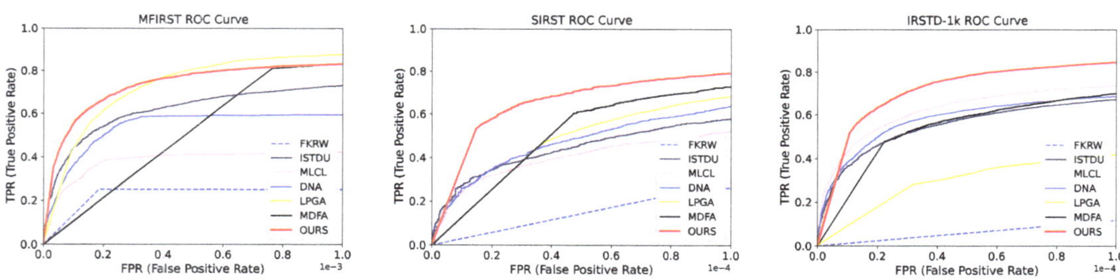

(**a**) (**b**) (**c**)

Figure 7. ROC curves of different algorithms. The ROC performance of IPI and MPCM was too poor to be shown in the figure. (**a**) ROC curves of different algorithms with the MFIRST dataset. (**b**) ROC curves of different algorithms with the SIRST dataset. (**c**) ROC curves of different algorithms with the IRSTD-1k dataset.

3.4.2. Visual Comparison

Some visualization examples of the MFIRST, SIRST, and IRSTD-1k datasets are shown in Figures 8–11, 12–15, 16–19, respectively. The yellow circles in the images indicate false alarms, and the red circle indicates leakage detection. We zoomed in on the target, which is displayed in the white box in the corner of the images, and when there were multiple targets, a blue dotted line is used to show the correspondence between the target and its zoomed-in image.

Among the traditional algorithms, IPI had a high detection rate, but the false alarm rate was also higher; the FKRW algorithm resulted in some leakage detection, and noise was introduced at the bottom edge of the image; the MPCM algorithm was very sensitive to boundary changes, had the highest false alarm rate, and had difficulty discriminating between the target and false alarms, so the detection effect figure for MPCM is not included. Overall, the traditional algorithms did not exhibit as good detection performance as the deep learning algorithms due to their reliance on a priori knowledge and lack of generalization ability.

Among the deep learning algorithms, MLCL had fewer false alarms but more false alarms; MDFA has few false alarms but many false alarms, even worse than the traditional IPI algorithm; ISTDU and DNA could detect all the targets but had false alarms to different degrees; LPNet could accurately detect all targets, but the target became blurred and less information was retained following subsequent processing. Thanks to the ability of our MCSAM to better localize the target area and our algorithm's advantages in extracting different features, our algorithm achieved the best detection results. Compared to other deep learning algorithms, our proposed algorithm could accurately detect all targets and achieved the lowest leakage and false alarm rates.

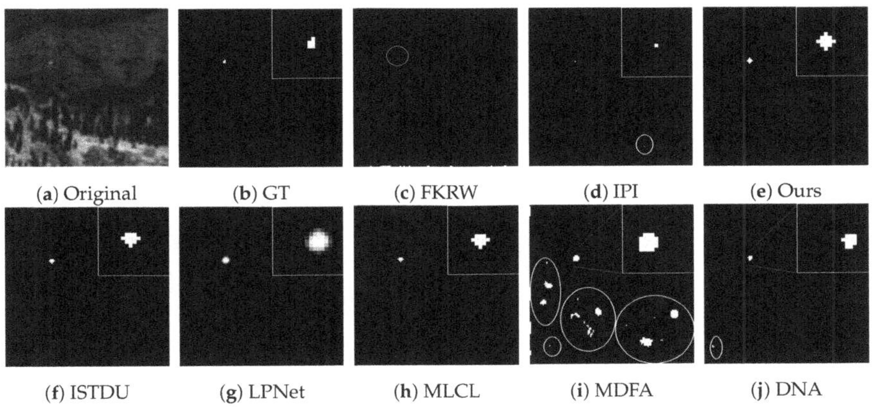

Figure 8. Visual example one of some representative methods for the MFIRST dataset.

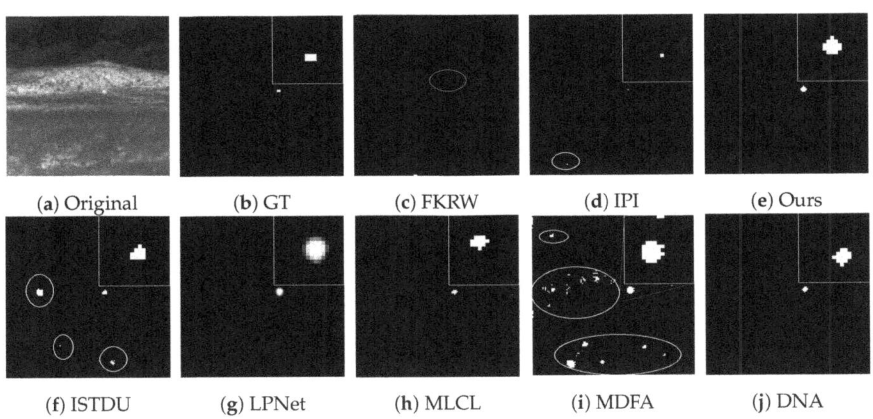

Figure 9. Visual example two of some representative methods for the MFIRST dataset.

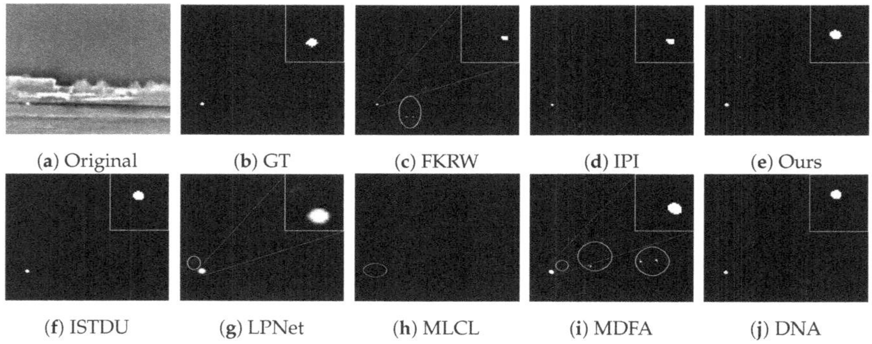

Figure 10. Visual example three of some representative methods for the MFIRST dataset.

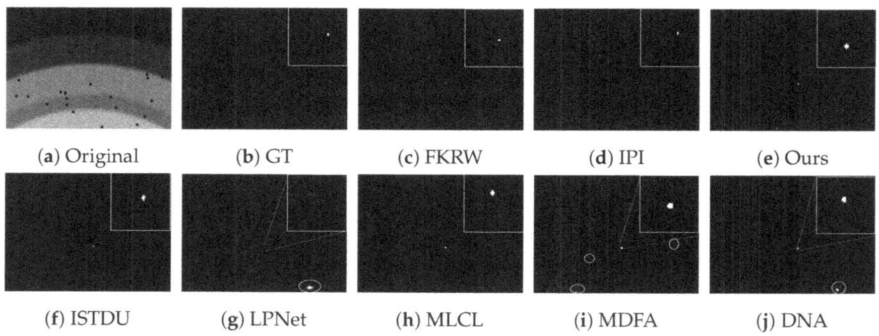

Figure 11. Visual example four of some representative methods for the MFIRST dataset.

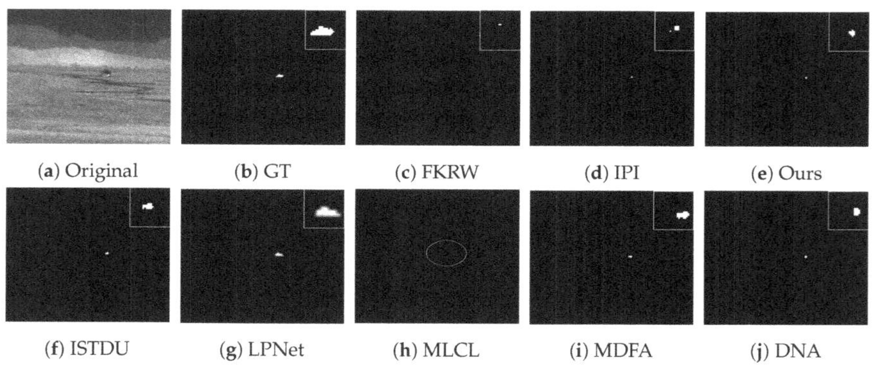

Figure 12. Visual example one of some representative methods for the SIRST dataset.

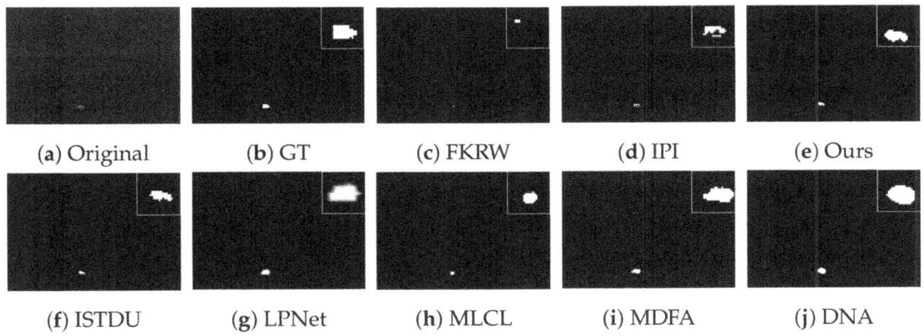

Figure 13. Visual example two of some representative methods for the SIRST dataset.

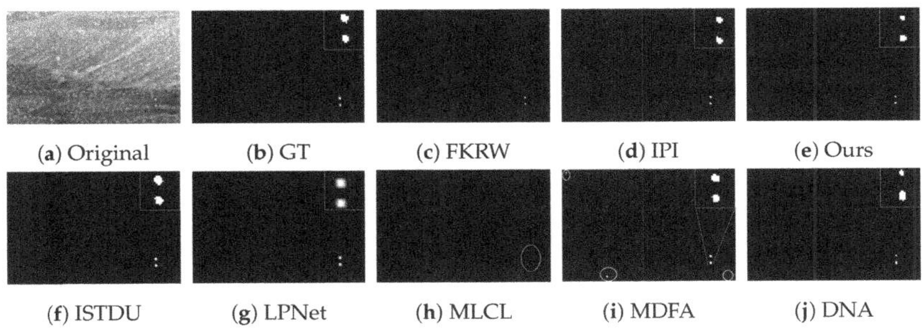

Figure 14. Visual example three of some representative methods for the SIRST dataset.

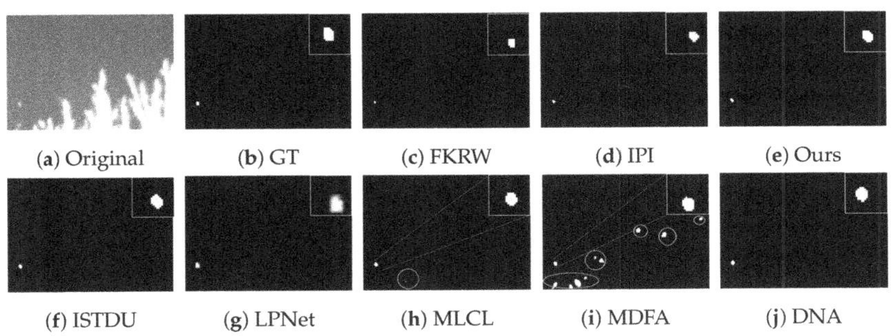

Figure 15. Visual example four of some representative methods for the SIRST dataset.

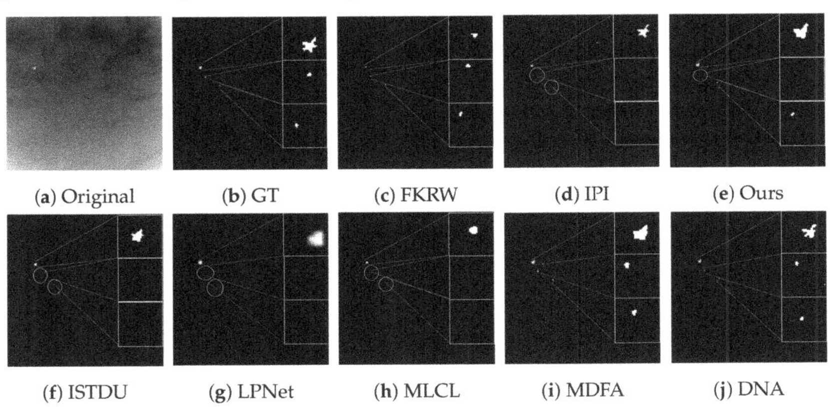

Figure 16. Visual example one of some representative methods for the IRSTD-1k dataset.

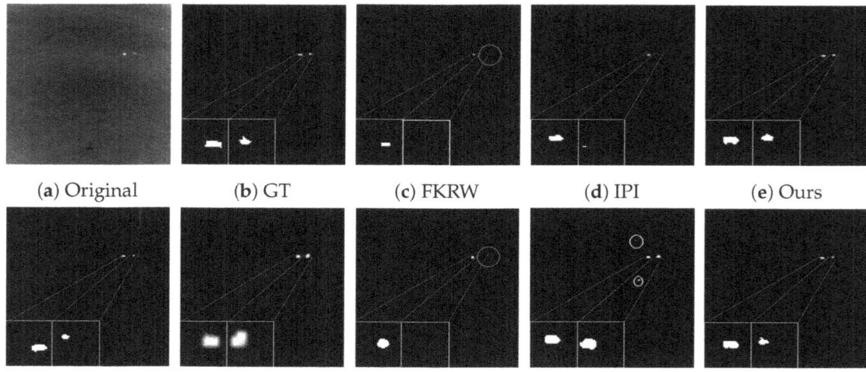

(**a**) Original (**b**) GT (**c**) FKRW (**d**) IPI (**e**) Ours

(**f**) ISTDU (**g**) LPNet (**h**) MLCL (**i**) MDFA (**j**) DNA

Figure 17. Visual example two of some representative methods for the IRSTD-1k dataset.

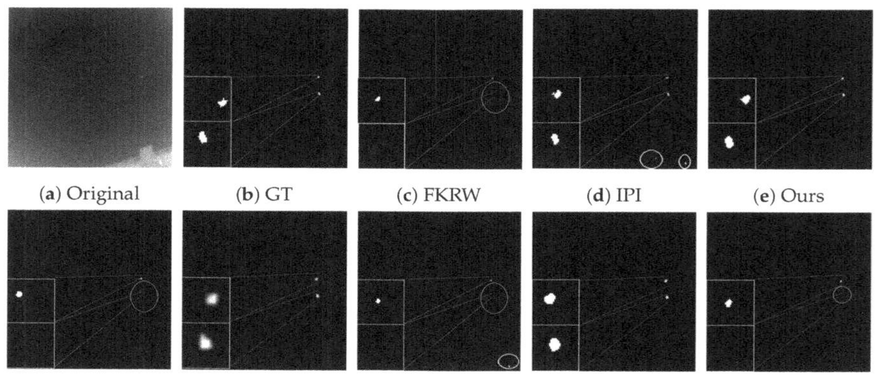

(**a**) Original (**b**) GT (**c**) FKRW (**d**) IPI (**e**) Ours

(**f**) ISTDU (**g**) LPNet (**h**) MLCL (**i**) MDFA (**j**) DNA

Figure 18. Visual example three of some representative methods for the IRSTD-1k dataset.

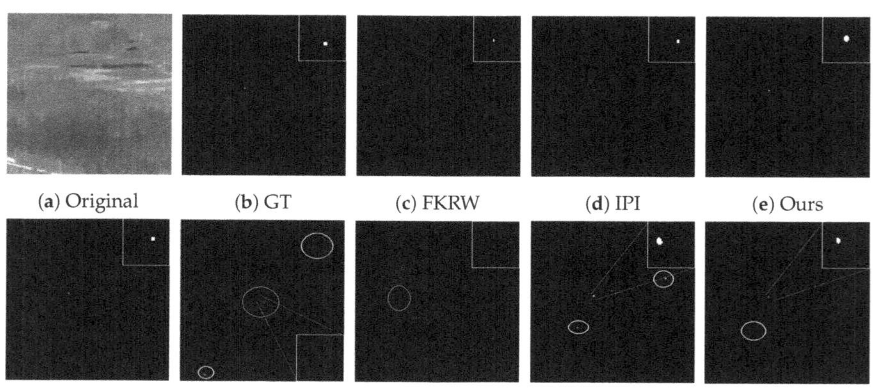

(**a**) Original (**b**) GT (**c**) FKRW (**d**) IPI (**e**) Ours

(**f**) ISTDU (**g**) LPNet (**h**) MLCL (**i**) MDFA (**j**) DNA

Figure 19. Visual example four of some representative methods for the IRSTD-1k dataset.

4. Discussion

Considering both the above quantitative and visual comparisons, the deep learning algorithms generally outperformed the traditional algorithms in terms of rates of detection and missed detection. Deep learning algorithms can extract rich feature information and automatically learn the features of a dataset through training, thus improving the detection efficacy of the algorithms, whereas traditional algorithms rely on a priori knowledge and can only be adapted to specific scenarios, making it difficult to detect targets in complex backgrounds.

Our algorithm can have different receptive fields through multiscale feature extraction, which improves its ability to adapt to targets of different sizes, including very small targets, and a high detection rate can be achieved. By using MCSAM, global information can be extracted and the target area can be made more prominent, thus improving detection in complex scenes and helping to achieve extremely low false alarm rates. In terms of quantitative metrics, our algorithm outperformed other state-of-the-art algorithms: we achieved the highest detection rate, the lowest false alarm rate, and the highest IoU values with different datasets; moreover, our ROC curve was closest to the upper left. Ablation and comparison experiments with different data demonstrate that our proposed amendments can effectively improve the detection performance of the algorithm.

5. Conclusions

In this paper, we present our proposed multiscale feature extraction U-Net network called MFEU-Net. MFEU-Net uses RSU and Inception as the encoder and decoder and extracts rich multiscale feature information through skip connections and a parallel branching structure, which enables the network to have different receptive field sizes at different layers. In addition, through MCSAM, weighting is performed in the channel and spatial domains separately, so the model can automatically learn the key patterns and features in the data, thereby focusing on the important regions in the feature map and thus improving its performance. In the experiments with different datasets, MFEU-Net achieved better detection results, demonstrating its effectiveness and that the changes result in an advancement.

Author Contributions: Conceptualization, X.W. (Xiaozhen Wang); methodology, X.W. (Xiaozhen Wang); software, X.W. (Xiaozhen Wang) and M.L.; validation, X.W. (Xiaofeng Wang), C.H.; writing—original draft preparation, X.W. (Xiaozhen Wang) and J.L.; writing—review and editing, X.W. (Xiaozhen Wang), T.N., M.L. and L.H. All authors have read and agreed to the published version of the manuscript.

Funding: This work was supported by the National Natural Science Foundation of China (No. 62105328).

Data Availability Statement: The SIRST, MFIRST, and IRSTD-1k image data used to support the research are available from the websites https://github.com/YimianDai/sirst, accessed on 29 July 2020, https://github.com/wanghuanphd/MDvsFAcGAN, accessed on 4 December 2019, https://github.com/RuiZhang97/ISNet, accessed on 20 March 2022.

Acknowledgments: The authors would like to thank D.Y., W.H., and Z.M. for providing the data.

Conflicts of Interest: The authors declare no conflicts of interest.

References

1. Zhang, K.; Ni, S.; Yan, D.; Zhang, A. Review of dim small target detection algorithms in single-frame infrared images. In Proceedings of the 2021 IEEE 4th Advanced Information Management, Communicates, Electronic and Automation Control Conference (IMCEC), Chongqing, China, 18–20 June 2021; Volume 4, pp. 2115–2120.
2. Wang, W.; Xiao, C.; Dou, H.; Liang, R.; Yuan, H.; Zhao, G.; Chen, Z.; Huang, Y. CCRANet: A Two-Stage Local Attention Network for Single-Frame Low-Resolution Infrared Small Target Detection. *Remote Sens.* **2023**, *15*, 5539. [CrossRef]
3. Eysa, R.; Hamdulla, A. Issues on Infrared Dim Small Target Detection and Tracking. In Proceedings of the 2019 International Conference on Smart Grid and Electrical Automation (ICSGEA), Xiangtan, China, 10–11 August 2019; pp. 452–456. [CrossRef]
4. Hao, X.; Liu, X.; Liu, Y.; Cui, Y.; Lei, T. Infrared Small-Target Detection Based on Background-Suppression Proximal Gradient and GPU Acceleration. *Remote Sens.* **2023**, *15*, 5424. [CrossRef]

5. Rawat, S.S.; Verma, S.K.; Kumar, Y. Review on recent development in infrared small target detection algorithms. *Procedia Comput. Sci.* **2020**, *167*, 2496–2505. [CrossRef]
6. Hou, Q.; Wang, Z.; Tan, F.; Zhao, Y.; Zheng, H.; Zhang, W. RISTDnet: Robust infrared small target detection network. *IEEE Geosci. Remote Sens. Lett.* **2021**, *19*, 7000805. [CrossRef]
7. Wang, Y.; Cao, L.; Su, K.; Dai, D.; Li, N.; Wu, D. Infrared Moving Small Target Detection Based on Space–Time Combination in Complex Scenes. *Remote Sens.* **2023**, *15*, 5380. [CrossRef]
8. Marvasti, F.S.; Mosavi, M.R.; Nasiri, M. Flying small target detection in IR images based on adaptive toggle operator. *IET Comput. Vis.* **2018**, *12*, 527–534. [CrossRef]
9. Chen, Y.; Li, L.; Liu, X.; Su, X. A multi-task framework for infrared small target detection and segmentation. *IEEE Trans. Geosci. Remote Sens.* **2022**, *60*, 5003109 . [CrossRef]
10. Kim, S.; Song, W.J.; Kim, S.H. Double weight-based SAR and infrared sensor fusion for automatic ground target recognition with deep learning. *Remote Sens.* **2018**, *10*, 72. [CrossRef]
11. Kwan, C.; Chou, B.; Yang, J.; Tran, T. Deep learning based target tracking and classification for infrared videos using compressive measurements. *J. Signal Inf. Process.* **2019**, *10*, 167. [CrossRef]
12. Ju, M.; Luo, J.; Liu, G.; Luo, H. ISTDet: An efficient end-to-end neural network for infrared small target detection. *Infrared Phys. Technol.* **2021**, *114*, 103659. [CrossRef]
13. Yao, J.; Xiao, S.; Deng, Q.; Wen, G.; Tao, H.; Du, J. An Infrared Maritime Small Target Detection Algorithm Based on Semantic, Detail, and Edge Multidimensional Information Fusion. *Remote Sens.* **2023**, *15*, 4909. [CrossRef]
14. Kim, J.H.; Hwang, Y. GAN-based synthetic data augmentation for infrared small target detection. *IEEE Trans. Geosci. Remote Sens.* **2022**, *60*, 5002512. [CrossRef]
15. Zhang, M.; Yang, H.; Yue, K.; Zhang, X.; Zhu, Y.; Li, Y. Thermodynamics-Inspired Multi-Feature Network for Infrared Small Target Detection. *Remote Sens.* **2023**, *15*, 4716. [CrossRef]
16. Zuo, Z.; Tong, X.; Wei, J.; Su, S.; Wu, P.; Guo, R.; Sun, B. AFFPN: Attention fusion feature pyramid network for small infrared target detection. *Remote Sens.* **2022**, *14*, 3412. [CrossRef]
17. Deshpande, S.D.; Er, M.H.; Venkateswarlu, R.; Chan, P. Max-mean and max-median filters for detection of small targets. In Proceedings of the Signal and Data Processing of Small Targets 1999, Denver, CO, USA, 4 October 1999; Volume 3809, pp. 74–83.
18. Starck, J.L.; Candès, E.J.; Donoho, D.L. The curvelet transform for image denoising. *IEEE Trans. Image Process.* **2002**, *11*, 670–684. [CrossRef] [PubMed]
19. Chen, C.P.; Li, H.; Wei, Y.; Xia, T.; Tang, Y.Y. A local contrast method for small infrared target detection. *IEEE Trans. Geosci. Remote Sens.* **2013**, *52*, 574–581. [CrossRef]
20. Han, J.; Moradi, S.; Faramarzi, I.; Liu, C.; Zhang, H.; Zhao, Q. A local contrast method for infrared small-target detection utilizing a tri-layer window. *IEEE Geosci. Remote Sens. Lett.* **2019**, *17*, 1822–1826. [CrossRef]
21. Wei, Y.; You, X.; Li, H. Multiscale patch-based contrast measure for small infrared target detection. *Pattern Recognit.* **2016**, *58*, 216–226. [CrossRef]
22. Sun, Y.; Yang, J.; An, W. Infrared dim and small target detection via multiple subspace learning and spatial-temporal patch-tensor model. *IEEE Trans. Geosci. Remote Sens.* **2020**, *59*, 3737–3752. [CrossRef]
23. Zhang, L.; Peng, L.; Zhang, T.; Cao, S.; Peng, Z. Infrared small target detection via non-convex rank approximation minimization joint l 2, 1 norm. *Remote Sens.* **2018**, *10*, 1821. [CrossRef]
24. Baili, N.; Moalla, M.; Frigui, H.; Karem, A.D. Multistage approach for automatic target detection and recognition in infrared imagery using deep learning. *J. Appl. Remote Sens.* **2022**, *16*, 048505. [CrossRef]
25. Wang, H.; Zhou, L.; Wang, L. Miss detection vs. false alarm: Adversarial learning for small object segmentation in infrared images. In Proceedings of the IEEE/CVF International Conference on Computer Vision, Seoul, Republic of Korea, 27 October–2 November 2019; pp. 8509–8518.
26. Li, B.; Xiao, C.; Wang, L.; Wang, Y.; Lin, Z.; Li, M.; An, W.; Guo, Y. Dense nested attention network for infrared small target detection. *IEEE Trans. Image Process.* **2022**, *32*, 1745–1758. [CrossRef] [PubMed]
27. Chen, F.; Gao, C.; Liu, F.; Zhao, Y.; Zhou, Y.; Meng, D.; Zuo, W. Local patch network with global attention for infrared small target detection. *IEEE Trans. Aerosp. Electron. Syst.* **2022**, *58*, 3979–3991. [CrossRef]
28. Hou, Q.; Zhang, L.; Tan, F.; Xi, Y.; Zheng, H.; Li, N. ISTDU-Net: Infrared Small-Target Detection U-Net. *IEEE Geosci. Remote Sens. Lett.* **2022**, *19*, 7506205. [CrossRef]
29. Yu, C.; Liu, Y.; Wu, S.; Hu, Z.; Xia, X.; Lan, D.; Liu, X. Infrared small target detection based on multiscale local contrast learning networks. *Infrared Phys. Technol.* **2022**, *123*, 104107. [CrossRef]
30. Li, R.; Shen, Y. YOLOSR-IST: A deep learning method for small target detection in infrared remote sensing images based on super-resolution and YOLO. *Signal Process.* **2023**, *208*, 108962. [CrossRef]
31. Zhang, M.; Zhang, R.; Zhang, J.; Guo, J.; Li, Y.; Gao, X. Dim2Clear network for infrared small target detection. *IEEE Trans. Geosci. Remote Sens.* **2023**, *61*, 5001714. [CrossRef]
32. Hu, J.; Shen, L.; Sun, G. Squeeze-and-excitation networks. In Proceedings of the IEEE Conference on Computer Vision and Pattern Recognition, Salt Lake City, UT, USA, 18–23 June 2018; pp. 7132–7141.

33. Wang, Q.; Wu, B.; Zhu, P.; Li, P.; Zuo, W.; Hu, Q. ECA-Net: Efficient channel attention for deep convolutional neural networks. In Proceedings of the IEEE/CVF Conference on Computer Vision and Pattern Recognition, Seattle, WA, USA, 13–19 June 2020; pp. 11534–11542.
34. Woo, S.; Park, J.; Lee, J.Y.; Kweon, I.S. Cbam: Convolutional block attention module. In Proceedings of the European Conference on Computer Vision (ECCV), Munich, Germany, 8–14 September 2018; pp. 3–19.
35. Qin, X.; Zhang, Z.; Huang, C.; Dehghan, M.; Zaiane, O.R.; Jagersand, M. U2-Net: Going deeper with nested U-structure for salient object detection. *Pattern Recognit.* **2020**, *106*, 107404. [CrossRef]
36. Szegedy, C.; Ioffe, S.; Vanhoucke, V.; Alemi, A. Inception-ResNet and the impact of residual connections on learning. *arXiv* **2016**, arXiv:1602.07261.
37. Li, X.; Sun, X.; Meng, Y.; Liang, J.; Wu, F.; Li, J. Dice loss for data-imbalanced NLP tasks. *arXiv* **2019**, arXiv:1911.02855.
38. Lin, T.Y.; Goyal, P.; Girshick, R.; He, K.; Dollár, P. Focal loss for dense object detection. In Proceedings of the IEEE International Conference on Computer Vision, Venice, Italy, 22–29 October 2017; pp. 2980–2988.
39. Cheng, B.; Girshick, R.; Dollár, P.; Berg, A.C.; Kirillov, A. Boundary IoU: Improving object-centric image segmentation evaluation. In Proceedings of the IEEE/CVF Conference on Computer Vision and Pattern Recognition, Nashville, TN, USA, 20–25 June 2021; pp. 15334–15342.
40. Dai, Y.; Wu, Y.; Zhou, F.; Barnard, K. Asymmetric contextual modulation for infrared small target detection. In Proceedings of the IEEE/CVF Winter Conference on Applications of Computer Vision, Waikoloa, HI, USA, 3–8 January 2021; pp. 950–959.
41. Zhang, M.; Zhang, R.; Yang, Y.; Bai, H.; Zhang, J.; Guo, J. ISNet: Shape Matters for Infrared Small Target Detection. In Proceedings of the 2022 IEEE/CVF Conference on Computer Vision and Pattern Recognition (CVPR), New Orleans, LA, USA, 18–24 June 2022; pp. 867–876. [CrossRef]
42. Gao, C.; Meng, D.; Yang, Y.; Wang, Y.; Zhou, X.; Hauptmann, A.G. Infrared patch-image model for small target detection in a single image. *IEEE Trans. Image Process.* **2013**, *22*, 4996–5009. [CrossRef] [PubMed]
43. Qin, Y.; Bruzzone, L.; Gao, C.; Li, B. Infrared small target detection based on facet kernel and random walker. *IEEE Trans. Geosci. Remote Sens.* **2019**, *57*, 7104–7118. [CrossRef]

remote sensing

MDPI

Article

Infrared Small Dim Target Detection Using Group Regularized Principle Component Pursuit

Meihui Li [1,2,3,4], Yuxing Wei [1,2,3,4], Bingbing Dan [1,2,3,4], Dongxu Liu [1,2,3] and Jianlin Zhang [1,2,3,4,*]

1. Institute of Optics and Electronics, Chinese Academy of Sciences, Chengdu 610209, China; limeihui@ioe.ac.cn (M.L.); weiyuxing@ioe.ac.cn (Y.W.); danbingbing20@mails.ucas.ac.cn (B.D.); liudongxu18@mails.ucas.ac.cn (D.L.)
2. National Key Laboratory of Optical Field Manipulation Science and Technology, Chinese Academy of Sciences, Chengdu 610209, China
3. Key Laboratory of Optical Engineering, Chinese Academy of Sciences, Chengdu 610209, China
4. University of Chinese Academy of Sciences, Beijing 100049, China

* Correspondence: jlin@ioe.ac.cn

Abstract: The detection of an infrared small target faces the problems of background interference and non-obvious target features, which have yet to be efficiently solved. By employing the non-local self-correlation characteristic of the infrared images, the principle component pursuit (PCP)-based methods are demonstrated to be applicable to infrared small target detection in a complex scene. However, existing PCP-based methods heavily depend on the uniform distribution of the background pixels and are prone to generating a high number of false alarms under strong clutter situations. In this paper, we propose a group low-rank regularized principle component pursuit model (GPCP) to solve this problem. First, the local image patches are clustered into several groups that correspond to different grayscale distributions. These patch groups are regularized with a group low-rank constraint, enabling an independent recovery of different background regions. Then, GPCP model integrates the group low-rank components with a global sparse component to extract small targets from the background. Different singular value thresholds can be exploited for image groups corresponding to different brightness and grayscale variance, boosting the recovery of background clutters and also enhancing the detection of small targets. Finally, a customized optimization approach based on alternating direction method of multipliers is proposed to solve this model. We set three representative detection scenes, including the ground background, sea background and sky background for experiment analysis and model comparison. The evaluation results show the proposed model has superiority in background suppression and achieves better adaptability for different scenes compared with various state-of-the-art methods.

Keywords: infrared small target detection; principle component pursuit; group low-rank regularization; infrared patch-image model

Citation: Li, M.; Wei, Y.; Dan, B.; Liu, D.; Zhang, J. Infrared Small Dim Target Detection Using Group Regularized Principle Component Pursuit. *Remote Sens.* **2024**, *16*, 16. https://doi.org/10.3390/rs16010016

Academic Editor: Paolo Tripicchio

Received: 2 November 2023
Revised: 5 December 2023
Accepted: 14 December 2023
Published: 20 December 2023

1. Introduction

The infrared search system has the merits of working in all weather, all day and at long ranges, which is applicable to many important fields such as early-warning systems, aerospace technology, remote sensing [1–3], etc. In the moving process of the infrared target, it is easy for it to be submerged in high brightness clutter such as clouds, sea-sky-line, etc. In addition, the search system usually needs to detect long-range targets [4], which means the target size is very small and the useful signal is very weak. To adapt to these real-world scenarios, the detection algorithms should be designed to handle the interference of background clutter and achieve the effective extraction of the small ($<9 \times 9$ pixels) and weak (<3 SNR) targets.

Over the years, a plethora of small dim target detection algorithms have been proposed. From the perspective of image characteristic utilization, these algorithms can be categorized into three types: target characteristic-based method [5–8], background characteristic-based method [9–11] and target/background characteristic integration-based method [12–15]. Generally, the infrared small targets appear to have large gray-scale values and are prone to distribution in high-frequency areas. These properties are usually adopted by the target characteristic-based methods for potential target region extraction, such as the local contrast measure [5,16], local entropy measure [3] and frequency-domain saliency region segmentation [17]. These methods are adequate for their relatively uniform background and targets with high brightness. However, due to the lack of background modeling, the strong edges or background clutters can easily be detected as false alarms in the target characteristic-based methods. The background characteristic-based methods can avoid the confusion of the targets and background interference to some extent, in which the background pixels of adjacent image area and successive frames are assumed to be spatially consistent. According to this, the target is detected by removing the predicted background from the original image. However, background characteristic-based methods are not suitable for handling complex backgrounds due to the difficulty of background estimation. It has been extensively shown that using single-target or background information is not effective for detecting small targets in complex situations.

The infrared patch-image (IPI) model [12] is a representative target and background integration-based method. By employing the non-local self-correlation of the background, IPI transfers the small target detection problem to the recovery of a low-rank matrix and a sparse matrix, which correspond to the background component, and target component respectively. The target components are regarded as outliers that increase the rank value of the data matrix and can be efficiently separated by the PCP model. Accurate approximation of the matrix rank is a major difficulty for the IPI model. Overlarge rank is good for background interference suppression but is more likely to cause miss detection. On the contrary, a lower rank will introduce an enormous number of false alarms. Recently, many works concentrate on the approximation of the matrix ranks. Zhang et al. [18] propose the modification of the low-rank constraint of IPI to a tighter-rank surrogate— $l_{2,1}$ norm to remove the unpredictable background residuals. Zhu et al. use a smooth but nonconvex surrogate of the rank based on the Log operator [19], which is closer to the rank minimization optimization than the nuclear norm. In the NIPPS model [20] and PSTNN model [14], by using the partial sum of nuclear norms, the matrix and tensor ranks are approximated by the energy ratio of the principle matrix to adapt to the changeable background. Liu et al. proposed a non-convex tensor low-rank approximation (NTLA) method to adaptively assign different weights to different singular values [21]. The above-mentioned methods focus more on using different surrogates to replace the low-rank constraint. However, due to the intrinsic diversity of different local image regions, using one single low-rank constraint is difficult to describe the whole background. For the sake of a more accurate description model, it is necessary to explore the complexity variations of different local regions and assign different rank thresholds in reconstruction.

Considering the fact that the infrared image is nonuniform and its complexity varies spatially, we establish a novel group regularized PCP model, named as GPCP for the small target detection in complex scenes. The proposed model employs a group low-rank constraint to replace the previous global low-rank constraint in recovering the background component, and enforces using different number of principle components for image data groups corresponding to different complexity and brightness. By minimizing the group low-rank constraints of the GPCP model, more image details can be reconstructed, so that the residual errors can be eliminated from the target components. The contributions of this article can be summarized as follows:

- We analyze the low-rank property of the global data matrix and grouped data matrix, and find there is a significant difference of principle component number in recovering the data matrix with different complexity.

- We propose a group low-rank constraint for background recovery and combine it with a global sparse regularization term for target recovery, which can remove the residual errors in the target component efficiently.
- A customized optimization algorithm is adopted to solve the proposed GPCP model, in which the group low-rank components are decoupled by the ADMM algorithm.

The rest of the sections are organized as follows. In Section 2, some related works on the small target detection are briefly reviewed. Section 3 introduces the algorithm flowchart and implementation details of the proposed detection model. Section 4 gives the experimental results on different background situations to demonstrate the effectiveness of our proposed method. Section 5 concludes the whole paper and discusses future works.

2. Related Work

We briefly review the related work on the small target detection methods using target characteristics, background characteristics and integration idea.

2.1. Target Characteristic-Based Method

The target characteristic-based method mainly focuses on the distinction between the target and its surrounding background. Many representative methods have been proposed in this research branch, such as the local contrast measure (LCM) [5,22], entropy contrast measure (ECM) [23,24], sparse representation-based methods [25], and so on. This type of method utilizes the shape or statistical characteristics of the small target for target detection. However, due to the similar image property of the small target and the strong background edges, the background clutters could easily be mistaken for a target. To address the issue of false detection, relative methods have been proposed to enhance the target intensity while suppressing the background region, such as the weighted local difference measure (WLDM) [6], multiscale local homogeneity measure (MLHM) [7], self-regularized weighted sparse model (SRWS) [26] etc. In the recent studies, saliency features are also utilized to associate the gray intensity with the entropy [19,27] or frequency domain [3,17], which have gained better results in small target detection. It is noticeable that these methods are sensitive to the settings of the target size and window size, which are hard to balance without prior information.

2.2. Background Characteristic-Based Method

The background characteristic-based method is usually based on the assumption that the background pixels are highly correlated, and targets are the parts that break this relationship. So, many background characteristic-based methods study the background estimation algorithms by using the neighboring pixels [9,11,28]. For example, the difference of Gaussian (DoG) filter uses the weighted sum of local neighborhood pixels as the background [29]. To cope with the problem of edge sensitivity, many methods propose to add the orientation information for background estimation, such as the max-mean and max-median filters [30], in which the maximum values of the mean or median arrays of different lines is taken as the background. The above mentioned background characteristic-based methods are all based on the local estimation model, in addition, the estimation strategy usually selects the maximum value of different orientations, which is not accurately designed. Aiming at this problem, some researchers propose to adopt the transform domain information for background suppression [9,31]. In [9], the whole infrared image is transformed into the frequency domain, and the background component is suppressed by removing the low frequency component from the original image. However, this type of method cannot suppress the complex background because the strong edges also belong to high-frequency subbands.

2.3. Target/Background Characteristic Integration-Based Method

Recently, the low-rank sparse model, which integrates the target characteristic and background characteristic by image data decomposition, has achieved considerable advances in the small target detection area. In [12], Gao et al. presented an infrared patch image (IPI) model, in which the target component and background component are assumed to be sparse and low rank, respectively. Considering that equally weighted singular values will restrict the description ability of low rank nature for the background patches, Zhang et al. proposed to modify the nuclear norm regularization to a weighted nuclear norm [13], which makes the model more flexible for complex background. Afterwards, Dai et al. pointed out that when facing extremely complex background, the low rank assumption of IPI model has a mismatching problem, which may lead the strong edges to be considered as outliers [20]. To solve this problem, they adopted the partial sum of singular values to constrain the low-rank background instead of the nuclear norm. Similar idea has also been mentioned in [14], where the partial sum minimization constraint of singular values is extended to the patch-tensor model. In order to transfer the NP-hard problem of PCP model into a non-convex optimization problem, Zhang et al. proposed to apply schatten q-norm and l_p-norm to the small target detection area, which is named as NOLC model [32]. To enhance the detection accuracy, in [26], an overlapping edge information is applied to mine the structure information of background. Multiple frames-based models [33] are also reported for small target detection in complex scenes. In [34], Aliha et al. built a block-matching patch-tensor model based on the spatial–temporal domain to extract inter-frame information. Hu et al. further used a simultaneous sampling in spatial and temporal domains to make full use of the information between multiple frames [26]. Considering the target's local continuity in the spatial–temporal domain, Li et al. [35] proposed a spatial regularized spatial–temporal twist tensor model, which can reduce the global noise to some extent.

Recently, convolutional neural network (CNN) began to appear in the infrared small target detection study area. Du et al. [36] proposed a shallow–deep feature-based detection model, which demonstrates that shallow features are important for small target detection. Regarding the feature lacking problem, Bai et al. used a cross-connection bidirectional pyramid network to provide more comprehensive target information [37]. To cope with the miss detection problem, Liu et al. adopted the transformer to learn the correlation of image features in a larger range [38]. Among existing deep learning methods, feature learning still remains challenging due to the small size and non-obvious image features of the infrared small targets.

3. Proposed Small Target Detection Using GPCP

In this section, we first analyze the low-rank property of the patch-image data matrix, including the global data matrix, the bright-uniform part, the dark-uniform part and the cluttered part. Then, a group-regularized principle component pursuit model (GPCP) is constructed according to the diverse characteristics of the local image parts. Finally, the sparse component which includes the small target is separated from the complex background using the GPCP model, as shown in Figure 1. The algorithm steps and results of the traditional PCP model and the proposed GPCP model are also illustrated in Figure 1. Next, we will elaborate the details of the proposed small target detection model.

3.1. Low-Rank Property of Image Groups

The existing PCP-based models mainly focus on the low-rank structure of the global data matrix and ignore the inhomogeneous information among local background regions, which makes these models not suitable to handle complex scenes. As illustrated in the upper part of Figure 1, there are many background clutters (labeled by blue boxes) remaining in the sparse component after the global PCP based decomposition process. It could also be observed that the residual clutters are mainly distributed in the image parts with strong edges or big gray level changes. That is to say, such a decomposition model forms a confusion of the small targets and some background clutters. Since the background component is recovered by a low-rank constraint, the key problem is then transferred into how to determine the rank threshold in the PCP process.

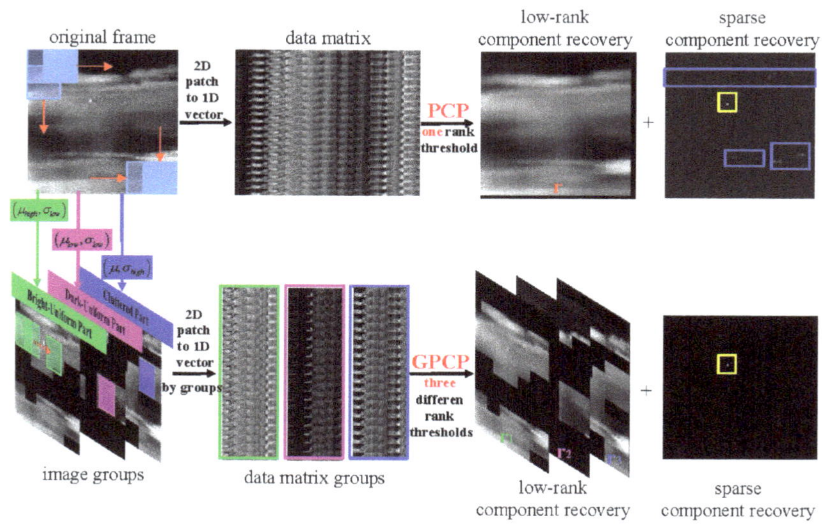

Figure 1. Illustration of the proposed GPCP model for small target detection.

To handle the aforementioned issue, we need to have a deeper understanding of the low-rank characteristics of different background parts. Figure 2 illustrates the eigenvalue curves of the global data matrix and the grouped local data matrix, which are generated by the distribution of data. To avoid the influence of matrix dimension on the result of eigenvalues, the column size of the global data matrix and the grouped data matrix is down-sampled to keep it the same. The X-axis of Figure 2c represents the number of principle components, which is defined as "rank threshold" in the optimization process. Y-axis represents the eigenvalues of data matrix. Here, we set 2 as the boundary of principle components, which means the eigenvectors whose corresponding eigenvalues are greater than 2 are regarded as principle components. From Figure 2c, we can see that the principle component of the global data matrix (red line) is concentrated in the top nine eigenvectors. For the uniform data matrix groups (green and pink lines), the number of principle components is about seven to eight. By comparison, the threshold value of the cluttered part (blue line) is 10, which is much larger than the other two uniform parts and is a bit larger than the global data matrix. This demonstrates that there is a significant difference on the low-rank characteristics among the bright-uniform part, the dark-uniform part and the cluttered part, which motivates us to consider whether we can use a group regularized PCP model to cope with the clutter interference problems in complex situations.

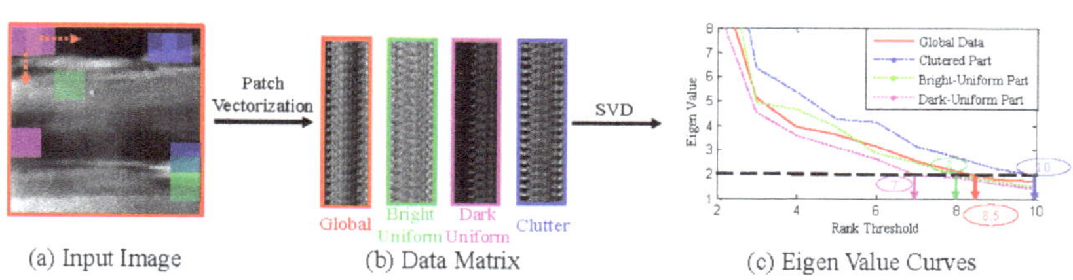

Figure 2. Low-rank property of entire data matrix and grouped data matrix. (**a**) Input images (**b**) data matrix (**c**) eigen value curves.

3.2. Construction of the GPCP Model

PCP is a convex model which aims to recover the low-rank matrix when the data matrix is corrupted by gross sparse errors [39] and is playing an important role in the recent patch-image based small target detection methods. Mathematically, it considers the data matrix $D \in R^{n_1 \times n_2}$ is composed of a low-rank component L and a sparse component S and solves the following convex optimization problem:

$$\arg\min_{L,S} \|L\|_* + \lambda \|S\|_0, s.t. \|D - L - S\| \leq \varepsilon \tag{1}$$

To recover L and S, the low-rank component L should be limited to the following three conditions:

$$\max_i \|U^* e_i\|^2 \leq \frac{\mu r}{n_1}, \max_i \|V^* e_i\|^2 \leq \frac{\mu r}{n_2}, \|UV^*\|_\infty \leq \sqrt{\frac{\mu r}{n_1 n_2}} \tag{2}$$

where $L = U\Sigma V^* = \sum_{i=1}^{r} \sigma_i \mu_i v_i^*$. By arranging an appropriate r, the L and S components can be efficiently separated after the PCP operation. Yet a unified r is not suitable to handle the overall data matrix since the image data always corresponds to different complexity. An extreme example is that the small target is located in the smooth background part, meanwhile strong edges exist in the other part of the background. When r is set to a small value, many residual errors will remain in the sparse component; when r is large, the real target will be regarded as the low-rank component. Therefore, the true reason causing missed detection and false detection lies in the data structure diversity of different background parts.

The newly designed GPCP model we consider in this paper assumes the low-rank component L satisfies a group low-rank structure, which is defined as follows:

$$rank_{group}(L) = \sum_k \mu_k \|L_k\|_* \tag{3}$$

where L_k represents the kth group of the low-rank component, μ_k is used to balance the image groups with different data number. In this way, each L_k is considered independent with each other and will correspond to different shrink thresholds r_k for decomposition. The eigen value curves of the data matrixes in Figure 2 also show that compared with the global data matrix, the group data matrix has a better property on the low-rank condition of PCP model. To recover the low-rank components $L_k (k = 1, 2, \ldots)$ and the spare component S, we need to solve the following GPCP model:

$$\begin{aligned} &\arg\min_{L,S} \sum_k \mu_k \|L_k\|_* + \lambda \|S\|_1 \\ &s.t. D = L + S + N, L = \{L_1, \ldots L_k, \ldots\} \end{aligned} \tag{4}$$

3.3. Small Target Detection Using GPCP

Typically, the small target detection model can be written as follows:

$$D = T + B + N \tag{5}$$

where D represents the input image, T, B and N represent the target, background and noise, respectively.

In this paper, we follow the basic idea of the infrared patch-image model [12] and denote D as a data matrix, which is composed of column-wise local patches of the input image. To explore the data structure of the background component, the image patch vectors with similar property on gray-scale variation should be clustered together. The complexity and gray level of an image are reflected by the variance value σ and the mean value μ, respectively. So, we employ (μ, σ) as the data feature descriptor.

Firstly, the image data can be divided into a clutter group and a uniform group according to σ. Then, for the uniform group, the bright part and the dark part also correspond to different image properties. The dissimilarity degree between two data samples is calculated by:

$$d_1 = |\mu_1 - \mu_1|, d_2 = |\sigma_1 - \sigma_1| \tag{6}$$

According to Equation (6), large d_1 and d_2 indicate a big difference between two samples. The k-means cluster algorithm is employed to divide the entire data into three groups: the bright-uniform part, the dark-uniform part and the cluttered part, which is shown in Figure 2. According to the previous discussion in Sections 3.1 and 3.2, the image data in different groups always corresponds to different low-rank structure and should be regularized separately. So, we propose to use the GPCP model to depict the background patch-image in complex scenes, which is defined as:

$$\begin{aligned} \|B\|_{g*} &= \sum_k \mu_k \|B_k\|_* \\ B &= \{B_1, B_2, \ldots, B_k\} \end{aligned} \tag{7}$$

where B_k represents the k-th group of the background data, μ_k is used to balance the image groups with different data number, which is defined as:

$$\mu_k = \frac{data\ \ number\ \ of\ \ group\ \ k}{total\ \ number} \tag{8}$$

Here, we use the group-regularized nuclear norm $\sum_k \mu_k \|B_k\|_*$ to approximate the rank property of the background component B, instead of $\|B\|_*$. So, the whole background is composed of the recovering of these separated image groups. Generally, the image groups containing strong edges and clutters will correspond to a large singular threshold, and the uniform image groups will correspond to a lower one. Compared with the previous detection method which uses one single low-rank constraint for the whole background component, the group low rank regularization can better explore the local structure of the image and lead to a more accurate decomposition result.

In the infrared images, small targets are usually randomly distributed in different groups. So, to keep the sparsity of the entire target component rather than the group component, we use a global sparse constraint for the whole target component T, which is defined as $\|T\|_1$. Therefore, the group IPI model is defined as follows:

$$\begin{aligned} &\arg\min_{L,S} \sum_k \mu_k \|B_k\|_* + \lambda \|T\|_1 \\ &s.t. D = B + T + N, B = \{B_1, B_2, \ldots, B_k\} \end{aligned} \tag{9}$$

3.4. Optimization Method of the GPCP Model

The objective function defined in Equation (9) is a convex problem which includes two variables B and T to be solved. It should be noticed that the background component B in Equation (9) is composed of several local groups and each group is independent of one another, which has a great difference compared with the traditional PCP model. In accordance with this complex situation, we adopt the ADMM algorithm to decouple the group principle component pursuit model into several sub-problems and alternatively optimize one variable while keeping others fixed. The augmented Lagrangian expression of Equation (9) can be rewritten as the following form:

$$\begin{aligned} L_\rho(B, T, F) = &\sum_k \mu_k \|B_k\|_* + \tfrac{\rho}{2} \|D - B - T\|_F^2 \\ &+ \lambda \|T\|_1 + \langle F, D - B - T \rangle \end{aligned} \tag{10}$$

where F represents the dual vector, $\rho > 0$ is the penalty parameter. The algorithm flow of ADMM is summarized in Algorithm 1.

Algorithm 1 ADMM (Alternating Direction Method of Multipliers) Algorithm for GPCP model

Input: group number K, regularization parameter λ, penalty parameter ρ, update factor for ρ: $\mu\rho$, maximum iteration max_iter, tolerance error tol.
while not converged **do**
1. Compute group background component B_k using $U_k diag\left(pos\left(\sigma_k - \frac{1}{\rho}\right)\right)V_k^T$;
2. Combine K group components $B_{1:K}$ into a global form B;
3. Compute target component T using $th_{\frac{2\lambda}{\rho}}\left(D - B + \frac{F}{\rho}\right)$;
4. Update dual factor F using $F^{t+1} = F^t + \rho^t(T^{t+1} + B^{t+1} - D)$;
5. Update penalty factor ρ using $\rho^{t+1} = \mu\rho \times \rho^t$;
6. Set termination condition:
(1) Compute reconstruction error $err = \|T + B - D\| < tol$;
(2) Target component not change $\sum\limits_{i,j} T^{t-1}(i,j) = \sum\limits_{i,j} T^t(i,j)$;
(3) Reach the maximum iteration number max_iter.
end while
Output: Sparse coefficient matrix $X^{(k)}$.

(1) Solution of background component B
The objective expression with regard to B can be summarized as:

$$L_\rho(B) = \sum_k \mu_k \|B_k\|_* + \frac{\rho}{2}\|D - B - T\|_F^2 + \langle F, D - B - T\rangle \tag{11}$$

The group members in B are independent with each other, so the minimization problem of $\frac{\rho}{2}\|D - B - T\|_F^2$ is equal to minimizing $\frac{\rho}{2}\sum\limits_k \|D_k - B_k - T_k\|_F^2$, and the minimization problem of $\langle F, D - B - T\rangle$ is equal to minimizing $\langle F, D_k - B_k - T_k\rangle$. According to this, Equation (10) can also be described as the following grouped summation form:

$$\begin{aligned} L_\rho(B_k) = &\sum_k \mu_k \|B_k\|_* + \langle F, D_k - B_k - T\rangle \\ &+ \frac{\rho}{2}\sum_k \|D_k - B_k - T_k\|_F^2 \end{aligned} \tag{12}$$

For each group, the objective function related to its corresponding background component B_k can be rewritten as the separated group form:

$$\begin{aligned} L_\rho(B_k) &= \mu_k \|B_k\|_* + \langle F, D_k - B_k - T\rangle \\ &\quad + \frac{\rho}{2}\|D_k - B_k - T_k\|_F^2 \\ &= \mu_k \|B_k\|_* + \frac{\rho}{2}\left\|B_k - \left(D_k - T_k + \frac{F}{\rho}\right)\right\|_F^2 \end{aligned} \tag{13}$$

The above problem can be solved by the singular value thresholding algorithm [40], which is defined as follows:

$$\begin{aligned} B_k^{t+1} &= SVD_{\frac{1}{\rho}}\left(D_k - T_k - \frac{F}{\rho}\right) \\ &= U_k diag\left(pos\left(\sigma_k - \frac{1}{\rho}\right)\right)V_k^T \\ pos\left(\sigma_k - \frac{1}{\rho}\right) &= \begin{cases} \sigma_k - \frac{1}{\rho}, if \sigma_k > \frac{1}{\rho} \\ 0, otherwise \end{cases} \end{aligned} \tag{14}$$

where U_k, V_k and σ_k are the left eigen-vector, right eigen-vector and singular values of matrix $D_k - T_k - \frac{F}{\rho}$, respectively.
(2) Solution of target component

The objective expression with regard to T can be summarized as:

$$\begin{array}{l} L_\rho(T) = \lambda\|T\|_1 + \frac{\rho}{2}\|D - B - T\|_F^2 + \langle F, D_k - B_k - T_k\rangle \\ B = \{B_1, \ldots B_k, \ldots\} \end{array} \tag{15}$$

Similar to Equation (13), the above expression can be rewritten as the following form:

$$L_\rho(T) = \lambda\|T\|_1 + \frac{\rho}{2}\left\|T - \left(D - B + \frac{F}{\rho}\right)\right\|_F^2 \tag{16}$$

According to reference [41], the solution of Equation (16) is given by the soft-thresholding function:

$$\begin{array}{l} T^{t+1} = th_{\frac{2\lambda}{\rho}}\left(D - B + \frac{F}{\rho}\right) \\ th_s(W) = \begin{cases} w - s, w > s \\ w + s, w < -s \\ 0, otherwise \end{cases} \end{array} \tag{17}$$

in which w represents the element of matrix W, T^{t+1} represents the updated target component in the next iteration.

(3) Update dual factor F and penalty factor ρ

The dual factor F and penalty factor ρ are all updated in a standard way as shown in the following:

$$\begin{array}{l} F^{t+1} = F^t + \rho^t(T^{t+1} + B^{t+1} - D) \\ \rho^{t+1} = \mu\rho \times \rho^t \end{array} \tag{18}$$

where $\mu\rho$ is the update factor for ρ.

4. Experimental Evaluations

4.1. Experiment Settings

4.1.1. Parameter Settings

In our experiment, the image is divided into 16×16 local patches with 10 pixel step size. The group number is set to 3. The regularization factor λ of the target component is $1/\left[\sqrt{\min(M,N)}\right]$, where M and N represent the patch size and patch number, respectively. The penalty factor ρ of the ADMM method is set to 0.001, and the update factor $\mu\rho$ is 1.05. The maximum iteration of ADMM method is set to 500.

4.1.2. Evaluation Metrics

We adopt three metrics to evaluate the performance of the detection algorithms. The first one is receiver operating characteristic (ROC) curve, which describes the sensitivity (or called saliency) of the target after detection operation. The false alarm ratio F_a and probability of detection P_d are employed to form the horizontal and vertical axis of the ROC curve, which are separately defined as below:

$$P_d = \frac{detected\ \ target\ \ number}{real\ \ target\ \ number} \tag{19}$$

$$F_a = \frac{falsely\ \ detected\ \ pixel\ \ number}{total\ \ pixel\ \ number} \tag{20}$$

For a randomly selected segmentation threshold, a good detection result should have a low false alarm ratio, while keeping a high target detection rate.

Another two metrics, signal-to-clutter ratio gain (SCRG) and background suppression factor (BSF) are used to measure the information change between the input images and output images. SCRG mainly reflects the enhanced capability to the target and BSF focuses on measuring the suppression effect on the background, which are separately defined as:

$$SCRG = \frac{SCRG_{out}}{SCRG_{in}}, SCR = \frac{|\mu_t - \mu_b|}{\sigma_b} \tag{21}$$

$$BSF = \frac{(\sigma_b)_{in}}{(\sigma_b)_{out}} \tag{22}$$

where μ_t and μ_b represent the mean value of the target part and background part, respectively. σ_b represents the standard deviation of the background part. Larger SCRG and BSF scores indicate a better detection performance.

4.1.3. Baseline Algorithms

To evaluate the performance of our proposed detection algorithm, several state-of-the-art methods are introduced as the comparison group, involving non-convex tensor low-rank approximation method (ASTTV-NTLA) [21], infrared patch image model (IPI) [12], partial sum of tensor nuclear norm-based detection model (PSTNN) [14], total variation regularization-based model (TVPCP) [42], reweighted image patch tensor model (RIPT) [43], non-convex rank approximation minimization joint $l_{1,2}$ norm-based model (NRAM) [18], multiscale patch-based contrast measure-based model (MPCM) [44] and sparse regularization-based spatial–temporal twist tensor (SRSTT) model [35]. Table 1 shows the detailed parameter settings of the compared methods in this paper.

Table 1. Detailed parameter settings for compared methods.

Methods	**Acronyms**	**Parameter Settings**
Non-convex tensor low-rank approximation method	ASTTV-NTLA	$L = 3, H = 6, \lambda_{tv} = 0.005, \lambda_s = \frac{H}{\sqrt{\max(M,N)*L}}, \lambda_3 = 100$
Infrared patch image model	IPI	Patchsize: 30×30, step: 10, $\lambda = \frac{1}{\sqrt{\min(m,n)}}, \varepsilon = 10^{-7}$
Partial sum of tensor nuclear norm-based detection model	PSTNN	Patchsize: 40×40, step: 40, $\lambda = \frac{0.6}{\sqrt{\max(n_1,n_2)*n_3}}, \varepsilon = 10^{-7}$
Total variation regularization-based model	TVPCP	$\lambda_1 = 0.005, \lambda_2 = \frac{1}{\sqrt{\max(M,N)}}, \beta = 0.025, \gamma = 1.5$
Reweighted image patch tensor model	RIPT	Patchsize: 50×50, step: 10, $\lambda = \frac{L}{\sqrt{\min(n_1,n_2,n_3)}}$ $L = 1, H = 10, \varepsilon = 10^{-7}$
Non-convex rank approximation minimization joint $l_{1,2}$ norm-based model	NRAM	Patchsize: 50×50, step: 10, $\gamma = 0.002, \lambda = \frac{1}{\sqrt{\max(M,N)}}$ $C = \frac{\sqrt{\min(M,N)}}{2.5}, \mu^0 = 3\sqrt{\min(M,N)}, \varepsilon = 10^{-7}$
Multiscale patch-based contrast measure-based model	MPCM	Mean filter size: 3×3, $N = 3,5,7,9$
Sparse regularization-based spatial–temporal twist tensor	SRSTT	$L = 30, \lambda_1 = 0.05, \lambda_2 = 0.1, \lambda_3 = 100, \varepsilon = 10^{-7}, \mu = 0.01$
Group-regularized principle component pursuit	GPCP	Patchsize: 30×30, step: 10, groupnum: 3, $\lambda = \frac{1}{\sqrt{\min(m,n)}}, \varepsilon = 10^{-7}$

4.1.4. Dataset

The full dataset contains 12 sequences. According to the type of detection scene, we have manually divided these sequences into 3 categories, including 3 ground-background sequences, 3 sea-background sequences and 6 sky-background sequences. The frame number, image size and signal-to-clutter information of each sequence are shown in Table 2.

Representative frames of each detection scene are shown in Figures 3–5. It is noticeable that the ground-background is the most complex compared with other two situations. The road surface with high gray-scale level leads to a very strong background edge, which causes great interference for detecting the real target. For the scene of sea-background, the warship target usually moves nearby the sea-level line. The clutters caused by the clouds and lighthouses will also increase the difficulty of small target detection. On the other hand, the imaging noise is very high in this situation, as shown in the sequence Sea-1. In the sky-background situation, the target energy is the lowest among these three situations. Specifically, the average signal-to-clutter ratio of sequence Sky-4 is less than zero, which indicates a very challenging task to detect the small target. In other sky-background sequences, the targets are submerged by the clouds from time to time.

Table 2. Dataset Information.

	Sequence Name	Frame Number	Image Size	Average SCR
Ground Background	Ground-1	200	256 × 256	2.21 dB
	Ground-2	200	256 × 256	3.41 dB
	Ground-3	200	256 × 256	5.01 dB
Sea Background	Sea-1	100	128 × 128	2.29 dB
	Sea-2	87	284 × 213	6.32 dB
	Sea-3	185	252 × 213	2.28 dB
Sky Background	Sky-1	60	320 × 240	6.86 dB
	Sky-2	67	320 × 240	0.87 dB
	Sky-3	400	256 × 172	4.14 dB
	Sky-4	200	256 × 208	−2.56 dB
	Sky-5	40	128 × 128	2.73 dB
	Sky-6	40	256 × 200	2.44 dB

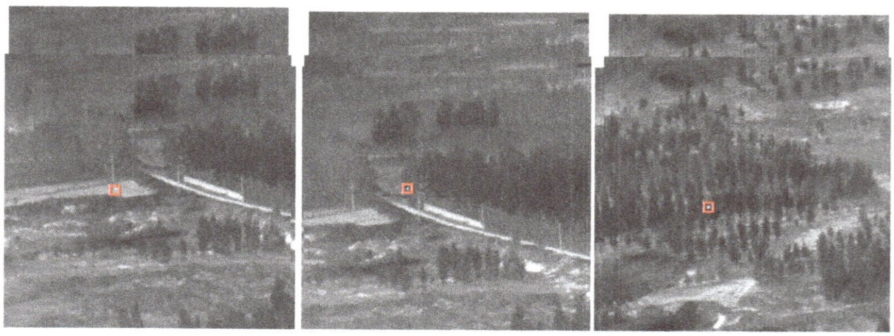

Figure 3. Ground-background sequences. (**a**) Ground-1 (**b**) Ground-2 (**c**) Ground-3. Targets are marked in red boxes.

(a) Sea-1 (b) Sea-2 (c) Sea-3

Figure 4. Sea-background sequences. (**a**) Sea-1 (**b**) Sea-2 (**c**) Sea-3. Targets are marked in red boxes.

4.2. Quantitative Comparison

To evaluate the detection performance of the proposed GPCP model, we first report the ROC curves of 9 infrared small target detection algorithms on the whole dataset, as shown in Figure 6. It can be observed that the curve of GPCP is the closest to the upper left corner, which means for any given false alarm rate, the proposed GPCP model achieves the highest accurate detection rate, and for any given detection rate, the proposed GPCP model achieves the lowest false alarm rate. The first line in Table 3 also shows the proposed GPCP model has the highest area under curve (AUC) value in all 9 algorithms, PSTNN is second only to our proposed model. That is to say, the proposed GPCP model has a relatively good detection performance on the whole dataset.

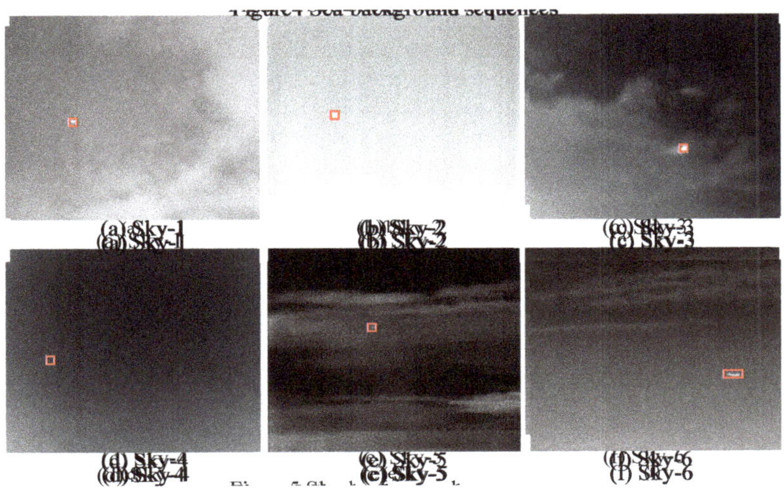

Figure 5. Sea-background sequences. (**a**) Sky-1 (**b**) Sky-2 (**c**) Sky-3 (**d**) Sky-4 (**e**) Sky-5 (**f**) Sky-6. Targets are marked in red boxes.

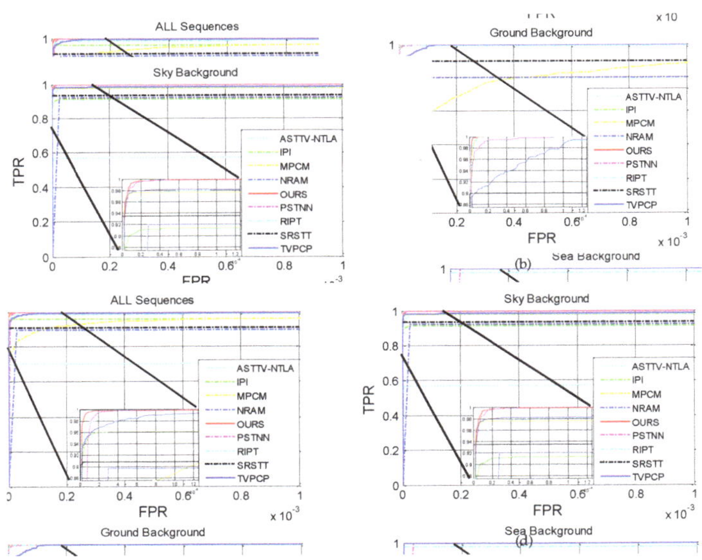

Figure 6. ROC curves of the whole dataset and 3 different background categories. (**a**) All sequences, (**b**) ground background, (**c**) sea background, (**d**) sky background.

Table 3. The table shows the AUC of 9 small target detection algorithms in the whole dataset and 3 different background categories. For each category, the best results are marked in the red color.

	OURS	ASTTV-NTLA	PSTNN	TVPCP	IPI	NRAM	RIPT	MPCM	SRSTT
All	0.999994	0.7391	0.999992	0.99979	0.9606	0.9485	0.9916	0.9945	0.9147
Ground	0.999999	0.9933	0.999998	0.999993	0.999999	0.9008	0.999999	0.9775	0.90
Sea	1	0.5355	0.999964	1	1	1	0.9785	1	0.8388
Sky	0.999998	0.6072	0.999998	0.9991	0.9132	0.9603	0.9913	0.9995	0.9346

We also report the ROC curves on 3 different background categories: the ground background, the sea background and the sky background, which are illustrated in Figure 6b, Figure 6c and Figure 6d, respectively. Combined with the AUC scores shown in Table 3, we can see that the GPCP curve is the closest to the upper left corner and corresponds to the largest AUC value, which indicates the proposed model has the best detection performance in the ground background. RIPT and IPI are the second- and third-best algorithms in this situation. For the sea background, most algorithms perform well. As the third line of Table 3 shows, the AUC values of IPI, MPCM, NRAM, GPCP and TVPCP are 1. Yet it is worth noting that the ASTTV-NTLA has a relatively small AUC value in this situation. For the sky background, from Figure 6d and the forth line of Table 3, we can see the proposed method and PSTNN have the best detection performance. The AUC values of these two models are the same. The main difference of these two methods lies in the GPCP model performs better in suppressing false alarms and PSTNN performs better in detection rate Figure 6d can prove this point, in the case of lower false-alarm rate, the proposed GPCP model has a higher accurate detection rate; in the case of a higher detection rate, the PSTNN achieves a lower false-alarm rate.

To analyze the algorithm performance more specifically, the signal-to-clutter gain (SCRG) and background suppression factor (BSF) of 9 algorithms on each individual sequence are also calculated, as shown in Tables 4–6. A good algorithm should achieve high SCRG and high BSF, which represent the performance on target enhancement and background clutter suppression, respectively. From Table 4, we can see that our proposed method achieves the highest SCRG on all three sequences in the ground background. In Ground−1, the proposed model also achieves the highest BSF value. In Ground−2 and Ground−3, BSF values of the proposed model are a bit lower than the NRAM and ASTTV-NTLA, which means the remained background pixel value of our model is a bit higher than the other two models. Based on the fact that the proposed model has the highest SCRG values in these two sequences, we can conclude that our model still achieves the largest contrast between the target and background. Table 5 shows the algorithm performance on three sequences in the sea background. GPCP model achieves the highest values of SCRG and BSF values in sequence Sea−1 and Sea−2. ASTTV-NTLA model achieves the highest values of SCRG and BSF in sequence Sea−3 due to its multi-frame and TV model, but has a poor performance in Sea−1 and Sea−2. That is because the ASTTV-NTLA model is not suitable for infrared small target detection with low moving speed. The same experiment results also appear in Sky−4 and Sky−5. ASTTV-NTLA model misses all the targets in these two sequences due to the low moving speed. PSTNN also performs well in sequence Sea−3, especially in the SCRG value. This is due in large part to the usage of structure tensor. A prior weight representing the corner feature is added to the target component and makes the extracted target brighter. By comparison, target intensity values of the proposed model are a little bit lower than PSTNN. However, from Table 3, we can see the AUC values of the proposed model is higher than PSTNN. In the sky background, the proposed GPCP model has the largest SCRG and BSF values in sequence Sky−1, Sky−2, Sky−5 and Sky−6. NRAM and PSTNN achieves the highest SCRG and BSF in Sky−3 and Sky−4, respectively. From Figure 5, we can see the targets in Sky−3 are relatively large and has a gray variance. The detection results of NRAM only reserve several pixels in the target center position. By comparison, the proposed model has more pixels of targets in the detection results and is more coincide with the real target. For sequence Sky−4, there are some residual pixels with low values remained in the proposed model compared with PSTNN. The reason lies that in this sequence, the gray-scale difference of the local background regions is not very great. Current group strategy which employs the complexity difference for patch grouping is disabled. Therefore, in sequence Sky−4, GPCP model is almost equal to its baseline IPI. The experiment results in Table 6 also shows the performance of GPCP model is similar to IPI model.

Table 4. The table shows signal-to-clutter ratio gain (SCRG) and background suppression factor (BSF) of 8 small target detection algorithms on 3 ground-background sequences. For each sequence, the best results are marked in the red color.

		OURS	ASTTV-NTLA	PSTNN	TVPCP	IPI	NRAM	RIPT	MPCM	SRSTT
Ground−1	SCRG	23.80	13.79	3.79	0.81	1.19	2.39	0.95	0.69	5.63
	BSF	5402	126.56	9.73	7.13	10.38	18.51	9.45	11.30	10.35
Ground−2	SCRG	0.3038	0.22	0.03	0.004	0.06	0.002	0.13	0.18	0.28
	BSF	10.16	8.54	2.64	2.48	4.22	10.54	7.15	3.38	4.35
Ground−3	SCRG	5.55	5.10	3.03	1.44	2.51	4.43	2.48	0.12	5.50
	BSF	9.04	9.32	5.45	2.44	6.00	7.56	9.29	2.93	2.34

Table 5. The table shows signal-to-clutter ratio gain (SCRG) and background suppression factor (BSF) of 8 small target detection algorithms on 3 sea-background sequences. For each sequence, the best results are marked in the red color.

		OURS	ASTTV-NTLA	PSTNN	TVPCP	IPI	NRAM	RIPT	MPCM	SRSTT
Sea−1	SCRG	30,498	0	42.07	943.90	3.82	25.53	4.69	1.38	6099
	BSF	15,784	0	124.55	3146	10.06	58.24	9.69	4.89	14,376
Sea−2	SCRG	15,659	0	12,144	16.60	18.36	11,399	13,648	4.89	0.25
	BSF	13,438	0	13,438	26.71	41.00	1	13,438	15.63	10,458
Sea−3	SCRG	97.74	2323	554.47	17.28	19.46	34.97	2318	1.15	6.48
	BSF	205.78	6461	211.13	14.15	14.46	24.57	1079	2.63	4.18

Table 6. The table shows signal-to-clutter ratio gain (SCRG) and background suppression factor (BSF) of 9 small target detection algorithms on 6 sky-background sequences. For each sequence, the best results are marked in the red color.

		OURS	ASTTV-NTLA	PSTNN	TVPCP	IPI	NRAM	RIPT	MPCM	SRSTT
Sky−1	SCRG	17.02	1.52	13.91	1.33	12.63	2.49	3.88	0.34	2.34
	BSF	17.26	3.56	15.39	1.27	15.08	17.58	11.92	8.91	2.31
Sky−2	SCRG	15.89	13.1	13.23	1.42	4.67	6.83	0.02	0.04	7.92
	BSF	20.28	8.67	8.46	1.17	14.18	10.79	8.93	5.91	5.23
Sky−3	SCRG	8.11	4.56	9.39	7.67	7.41	18.57	0.01	0.15	0.98
	BSF	10.99	1.16	16.79	10.74	10.42	1066	172.93	19.22	11.66
Sky−4	SCRG	25.46	0	7073	27.23	30.79	4105	28.13	0.17	59.50
	BSF	28.55	0	1078	16.19	26.63	1023	20.37	4.95	17.33
Sky−5	SCRG	1991	0	125	0.01	8.16	29.68	8.77	0.69	7.13
	BSF	1735	0	95.22	3.80	9.36	38.33	7.68	0.88	32.59
Sky−6	SCRG	15.87	0.006	12.09	13.90	13.66	14.96	6.59	2.19	3.45
	BSF	20.76	8.42	8.53	10.63	10.76	17.57	8.79	9.81	2.71

4.3. Qualitative Comparison

To have a more direct and deeper impression on the effect of each method, we select several representative results as well as the corresponding three-dimensional surface results from each type of detection scene for illustration. The small target detection task in the ground background is the most difficult situation. Three representative examples are shown in Figure 7. The detection for sequence Ground−1 is relatively simple because there is a large contrast between the target and its surrounding background. The most challenging factor is the interference caused by the road edge. The proposed GPCP model achieves the best detection performance in this situation. Meanwhile, from the three-dimensional surface results, we can see the proposed method gets the best performance on background suppression among all the 9 algorithms. The detection task for sequence Ground−2 and Ground−3 is more difficult compared with Ground−1. The original images of these two sequences show the targets have been basically submerged into the background, in addition, there are many background clutters with similar appearance to the small dim targets. In these two sequences, only the proposed GPCP model successfully detects the target while suppressing the background clutters. Other methods leave many false alarms in the detection results, as the green box and three-dimensional surfaces show.

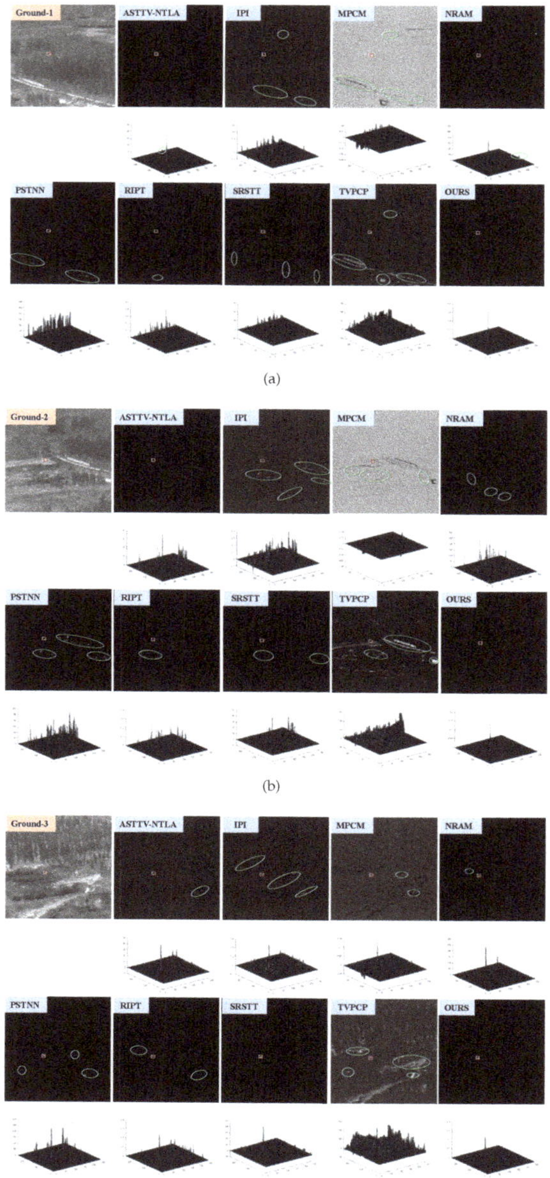

Figure 7. Representative results in the ground background. (**a**) Ground−1 (**b**) Ground−2 (**c**) Ground−3. Targets are labeled by the red box. The remained background clutters are labeled by the green box.

Figure 8 illustrates the detection results in the scene of sea background, in which the most challenging factor lies in the background interference caused by the lighthouse and water grass shelter. In sequence Sea−1, it can be seen that the ASTTV-NTLA, IPI, MPCM and PSTNN models have a poor detection performance, where the gray-scale value of the lighthouse outline is even larger than the real target after detection. For the RIPT method, the imaging noise has a certain impact on the detection performance, in which many false alarms are remained in the background part. By comparison, the NRAM, TVPCP and the proposed methods are good at suppressing the strong background edges as well as the imaging noise. In sequence Sea−3, the gray value of the water grass is larger than the target, in addition, both of these two parts have sharp forms in appearance, making the small target hard to be distinguished from the background. In the detection results of SRSTT and TVPCP methods, there are many residual clutters remaining in the background, as the green boxes show. By comparison, the PSTNN, RIPT, NRAM and the proposed methods achieve satisfactory results.

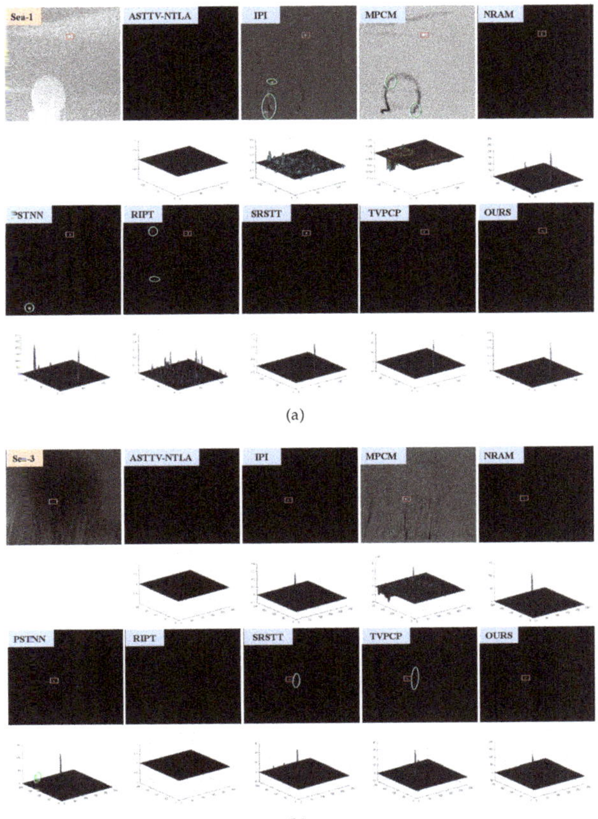

Figure 8. Representative results in the sea background. (**a**) Sea−1 (**b**) Sea−3. Targets are labeled by the red box. The remained background clutters are labeled by the green box.

Three representative detection results in the scene of sky background are shown in Figure 9. The target in sequence Sky-2 has a very low contrast compared with the background. In this situation, the ASTTV-NTLA and TVPCP methods fail to suppress the background noise and cannot find the real target. By comparison, the NRAM, PSTNN, RIPT, SRSTT and the proposed GPCP model achieve a better performance on target enhancement. We can see that the decomposition results of these four methods all correspond to a low level background noise. Sequence Sky-5 shows the small target detection results of 9 methods in the case of bright heavy cloud. There are many strong edges in the background part, especially in the top and right side of the image. As shown in Figure 9, the detection results of the IPI, MPCM, PSTNN, RIPT, Tophat and TVPCP methods remain having many background clutters, which are easily to confuse with the real small target. Only the NRAM and the proposed GPCP methods can extract the target while suppressing the clutters simultaneously. The IPI, PSTNN, RIPT and NRAM models are all based on the PCP theory and carry out a global low-rank decomposition to remove the background clutters. By comparison, the proposed group low-rank and sparse decomposition model has a significant effect to cope with the situation with strong background clutters.

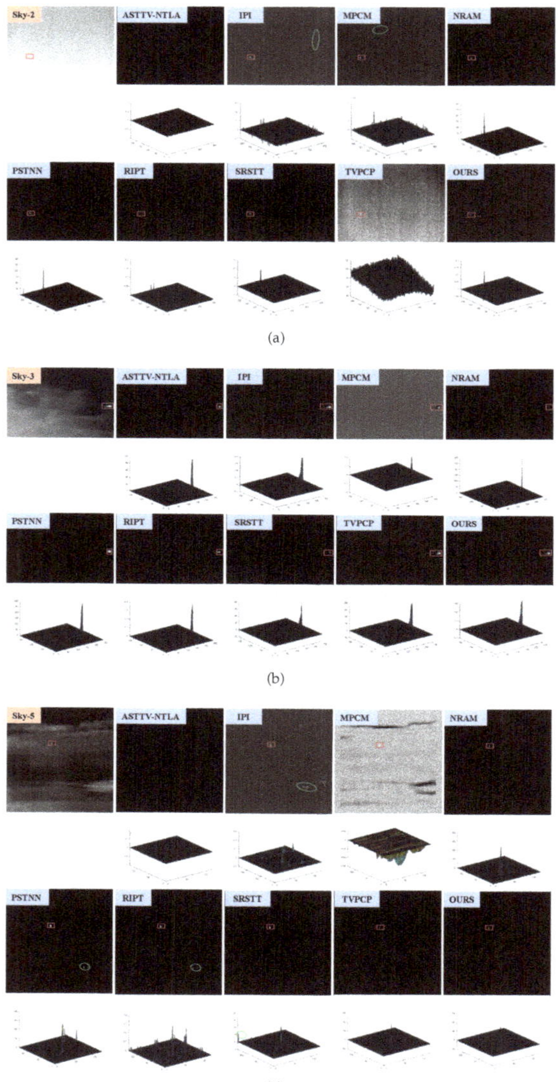

Figure 9. Representative results in the sky background. (**a**) Sky−2 (**b**) Sky−3 (**c**) Sky−5. Targets are labeled by the red box. The remained background clutters are labeled by the green box.

4.4. *Influence of Grouping Criteria on Our Method*

As mentioned above, the proposed model needs to divide the full image into several groups for image decomposition. In this part, an ablation experiment is carried out to discuss the effectiveness of the grouping criteria of the customized group low-rank strategy. The proposed model takes both of the gray-scale level and the clutter level into consideration and divides the data matrix into three groups. By comparison, the first contrast experiment is designed to only use the gray-scale information and divide the data matrix into a bright part and a dark part, which is named as GPCP−Gray. The second contrast experiment employs the variance information and divides the data matrix into a uniform part and a cluttered part, which is named as GPCP−Var. The IPI model, which decomposes the entire image data into a low-rank component and a sparse component plays as the baseline method. The ROC curves of these four experiments are shown in Figure 10.

In this experiment, the GPCP−Var model has the worst performance, especially in the sky background situation. By comparison, the GPCP−Gray model has a slight decline in the ROC curves compared with the proposed GPCP model, while performs better than the global regularized low rank and sparse decomposition model (IPI). This suggests that dividing the data matrix into two parts with different brightness level has a positive influence on background suppression in the PCP process. In addition, from the ROC results of the GPCP−Gray model and the proposed GPCP model, we can see that extracting the cluttered image data and decomposing this part independently can further improve the detection performance.

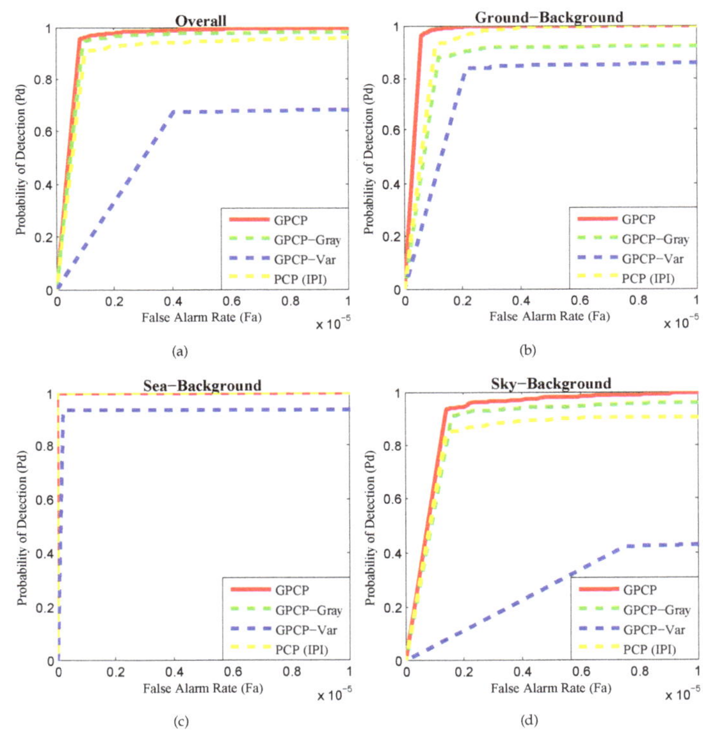

Figure 10. ROC curves of the overall dataset and 3 different background categories. (**a**) Overall (**b**) ground background (**c**) sea background (**d**) sky background.

4.5. Computation Complexity Analysis

The computation complexity of each comparison model is shown in Table 7. Suppose the image size is $M \times N$ and the patch image size is $m \times n$. The computation cost is $O(L^3MN)$. The major time-consuming part is saliency map calculation, in each scale, the computation cost is $O(L^2MN)$. The total cost in all scales is $O(L^3MN)$. For the patch-based models, including IPI, NRAM and the proposed GPCP model, its computation complexity mainly comes from the SVD decomposition. For an $m \times n$ patch matrix, the computation complexity of SVD is $O(mn^2)$. For the patch-tensor models, including

RIPT, PSTNN, ASTTV-NTLA and SRSTT, the main time-consuming part is the SVD decomposition progress in the frequency domain. For a tensor size with $n_1 \times n_2 \times n_3$, the computation complexity of SVD is $O(n_1 n_2 n_3 \log(n_1 n_2 n_3))$, the computation complexity of FFT is $O(n_1 n_2^2 [(n_3 + 1)/2])$. They are faster than the SVD decomposition of patch-based models, which are calculated in the spatial domain. The TVPCP model is a little time consuming due to the matrix inversion calculation. It is worth noting that the proposed GPCP model is faster than its baseline IPI model. From the grouping criteria, we can see that $n_1 + n_2 + n_3 = n$. Based on the fact that $n_1^2 + n_2^2 + n_3^2 < n^2$, the computation cost of GPCP is lower than IPI. In our experiments, for a 256 × 256 image, GPCP needs 5.9 s to obtain the detection result. By comparison, IPI needs 11.9s. The speed increases doubly.

Table 7. The table shows the computation complexity of 9 small target detection algorithms.

	OURS	ASTTV-NTLA	PSTNN	TVPCP	IPI	NRAM	RIPT	MPCM	SRSTT
Complexity	$O(m(n_1^2 + n_2^2 + n_3^2))$	$O(MNL\log(MNL))$	$O(n_1 n_2 n_3 \log(n_1 n_2 n_3))$ $+O(n_1 n_2^2[(n_3+1)/2])$	$O(MN^2 + N^4)$	$O(mn^2)$	$O(mn^2)$	$O(n_1 n_2 n_3(n_1 n_2 + n_2 n_3 + n_1 n_3))$	$O(L^3 MN)$	$O(Ln_1(n_2^2 + n_3\log((n_2+1)/2)))$

5. Conclusions

In this paper, a novel group regularized low-rank and sparse decomposition model is proposed for infrared small dim target detection. The traditional decomposition-based models are usually sensitive to strong edges and background clutters due to the ignorance of data structure diversity. The proposed method is able to solve this problem by using a customized group low-rank strategy. Firstly, it exploits different singular value thresholds for the low-rank decomposition of image groups corresponding to different complexity. Then, the newly designed group low-rank regularization is integrated with the sparse constraint for background and target separation, in which more prior information related to data structure can be utilized in the decomposition process. Experimental results on 3 different detection scenes, which includes 12 sequences, have shown the priority of the proposed in terms of probability of detection, false alarm rates, target enhancement and background suppression factors.

There also exist some issues worth considering. For example, we use the brightness and gray-scale variance to divide patches into groups, other strategies such as image feature-based methods can be further considered for patch grouping. This method is also time consuming, especially in the background solving process, other background modeling methods need to be explored in the future work.

Author Contributions: Conceptualization, M.L. and J.Z.; methodology, M.L. and Y.W.; software, M.L.; validation, M.L. and B.D.; formal analysis, M.L.; investigation, M.L.; resources, Y.W.; data curation, M.L.; writing—original draft preparation, M.L.; writing—review and editing, M.L. and D.L.; visualization, M.L.; project administration, M.L.; funding acquisition, M.L. All authors have read and agreed to the published version of the manuscript.

Funding: This research was funded by the National Natural Science Foundation of China grant number 62101529.

Data Availability Statement: The data presented in this study are available on request from the corresponding author.

Conflicts of Interest: The authors declare no conflict of interest.

References

1. Nasiri, M.; Chehresa, S. Infrared small target enhancement based on variance difference. *Infrared Phys. Technol.* **2017**, *82*, 107–119. [CrossRef]
2. Liu, H.K.; Zhang, L.; Huang, H. Small Target Detection in Infrared Videos Based on Spatio-Temporal Tensor Model. *IEEE Trans. Geosci. Remote Sens.* **2020**, *58*, 8689–8700. [CrossRef]
3. Deng, H.; Sun, X.; Liu, M.; Ye, C. Infrared small-target detection using multiscale gray difference weighted image entropy. *IEEE Trans. Aerosp. Electron. Syst.* **2016**, *52*, 60–72. [CrossRef]

4. Xiong, B.; Huang, X.; Wang, M.; Peng, G. Small target detection for infrared image based on optimal infrared patch-image model by solving modified adaptive RPCA problem. *Int. J. Pattern Recognit. Artif. Intell.* **2021**, *35*, 2150007. [CrossRef]
5. Chen, C.L.P.; Li, H.; Wei, Y.; Xia, T.; Tang, Y.Y. A Local Contrast Method for Small Infrared Target Detection. *IEEE Trans. Geosci. Remote Sens.* **2014**, *52*, 574–581. [CrossRef]
6. Deng, H.; Sun, X.; Liu, M.; Ye, C.; Zhou, X. Small Infrared Target Detection Based on Weighted Local Difference Measure. *IEEE Trans. Geosci. Remote Sens.* **2016**, *54*, 4204–4214. [CrossRef]
7. Nie, J.; Qu, S.; Wei, Y.; Zhang, L.; Deng, L. An Infrared Small Target Detection Method Based on Multiscale Local Homogeneity Measure. *Infrared Phys. Technol.* **2018**, *90*, 186–194. [CrossRef]
8. Qu, X.; He, C.; Peng, G. Novel detection method for infrared small targets using weighted information entropy. *J. Syst. Eng. Electron.* **2012**, *23*, 838–842. [CrossRef]
9. Gu, Y.; Wang, C.; Liu, B.X ; Zhang, Y. A Kernel-Based Nonparametric Regression Method for Clutter Removal in Infrared Small-Target Detection Applications. *IEEE Geosci. Remote Sens. Lett.* **2010**, *7*, 469–473. [CrossRef]
10. Bin, Y.E.; Jiaxiong, P. Small Target Detection Method Based on Morphology Top-Hat Operator. *J. Image Graph.* **2002**, *7*, 638–642.
11. Han, J.; Liu, C.; Liu, Y.; Luo, Z.; Niu, Q. Infrared Small Target Detection Utilizing the Enhanced Closest-Mean Background Estimation. *IEEE J. Sel. Top. Appl. Earth Obs. Remote Sens.* **2020**, *14*, 645–662. [CrossRef]
12. Gao, C.; Meng, D.; Yang, Y.; Wang, Y.; Zhou, X.; Hauptmann, A.G. Infrared Patch-Image Model for Small Target Detection in a Single Image. *IEEE Trans. Image Process.* **2013**, *22*, 4996–5009. [CrossRef] [PubMed]
13. Zhang, C.; Wang, H.; Lou, J. Infrared small and dim target detection based on weighted nuclear norm minimization. *J. Huazhong Univ. Sci. Technol.* **2017**, *45*, 31–37.
14. Zhang, L.; Peng, Z. Infrared Small Target Detection Based on Partial Sum of the Tensor Nuclear Norm. *Remote Sens.* **2019**, *11*, 382. [CrossRef]
15. Kong, X.; Yang, C.; Cao, S.; Li, C.; Peng, Z. Infrared Small Target Eetection via Nonconvex Tensor Fibered Rank Approximation. *IEEE Trans. Geosci. Remote Sens.* **2021**, *60*, 5000321.
16. Shi, Y.; Wei, Y.; Yao, H.; Pan, D.; Xiao, G. High-Boost-Based Multiscale Local Contrast Measure for Infrared Small Target Detection. *IEEE Geoence Remote Sens. Lett.* **2017**, *15*, 33–37. [CrossRef]
17. Tang, W.; Zheng, Y.; Lu, R.; Huang, X. A novel infrared dim small target detection algorithm based on frequency domain saliency. In Proceedings of the 2016 IEEE Advanced Information Management, Communicates, Electronic and Automation Control Conference (IMCEC), Xi'an, China, 3–5 October 2016; pp. 1053–1057.
18. Zhang, L.; Peng, L.; Zhang, T.; Cao, S.; Peng, Z. Infrared Small Target Detection via Non-Convex Rank Approximation Minimization Joint l2,1 Norm. *Remote Sens.* **2018**, *10*, 1821. [CrossRef]
19. Zhu, H.; Ni, H.; Liu, S.; Xu, G.; Deng, L. TNLRS: Target-Aware Non-local Low-Rank Modeling with Saliency Filtering Regularization for Infrared Small Target Detection. *IEEE Trans. Image Process.* **2020**, *29*, 9546–9558. [CrossRef]
20. Dai, Y.; Wu, Y.; Song, Y.; Guo, J. Non-negative infrared patch-image model: Robust target-background separation via partial sum minimization of singular values. *Infrared Phys. Technol.* **2017**, *81*, 182–194. [CrossRef]
21. Liu, T.; Yang, J.; Li, B.; Xiao, C.; Sun, Y.; Wang, Y.; An, W. Nonconvex tensor low-rank approximation for infrared small target detection. *IEEE Trans. Geosci. Remote Sens.* **2021**, *60*, 5614718. [CrossRef]
22. Liu, J.; Wang, H.; Lei, L.; He, J. Infrared Small Target Detection Utilizing Halo Structure Prior-Based Local Contrast Measure. *IEEE Geosci. Remote Sens. Lett.* **2022**, *19*, 6508205. [CrossRef]
23. Bai, X.; Bi, Y. Derivative dntropy-based contrast measure for infrared small-target detection. *IEEE Trans. Geosci. Remote Sens.* **2018**, *56*, 2452–2466. [CrossRef]
24. Deng, L.; Zhang, J.; Xu, G.; Zhu, H. Infrared small target detection via adaptive M-estimator ring top-hat transformation. *Pattern Recognit. J. Pattern Recognit. Soc.* **2021**, *112*, 107729. [CrossRef]
25. Depeng, L.; Zhengzhou, L.; Bing, L.; Wenhao, C.; Tianmei, L.; Lei, C. Infrared small target detection in heavy sky scene clutter based on sparse representation. *Infrared Phys. Technol.* **2017**, *85*, 13–31.
26. Zhang, T.; Peng, Z.; Wu, H.; He, Y.; Li, C.; Yang, C. Infrared small target detection via self-regularized weighted sparse model. *Neurocomputing* **2021**, *420*, 124–148. [CrossRef]
27. Zhang, H.; Zhou, Z. Small target detection based on automatic ROI extraction and local directional gray and entropy contrast map. *Infrared Phys. Technol.* **2020**, *107*, 103290. [CrossRef]
28. Barnett, J.T. Statistical Analysis of Median Subtraction Filtering with Application to Point Target Detection in Infrared Backgrounds. In Proceedings of the SPIE—The International Society for Optical Engineering, Infrared Systems and Components III, Los Angeles, CA, USA, 15–20 January 1989; Volume 1050.
29. Dong, X.; Huang, X.; Zheng, Y.; Bai, S.; Xu, W. A novel infrared small moving target detection method based on tracking interest points under complicated background. *Infrared Phys. Technol.* **2014**, *65*, 36–42. [CrossRef]
30. Deshpande, S.D.; Meng, H.E.; Ronda, V.; Chan, P. Max-Mean and Max-Median Filters for Detection of Small-Targets. In Proceedings of the SPIE—The International Society for Optical Engineering, Signal and Data Processing of Small Targets, Denver, CO, USA, 18–23 July 1999; Volume 3809, pp. 74–83.
31. Sun, Y.Q.; Tian, J.W.; Liu, J. Background suppression based-on wavelet transformation to detect infrared target. In Proceedings of the 2005 International Conference on Machine Learning and Cybernetics, Guangzhou, China, 18–21 August 2005; Volume 8, pp. 4611–4615.

32. Zhang, T.; Wu, H.; Liu, Y.; Peng, L.; Yang, C.; Peng, Z. Infrared small target detection based on non-convex optimization with Lp-norm constraint. *Remote Sens.* **2019**, *11*, 559. [CrossRef]
33. Kwan, C.; Budavari, B. A high-performance approach to detecting small targets in long-range low-quality infrared videos. *Signal Image Video Process.* **2022**, *16*, 93–101. [CrossRef]
34. Aliha, A.; Liu, Y.; Ma, Y.; Hu, Y.; Pan, Z.; Zhou, G. A Spatial and Temporal Block-Matching Patch-Tensor Model for Infrared Small Moving Target Detection in Complex Scenes. *Remote Sens.* **2023**, *15*, 4316. [CrossRef]
35. Li, J.; Zhang, P.; Zhang, L.; Zhang, Z. Sparse Regularization-Based Spatial–Temporal Twist Tensor Model for Infrared Small Target Detection. *IEEE Trans. Geosci. Remote Sens.* **2023**, *61*, 5000417. [CrossRef]
36. Du, J.; Lu, H.; Hu, M.; Zhang, L.; Shen, X. CNN-based infrared dim small target detection algorithm using target-oriented shallow-deep features and effective small anchor. *IET Image Process.* **2021**, *15*, 1–15. [CrossRef]
37. Bai, Y.; Li, R.; Gou, S.; Zhang, C.; Chen, Y.; Zheng, Z. Cross-connected bidirectional pyramid network for infrared small-dim target detection. *IEEE Geosci. Remote Sens. Lett.* **2022**, *19*, 7506405. [CrossRef]
38. Liu, F.; Gao, C.; Chen, F.; Meng, D.; Zuo, W.; Gao, X. Infrared Small and Dim Target Detection With Transformer Under Complex Backgrounds. *IEEE Trans. Image Process.* **2023**, *32*, 5921–5932. [CrossRef] [PubMed]
39. Yan, Z.; Chen, C.Y.; Luo, L.; Yao, Y. Stable principal component pursuit-based thermographic data analysis for defect detection in polymer composites. *J. Process. Control.* **2017**, *49*, 36–44. [CrossRef]
40. Cai, J.F.; Candès, E.J.; Shen, Z. A Singular value thresholding algorithm for matrix completion. *SIAM J. Optim.* **2010**, *20*, 1956–1982 [CrossRef]
41. Wright, S.J.; Nowak, R.D.; Figueiredo, M.A.T. Sparse reconstruction by separable approximation. In Proceedings of the 2008 IEEE International Conference on Acoustics, Speech and Signal Processing, Las Vegas, NV, USA, 31 March–4 April 2008; pp. 3373–3376
42. Wang, X.; Peng, Z.; Kong, D.; Zhang, P.; He, Y. Infrared dim target detection based on total variation regularization and principal component pursuit. *Image Vis. Comput.* **2017**, *63*, 1–9. [CrossRef]
43. Dai, Y.; Wu, Y. Reweighted Infrared Patch-Tensor Model With Both Nonlocal and Local Priors for Single-Frame Small Target Detection. *IEEE J. Sel. Top. Appl. Earth Obs. Remote Sens.* **2017**, *10*, 3752–3767. [CrossRef]
44. Wei, Y.; You, X.; Li, H. Multiscale patch-based contrast measure for small infrared target detection. *Pattern Recognit.* **2016**, *58*, 216–226. [CrossRef]

remote sensing

MDPI

Article

Infrared Small-Target Detection Based on Background-Suppression Proximal Gradient and GPU Acceleration

Xuying Hao [1,2,3], Xianyuan Liu [1,2,3], Yujia Liu [1,2,3], Yi Cui [1,2,3] and Tao Lei [1,2,3,*]

1 National Key Laboratory of Optical Field Manipulation Science and Technology, Chinese Academy of Sciences, Chengdu 610209, China; haoxuying20@mails.ucas.ac.cn (X.H.); liuxianyuan16@mails.ucas.ac.cn (X.L.); liuyujia20@mails.ucas.ac.cn (Y.L.); cuiyi@ioe.ac.cn (Y.C.)
2 Institute of Optics and Electronics, Chinese Academy of Sciences, Chengdu 610209, China
3 School of Electronic, Electrical and Communication Engineering, University of Chinese Academy of Sciences, Beijing 101408, China
* Correspondence: taoleiyan@ioe.ac.cn

Abstract: Patch-based methods improve the performance of infrared small target detection, transforming the detection problem into a Low-Rank Sparse Decomposition (LRSD) problem. However, two challenges hinder the success of these methods: (1) The interference from strong edges of the background, and (2) the time-consuming nature of solving the model. To tackle these two challenges, we propose a novel infrared small-target detection method using a Background-Suppression Proximal Gradient (BSPG) and GPU parallelism. We first propose a new continuation strategy to suppress the strong edges. This strategy enables the model to simultaneously consider heterogeneous components while dealing with low-rank backgrounds. Then, the Approximate Partial Singular Value Decomposition (APSVD) is presented to accelerate solution of the LRSD problem and further improve the solution accuracy. Finally, we implement our method on GPU using multi-threaded parallelism, in order to further enhance the computational efficiency of the model. The experimental results demonstrate that our method out-performs existing advanced methods, in terms of detection accuracy and execution time.

Keywords: infrared small target detection; proximal gradient; approximate partial SVD; GPU acceleration

Citation: Hao, X.; Liu, X.; Liu, Y.; Cui, Y.; Lei, T. Infrared Small-Target Detection Based on Background-Suppression Proximal Gradient and GPU Acceleration. *Remote Sens.* **2023**, *15*, 5424. https://doi.org/10.3390/rs15225424

Academic Editor: Paolo Tripicchio

Received: 1 October 2023
Revised: 5 November 2023
Accepted: 15 November 2023
Published: 20 November 2023

1. Introduction

Infrared Small-Target Detection (ISTD) is an important component of infrared search and tracking, aiming to exploit the thermal radiation difference between a target and its background to achieve long-range target detection. According to the definition by the Society of Photo-Optical Instrumentation Engineers (SPIE), small targets typically refers to objects in a 256 × 256 image with an area of fewer than 80 pixels, accounting for approximately 0.12% of the total image area [1]. These small targets usually appear as faint, tiny points, characterized by their diminutive size and a lack of clear texture and shape features. Moreover, the background in infrared images is often affected by random noise, clutter, and environmental factors, making small targets vulnerable to interference. Furthermore, some practical applications have strict requirements on the real-time performance of detection algorithms. Therefore, the rapid and accurate detection of small targets in complex backgrounds poses a significant challenge.

Two primary methods are employed in ISTD for target detection: Tracking-Before-Detection (TBD) and Detection-Before-Tracking (DBT). TBD relies on the temporal information of consecutive frames to capture the movement and features of potential targets. It struggles with stationary or sporadically moving targets and is constrained by computational resources. On the other hand, DBT applies single-frame ISTD to infrared data, identifying potential targets based on features such as contrast and low-rank sparsity. Single-frame infrared small target detection has been widely concerned because of its simple data acquisition, low computational complexity, not affected by target motion, and wide applicability.

The categorization of single-frame ISTD can be determined by the structure of the image; that is, whether (1) the original image or (2) the patch image is used [2]. The first category detects the target directly from the original image; for example, by filtering or Human Vision System (HVS). Filter-based methods [3–6] have limited utility in ISTD, due to their strict requirements on the background variation and prior knowledge. Meanwhile, HVS-based methods [7–11] use the contrast mechanism to quantify the difference between the target and the background, thereby enhancing small targets. However, these methods are limited by the local saliency of the target, rendering them ineffective when detecting targets that are dark or similar to the background. Some deep learning technologies [12–15] have recently been applied to this category, but a lack of large data sets limits their performance.

The other category—namely, patch-based methods—transforms small target detection into a low-rank matrix recovery problem [16]. This transformation can circumvent the aforementioned limitations, such as the dependence on prior knowledge and target saliency, as well as the false detection of dark targets. The most popular method is Infrared Patch-Image (IPI) [17], which uses a sliding window technique to generate a corresponding patch image from the original image. Due to its outstanding performance, many studies [18–26] have been conducted on IPI, which typically yields superior results. However, patch-based methods still have two problems: (1) The misclassification of strong edges as sparse target components, and (2) the time-consuming nature of the method.

The above-mentioned misclassification arises from the limited ability of the model to distinguish strong edges from sparse components. To address this issue, we propose a Background Suppression Proximal Gradient (BSPG) method, incorporating a novel continuation strategy during the alternating updating of low-rank and sparse components. Our proposed continuation strategy can preserve more components while updating the low-rank matrix, while also reducing the update rate of sparse matrix. As strong edges frequently correspond to larger singular values than targets, the former facilitates the transition of strong edges from sparse components to low-rank components, thereby enabling the model to eliminate the affect of strong edges. Meanwhile, the latter ensures the convergence of the algorithm.

The time-consuming nature of patch-based methods is due to the complex nature of solving the method, mainly including solving the LRSD problem and constructing/reconstructing patch images. To address this issue, we utilize both algorithmic optimizations and hardware enhancements. At the algorithmic level, we propose an approximate partial SVD (APSVD) for efficiently solving the LRSD problem and use a rank estimation method to ensure the accuracy of the solution. At the hardware level, we propose the use of GPU multi-threaded parallelism strategies to expedite the construction and reconstruction modules, as these modules can be decomposed into repetitive and independent sub-tasks.

Our main contributions can be summarized as follows:

- A novel continuation strategy based on the Proximal Gradient (PG) algorithm is introduced to suppress strong edges. This continuation strategy preserves heterogeneous backgrounds as low-rank components, hence reducing false alarms.
- The APSVD is proposed for solving the LRSD problem, which is more efficient than the original SVD. Subsequently, parallel strategies are presented to accelerate the construction and reconstruction of patch images. These designs can reduce the computation time at the algorithmic and hardware levels, facilitating rapid and accurate solution.
- Implementation of the proposed method on GPU is executed and experimentally validate its effectiveness with respect to the detection accuracy and computation time. The obtained results demonstrate that the proposed method out-performs nine state-of-the-art methods.

The remainder of this paper is organized as follows: Section 2 presents the related work. Section 3 details our proposed BSPG algorithm and GPU acceleration strategies. Section 4 introduces the data set and experimental settings, as well as providing the experimental results and analysis. Section 5 discusses the results of the experiments. Section 6 gives a summary of our method.

2. Related Work

2.1. HVS-Based Methods

HVS-based methods detect small targets by utilizing the contrast differences between the target region and its surrounding background. These methods can be categorized based on the type of information they use: grey scale information, gradient information, and a combination of both grey scale and gradient information. Local Contrast Measure (LCM) [1] proposes a novel method for detecting small targets by leveraging grey scale contrast. This method uses a contrast mechanism designed to enhance small targets while effectively suppressing the background noise. Based on the improvement of LCM algorithm, Relative Local Contrast Measure (RLCM) [8], Multiscale Patch-based Contrast Measure (MPCM) [9], Weighted Local Difference Measure (WLDM) [27] and other methods were proposed. Gradient-based contrast methods use first-order or second-order derivatives of the image to extract gradient information. They then utilize this information to design a gradient difference measure that effectively discriminates between small targets and the surrounding background. Building on this concept, Derivative Entropy-based Contrast Measure (DECM) [28] and Local Contrast-Weighted Multidirectional Derivative (LCWMD) [29] propose the use of multidirectional derivative to incorporate more gradient information. In addition, Local Intensity and Gradient (LIG) [30], Gradient-Intensity Joint Saliency Measure (GISM) [31] fuse gradient and intensity information to further highlight small targets. Although HVS-based methods can be effective in many scenarios, they are susceptible to missed detections and false positives in images characterized by low signal-to-clutter ratios and high-intensity backgrounds.

2.2. Deep Learning-Based Methods

In recent years, there has been a significant research focus on deep learning-based methods for infrared small target detection, which seek to achieve high-accuracy detection rates. These deep learning models are trained to discern features within infrared images using vast datasets, thereby enhancing their detection capabilities. To address the problem that infrared small target features are easily lost in deep neural networks, Attention Local Contrast Network (ALCNet) [32] proposes asymmetric contextual modulation to interact the feature information between the high and low levels. Dense Nested Attention Network (DNANet) [15] adequately fuses feature information through densely nested interaction modules to maintain small targets in deep layers. Miss Detection vs. False Alarm (MDvsFA) [33] proposes dual generative adversarial network models, trained inversely to decompose the detection challenge into sub-problems, aiming to strike a balance between miss detections and false alarms. While publicly available datasets have advanced deep learning for infrared small target detection, the scant features of small targets and the dependency on training samples limit the applicability of the model in varied real-world scenarios.

2.3. Patch-Based Methods

A significant amount of research has been conducted to improve the detection ability of IPI [17]. On one hand, some methods have used prior constraints, including Column-Weighted IPI (WIPI) [18], Non-negative IPI with Partial Sum (NIPPS) [20], and Re-Weighted IPI (ReWIPI) [21]. On the other hand, some studies have identified limitations in the nuclear norm and L1 norm and, so, alternative norms to achieve improved target representation and background suppression have been proposed; for example, Non-convex Rank Approximation Minimization (NRAM) [22] and Non-convex Optimization with Lp norm Constraint (NOLC) [23] introduce non-convex matrix rank approximation coupled with L2,1 norm and Lp norm regularization, while Total Variation Weighted Low-Rank (TVWLR) [24], Kernel Robust Principal Component Analysis (KRPCA) [25] introduce total variation regularization, High Local Variance (HLV) [26] method present LV* norm to constrain the background's local variance. Patch-based methods mainly consider the low-rank nature of the background, affecting their performance in the presence of strong edges. However, our method pays additional attention to heterogeneous background suppression in low-rank constraints, in order to avoid this problem.

2.4. Acceleration Strategies for Patch-Based Methods

Acceleration strategies for patch-based methods can be categorized into algorithm-level and hardware-level acceleration. The first category mainly relies on the strategy of reducing the number of iterations. Self-Regularized Weighted Sparse (SRWS) [34] and NOLC [23] improve the iteration termination condition for acceleration, but still suffer from the time consumption associated to decomposing large matrices. The other category (i.e., hardware acceleration) relies on the use of computationally powerful hardware and efficient parallelization strategies. In [35], the researchers proposed Separable Convolutional Templates (SCT); however, this method has poor performance under complex backgrounds. In addition, extending the patch model to tensor space can also achieve acceleration [36–41]. Representative methods in this direction include Re-weighted Infrared Patch-Tensor (RIPT) [36], LogTFNN [39] and the Pareto Frontier Algorithm (PFA) [37] However, unfolding the tensor into a two-dimensional matrix before decomposition increases the algorithm's complexity. Partial Sum of the Tensor Nuclear Norm (PSTNN) [38] and Self-Adaptive and Non-Local Patch-Tensor Model (ANLPT) [42] utilize the t-SVD speed up tensor decomposition with t-SVD. However, these methods are limited by the complexity of finding the applicable constrained kernel norm. Our work investigates accelerated patch-based methods at both the algorithmic and hardware levels.

3. Method

In this section, we present the details and principles of the proposed method. First, a novel continuous strategy is proposed for the suppression of strong edges. Then, APSVD is used to accelerate solution of the LRSD problem. Finally, the integration of our proposed method on GPU is presented. The overall framework is shown in Figure 1.

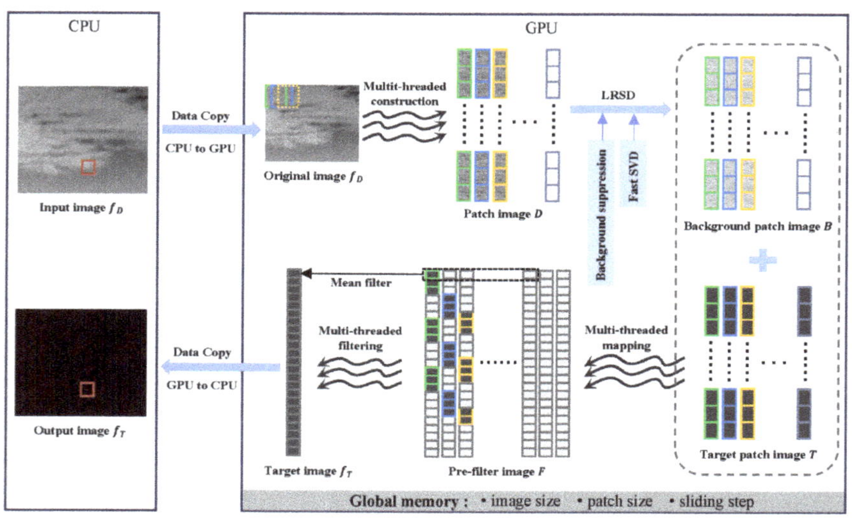

Figure 1. Framework of the proposed infrared small-target detection method. Targets in the input and output images are highlighted with red boxes.

3.1. BSPG Model

The infrared image is considered to be composed of the target image, background image, and noise image, formulated as

$$f_D = f_B + f_T + f_N, \tag{1}$$

where f_D, f_B, f_T, and f_N represent the original infrared, background, target, and noise images, respectively. The IPI model uses a sliding window (from top-left to bottom-right) to convert the original image into a patch image. The IPI model can be formulated as

$$D = B + T + N, \tag{2}$$

where D, B, T, and N represent the patch image, background patch image, target patch image, and noise patch image, respectively. Then, we can transform the small-target detection problem into the following convex optimization under broad conditions; that is,

$$\underset{B,T}{min} \|B\|_* + \lambda\|T\|_1, \quad s.t. \ \|D - B - T\|_F \leq \delta, \tag{3}$$

where $\|.\|_*$ represents the nuclear norm, $\|.\|_1$ represents the column sum norm, λ is a positive weighting parameter, and $\delta > 0$. This convex optimization problem is called robust principle component analysis (RPCA), which can recover low-rank and sparse parts of the data matrix even when a fraction of the entries are missing. Let $f(X) = \frac{1}{2}\|D - B - T\|_F^2$, $P(X) = \mu(\|B\|_* + \lambda\|T\|_1)$, where μ is a relaxation parameter. Hence, we can express Equation (3) as

$$minF(X) = f(X) + P(X). \tag{4}$$

The PG algorithm is an efficient method to solve the RPCA problem, which estimates the background image and the target image by minimizing the separable quadratic approximation sequence of Equation (4); that is,

$$\begin{aligned} Q(X,Y) &\doteq f(Y) + \langle \triangledown f(Y), X - Y\rangle + \frac{\tau}{2}\|X - Y\|_F^2 + P(X) \\ &= \frac{\tau}{2}\|(X - G)\|_F^2 + P(X) + f(Y) - \frac{1}{2\tau}\|\triangledown Y\|_F^2, \end{aligned} \tag{5}$$

where $G = Y - \frac{1}{L_f}\triangledown f$, L_f is the Lipschitz constant (which is set to 2 in this problem), and $\tau > 0$ is a given parameter. The following function has a unique optimal solution as Equation (5) is convex:

$$arg\,min\{Q_\tau(X, Y|X \in dom(P))\}, \tag{6}$$

where $dom(P) = \{X|P(X) < +\infty\}$. In our method, Equation (5) can be expressed as:

$$\begin{aligned} Q(B,T,\mu,Y_B,Y_T) = &\frac{\tau}{2}\|B - Y_B|_F^2 + \frac{\tau}{2}\|T - Y_T|_F^2 + f(Y_B,Y_T) \\ &+ \mu P(B,T) + \frac{1}{2\tau}|\triangledown f(Y_B,Y_T)|_F^2. \end{aligned} \tag{7}$$

To solve Equation (6), the iterative process of the PG algorithm repeatedly sets $X_{k+1} = arg\,minQ(X, Y_k)$, and Y_k is obtained from $X_0, X_1, ...X_k$. In our method, X_K to be solved are ordered pairs (B_k, T_k). Therefore, we set

$$Y_k = X_k + \frac{t_{k-1} - 1}{t_k}(X_k - X_{k-1}), \tag{8}$$

where t_k is the sequence satisfying ${t_{k+1}}^2 - t_{k+1} \leq {t_k}^2$.

The closed-form expression of X_{k+1} can be obtained by soft-thresholding the singular values. The soft-threshold operation is defined as

$$S_\varepsilon[x] \doteq \begin{cases} x - \varepsilon, & if \ x > \varepsilon, \\ x + \varepsilon, & if \ x < -\varepsilon, \\ 0, & otherwise. \end{cases} \tag{9}$$

Then,

$$\begin{aligned} T_{k+1} &= S_\tau[G_k^T] = S_{\frac{\lambda\mu}{L_f}}[G_k^T], \\ B_{k+1} &= S_\tau[G_k^B] = US_{\frac{\mu}{L_f}}V^T, \end{aligned} \tag{10}$$

where USV^T is the SVD of G_k^B.

The continuation strategy in [43] can speed up the convergence of the PG. This technique employs a decreasing sequence to derive $\bar{\mu}$, where μ_k is updated as follows:

$$\mu_k = max\{\eta\mu_{k-1}, \bar{\mu}\}, \quad 0 < \eta < 1. \tag{11}$$

The value of η affects the convergence of the algorithm. A larger decrease results in more G_k components to be retained, while fewer iterations to update $\bar{\mu}$ results in an inability to separate the target. Conversely, a smaller η decrease has the opposite effect. Figure 2 shows examples of the target images at different iterations. It can be seen that, in the early iterations, the strong edges are separated first as the low component of the background is retained at the highlights. This leads to many false alarms at the strong edges when the target is separated. This phenomenon motivates us to use different decrement rates for G_k^B and G_k^T. We set a higher rate for μ^B and set G_{k+1}^B to compute more singular values, in order to retain strong edges. We also set a lower rate for μ^T, ensuring that the target is decomposed into sparse parts when the algorithm converges. Thus, μ_k is updated as follows:

$$\begin{aligned} \mu_k^T &= max\{\alpha\mu_{k-1}^T, \bar{\mu}\}, \\ \mu_k^B &= max\{\beta\mu_k^T, \bar{\mu}\}, \end{aligned} \tag{12}$$

where $0 < \alpha, \beta < 1$. The solution of the BSPG algorithm is described in Algorithm 1.

The upper bound of the algorithm is discussed below. By denoting $\{X^k, Y^k, t^k\}$ as the sequence obtained by the algorithm with $t^k \geq \frac{k+2}{2}$, according to [44], for any $k \geq 1$, we have

$$F(X^k) - F(X^*) \leq \frac{2L_f\|X^* - X^0\|_F^2}{(k+1)^2}, \quad X^* \neq 0. \tag{13}$$

Then,

$$F(X^k) - F(X^*) \leq \epsilon, \tag{14}$$

when $k > k_0 + \sqrt{\frac{L_f}{\epsilon}}\|X_{k_0} - X^*\|_F$, where the convergence accuracy is $\epsilon > 0$, yielding that the algorithm has $O\left(\sqrt{\frac{L_f}{\epsilon}}\right)$ iteration complexity.

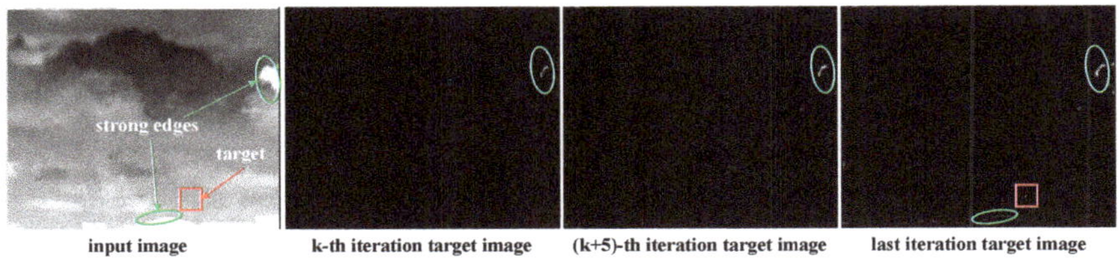

Figure 2. IPI target images at different iterations. Strong edges are preserved when the target is detected. Targets are shown in red boxes and strong edges are denoted by green circles.

Algorithm 1: BSPG solution via APSVD

Input : Patch image: $D \in \mathbb{R}^{m \times n}$, weighting parameters: $\lambda, \overline{\mu}, \alpha, \beta$
Output: $B \leftarrow B_k, T \leftarrow T_k$

Initialization: $k = 0, Y_B^0, Y_T^0, B_0, T_0, t_0, \mu_0$;
while *not converged* **do**
 update Y_B^k, Y_T^k by Equation (8);
 $G_k^B \leftarrow Y_k^B - 1/2(Y_k^B + Y_k^T - D); G_k^T \leftarrow Y_k^T - 1/2(Y_k^B + Y_k^T - D)$;
 update rank estimation quantity sv_k by Equation (15);
 //Approximate Partial SVD:
 $G_k^{B'} \leftarrow G_k^{B^T} \times G_k^B$;
 $(S, V)_{sv_k} \leftarrow partial\_eig(G_k^{B'}, sv_k)$;
 $U_{sv_k} \leftarrow G_k^B V_{sv_k} S_{sv_k}{}^{-1}$;
 update B_{k+1}, T_{k+1} by Equation (10);
 compute the current rank quantity:
 $sv_{k+1} = length\Big(find\big(diagS > \mu_k^B / L_f\big)\Big)$;
 update μ_{k+1}^T, μ_{k+1}^B by Equation (12);
 $k \leftarrow k+1$
end

3.2. APSVD

The most time-consuming step in each iteration of the PG algorithm is the execution of the full SVD. It is worth mentioning that the soft-threshold operation only leaves a portion of the singular values and vectors to participate in the subsequent calculations. In particular, few singular values are needed in early iterations. Therefore, it is feasible to replace full SVD with partial SVD. The crucial step of this strategy is rank estimation, which involves estimating the number of singular values and singular vectors participating in the computation after truncation. As the noise in infrared images is often not simply Gaussian distributed, estimation functions such as minimax estimator or the simple quantitative increase method proposed in [43] are not feasible. Due to the low-rank nature of the patch image, the singular value matrix has a clear trend of change, as shown in Figure 3. Thus, we estimate the rank by evaluating the degree of variation of the singular values. In the kth iteration, the pre-determined rank sv_{k-1} is initialized by the number of singular values in S greater than μ_k / L_f. We update sv_k as follows:

$$sv_k = \begin{cases} sv_{k-1} + 5, & if \ \dfrac{\sigma_{sv_k}}{\sigma_{sv_{k-1}}} < \delta, \\ sv_{k-1} + \lceil \gamma N \rceil, & otherwise, \end{cases} \tag{15}$$

where δ is the threshold for measuring the degree of singular value variation, N is the width of the patch image, and δ and γ are empirically set to 0.95 and 0.1, respectively.

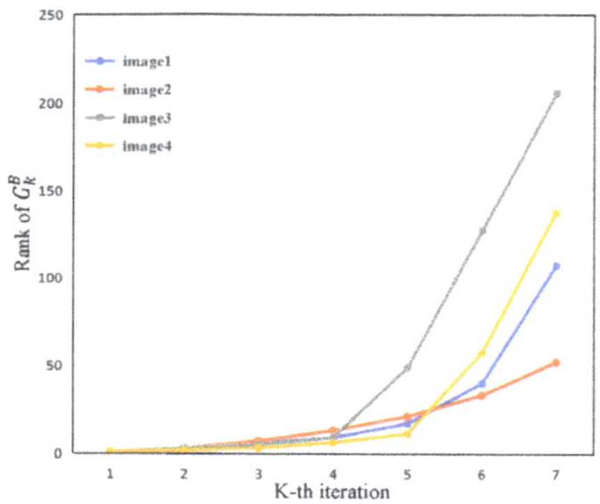

Figure 3. Trend of rank with an increasing number of iterations. Image 1 to image 4 represent the patch images corresponding to four different infrared images from the used data set.

The truncated SVD of the patch image needs to satisfy two requirements: (1) Only a small number of large singular values are retained, and (2) the singular vectors (including the left and right) need to be computed. We approximate the SVD of the patch image A by solving the eigenvalue decomposition of its covariance matrix $A^T A$. Given the slender nature of the patch matrix, the latter computation is considerably more straightforward compared to the former. Moreover, the symmetry of $A^T A$ guarantees its suitability for eigenvalue decomposition, leading to

$$\begin{cases} A = USV^T, \\ A^T A = VS^2V^T. \end{cases} \tag{16}$$

It is apparent, from Equation (16), that the eigenvalues of $A^T A$ correspond to the squares of the singular values of A. Additionally, both the right singular vectors of A and the eigenvectors of $A^T A$ are unitary. This relationship further implies that the left singular vectors U follow the equation $U = AVS^{-1}$. The approximate SVD of matrix A satisfies

$$U^T AV = S + \xi, \tag{17}$$

where ξ depends on the rounding errors. The singular values and singular vectors of the eigendecomposition in CUDA can approximate the accuracy of SVD to machine zero [45]. Furthermore, the larger the singular value, the smaller the error of U.

3.3. *GPU Parallel Implementation*

The GPU implementation consists of three main parts: Constructing the patch image, solving the LRSD problem, and reconstructing the patch image. Our method uses GPU for implementation purposes, and CPU only for data transfer and GPU control.

3.3.1. Construction

In order to reduce the number of data transfers between the host memory and the device's global memory, we first copy all the image data and hyperparameters read by the CPU to the GPU via PCI Express, and then execute the parallelization kernel functions. GPU parallelism mainly relies on data parallelism, i.e., performing the same operation on multiple data elements. Correspondingly, a patch image is constructed to change the storage location of each pixel in the original image. Therefore we build an index mapping between the original image and the patch image. This mapping allows a thread to manage the correspondence of a pixel position, facilitating the parallel processing of all pixels.

Let dw, dh denote the width and height of the sliding window, s_x, s_y denote the sliding step in the x, y directions, and p denotes the number of patches in the x direction. The mapping of the patch image pixel index I_x^p to the original image pixel index I_x^o is

$$
\begin{aligned}
I_x^o &= I_x^p \,\%\, p \times s_y + \frac{I_y^p}{dh}, \\
I_y^o &= \frac{I_x^p}{p} \times s_x + I_y^p \,\%\, dh.
\end{aligned}
\tag{18}
$$

The execution of the kernel function requires the determination of the thread block and grid size, where the grid size is determined based on the number of processing subtasks and the thread block size. Suppose the output image size n of the kernel function is $n_x \times n_y$ and the thread block size k is set to $k_x \times k_y$ (where n_x, n_y denote the size of the image in the x and y direction, respectively, and k_x, k_y denote the number of threads per block in the x and y direction, respectively), then the grid size is determined as $((n_x)_{k_x}, (n_y)_{k_y})$, where operator $(*)_k$ is defined as

$$
(n)_k = k\lceil * \rceil \frac{n}{k}, \quad k \in \mathbb{N}, \quad n \in \mathbb{R}. \tag{19}
$$

We set the thread block size for the construction kernel function based on the size of the patch image. Assuming that the patch image dimensions are $p_x \times p_y$, we configure the thread block size as $(p_x, \lfloor 1024/p_x \rfloor)$. For instance, given an image with a size of 200×150, a sliding window with a size of 50×50, and a sliding step of 10, the size of the resulting patch image would be 176×2500. Accordingly, the thread block size is set to $(176, 5)$ and the grid dimensions are set to $(1, 500)$. Consequently, a row of threads in the x direction corresponds to a row within the patch image. The construction process is carried out pixel-by-pixel. The processing time of each pixel is assumed to be t, for an $M \times N$ patch image, serial execution of the construction module takes $M \times N \times t$. In contrast, our method operates $M \times N$ threads in parallel, completing the process in time t. The theoretical speedup ratio is $M \times N$.

3.3.2. Reconstruction

Figure 4 shows the steps for reconstructing the background and target patch images after LRSD. First, the target patch image is transformed into the pre-filtered image. We provide the pseudo-code for this transformation in Algorithm 2. Then, the indices of the first and last patches containing valid information are determined. Finally, filtering is performed on the valid portions of each row to obtain the target image. In summary, the reconstruction includes two parallel processes: One involving mapping from the patch image to the pre-filter image, and another entailing filtering.

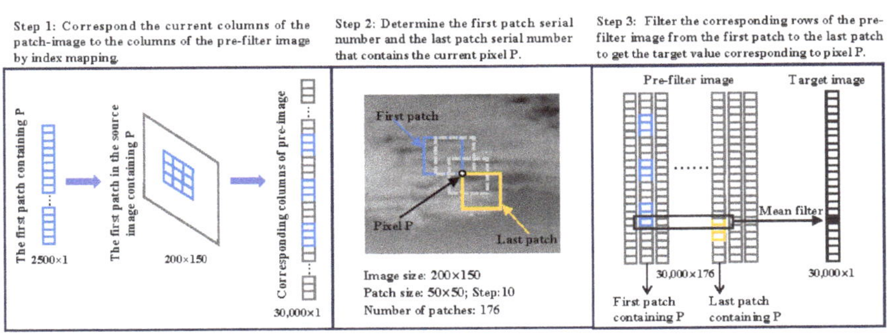

Figure 4. Three steps of the proposed reconstruction method.

Algorithm 2: The mapping of patch image and pre-filter image

Input : Patch image D, original image size w and h, patch size dw and dh, step s, patch number of per row p_r

Output: pre-filter image F

Step I: Compute the index I_D in the patch image by using the row and column numbers R_D, C_D.

$R_D = blockIdx.y \times blockDim.y + threadIdx.y;$

$C_D = blockIdx.x \times blockDim.x + threadIdx.x;$

$I_D = C_D \times dw \times dh + R_D;$

Step II: Compute the row index I_p^r and column index I_p^c of the patch by using patch index I_p.

$I_p = I_D/(dw \times dh);$

$I_p^r = I_p/p_r;\ I_p^c = I_p\%p_r;$

Step III: Compute the index I_F in F by using the row number R_O and column number C_O in the original image O.

$R_O = I_p^r \times s + R_D\%dh;$

$C_O = I_p^c \times s + R_D/dh;$

$I_F = C_D \times w \times h + C_O \times h + R_O;$

$F[I_F] = D[I_D].$

We set the thread block and grid of the mapping kernel to the same size as the construction kernel. The filter kernel handles a much larger matrix, and in order to improve the resource usage, the thread block size is set to $(32, 32)$, and the grid size is obtained according to Equation (19). Then, the indices of the first patch and the last patch I_p^f, I_p^l containing valid information can be expressed as:

$$\begin{aligned} I_p^f &= \frac{I_x^o - dw}{s_x} + 1 + (\frac{I_y^o - dh}{s_y} + 1) \times p, \\ I_p^l &= \frac{I_x^o}{s_x} + 1 + (\frac{I_y^o}{s_y} + 1) \times p. \end{aligned} \tag{20}$$

In NVIDIA's GPU architecture, one warp typically consists of 32 threads, while our thread block contains 32×32 threads. This means that each warp can effectively execute an entire thread block. This high warp occupancy rate reaches 100%, efficiently harnessing the performance of the GPU.

3.3.3. APSVD Using CUDA

The key to implementing the APSVD is the exact eigendecomposition, which can be achieved using the Symmetric Eigenvalue Divide (SYEVD) function based on QR decomposition or the Symmetric Eigenvalue Jacobi (SYEVJ) function based on Jacobi decomposition [46]. SYEVD employs a divide-and-conquer method to decompose a symmetric matrix into smaller sub-problems and solves them recursively. Its runtime is primarily attributed to QR decomposition. QR decomposition can be expressed as

$$A_{m\times n} = Q_{m\times n} R_{n\times n}, \tag{21}$$

where Q is an orthogonal matrix and R is an upper triangular matrix. QR decomposition usually requires Householder transformations for multiple iterations, and each transformation needs to manipulate all elements of the matrix, which becomes redundant for small matrices.

SYEVJ transforms the symmetric matrix A into a diagonal matrix D by performing a rotational transformation via the bilateral Jacobi method.

$$D = \cdots J_3^T(J_2^T(J_1^T A J_1)J_2)J_3 \cdots = (\cdots J_3^T J_2^T J_1^T) A (J_1 J_2 J_3 \cdots), \tag{22}$$

where J is denoted as $J(i, j, \theta)$, contains the rotation angle θ and an index pair (i, j), and satisfies $J(i, j, \theta)^T J(i, j\theta) = E$. The Jacobi method, with its element-wise rotations, offers lower computational complexity, localized memory access, and parallelization potential, making it more efficient for small matrices.

To quantitatively analyze both methods, we introduce arithmetic intensity [46] which is a metric used to evaluate the performance of parallel computational tasks. Specifically, the arithmetic intensity I is defined as

$$I = \frac{\text{FLOPs}}{\text{bytes loaded}}, \tag{23}$$

where FLOPs represents the number of floating point operations and can measure the complexity of an algorithm, bytes loaded represents the number of bytes loaded from memory during kernel execution. Given an $N \times N$ matrix (single-precision), a Givens rotation typically requires 8 floating-point operations (2 trigonometric functions, 4 multiplications, 2 additions) and loads $2N$ elements. This means that the number of FLOPS per iteration is 8, and the memory access requires loading $8N$ bytes. In QR decomposition, the computational complexity of each iteration, which involves Householder transformations, is $2N^3$, and it loads the entire matrix, including $4N^2$ bytes. The arithmetic intensity of the Jacobi kernel I_J and the arithmetic intensity of the QR decomposition kernel I_{QR} can be expressed as

$$I_J = \frac{8}{2N}, \quad I_{QR} = \frac{2N^3}{4N^2}. \tag{24}$$

Therefore, from the perspective of arithmetic intensity, we choose the more efficient Jacobi method to perform eigenvalue decomposition. The Jacobi method typically has lower arithmetic intensity and is relatively memory-access efficient, whereas QR decomposition involves orthogonal transformations and matrix updates, often requiring more memory bandwidth and computation. QR decomposition has a higher arithmetic intensity, making it perform better on larger matrices where the high arithmetic intensity can be fully utilized, but not on small matrices.

Furthermore, the implementation of APSVD requires an efficient matrix multiplication function. The General Element-wise Matrix Multiply (GEMM) function in CUDA takes advantage of the GPU's parallel computing capabilities and efficiently processes substantial amounts of data, thereby enhancing computational performance. To reduce routing errors, we use the double precision-controlled GEMM function, DGEMM, which has a time complexity of $2MN^2$. Notably, other functions utilize single precision to strike a balance between instruction throughput and accuracy. Figure 5 illustrates the runtime ratios for each component of APSVD on matrices of varying size. Notably, the efficiency of SYEVJ is demonstrated, as it is unaffected by the matrix height and exhibits a decreasing ratio of time spent on eigendecomposition as the matrix size increases.

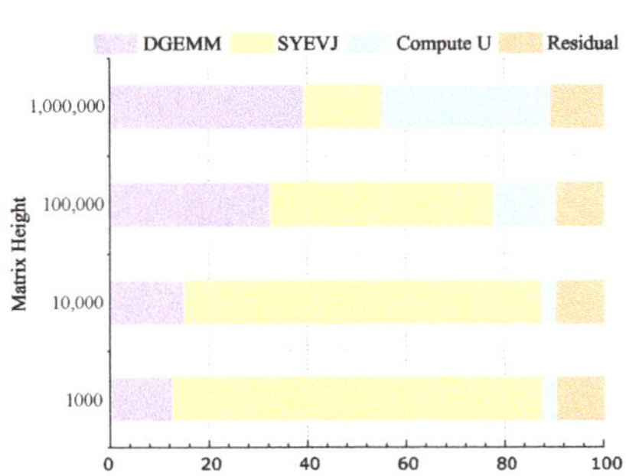

Figure 5. The running time ratio of each component in APSVD. The matrix is taken from SIR_1 in the experiment with a fixed width of 32.

4. Experiments and Analysis

In this section, we provide an evaluation of our method in terms of detection accuracy and execution time.

4.1. Experimental Setup

Data. The experiments used the real single-frame infrared images provided in [32] selected from infrared sequences in a variety of scenes, including ocean, cloud, sky, and urban areas, as shown in Figure 6. The targets are marked with red boxes and magnified for convenient viewing in the bottom left corner of each image. It can be seen that the targets occupy very few pixels; most small targets lack shape and texture information and have low intensity. Images with poor imaging quality exhibit strong noise, such as SIR_2 and SIR_3. Some small targets are submerged in cloud or sea clutter as SIR_11 and SIR_13 and suffered from highlight backgrounds with strong edges as SIR_8 to SIR_14. Furthermore we provide detailed information of the test images, including the background type, target type, Signal Clutter Ratio (SCR), target size, and detection challenges, in Table 1.

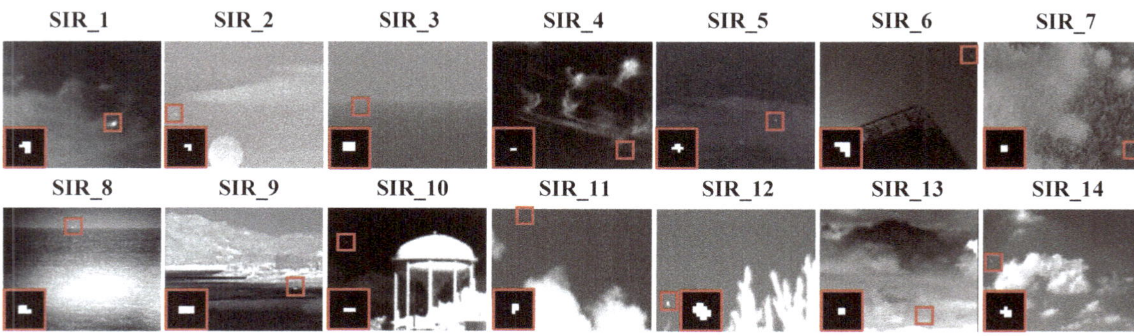

Figure 6. Test images SIR_1 to SIR_14. The targets are highlighted with red boxes and the binary mask of the target is given in the lower left corner of each image.

Table 1. Detailed information of the test images. The target size is expressed as the number of pixels.

Data	Image Size	Target Size	SCR	Background Type	Target Type	Detection Challenges			
						Strong Edge	Low Contrast	Heavy Noise	Cloud Clutter
SIR_1	256 × 172	11	6.52	cloud + sky	Irregular shape			✓	
SIR_2	256 × 239	3	8.63	building + sky	Weak		✓	✓	✓
SIR_3	300 × 209	12	1.04	sea + sky	Low intensity		✓	✓	
SIR_4	280 × 228	2	3.09	cloud + sky	Weak, hidden		✓		✓
SIR_5	320 × 240	7	11.11	cloud + sky	Hidden			✓	
SIR_6	359 × 249	6	6.14	building + sky	Irregular shape		✓		
SIR_7	640 × 512	4	10.52	cloud + sky	Weak, hidden	✓			✓
SIR_8	320 × 256	5	5.36	sea + sky	Weak	✓			
SIR_9	283 × 182	8	1.59	cloud + sea	Hidden	✓			✓
SIR_10	379 × 246	3	10.57	building + sky	Low intensity	✓	✓		
SIR_11	315 × 206	5	9.61	cloud + sky	Low intensity	✓			✓
SIR_12	305 × 214	17	8.43	tree + sky	Irregular shape	✓			
SIR_13	320 × 255	4	4.12	cloud + sky	Low intensity	✓	✓		✓
SIR_14	377 × 261	6	2.38	cloud + sky	Low intensity	✓			✓

Hardware. We implemented our method on the embedded GPU Jetson AGX Xavier, which has 7764 MB of global memory and 48 KB of shared memory. The version of CUDA was 10.2. The experiments in MATLAB were based on an Intel(R) Core(TM) i7-8750H CPU with 8 GB RAM.

Baselines and parameter settings. We compared our proposed method to other state-of-the-art patch-based methods, including IPI [17], NIPPS [20], NRAM [22], NOLC [23], SRWS [34] and HLV [26]. As tensor-based methods have better performance in terms of computational efficiency, we also included three tensor-based methods for comparison, including RIPT [36], PSTNN [38], PFA [37], LogTFNN [39] and ANLPT [42]. The parameter settings are provided in Table 2. We employed a sliding window with a size of 100×100 and a step size of 30 on images with resolution equal to or exceeding 640×512. This setting ensured that the execution time remained within the desired range.

Table 2. Parameter settings. All methods used their original settings.

Method	Patch Size	Step	Parameter
IPI [17]	50 × 50	10	$L = 1, \lambda = L/\sqrt{\min(m,n)}, \varepsilon = 10^{-7}$
RIPT [36]	50 × 50	10	$L = 1, \lambda = L/\sqrt{\min(n_1,n_2,n_3)}, \varepsilon = 10^{-7}$
NIPPS [20]	50 × 50	10	$L = 1, \lambda = L/\sqrt{\min(m,n)}, \varepsilon = 10^{-7}$
NRAM [22]	50 × 50	10	$L = 1, \lambda = L/\sqrt{\min(m,n)}, \varepsilon = 10^{-7}$
NOLC [23]	50 × 50	10	$L = 1, \lambda = L/\sqrt{\min(size(D))}, p = 0.5, \varepsilon = 10^{-7}$
PSTNN [38]	40 × 40	40	$L = 0.7, \lambda = L/\sqrt{\min(n_1,n_2) * n_3}, \varepsilon = 10^{-7}$
SRWS [34]	50 × 50	10	$L = 1, \lambda = L/\sqrt{\min(m,n)}, \gamma = 0.09/\sqrt{min(m,n)}, \varepsilon = 10^{-7}$
PFA [37]	25 × 25	25	$\kappa = 30, \tau_0 = 1e+5, \varepsilon = 10^{-5}$
LogTFNN [39]	40 × 40	40	$L = 1, \lambda = L/\sqrt{\min(n_1,n_2) \times n_3}, \beta = 0.01, \mu = 200$
HLV [26]	50 × 50	10	$L = 1, \lambda = L/\sqrt{\max(m,n)}, \alpha = 1.3, \beta = 2.5, C = 8$
ANLPT [42]	50 × 50	10	$\lambda = sigmoid(E/n_3)/\sqrt{\min(n_1,n_2) \times n_3}, E = entropy(T)$
Ours	50 × 50	10	$L = 1, \lambda = L/\sqrt{\max(m,n)}, \varepsilon = 10^{-7}$

Evaluation metrics. We used two quantitative analysis evaluation indicators commonly used for small-target detection to evaluate our method: Signal Clutter Ratio Gain (SCRG) and Background Suppress Factor (BSF). SCRG reflects the effect of increasing target saliency, and is defined as follows:

$$SCRG = \frac{SCR_{out}}{SCR_{in}}, SCR = \frac{|\mu_t - \mu_b|}{\sigma_b}, \tag{25}$$

where SCR_{in} and SCR_{out} represent the signal-to-clutter ratio of the input and output images, respectively, μ_t represents the average pixel gray value of the target, μ_b represents the average pixel gray value of the local background around the target, and σ_b represents the standard deviation of the gray pixel value of the local background around the target. BSF reflects the effect of suppressing background interference and is defined as follows:

$$BSF = \frac{\sigma_{in}}{\sigma_{out}}, \tag{26}$$

where σ_{out} and σ_{in} are the standard deviation values of the local background around the target in the output image and the original image, respectively.

We also analyzed the results using Receiver Operating Characteristic (ROC) curves. The ROC curve is plotted by assessing the True Positive Rate (TPR) and the False Positive Rate (FPR) at different classification thresholds, as defined below:

$$\begin{aligned} TPR &= \frac{number\ of\ real\ targets\ detected}{number\ of\ real\ targets}, \\ FPR &= \frac{numer\ of\ false\ targets\ detected}{number\ of\ real\ targets}. \end{aligned} \tag{27}$$

To quantitatively compare the ROC curves, the area under the curve (AUC) can be used as an evaluation criterion; the larger the AUC, the more accurate the detection performance.

4.2. Visual Comparison with Baselines

The visualization results of twelve detection methods are shown in Figures 7 and 8. IPI, RIPT, and NIPPS only successfully detect targets under simple backgrounds with high local patch similarity. In images with complex backgrounds, NIPPS present noticeable clutter, while IPI and RIPT exhibit noise in highlighted backgrounds. NRAM and PFA can detect the majority of targets, but they are prone to generating false alarms in regions with strong edges. NOLC and SRWS suffer from the sea surface background and can not detect weak dark targets. PSTNN has similar poor performance, with many false alarms under clutter. LogTFNN is poorly detected under high-intensity backgrounds, leaving a large amount of background residue. HLV and ANLPT perform poorly when detecting targets that are dark or have low contrast with the neighboring background. In the case of images SIR_7 to SIR_14, the highlighted backgrounds cause most methods to produce false alarms near strong edges, leading to inaccurate detection. However, our method excels in terms of effectively suppressing strong edges under such conditions. From the visualized detection results, our method exhibits robust detection performance.

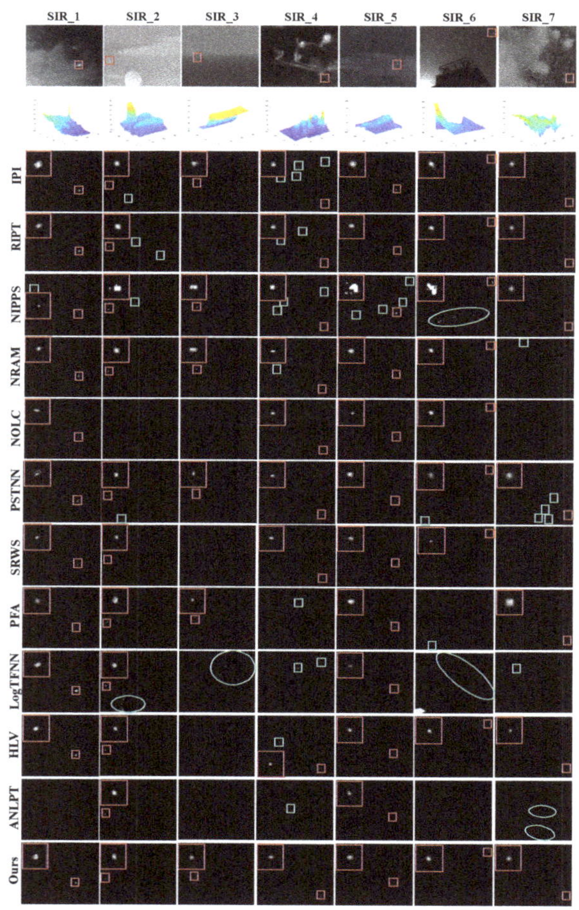

Figure 7. Partial detection results for different methods. The correctly detected targets are highlighted with red boxes and enlarged in the top left corner of each target image. The incorrect targets are highlighted with green boxes and circles.

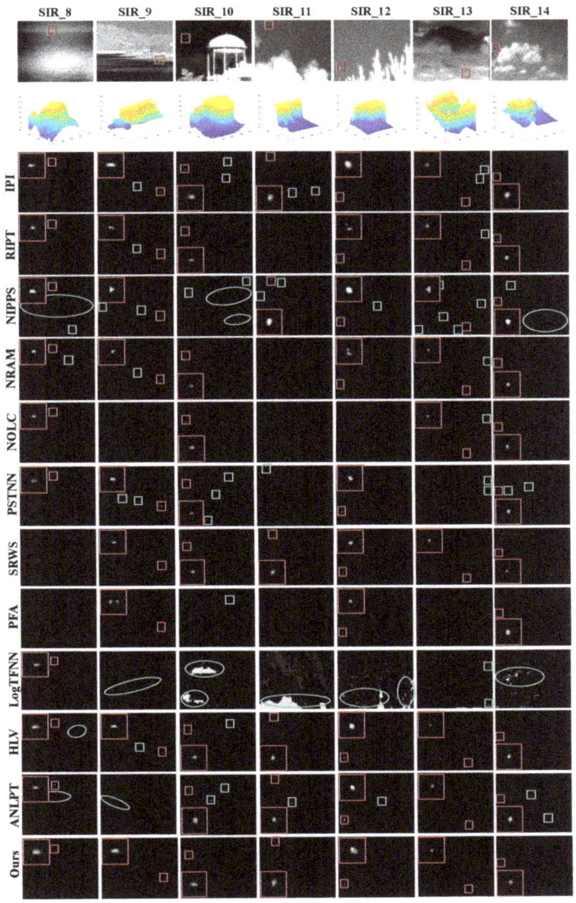

Figure 8. Partial detection results for different methods. The correctly detected targets are highlighted with red boxes and enlarged in the top left corner of each target image. The incorrect targets are highlighted with green boxes and circles.

4.3. Quantitative Evaluation and Analysis

The results of the quantitative comparison of the various methods on the test images are shown in Table 3. From the definitions of the two evaluation metrics, the larger the SCRG and BSF values, the better the detection results. In terms of BSF, our method performs better than the other methods on all images, indicating that our method excels in suppressing the background. In terms of SCRG, our method outperforms the other methods on most images. When the target is missed, the local background standard deviation is 0. Consequently, this leads to the SCRG appearing as a NaN result and BSF tending to infinity. In addition, we plotted the ROC curves corresponding to the experiments, in order to further validate the effectiveness of our method. The results in Figure 9 demonstrate that our method outperforms the other comparative methods on the test images. IPI and NIPPS are sensitive to clutter and high-intensity backgrounds due to the difficulty in distinguishing between targets and strong edges. NRAM and NOLC exhibit instability in detecting dim and weak targets. HLV experiences an increase in false alarms when dealing with strong clutter interference on the sea surface. Tensor-based detection methods show stable performance in simple backgrounds. However, RIPT and LogTFNN exhibit low detection accuracy in high-intensity backgrounds due to their high requirements for sparsity. PSTNN and PFA tend to erroneously reject targets in the presence of background clutter. SRWS demonstrates effective clutter suppression in high-intensity backgrounds but struggles to detect low-contrast weak targets. ANLPT exhibits weaker clutter suppression capabilities in high-intensity structured backgrounds. In contrast our method achieves remarkable results in terms of detection accuracy and false alarm rate in a wide range of scenarios.

To evaluate the execution speed of our method, we compared our method with other patch-based methods. For fair comparison, we set these methods to share the same patch and step configurations. We implemented our method in two versions; that is, CPU and GPU versions. The GPU execution time includes the data transfer time between the host and the device. To ensure the reliability of the time statistics, we took the average time of 10 executions for each method. The comparison results are presented in Table 4. Most patch-based methods are time-consuming due to iteration and complex matrix decomposition including IPI, NIPPS, and NRAM. RIPT and ANLPT relatively improve the detection efficiency, but the tensor decomposition is still complex. SRWS and HLV optimise the iterative termination conditions to achieve a faster detection speed. The results demonstrate that our method achieved impressive speed, particularly with significant acceleration when using the GPU. Combined with the previous detection accuracy evaluation, our method was found to achieve faster detection while maintaining higher accuracy.

Table 3. Comparison of SCRG and BSF under various methods. The best performance is indicated in **bold**.

Methods		IPI [17]	RIPT [17]	NIPPS [20]	NRAM [22]	NOLC [23]	PSTNN [38]	SRWS [34]	PFA [37]	LogTFNN [39]	HLV [26]	ANLPT [42]	Ours
SIR_1	SCRG	2.08	2.55	0.05	2.76	2.58	1.81	2.78	0.03	1.56	2.85	NaN	**20.67**
	BSF	1.51	2.26	3.45	2.82	1.98	1.31	5.54	4.50	1.14	2.14	Inf	**32.40**
SIR_2	SCRG	3.29	2.38	1.17	2.89	NaN	3.13	5.20	0.91	1.82	4.24	3.40	**23.50**
	BSF	1.05	0.59	0.26	0.75	Inf	0.80	2.48	0.40	0.48	1.08	0.83	**7.20**
SIR_3	SCRG	137.56	NaN	102.47	**235.38**	NaN	90.23	NaN	32.40	11.80	NaN	NaN	151.21
	BSF	11.39	Inf	5.99	17.02	Inf	18.42	Inf	12.10	1.28	Inf	Inf	**19.48**
SIR_4	SCRG	16.36	15.36	9.46	**Inf**	39.94	**Inf**	60.86	NaN	NaN	16.74	NaN	**Inf**
	BSF	3.55	3.47	2.04	**Inf**	8.80	**Inf**	13.90	Inf	Inf	3.61	Inf	**Inf**
SIR_5	SCRG	2.18	5.60	0.68	4.79	4.96	1.53	6.57	2.39	1.41	1.82	0.01	**7.81**
	BSF	0.77	2.07	0.16	1.63	1.72	0.49	2.49	0.80	0.71	0.61	0.68	**3.59**
SIR_6	SCRG	28.99	17.08	7.77	**Inf**	26.84	NaN	**Inf**	NaN	NaN	2.56	NaN	**Inf**
	BSF	32.21	6.18	1.96	**Inf**	8.08	Inf	**Inf**	Inf	Inf	0.90	Inf	**Inf**
SIR_7	SCRG	275.57	**Inf**	**Inf**	NaN	NaN	5.36	NaN	2.13	3.42	351.29	NaN	**Inf**
	BSF	169.41	**Inf**	**Inf**	Inf	Inf	3.30	Inf	1.69	2.47	215.97	Inf	**Inf**
SIR_8	SCRG	7.53	32.22	7.89	17.04	41.07	6.16	NaN	2.40	3.48	8.75	4.76	**90.97**
	BSF	3.98	25.67	3.28	9.77	43.57	4.50	Inf	192.28	1.82	4.88	2.39	**69.74**
SIR_9	SCRG	24.34	25.51	11.86	**Inf**	NaN	14.85	**Inf**	5.41	18.08	23.11	NaN	**Inf**
	BSF	12.92	24.33	9.04	**Inf**	Inf	7.95	**Inf**	10.00	9.44	12.42	Inf	**Inf**
SIR_10	SCRG	1.94	**Inf**	0.38	**Inf**	3.37	**Inf**	4.31	NaN	NaN	2.39	2.02	**Inf**
	BSF	1.04	**Inf**	0.16	**Inf**	1.85	**Inf**	2.47	Inf	Inf	1.30	1.36	**Inf**
SIR_11	SCRG	2.57	NaN	0.87	NaN	NaN	NaN	10.58	NaN	0.06	**Inf**	1.73	**Inf**
	BSF	0.28	Inf	0.07	Inf	Inf	Inf	1.46	Inf	0.05	**Inf**	0.18	**Inf**
SIR_12	SCRG	1.47	**Inf**	1.42	**Inf**	NaN	1.91	1.11	**Inf**	0.52	1.75	1.14	**Inf**
	BSF	0.73	**Inf**	0.62	**Inf**	Inf	1.02	1.75	**Inf**	0.25	0.91	0.55	**Inf**
SIR_13	SCRG	1.58	**Inf**	0.30	**Inf**	**Inf**	5.67	**Inf**	NaN	NaN	31.94	**Inf**	**Inf**
	BSF	0.53	**Inf**	0.08	**Inf**	**Inf**	3.40	**Inf**	Inf	Inf	6.23	**Inf**	**Inf**
SIR_14	SCRG	4.28	7.69	1.87	8.26	**Inf**	3.25	**Inf**	1.58	0.52	7.25	5.73	**Inf**
	BSF	1.55	2.79	0.48	3.09	**Inf**	1.14	**Inf**	0.63	0.19	2.88	2.06	**Inf**

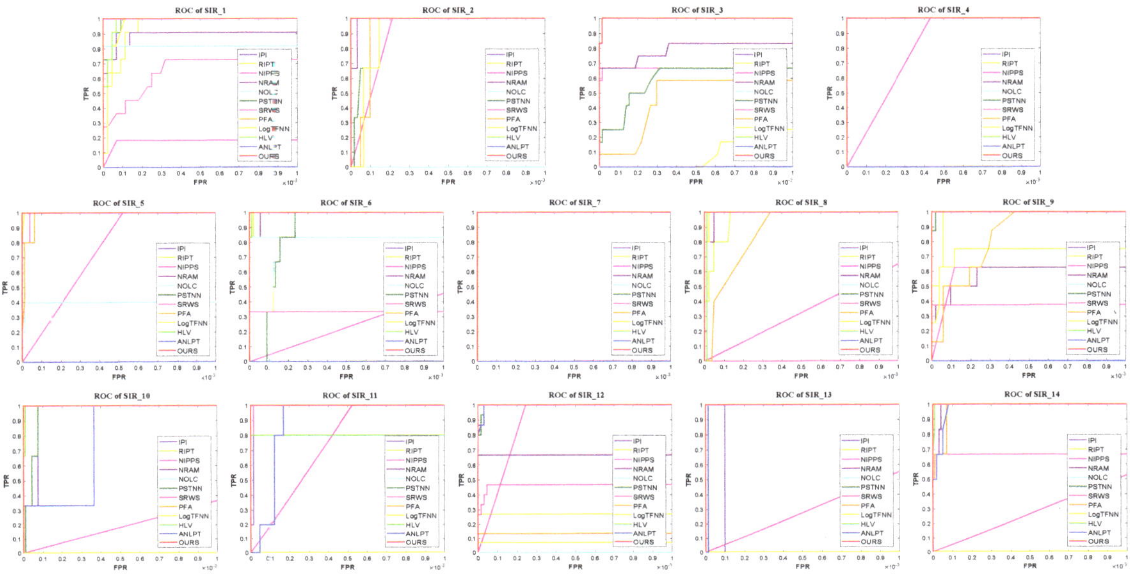

Figure 9. ROC curves for the twelve methods on test images SIR_1 to SIR_14.

Table 4. Comparison of the execution time (s) across various patch-based methods. All baselines were tested on the CPU, while our methods were tested on both the CPU and GPU. The best time is denoted in **bold**, while the second-best is underlined.

Image id	SIR_1	SIR_2	SIR_3	SIR_4	SIR_5	SIR_6	SIR_7	SIR_8	SIR_9	SIR_10	SIR_11	SIR_12	SIR_13	SIR_14
IPI [17]	3.28	5.23	7.63	6.45	12.52	12.93	12.67	11.28	4.12	15.32	7.88	7.29	14.87	18.72
RIPT [36]	1.17	2.76	2.02	2.82	4.70	2.88	8.01	4.35	0.96	1.85	1.02	1.40	2.12	2.14
NIPPS [20]	1.88	3.34	3.60	3.56	5.51	6.82	7.11	6.71	2.84	9.18	3.95	3.99	7.51	9.96
NRAM [22]	2.17	2.14	1.55	2.61	2.99	3.88	2.38	4.79	1.44	4.20	2.09	2.27	3.94	4.20
NOLC [23]	0.72	0.86	1.11	1.15	1.24	1.67	3.62	1.64	0.94	3.17	1.55	1.28	1.33	2.11
SRWS [34]	2.01	2.01	1.10	3.12	2.12	2.60	3.65	1.63	0.78	1.57	1.01	1.29	1.46	1.77
HLV [26]	1.13	1.76	2.32	1.55	2.86	4.51	4.26	3.54	1.44	4.47	2.30	2.27	4.01	6.09
ANLPT [42]	1.53	1.79	1.91	1.73	2.05	2.18	8.07	2.57	1.53	2.29	1.99	2.15	2.52	2.80
Ours (CPU)	0.49	0.76	0.94	0.93	1.55	1.94	2.10	1.29	0.53	1.77	0.86	0.87	1.64	1.89
Ours (GPU)	**0.34**	**0.42**	**0.54**	**0.52**	**0.87**	**0.98**	**0.54**	**0.90**	**0.36**	**0.84**	**0.47**	**0.42**	**0.82**	**0.85**

4.4. Ablation Study

The effect of relaxation parameters. We explored the effects of the relaxation parameters α and β on the detection accuracy of our method using ROC curves. Figure 10 shows that excessively large or small values of α and β led to a decrease in the AUC value. Small α can retain more background components, but overly small values result in insufficient iteration, thereby failing to separate targets. Meanwhile, large α and β values can cause false alarms by decomposing strong edges into sparse target portions. Therefore, we set α and β to 0.4 and 0.7, respectively.

Comparison with other tensor-based methods. For a fair comparison, we studied the tensor-based methods under the same settings as used for ours. We evaluated their detection accuracy and execution time under various patch and step configurations. Table 5 shows the execution time results. When using the same patch and step settings, our method on GPU is faster than PFA, PSTNN, and LogTFNN. As illustrated in Figure 11, our method consistently achieves the best detection accuracy across most scenarios. However, the detection accuracy of PFA, PSTNN, and LogTFNN significantly diminishes between images with variations in the patch size and step values.

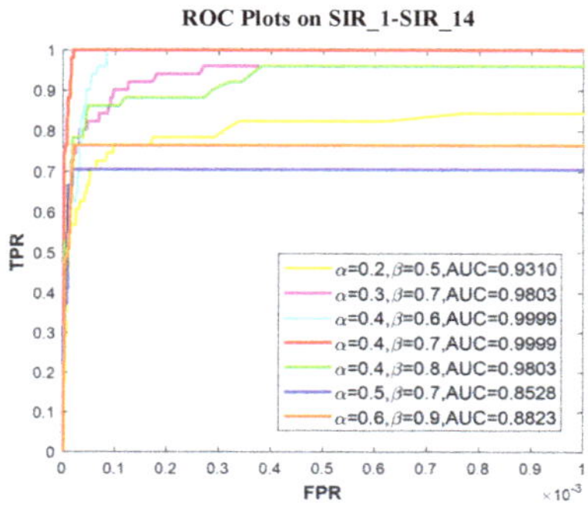

Figure 10. ROC curves of our method under different α and β values.

Table 5. Execution time (s) of PFA, PSTNN, LogTFNN and the proposed method (Ours) under varying patch size and number of steps. The best is denoted in **bold**, while the second-best is underlined.

Method	SIR_1			SIR_2			SIR_3		
(Patch, Step)	(25,25)	(40,40)	(50,10)	(25,25)	(40,40)	(50,10)	(25,25)	(40,40)	(50,10)
PFA [37]	9.96	0.33	1.39	12.68	0.26	1.69	0.33	0.26	2.19
PSTNN [38]	0.04	0.05	1.15	0.06	0.07	3.90	0.16	0.06	1.44
LogTFNN [39]	0.89	1.33	15.06	1.22	1.81	11.63	1.27	1.38	26.92
Ours(CPU)	0.12	0.13	0.49	0.19	0.17	0.76	0.16	0.14	0.94
Ours(GPU)	**0.02**	**0.02**	**0.34**	**0.04**	**0.02**	**0.42**	**0.02**	**0.01**	**0.54**

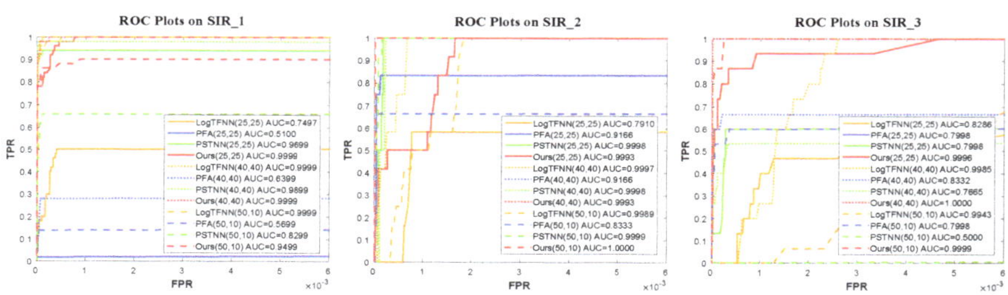

Figure 11. ROC curves for PFA, PSTNN, LogTFNN and the proposed method (Ours). The patch size and step are labeled in the figure; for example, (25,25) means that the patch size was set to 25×25 and the step was set to 25.

The acceleration effect of different strategies. To validate the speed enhancement due to our proposed APSVD, we compared its implementations on both MATLAB 2017b and CUDA 10.2 platforms with various SVD functions, as shown in Table 6. It can be observed that APSVD exhibits high efficiency on slender matrices. To explore the effectiveness of the proposed acceleration strategies, we conducted tests on their cumulative acceleration effects over the baseline method IPI. Table 7 demonstrates that our proposed acceleration strategies yield commendable speed increases. Notably, PASVD avoids the intricate decomposition of large matrices, thus significantly saving time; the new continuation strategy employs a greater decrement for the relaxation parameters, reducing the number of iterations and consequently accelerating the overall speed; and the GPU parallel strategies provide significant acceleration, especially for larger images.

Table 6. Execution times (ms) of SVD functions in MATLAB and CUDA on different matrices, with a fixed matrix width of 32 and a rank of 10 for partial SVD. The fastest time on MATLAB and CUDA is marked in **bold**.

Matrix Height	MATLAB					CUDA		
	SVD	SVDS	Lanczos	RSVD	APSVD	SGESVD	SGESVDJ	APSVD
1000	1.03	6.68	4.59	7.67	**0.53**	9.07	5.75	**1.06**
10,000	6.27	19.93	22.08	10.05	**1.83**	16.71	7.36	**1.24**
100,000	280.12	406.70	298.77	50.82	**11.82**	/	24.06	**9.58**

Table 7. Cumulative acceleration effects at different sizes of SIR_1, obtained using the resize function.

Image Size	Base	+PASVD	+New Continuation	+GPU Parallelism
200 × 150	1.42	0.86	0.29	0.09
280 × 228	6.23	4.91	1.12	0.41
320 × 256	12.77	9.24	1.99	0.74
640 × 512	13.60	7.33	2.31	0.59
1020 × 750	57.79	34.6	7.32	2.38
1260 × 1024	207.13	116.58	22.20	3.65

The acceleration effects on images with different attributes. To validate the acceleration effect of the proposed method, we conducted experiments on images with varying attributes (i.e., resolution and background complexity). As shown in Figure 12, the acceleration effect of our method becomes more pronounced with increasing image resolution. On an image with a resolution of 1024 × 1020, the execution speed of the proposed method is nearly 60 times faster than that of IPI. Due to the influence of image complexity on execution time, the acceleration effect varies slightly at the same resolution. Furthermore, we conducted a comparative analysis of the three stages of our method—namely, constructing a patch image, solving the LRSD problem, and reconstructing a patch image—as shown in Figure 13. It is evident that multi-threading parallelism and optimized memory access significantly reduce the time required for the construction and reconstruction modules. Additionally, the new continuation strategy and APSVD greatly contribute to reducing the time required to solve the LRSD problem.

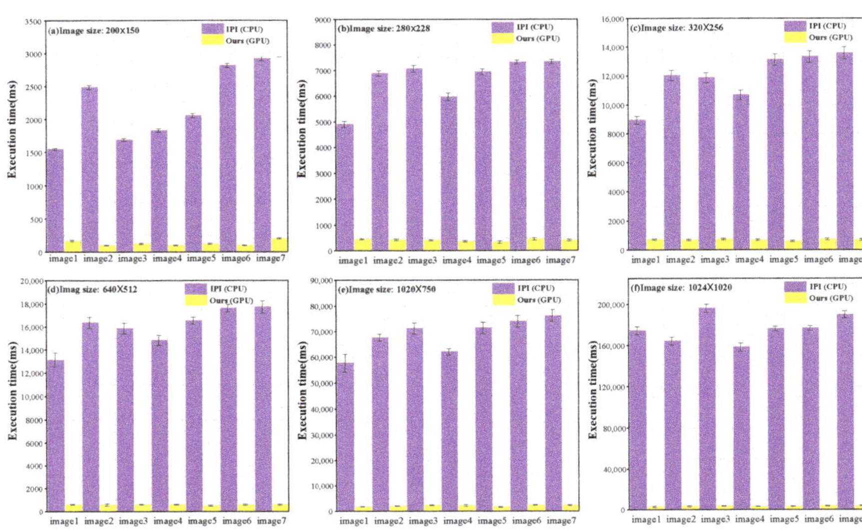

Figure 12. Comparison of execution time between IPI and the proposed method (Ours) for images of different resolution and complexity.

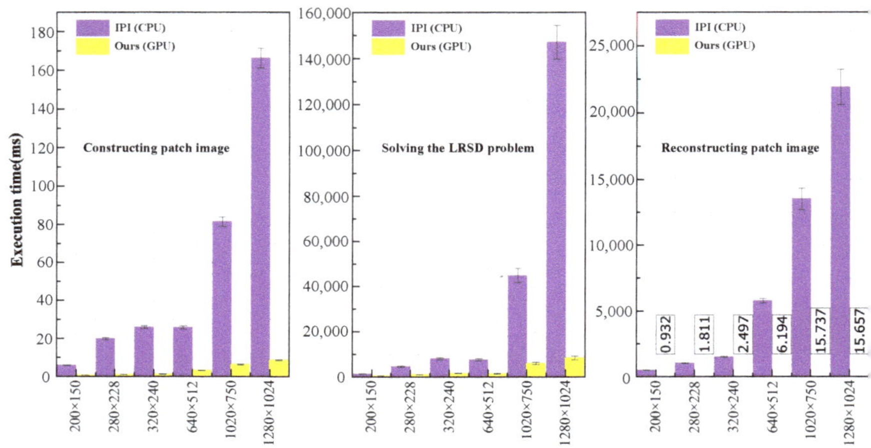

Figure 13. Comparison of execution time between IPI and the proposed method (Ours) for the three parts.

5. Discussion

Patch-based methods are well-studied in single-frame infrared small target detection for their reliability. The classical IPI algorithm is a notable example. It transforms the original image into a patch image and leverages the non-local self-similarity of the background to enhance the low-rank property of the patch image. This method allows for effective infrared small target detection through low-rank and sparse decomposition.

In our comparison of small target detection methods, we observed that methods like NIPPS, NRAM, and NOLC aim to improve detection accuracy by enhancing the nuclear norm and l1 norm. However, these methods involve complex matrix decomposition and iterative processes, leading to time-consuming issues. These methods also struggle to differentiate between the edges and the actual targets due to the local sparsity of strong edges. Conversely, SRWS and HLV, with their proposed multi-subspace assumptions and high local variance constraints, generally perform well in most cases. They effectively suppress strong edges but may miss dark and weak targets. Additionally, these methods require complex matrix decomposition, making them time-consuming. RIPT expands the patch model into tensor space, adding to the computational burden as tensors are unfolded and decomposed. While tensor-based methods such as PSTNN, PFA, and logTFNN have accelerated detection somewhat, their effectiveness is limited by the challenges of accurately approximating nuclear norms within tensor models.

This paper aims to strike a balance between detection performance and time consumption. To address interference from strong edges, the BSPG method proposed in this paper introduces a novel continuous strategy in the alternating update process of low-rank and sparse components. This allows the model to mitigate the influence of strong edges by preserving more components while updating the low-rank matrix. For algorithm acceleration, a combined approach involving algorithm optimization and hardware enhancement is presented. On the algorithmic front, APSVD is introduced to expedite solving the LRSD problem. On the hardware front, we suggest utilizing GPU multi-thread parallel strategies to accelerate the construction and reconstruction of modules. This is possible as these modules can be decomposed into repetitive and independent subtasks. Visual and quantitative results from experiments demonstrate that our method outperforms other state-of-the-art methods. However, there is still room for improvement in terms of time performance, and in the future, we plan to explore even faster methods.

6. Conclusions

In this paper, we proposed a novel infrared small-target detection method using background-suppression proximal gradient and GPU parallelism. Considering that patch-based methods often result in false alarms at strong edges, we first proposed a novel continuation strategy to suppress such background interference. Then, we presented APSVD to accelerate the solution of the LRSD problem, which involves complex and time-consuming large matrix decomposition. Moreover, we employed GPU multi-threading parallelism to accelerate the construction and reconstruction of patch images. Finally, we optimized the proposed method on the GPU, ultimately achieving outstanding performance. The obtained experimental results demonstrated that our method out-performs nine state-of-the-art methods in terms of both detection accuracy and computational efficiency. The proposed GPU parallelism strategy can be applied to infrared motion sensors and other patch-based infrared small-target detection methods, thus facilitating their application in practical engineering.

Author Contributions: Conceptualization, X.H.; data curation, X.H.; investigation, T.L.; methodology, X.H.; software, X.H. and Y.L.; writing—original draft, X.H.; writing—review & editing, X.L. and Y.C. All authors have read and agreed to the published version of the manuscript.

Funding: This work was funded by the CAS "Light of West China" Program.

Data Availability Statement: The data presented in this study are cited within the article.

Conflicts of Interest: The authors declare no conflict of interest.

References

1. Chen, C.P.; Li, H.; Wei, Y.; Xia, T.; Tang, Y.Y. A local contrast method for small infrared target detection. *IEEE Trans. Geosci. Remote Sens.* **2013**, *52*, 574–581. [CrossRef]
2. Zhang, C.; He, Y.; Tang, Q.; Chen, Z.; Mu, T. Infrared Small Target Detection via Interpatch Correlation Enhancement and Joint Local Visual Saliency Prior. *IEEE Trans. Geosci. Remote Sens.* **2021**, *60*, 5001314. [CrossRef]
3. Bai, X.; Zhou, F. Analysis of new top-hat transformation and the application for infrared dim small target detection. *Pattern Recognit.* **2010**, *43*, 2145–2156. [CrossRef]
4. Zhao, Y.; Pan, H.; Du, C.; Peng, Y.; Zheng, Y. Bilateral two-dimensional least mean square filter for infrared small target detection. *Infrared Phys. Technol.* **2014**, *65*, 17–23. [CrossRef]
5. Deshpande, S.D.; Er, M.H.; Venkateswarlu, R.; Chan, P. Max-mean and max-median filters for detection of small targets. In Proceedings of the Signal and Data Processing of Small Targets, Denver, CO, USA, 20–22 July 1999; Volume 3809, pp. 74–83.
6. Liu, X.; Li, L.; Liu, L.; Su, X.; Chen, F. Moving dim and small target detection in multiframe infrared sequence with low SCR based on temporal profile similarity. *IEEE Geosci. Remote Sens. Lett.* **2022**, *19*, 7507005. [CrossRef]
7. Qin, Y.; Li, B. Effective infrared small target detection utilizing a novel local contrast method. *IEEE Geosci. Remote Sens. Lett.* **2016**, *13*, 1890–1894. [CrossRef]
8. Han, J.; Liang, K.; Zhou, B.; Zhu, X.; Zhao, J.; Zhao, L. Infrared small target detection utilizing the multiscale relative local contrast measure. *IEEE Geosci. Remote Sens. Lett.* **2018**, *15*, 612–616. [CrossRef]
9. Wei, Y.; You, X.; Li, H. Multiscale patch-based contrast measure for small infrared target detection. *Pattern Recognit.* **2016**, *58*, 216–226. [CrossRef]
10. Chen, Y.; Zhang, G.; Ma, Y.; Kang, J.U.; Kwan, C. Small infrared target detection based on fast adaptive masking and scaling with iterative segmentation. *IEEE Geosci. Remote Sens. Lett.* **2021**, *19*, 7000605. [CrossRef]
11. Cui, H.; Li, L.; Liu, X.; Su, X.; Chen, F. Infrared small target detection based on weighted three-layer window local contrast. *IEEE Geosci. Remote Sens. Lett.* **2021**, *19*, 7505705. [CrossRef]
12. Du, J.; Lu, H.; Hu, M.; Zhang, L.; Shen, X. CNN-based infrared dim small target detection algorithm using target-oriented shallow-deep features and effective small anchor. *IET Image Process.* **2021**, *15*, 1–15. [CrossRef]
13. Du, J.; Lu, H.; Zhang, L.; Hu, M.; Chen, S.; Deng, Y.; Shen, X.; Zhang, Y. A spatial-temporal feature-based detection framework for infrared dim small target. *IEEE Trans. Geosci. Remote Sens.* **2021**, *60*, 3000412. [CrossRef]
14. Zhang, M.; Dong, L.; Ma, D.; Xu, W. Infrared target detection in marine images with heavy waves via local patch similarity. *Infrared Phys. Technol.* **2022**, *125*, 104283. [CrossRef]
15. Li, B.; Xiao, C.; Wang, L.; Wang, Y.; Lin, Z.; Li, M.; An, W.; Guo, Y. Dense nested attention network for infrared small target detection. *IEEE Trans. Image Process.* **2022**, *32*, 1745–1758. [CrossRef]
16. Zhong, S.; Zhou, H.; Cui, X.; Cao, X.; Zhang, F. Infrared small target detection based on local-image construction and maximum correntropy. *Measurement* **2023**, *211*, 112662. [CrossRef]

17. Gao, C.; Meng, D.; Yang, Y.; Wang, Y.; Zhou, X.; Hauptmann, A.G. Infrared patch-image model for small target detection in a single image. *IEEE Trans. Image Process.* **2013**, *22*, 4996–5009. [CrossRef]
18. Dai, Y.; Wu, Y.; Song, Y. Infrared small target and background separation via column-wise weighted robust principal component analysis. *Infrared Phys. Technol.* **2016**, *77*, 421–430. [CrossRef]
19. Wang, X.; Peng, Z.; Kong, D.; Zhang, P.; He, Y. Infrared dim target detection based on total variation regularization and principal component pursuit. *Image Vis. Comput.* **2017**, *63*, 1–9. [CrossRef]
20. Dai, Y.; Wu, Y.; Song, Y.; Guo, J. Non-negative infrared patch-image model: Robust target-background separation via partial sum minimization of singular values. *Infrared Phys. Technol.* **2017**, *81*, 182–194. [CrossRef]
21. Guo, J.; Wu, Y.; Dai, Y. Small target detection based on reweighted infrared patch-image model. *IET Image Process.* **2018**, *12*, 70–79. [CrossRef]
22. Zhang, L.; Peng, L.; Zhang, T.; Cao, S.; Peng, Z. Infrared small target detection via non-convex rank approximation minimization joint l 2, 1 norm. *Remote Sens.* **2018**, *10*, 1821. [CrossRef]
23. Zhang, T.; Wu, H.; Liu, Y.; Peng, L.; Yang, C.; Peng, Z. Infrared small target detection based on non-convex optimization with Lp-norm constraint. *Remote Sens.* **2019**, *11*, 559. [CrossRef]
24. Chen, X.; Xu, W.; Tao, S.; Gao, T.; Feng, Q.; Piao, Y. Total Variation Weighted Low-Rank Constraint for Infrared Dim Small Target Detection. *Remote Sens.* **2022**, *14*, 4615. [CrossRef]
25. Yan, F.; Xu, G.; Wu, Q.; Wang, J.; Li, Z. Infrared small target detection using kernel low-rank approximation and regularization terms for constraints. *Infrared Phys. Technol.* **2022**, *125*, 104222. [CrossRef]
26. Liu, Y.; Liu, X.; Hao, X.; Tang, W.; Zhang, S.; Lei, T. Single-Frame Infrared Small Target Detection by High Local Variance, Low-Rank and Sparse Decomposition. *IEEE Trans. Geosci. Remote Sens.* **2023**, *61*, 5614317. [CrossRef]
27. Deng, H.; Sun, X.; Liu, M.; Ye, C.; Zhou, X. Entropy-based window selection for detecting dim and small infrared targets. *Pattern Recognit.* **2017**, *61*, 66–77. [CrossRef]
28. Bai, X.; Bi, Y. Derivative entropy-based contrast measure for infrared small-target detection. *IEEE Trans. Geosci. Remote Sens.* **2018**, *56*, 2452–2466. [CrossRef]
29. Xu, Y.; Wan, M.; Zhang, X.; Wu, J.; Chen, Y.; Chen, Q.; Gu, G. Infrared Small Target Detection Based on Local Contrast-Weighted Multidirectional Derivative. *IEEE Trans. Geosci. Remote Sens.* **2023**, *61*, 5000816. [CrossRef]
30. Zhang, H.; Zhang, L.; Yuan, D.; Chen, H. Infrared small target detection based on local intensity and gradient properties. *Infrared Phys. Technol.* **2018**, *89*, 88–96. [CrossRef]
31. Li, Y.; Li, Z.; Li, W.; Liu, Y. Infrared Small Target Detection Based on Gradient-Intensity Joint Saliency Measure. *IEEE J. Sel. Top. Appl. Earth Obs. Remote Sens.* **2022**, *15*, 7687–7699. [CrossRef]
32. Dai, Y.; Wu, Y.; Zhou, F.; Barnard, K. Attentional local contrast networks for infrared small target detection. *IEEE Trans. Geosci. Remote Sens.* **2021**, *59*, 9813–9824. [CrossRef]
33. Wang, H.; Zhou, L.; Wang, L. Miss detection vs. false alarm: Adversarial learning for small object segmentation in infrared images. In Proceedings of the IEEE/CVF International Conference on Computer Vision, Seoul, Korea, 27 October–2 November 2019; pp. 8509–8518.
34. Zhang, T.; Peng, Z.; Wu, H.; He, Y.; Li, C.; Yang, C. Infrared small target detection via self-regularized weighted sparse model. *Neurocomputing* **2021**, *420*, 124–148. [CrossRef]
35. Wu, X.; Zhang, J.Q.; Huang, X.; Liu, D.L. Separable convolution template (SCT) background prediction accelerated by CUDA for infrared small target detection. *Infrared Phys. Technol.* **2013**, *60*, 300–305. [CrossRef]
36. Dai, Y.; Wu, Y. Reweighted infrared patch-tensor model with both nonlocal and local priors for single-frame small target detection. *IEEE J. Sel. Top. Appl. Earth Obs. Remote Sens.* **2017**, *10*, 3752–3767. [CrossRef]
37. Xu, L.; Wei, Y.; Zhang, H.; Shang, S. Robust and fast infrared small target detection based on pareto frontier optimization. *Infrared Phys. Technol.* **2022**, *123*, 104192. [CrossRef]
38. Zhang, L.; Peng, Z. Infrared small target detection based on partial sum of the tensor nuclear norm. *Remote Sens.* **2019**, *11*, 382. [CrossRef]
39. Kong, X.; Yang, C.; Cao, S.; Li, C.; Peng, Z. Infrared small target detection via nonconvex tensor fibered rank approximation. *IEEE Trans. Geosci. Remote Sens.* **2021**, *60*, 5000321. [CrossRef]
40. Wang, G.; Tao, B.; Kong, X.; Peng, Z. Infrared Small Target Detection Using Nonoverlapping Patch Spatial—Temporal Tensor Factorization With Capped Nuclear Norm Regularization. *IEEE Trans. Geosci. Remote Sens.* **2021**, *60*, 5001417. [CrossRef]
41. Li, J.; Zhang, P.; Zhang, L.; Zhang, Z. Sparse Regularization-Based Spatial-Temporal Twist Tensor Model for Infrared Small Target Detection. *IEEE Trans. Geosci. Remote Sens.* **2023**, *61*, 5000417. [CrossRef]
42. Zhang, Z.; Ding, C.; Gao, Z.; Xie, C. ANLPT: Self-Adaptive and Non-Local Patch-Tensor Model for Infrared Small Target Detection. *Remote Sens.* **2023**, *15*, 1021. [CrossRef]
43. Toh, K.C.; Yun, S. An accelerated proximal gradient algorithm for nuclear norm regularized linear least squares problems. *Pac. J. Optim.* **2010**, *6*, 15.
44. Lin, Z.; Ganesh, A.; Wright, J.; Wu, L.; Chen, M.; Ma, Y. Fast Convex Optimization Algorithms for Exact Recovery of a Corrupted Low-Rank Matrix. *Coordinated Science Laboratory Report no. UILU-ENG-09-2214, DC-246*. 2009. Available online: https://people.eecs.berkeley.edu/~yima/matrix-rank/Files/rpca_algorithms.pdf (accessed on 30 September 2023).

45. Crdóñez, Á.; Argüello, F.; Heras, D.B.; Demir, B. GPU-accelerated registration of hyperspectral images using KAZE features. *J. Supercomput.* **2020**, *76*, 9478–9492. [CrossRef]
46. Seznec, M.; Gac, N.; Orieux, F.; Naik, A.S. Real-time optical flow processing on embedded GPU: An hardware-aware algorithm to implementation strategy. *J. Real Time Image Process.* **2022**, *19*, 317–329. [CrossRef]

remote sensing

MDPI

Article

Multi-Dimensional Low-Rank with Weighted Schatten *p*-Norm Minimization for Hyperspectral Anomaly Detection

Xi'ai Chen [1,2,*], Zhen Wang [1,2], Kaidong Wang [3], Huidi Jia [1,2,4], Zhi Han [1,2] and Yandong Tang [1,2]

1 State Key Laboratory of Robotics, Shenyang Institute of Automation, Chinese Academy of Sciences, Shenyang 110016, China; wangzhen2@sia.cn (Z.W.); jiahuidi@sia.cn (H.J.); hanzhi@sia.cn (Z.H.); ytang@sia.cn (Y.T.)
2 Institutes for Robotics and Intelligent Manufacturing, Chinese Academy of Sciences, Shenyang 110016, China
3 Center for Intelligent Decision-Making and Machine Learning, School of Management, Xi'an Jiaotong University, Xi'an 710049, China; wangkd13@gmail.com
4 University of Chinese Academy of Sciences, Beijing 100049, China
* Correspondence: chenxiai@sia.cn

Abstract: Hyperspectral anomaly detection is an important unsupervised binary classification problem that aims to effectively distinguish between background and anomalies in hyperspectral images (HSIs). In recent years, methods based on low-rank tensor representations have been proposed to decompose HSIs into low-rank background and sparse anomaly tensors. However, current methods neglect the low-rank information in the spatial dimension and rely heavily on the background information contained in the dictionary. Furthermore, these algorithms show limited robustness when the dictionary information is missing or corrupted by high level noise. To address these problems, we propose a novel method called multi-dimensional low-rank (MDLR) for HSI anomaly detection. It first reconstructs three background tensors separately from three directional slices of the background tensor. Then, weighted schatten p-norm minimization is employed to enforce the low-rank constraint on the background tensor, and $L_{F,1}$-norm regularization is used to describe the sparsity in the anomaly tensor. Finally, a well-designed alternating direction method of multipliers (ADMM) is employed to effectively solve the optimization problem. Extensive experiments on four real-world datasets show that our approach outperforms existing anomaly detection methods in terms of accuracy.

Keywords: anomaly detection; multi-dimensional; low-rank

Citation: Chen, X.; Wang, Z.; Wang, K.; Jia, H.; Han, Z.; Tang, Y. Multi-Dimensional Low-Rank with Weighted Schatten *p*-Norm Minimization for Hyperspectral Anomaly Detection. *Remote Sens.* **2024**, *16*, 74. https://doi.org/10.3390/rs16010074

Academic Editor: Paolo Tripicchio

Received: 30 September 2023
Revised: 18 December 2023
Accepted: 21 December 2023
Published: 24 December 2023

1. Introduction

Compared with conventional images such as RGB images, multispectral images, SAS images [1], and delay-Doppler images [2], hyperspectral images (HSIs) offer the advantage of capturing hundreds of contiguous spectral bands of the same scene. This unique characteristic of HSI proves to be beneficial for target detection and finds wide applications in various fields such as land cover classification [3–5], mineral survey [6–8], environmental protection [9–11], and other applications [12–18]. In hyperspectral target detection, when the target information is unknown, the unsupervised processing of the target detection is called anomaly detection. However, in practical applications, it is often difficult to obtain the prior information of the target, so hyperspectral anomaly detection is more suitable. In essence, hyperspectral anomaly detection can be viewed as an unsupervised binary classification problem that separates an image into background and anomalies, where anomalies typically represent rare targets that occupy only a small number of pixels.

Over the past two decades, there has been a growing interest in hyperspectral anomaly detection, leading to the development of numerous detection algorithms. The Reed–Xiaoli (RX) algorithm [19] is a classical statistical modelling method for anomaly detection, assuming that the background follows a multivariate Gaussian distribution. The main objective of the RX algorithm is to compute the Mahalanobis distance between the measured pixel and the background [20], which involves estimating the mean vector and the covariance matrix of the background. Two commonly studied extended versions of the RX algorithm are the global RX (GRX) [21] and the local RX (LRX) [22], where the former calculates the distance between the measured pixel and all background pixels, and the latter calculates the distance between the measured pixel and the surrounding background pixels. However, in hyperspectral applications, it is crude to describe the background with a single Gaussian distribution, and the mean vector and covariance matrix of the background are susceptible to the noisy pixels and anomalies.

In general, the HSI can be represented as a three-order tensor with two spatial dimensions and one spectral dimension. Taking into account the similarity between spectral bands, the HSI can be transformed into a matrix along the spectral dimension, which inspires the matrix-based anomaly detection methods. Anomalies are assumed to be randomly distributed in the background and to have sparse properties. By formulating a constrained convex optimization problem that incorporates the characteristics of both the background and the anomalies, successful separation of the anomalies from the background can be achieved. Consequently, the low-rank and sparse matrix decomposition (LRaSMD) algorithms [23–25] have been used to separate the HSI data into low-rank background and sparse anomalies and have demonstrated their effectiveness in previous studies [26–28]. According to the LRaSMD approach, the spectral response of a pixel $y_i (i \in \{1, \dots, N\})$ in d bands of the HSI can be represented as a spectral vector $\mathbf{y}_i \in \mathbb{R}^d$ with the decomposition.

$$\mathbf{y}_i = \mathbf{x}_i + \mathbf{s}_i, \quad \begin{cases} \mathbf{s}_i = 0, & \text{if } \mathbf{y}_i \text{ is part of background,} \\ \mathbf{s}_i \neq 0, & \text{if } \mathbf{y}_i \text{ is part of anomalies,} \end{cases} \tag{1}$$

which can further be written in matrix form as:

$$\mathbf{Y} = \mathbf{X} + \mathbf{S}, \tag{2}$$

where $\mathbf{X} = [\mathbf{x}_1, \mathbf{x}_2, \dots, \mathbf{x}_N]^T$ and $\mathbf{S} = [\mathbf{s}_1, \mathbf{s}_2, \dots, \mathbf{s}_N]^T \in \mathbb{R}^{N \times d}$ represent the background and anomaly components of the HSI matrix $\mathbf{Y} = [\mathbf{y}_1, \mathbf{y}_2, \dots, \mathbf{y}_N]^T \in \mathbb{R}^{N \times d}$, where N represents the number of pixels in the HSI, and d represents the number of spectral bands. Furthermore, to address the attention imbalance between anomalies and the background observed in LRaSMD, Zhang et al. [29] proposed the LRaSMD-based Mahalanobis distance (LSMAD) method. Xu et al. [30] integrated cooperative representation and Euclidean distance into the LRaSMD framework. Li et al. [31] investigated LRaSMD under the assumption of a mixture-of-Gaussian (MoG) distribution and developed a global detector based on the Manhattan distance. To further exploit the intrinsic information of the background, low-rank representation (LRR) [32,33] was proposed, which maps the HSI to multiple linear subspaces using a dictionary. Xu et al. [34] proposed a new anomaly detection method called low-rank and sparse representation (LRASR), which employed a dictionary construction strategy and a sparsity-inducing regularization term to reconstruct the background matrix. To preserve the local geometric structure and spatial relationships of the background, the graph and total variation regularized low-rank representation (GTVLRR) [35] method was introduced for HSI anomaly detection. Fu et al. [36] used convolutional neural network (CNN) denoisers [37] as priors for the coefficients of the dictionary.

The aforementioned matrix-based anomaly detection methods tend to destroy the spatial structure of HSI and fail to effectively exploit the inherent spatial information [38,39]. In recent years, tensor-based methods have emerged as a promising approach to HSI anomaly detection, allowing the decomposition of HSI data into low-rank and sparse components. Sun et al. [23] used Tucker decomposition to obtain the low-rank background, using an unmixing method to extract the spectral features of the anomaly. Li et al. [40] embedded priors into the dimensions of a tensor with different regularizations. Song et al. [41] proposed a dictionary construction strategy based on Tucker decomposition, which improved the inclusion of spectral segment information in the dictionary. Shang et al. [42] found a new prior that describes the sparsity of the core tensor of a gradient map (GCS) under Tucker decomposition. However, Tucker decomposition has inherent limitations in terms of rank. To address this issue, Wang et al. [43] extended the concept of LRR from matrix to tensor, taking into account the three-dimensional structure of HSI. Sun et al. [44] represented the background tensor as the product of a transformed tensor and a low-rank matrix. However, these methods pay primary attention to the low-rank of the spectral dimension of the tensor, neglecting the low-rank information in the spatial dimensions. The dictionary, which maps the HSI into multiple linear subspaces, plays a crucial role in the reconstruction of the background component. To achieve an effective separation of background and anomalies, the dictionary should primarily contain background information. Although some methods choose the original data themselves as the dictionary, they may still contain anomalies that can adversely affect the background reconstruction process.

To address these issues, we propose a multi-dimensional low-rank (MDLR) strategy for HSI anomaly detection. Unlike the existing tensor-based methods that construct one background tensor, our approach constructs three background tensors, two capturing the spatial dimension and one representing the spectral dimension. Using the tensor singular value decomposition (t-SVD) technique, we obtain the f-diagonal tensor $\mathcal{S}$, characterizing the background. To enforce low-rankness in the background tensor, we apply the weighted Schatten p-norm minimization (WSNM) to the slices of $\mathcal{S}$. Finally, the three background tensors are merged into a single background tensor. In addition, anomalies in the HSI tend to occur at consistent spatial locations across all spectral bands and exhibit a slight spectral density. To capture this property, we impose a joint spectral–spatial sparsity on the anomaly tensor using the $L_{F,1}$ norm. The main contributions of this work can be summarized as follows:

1. Low-rankness along three dimensions in the frequency domain is exploited. Through the low-rank property analysis of the tensor along different dimensions, we found that it is not sufficient to measure the low-rankness along only one dimension. Therefore, multi-dimensional low-rankness is embedded into different tensors with t-SVD along different slices. These tensors are then fused to form a background tensor that captures the low-rank characteristics across all three dimensions and enables the MDLR method to effectively explore more comprehensive background information.
2. To enforce low-rank in the background tensor, WSNM is applied to the frontal slices of the f-diagonal tensor, which enhances the preservation of the low-rank structure in the background tensor.

The rest of this paper is organized as follows. In Section 2, notations and preliminaries are introduced. The proposed multidimensional low-rank model is presented in detail in Section 3. The experimental results are demonstrated in Section 4. The conclusion is given in Section 5.

2. Notations and Preliminaries

In this section, we introduce the notations and preliminaries used in this paper. The column vectors are represented by lowercase letters, e.g., $\mathbf{x}$. The matrix is represented by bold capital letters, e.g., $\mathbf{X}$. An HSI with w rows, h columns, and d spectral bands can be naturally represented as a third-order tensor, $\mathcal{X} \in \mathbb{R}^{w \times h \times d}$. The discrete Fourier transform (DFT) of $\mathcal{X}$ along the spectral dimension can be written as $\hat{\mathcal{X}}$ = fft($\mathcal{X}$, [], 3). The inverse DFT of $\hat{\mathcal{X}}$ is written as $\mathcal{X}$= ifft($\hat{\mathcal{X}}$, [], 3). $\mathcal{X}^*$ represents the conjugate transpose of $\mathcal{X}$. $\mathcal{X}^{(i)}$ is the i-th frontal slice of $\mathcal{X}$. The block circulant matrix bcirc($\mathcal{N}$) of $\mathcal{N} \in \mathbb{R}^{w \times h \times d}$ is defined as follows:

$$\mathrm{bcirc}(\mathcal{N}) = \begin{bmatrix} \mathcal{N}^{(1)} & \mathcal{N}^{(d)} & \mathcal{N}^{(d-1)} & \cdots & \mathcal{N}^{(2)} \\ \mathcal{N}^{(2)} & \mathcal{N}^{(1)} & \mathcal{N}^{(d)} & \cdots & \mathcal{N}^{(3)} \\ \vdots & \ddots & \ddots & \ddots & \vdots \\ \mathcal{N}^{(d)} & \mathcal{N}^{(d-1)} & \ddots & \mathcal{N}^{(2)} & \mathcal{N}^{(1)}. \end{bmatrix}$$

The block vectorization operation bvec$(\cdot)$ of $\mathcal{N}$ and its inverse operation bvfold$(\cdot)$ are denoted as:

$$\text{bvec}(\mathcal{N}) = \begin{bmatrix} \mathcal{N}^{(1)} \\ \mathcal{N}^{(2)} \\ \vdots \\ \mathcal{N}^{(d)} \end{bmatrix}, \quad \text{bvfold}(\text{bvec}(\mathcal{N})) = \mathcal{N}.$$

Definition 1 (Tensor product). *The product of three-order tensor $\mathcal{N} \in \mathbb{R}^{n_1 \times n_2 \times n_3}$ and $\mathcal{M} \in \mathbb{R}^{n_2 \times n_4 \times n_3}$ is $\mathcal{A} \in \mathbb{R}^{n_1 \times n_4 \times n_3}$ defined as follows:*

$$\mathcal{A} = \mathcal{N} * \mathcal{M} = bvfold(bcirc(\mathcal{N}) * bvec(\mathcal{M})). \tag{3}$$

Definition 2 (Slices of Tensor). *There are three types of slices in a tensor: that is, horizontal slices $\mathcal{X}_{i::}$, lateral slices $\mathcal{X}_{:j:}$, and frontal slices $\mathcal{X}_{::k}$.*

Definition 3 (Identity Tensor). *The identity tensor $\mathcal{I} \in \mathbb{R}^{n_1 \times n_2 \times n_3}$ is defined by $\mathcal{I}(:,:,1) = eye(n_1, n_2)$, $\mathcal{I}(:,:,2:n_3) = 0$, where eye (n_1, n_2) is an identity matrix $(n_1 \times n_2)$.*

Definition 4 (Conjugate Transpose). *The conjugate transpose of a tensor $\mathcal{X} \in \mathbb{R}^{n_1 \times n_2 \times n_3}$ is denoted as $\mathcal{X}^*$ with*

$$\widehat{\mathcal{X}^*}^{(i)} = (\hat{\mathcal{X}}^{(i)})^T, i = 1, 2, ..., n_3. \tag{4}$$

Definition 5 (Orthogonal Tensor). *The orthogonal tensor $\mathcal{D}$ satisfies $\mathcal{D}^* * \mathcal{D} = \mathcal{D} * \mathcal{D}^* = \mathcal{I}$.*

Definition 6 (t-SVD). *The singular value decomposition of a tensor $\mathcal{X} \in \mathbb{R}^{w \times h \times d}$ can be decomposed into the product of three three-order tensors.*

$$\mathcal{X} = \mathcal{U} * \mathcal{S} * \mathcal{V}^*, \tag{5}$$

where $\mathcal{U} \in \mathbb{R}^{w \times w \times d}$ and $\mathcal{V} \in \mathbb{R}^{h \times h \times d}$ are orthogonal tensors and $\mathcal{S} \in \mathbb{R}^{w \times h \times d}$ is an f-diagonal tensor. The procedure of t-SVD is described in Algorithm 1.

Definition 7 (Tensor Tubal Rank). *For a tensor $\mathcal{X} \in \mathbb{R}^{n_1 \times n_2 \times n_3}$ with t-SVD $\mathcal{X} = \mathcal{U} * \mathcal{S} * \mathcal{V}^*$, its tubal rank is the number of non-zero tubes of $\mathcal{S}$:*

$$rank_t(\mathcal{X}) = \#\{k : \mathcal{S}(k,k,:) \neq 0\}. \tag{6}$$

Definition 8 (Tensor Nuclear Norm(TNN)). *The TNN of a tensor $X \in \mathbb{R}^{n_1 \times n_2 \times n_3}$ is the sum of singular values of all front slices of $\hat{\mathcal{X}}$, that is,*

$$\|\mathcal{X}\|_* := \sum_{k=1}^{n_3} \|\hat{\mathcal{X}}^{(k)}\|_*. \tag{7}$$

3. Proposed Method

An illustration of the proposed model is shown in Figure 1. Figure 1a illustrates the different dimension low-rank property of the HSI in the frequency domain. To exploit the low-rankness along different dimensions, we combine these three different dimensional tensors to form the background tensor and apply tensor low-rank and sparse decomposition to extract the sparse anomaly object from the low-rank background. Final detection map M can be obtained via the sparse $\mathcal{S}$ by computing $\sqrt{\sum_{k=1}^{d} |\mathcal{S}(i,j,k)|^2}$. We will introduce each part in detail in the following subsections.

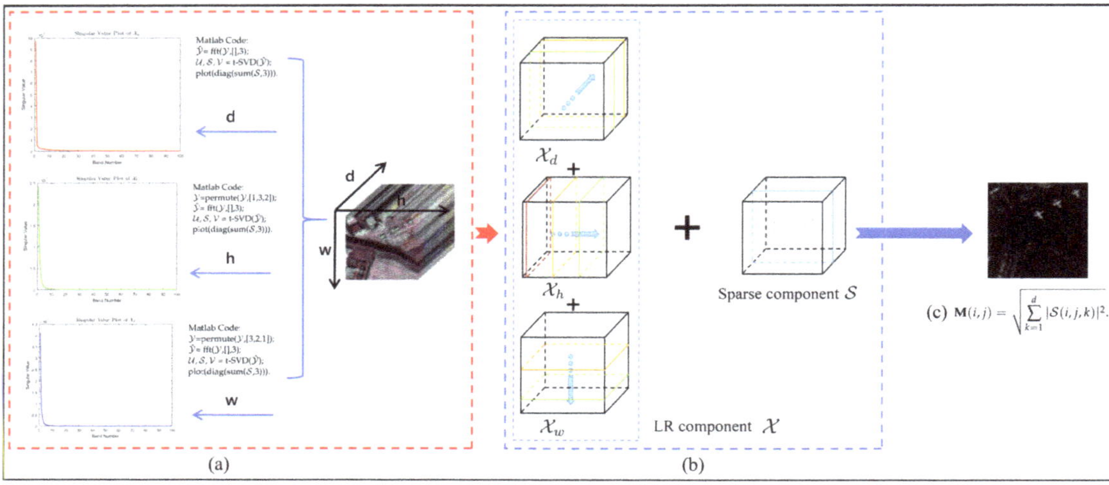

Figure 1. Illustration of the proposed model for HSI anomaly detection. (**a**) Multi-dimensional low-rank in frequency domain. (**b**) Tensor low-rank and sparse decomposition. (**c**) Detection map.

3.1. Tensor Low-Rank Linear Representation

LRR uses a dictionary to explore low-rank linear representations of HSI, but the matrix-based approach breaks the tensor structure inherent in HSI. To overcome this limitation, tensor LRR is proposed, which incorporates the t-product to preserve the spatial structure of the tensor. Given a tensor $\mathcal{Y} \in \mathbb{R}^{w \times h \times d}$, it can be decomposed using the tensor LRR formulation as follows:

$$\mathcal{Y} = \mathcal{A} * \mathcal{X} + \mathcal{S}, \tag{8}$$

where $\mathcal{X}$ is the low-rank background tensor, $\mathcal{S}$ is the sparse anomaly tensor, and $\mathcal{A}$ is the dictionary. Equation (8) aims to construct the low-rank and sparse components exactly and efficiently by dictionary $\mathcal{A}$ from HSI data.

$$\begin{aligned} \min_{\mathcal{X},\mathcal{S}} \quad & \text{rank}_t(\mathcal{X}) + \lambda \text{sparse}(\mathcal{S}) \\ s.t. \quad & \mathcal{Y} = \mathcal{A} * \mathcal{X} + \mathcal{S}, \end{aligned} \tag{9}$$

where $\text{rank}_t(\mathcal{X})$ denotes the tensor tubal rank function [45], λ is a regularization parameter of the $\mathcal{S}$, $\text{sparse}(\mathcal{S})$ is the sparse norm.

3.2. Weighted Schatten p-Norm Minimization

The problem of determining the $\text{rank}_t(\mathcal{X})$ is known to be NP-hard. To approximate the rank of a matrix, a commonly used method is nuclear norm minimization (NNM), which calculates the sum of the singular values of the matrix $\mathcal{X}$. NNM is typically solved using a singular value thresholding algorithm. However, to obtain a more accurate low-rank approximation, other methods [46–48] have been developed. These methods treat different singular values individually rather than uniformly as in NNM, resulting in improved performance. In WSNM, each singular value is assigned a specific weight, and the optimization problem aims to minimize the weighted Schatten p-norm of the matrix $\mathbf{X} \in \mathbb{R}^{h \times w}$.

$$\|\mathbf{X}\|_{w,S_p} = \left(\sum_{i=1}^{\min\{n,m\}} w_i \sigma_i^p \right)^{\frac{1}{p}}, \tag{10}$$

where σ_i is the i-th singular value of $\mathbf{X}$, w_i is the weight of σ_i, $\mathbf{w}$=[w_1,..., $w_{min(n,m)}$] is a non-negative vector to constrain the single value of $\mathbf{X}$. The weighted Schatten p norm minimization problem can be effectively solved by the generalized soft thresholds. Given p and w_i, the specific threshold can be obtained by:

$$GST(w_i, p) = (2w_i(1-p))^{\frac{1}{2-p}} + w_i p(2w_i(1-p))^{\frac{p-1}{2-p}}. \quad (11)$$

The main procedures of this approach are shown in Algorithm 1. In this work, the low-rank problem of HSI is solved in tensor form and the nuclear norm of the matrix is converted to the tensor nuclear norm (TNN). The WSNM is applied to the forward slices of $\mathcal{S}$.

3.3. Mutil-Dimensional Tensor Low-Rank Norm

According to the tensor LRR, tensor $\mathcal{X}$ can be expressed as the linear combination of the tensor dictionary $\mathcal{A}$. The choice of the dictionary plays a crucial role in the background tensor reconstruction. Conventional dictionary construction methods are often sensitive to noise and require separate construction for different datasets, making the anomaly detection process complicated. When the dictionary is an identity tensor, the tensor LRR is converted to tensor robust principal component analysis (TRPCA) [49]. By combining WSNM and TRPCA, we have the following:

$$\min_{\mathcal{X},\mathcal{S}} \quad \|\mathcal{X}\|_{w,S_p} + \lambda\|\mathcal{S}\|_{F,1}$$
$$s.t. \quad \mathcal{Y} = \mathcal{X} + \mathcal{S}. \quad (12)$$

In the field of HSI unmixing, a latent low-rank representation theory (LatLRR) has been proposed [50]. LatLRR treats itself as a dictionary and learns its own rows and columns separately to obtain two different background representations while incorporating low-rank constraints. Motivated by this concept, we aim to explore the background tensor and reorganize it from different directions of slices. To achieve this, we introduce three background tensors: $\mathcal{X}_w, \mathcal{X}_h, \mathcal{X}_d \in \mathbb{R}^{w\times h\times d}$. We run WSNM separately on these three tensors along different dimensional frontal slices. The proposed tensor-based method, called multi-dimensional low-rank (MDLR), can be expressed as follows:

$$\|\mathcal{X}\|_{msp,*} = \mu_w\|\mathcal{X}_w\|_{w,S_p} + \mu_h\|\mathcal{X}_h\|_{w,S_p} + \mu_d\|\mathcal{X}_d\|_{w,S_p}, \quad (13)$$

where $0 \le \mu_w \le 1$, $0 \le \mu_h \le 1$, and $\mu_d = 1 - \mu_w - \mu_h$ balance the contributions of $\mathcal{X}_w$, $\mathcal{X}_h$ and $\mathcal{X}_d$. We call the reconstruction of the background tensor $\mathcal{X}_w$, the reconstruction of the background from the w dimension, $\mathcal{X}_h$, the reconstruction of the background from the h dimension, $\mathcal{X}_d$ the reconstruction of the background from the d dimension. Finally, our model formulation can be written as:

$$\min_{\mathcal{X},\mathcal{S}} \quad \|\mathcal{X}\|_{msp,*} + \lambda\|\mathcal{S}\|_{F,1}$$
$$s.t. \quad \mathcal{Y} = \mathcal{X} + \mathcal{S}. \quad (14)$$

3.4. Optimization Procedure

By introducing auxiliary $\mathcal{X}_w, \mathcal{X}_h, \mathcal{X}_d$, Equation (12) can be written as the following equivalent problem:

$$\min_{\mathcal{X}_w,\mathcal{X}_h,\mathcal{X}_d,\mathcal{S}} \mu_w\|\mathcal{X}_w\|_{w,S_p} + \mu_h\|\mathcal{X}_h\|_{w,S_p} + \mu_d\|\mathcal{X}_d\|_{w,S_p} + \lambda\|\mathcal{S}\|_{F,1}$$
$$s.t.\ \mathcal{X} = \mathcal{X}_w,\ \mathcal{X} = \mathcal{X}_h,\ \mathcal{X} = \mathcal{X}_d,\ \mathcal{Y} = \mathcal{X} + \mathcal{S}. \quad (15)$$

The Lagrange multipliers $\mathcal{E}$, $\mathcal{Q}_{1,2,3}$ are introduced and we use the ADMM to solve the augmented Lagrange function. The optimization problem above is written as follows:

$$\begin{aligned}\min_{\mathcal{X}_w,\mathcal{X}_h,\mathcal{X}_d,\mathcal{S},\mathcal{E}} &\mu_w\|\mathcal{X}_w\|_{w,S_p} + \mu_h\|\mathcal{X}_h\|_{w,S_p} + \mu_d\|\mathcal{X}_d\|_{w,S_p} - \lambda\|\mathcal{S}\|_{F,1} \\ &+\frac{\alpha}{2}\|\mathcal{X} - \mathcal{X}_w + \frac{\mathcal{Q}_1}{\alpha}\|_F^2 + \frac{\alpha}{2}\|\mathcal{X} - \mathcal{X}_h + \frac{\mathcal{Q}_2}{\alpha}\|_F^2 \\ &+\frac{\alpha}{2}\|\mathcal{X} - \mathcal{X}_d + \frac{\mathcal{Q}_3}{\alpha}\|_F^2 + \frac{\alpha}{2}\|\mathcal{Y} - \mathcal{X} - \mathcal{S} + \frac{\mathcal{E}}{\alpha}\|_F^2 \end{aligned} \tag{16}$$

(1) **Update** $\mathcal{X}$

$$\begin{aligned}\mathcal{X} = \underset{\mathcal{X}}{\mathrm{argmin}}\, &\frac{\alpha}{2}\|\mathcal{X} - \mathcal{X}_w + \frac{\mathcal{Q}_1}{\alpha}\|_F^2 + \frac{\alpha}{2}\|\mathcal{X} - \mathcal{X}_h + \frac{\mathcal{Q}_2}{\alpha}\|_F^2 \\ &+\frac{\alpha}{2}\|\mathcal{X} - \mathcal{X}_d + \frac{\mathcal{Q}_3}{\alpha}\|_F^2 + \frac{\alpha}{2}\|\mathcal{Y} - \mathcal{X} - \mathcal{S} + \frac{\mathcal{E}}{\alpha}\|_F^2. \end{aligned} \tag{17}$$

The closed-form solution of $\mathcal{X}$ can be obtained by taking the derivative of the above objective function and setting it to zero, as follows:

$$\mathcal{X} = (\mathcal{Y} - \mathcal{S} + \frac{\mathcal{E}}{\alpha} + \mathcal{X}_w + \mathcal{X}_h + \mathcal{X}_d - \sum_{i=1}^{3} \frac{\mathcal{Q}_i}{\alpha})/4 \tag{18}$$

(2) **Update** $\mathcal{X}_w$

$$\mathcal{X}_w = \arg\min_{\mathcal{X}_w} \quad \mu_w\|\mathcal{X}_w\|_{w,s_p} + \frac{\alpha}{2}\|\mathcal{X} - \mathcal{X}_w + \frac{\mathcal{Q}_1}{\alpha}\|_F^2 \tag{19}$$

(3) **Update** $\mathcal{X}_h$

$$\mathcal{X}_h = \arg\min_{\mathcal{X}_h} \quad \mu_h\|\mathcal{X}_h\|_{w,s_p} + \frac{\alpha}{2}\|\mathcal{X} - \mathcal{X}_h + \frac{\mathcal{Q}_3}{\alpha}\|_F^2 \tag{20}$$

(4) **Update** $\mathcal{X}_d$

$$\mathcal{X}_d = \arg\min_{\mathcal{X}_d} \quad \mu_d\|\mathcal{X}_d\|_{w,s_p} + \frac{\alpha}{2}\|\mathcal{X} - \mathcal{X}_d + \frac{\mathcal{Q}_3}{\alpha}\|_F^2. \tag{21}$$

The subproblem $\mathcal{X}_{w,h,d}$ can be solved using generalized soft-thresholding as shown in Algorithm 1. Before applying the Algorithm 1, $\mathcal{X}_w$ should be converted to $\mathcal{X}_w \in \mathbb{R}^{d\times h\times w}$, and then it must be reshaped again as $\mathcal{X}_w \in \mathbb{R}^{w\times h\times d}$ after Algorithm 1. Similarly, $\mathcal{X}_h$ should be converted to $\mathcal{X}_w \in \mathbb{R}^{w\times d\times h}$ before Algorithm 1 and back to $\mathcal{X}_h \in \mathbb{R}^{w\times h\times d}$ after Algorithm 1.

(5) **Update** $\mathcal{S}$

$$\mathcal{S} = \arg\min_{\mathcal{S}} \lambda\|\mathcal{S}\|_{F,1} + \frac{\alpha}{2}\|\mathcal{Y} - \mathcal{X} - \mathcal{S} + \frac{\mathcal{E}}{\alpha}\|_F^2. \tag{22}$$

Then, we have the following closed solution:

$$\mathcal{S}(:,:,k) = \begin{cases} \frac{\|\mathcal{M}(:,:,k)\|_F+\lambda}{\mathcal{M}(:,:,k)\|_F}\mathcal{M}(:,:,k), & \lambda < \|\mathcal{M}(:,:,k)\|_F \\ 0, & \text{otherwise} \\ , & \end{cases} \tag{23}$$

where $\mathcal{M} = \mathcal{Y} - \mathcal{X} + \frac{\mathcal{E}}{\alpha}$.

(6) **Update** Lagrange multiplier $\mathcal{E}$ and $\mathcal{Q}_{1,2,3}$

$$\mathcal{E} = \mathcal{E} + \alpha * (\mathcal{Y} - \mathcal{X} - \mathcal{S})$$

$$\mathcal{Q}_1 = \mathcal{Q}_1 + \alpha * (\mathcal{X} - \mathcal{X}_w)$$
$$\mathcal{Q}_2 = \mathcal{Q}_2 + \alpha * (\mathcal{X} - \mathcal{X}_h)$$
$$\mathcal{Q}_3 = \mathcal{Q}_3 + \alpha * (\mathcal{X} - \mathcal{X}_d). \quad (24)$$

The overall process of the proposed method is concluded in Algorithm 2. When the optimization process is complete, the anomaly detection map **M** of HSI data can obtain by the sparse anomaly tensor S as follows:

$$\mathbf{M}(i,j) = \sqrt{\sum_{k=1}^{d} |\mathcal{S}(i,j,k)|^2}. \quad (25)$$

Due to the WSNM regularization, the solution process of the problem in Equations (19)–(21) is actually not a convex optimization problem. Nevertheless, Xie et al. [51] prove that WSNM is not convex, and if the weights satisfy $0 \leq w_1 \leq w_2 \leq \ldots w_i$, at least one accumulation point satisfies (26). A convergence analysis can be found in Theorem 3 of WSNM.

$$\lim_{k \to \infty} \|\mathcal{X}_{k+1} - \mathcal{X}_k\|_F^2 + \|\mathcal{S}_{k+1} - \mathcal{S}_k\|_F^2. \quad (26)$$

Algorithm 1 WSNM based on t-SVD.

Input: $\mathcal{X}, \mathcal{Q}, p, \alpha, \tau$
$\mathcal{P} = \mathcal{X} + \frac{\mathcal{Q}}{\alpha}$
$\hat{\mathcal{P}} = \text{fft}(\mathcal{P}, [], 3)$
for i = 0,1,..., $[\frac{d+1}{2}]$
  $[\hat{\mathcal{U}}(:,:,i), \hat{\mathcal{S}}(:,:,i), \hat{\mathcal{V}}(:,:,i)] = \text{SVD}(\hat{\mathcal{P}}(:,:,i))$;
  $\hat{\mathcal{S}} = \hat{\mathcal{S}}(:,:,i)$;
  for $j = 1$:size(diag($\hat{\mathcal{S}}$))
    $w_j = \tau(2\sqrt{2}((\frac{1}{\alpha^2})\sqrt{wh})/(\text{diag}(\hat{\mathcal{S}}(j))^{1/p} + 1^{-6})$;
    get t by calculating Equation (11);
    if $|\text{diag}(\hat{\mathcal{S}}(j))| \leq t$, **then**
      $\text{diag}(\hat{\mathcal{S}}(j)) = 0$;
      else
      $k = 0, \mu_k = |\text{diag}(\hat{\mathcal{S}}(j))|$
      for $k = 0, 1, ..., J$ **do**
        $\mu_{k+1} = |\text{diag}(\hat{\mathcal{S}}(j))| - w_j p(\mu_k)^{p-1}$;
        $k = k + 1$;
      end
      $\text{diag}(\hat{\mathcal{S}}(j))$ =sgn$(\text{diag}(\hat{\mathcal{S}}(j))\mu_k$;
      $\hat{S}_{new}(:,:,i)$ =diag$(\hat{\mathcal{S}}(j))$;
    end
  end
end
for $i = [\frac{z+1}{2}] + 1, ..., z$
  $\hat{\mathcal{U}}(:,:,i) = \text{conj}(\hat{\mathcal{U}}(:,:,z-i+2))$
  $\hat{\mathcal{S}_{new}}(:,:,i) = \text{conj}(\hat{\mathcal{S}_{new}}(:,:,i))$
  $\hat{\mathcal{V}}(:,:,i) = \text{conj}(\hat{\mathcal{V}}(:,:,z-i+2))$
$\mathcal{U} = \text{ifft}(\hat{\mathcal{U}}, [], 3), \mathcal{S}_{new} = \text{iftt}(\hat{\mathcal{S}}_{new}, [], 3), \mathcal{V} = \text{ifft}(\hat{\mathcal{V}}, [], 3)$
end
Output: $\mathcal{Z} = \mathcal{U} * \mathcal{S}_{new} * \mathcal{V}^*$;

Algorithm 2 MDLR for HSI anomaly detection.

Input: HSI tensor $\mathcal{Y}$, $\mu_{w,h,d}$, λ, α
Initializtion: $\mathcal{X}, \mathcal{S}, \mathcal{X}_{w,h,d} = 0$, $\mathcal{Q}_{1,2,3}, \mathcal{E} = 0, i = 0$.
If i<maxiter or satisfy Equation (11).
  Update $\mathcal{X}$ by Equation (18);
  Update $\mathcal{X}_{w,j,d}$ by algorithm (2);
  Update $\mathcal{S}$ by Equation (23);
  Update $\mathcal{E}$ and $\mathcal{Q}_{1,2,3}$ by Equation (24);
End
compute the anomaly detection map **M** by Equation (25);
Output: anomaly detection map **M**.

3.5. Computational Complexity

Given a tensor $\mathcal{X} \in \mathbb{R}^{w \times h \times d}$, the computational complexity of our model mainly consists of the following two parts: (1) solving the subproblem of Equations (19)–(21) depends on the t-SVD, and the complexity is approximately $O(h^3d + d^3h + h^3w)$; (2) for Equation (22), $O(wh^2d + h(w+h)d\log(d)$ is required. Therefore, the total main cost of the proposed model is $O(h^3d + d^3h + h^3w + wh^2d + h(w+d)d\log(d))$. The computational complexity is studied by comparing their running time on San Diego data as shown in Table 1.

Table 1. Running time of all compared algorithms on San Diego.

HSI Data	RX	LSMAD	LRASR	GTVLRR	DeCNN-AD	PTA	PCA-TLRSR	MDLR
San Diego	2.054	38.46	56.394	214.343	256.589	34.344	8.312	132.46

4. Experimental Results

In this section, we verify the effectiveness of our method on an extensive dataset compared with the SOTA methods. Standard metrics, such as the 2D receiver operating characteristic (ROC) curve [52] and the area under the curve (AUC) metric [53], are used to quantify the results quantitatively. The ROC curve plots the probability of detection (PD) against the false alarm rate (FAR) for all possible thresholds. The AUC is calculated by integrating the area under the ROC curve. To effectively evaluate the detection performance, 3D ROCs [54] generated from 2D ROC and separability maps are also used for quantitative comparison. All the experimental algorithms are performed in MATLAB 2020a on a computer with Core i9-11900KF 3.50GHz CPU and 32-GB of RAM in Windows 11.

4.1. HSI Datasets

San Diego: The dataset is part of a collection captured by the AVIRIS sensor [55], which measures 100 × 100 × 189 and consists mainly of roofs, shadows, and grass, of which aircrafts are considered anomalies to be detected.

HYDICE-Urban: The dataset is collected by a hyperspectral digital imagery collection experiment (HYDICE) sensor [56] for an urban area, including one vegetation area, one built-up area, and several built-up areas of roads and some vehicles. Its spatial resolution is 1 meter. The size of the crop plus the water vapor removal is 80 × 100 × 175. The 21 pixels occupied by vehicles and roofs of different sizes are used as anomalies.

Airport 1–4: The dataset consists of four images of 100 by 100 pixels in 205 bands taken by the airborne visible/infrared imaging spectrometer (AVIRIS) sensor [57]. As above, they include surface vegetation, roads, and buildings as background. Aircraft flying at different altitudes are treated as anomalies.

Urban 1–4: This dataset of four urban scenes is obtained from a class of sensors as with the airport dataset, with pixels of 100 × 100 and a band number between 190 and 210.

4.2. Compared Methods and Parameter Setting

In this section, we briefly introduce the compared anomaly detection methods and their parameter settings. The parameter values of the compared methods in our experiments are tuned according to the corresponding references.

- **RX [19]:** The classical anomaly detection algorithm calculates the Mahalanobis distance between the pixel under test and the background pixels. The parameter λ of RX is set to 1/min(w,h).
- **LSMAD [29]:** A method based on low-rank sparse matrix decomposition (LRaSAM) with Mahalanobis distance. We set $r = 3$, $k = 0.8$.
- **LRASR [34]:** Learn low-rank linear representation (LRR) of backgrounds by constructing dictionaries. The parameters λ and β of LRASR are set to 0.1 and 0.1 in LRASR.
- **GTVLRR [35]:** Adding total variation (TV) and graph regularization to the restructuring of the background in the LRR-based method, we set $\lambda = 0.5$, $\beta = 0.2$, and $\omega = 0.05$ according to the GTVLRR.
- **PTA [40]:** According to the properties of the spatial and spectral dimensions of the HSI, PTA adds TV into spatial dimensions and low-rank into spectral dimensions. The parameters α, τ ,β of PTA are set to 1, 1, and 0.01 separately.
- **DeCNN-AD [36]:** Using convolutional neural network (CNN)-based denoisers as the prior for the dictionary representation coefficients, the cluster number of DeCNN-AD is set to 8 and λ, β are set to 0.01.
- **PCA-TLRSR [43]:** The first method extends LRR to tensor LRR for HSI anomaly detection. The reduced dimensions of PCA are tuned according to PCA-TLRSR and parameter λ is set to 0.4.

4.3. Detection Performance

4.3.1. San Diego

The false-color image, ground-truth map, and detection maps of all compared methods are shown in Figure 2. In the San Diego detection maps, methods such as RX, LSMAD, and LRASR fail to accurately detect the three aircrafts in the upper right corner of the HSI data. DeCNN-AD and GTVLRR have difficulty clearly identifying the outline of the aircrafts. PTA can observe the aircrafts, but it contains some background information, such as road buildings. PCA-TLRSR obtains a relatively good performance by recovering the outline of the aircrafts with less background information. In addition, our method can also capture more detailed features of the aircrafts. Figure 3 shows the anomaly detection evaluation metrics of different anomaly detectors for the San Diego dataset, including 3D ROC curves, 2D ROC curves, and separability maps. The proposed method has a slightly higher detection probability than PCA-TLRSR in 3D and 2D ROC curves. The gap between the background box and the anomaly box shows the degree of separation between background and anomaly on the separability maps. The separability maps on the San Diego dataset are shown in Figure 3. The proposed MDLR obtains a larger gap between the background and the anomaly box over all the other methods compared, which indicates that it has a better ability to separate the background and anomaly. The AUC values in the second row of Table 2 provide further evidence that our method achieves the highest performance on the San Diego dataset.

4.3.2. HYDICE-Ubran

The false-color image, ground-truth map, and detection maps of the competitive methods are visually shown in Figure 4. In the detection maps of all compared methods on the HYDICE-Urban dataset, RX and LSMAD have difficulty in clearly recovering the anomaly information. PTA and PCA-TLRSR can observe the anomaly information, but they also contain a significant amount of background information, such as roads. DeCNN-AD, LRASR, GTVLRR, and our proposed method can clearly identify the anomaly information. However, both LRASR and GTVLRR struggle to detect the anomaly in the lower left corner of the image. Our proposed method shows an improvement in terms of visual quality and achieves a higher AUC value compared to other methods, as shown in Table 2.

Table 2. AUC values of all compared algorithms on different datasets.

HSI Datasets	RX	LSMAD	LRASR	GTVLRR	DeCNN-AD	PTA	PCA-TLRSR	MDLR
San Diego	0.8885	0.9773	0.9853	0.9795	0.9901	0.9946	<u>0.9957</u>	**0.9976**
HYDICE-Urban	0.9856	0.9901	0.9918	0.9856	0.9935	<u>0.9953</u>	0.9941	**0.9975**
Airport-1	0.8220	0.8334	0.7854	0.9013	0.8503	0.9207	<u>0.9478</u>	**0.9538**
Airport-2	0.8403	0.9189	0.8657	0.8695	0.9204	0.9428	<u>0.9697</u>	**0.9738**
Airport-3	0.9228	0.9401	0.9408	0.9295	0.9434	0.9355	<u>0.9574</u>	**0.9590**
Airport-4	0.9526	0.9862	0.9723	0.9875	0.9897	0.9875	<u>0.9943</u>	**0.9953**
Urban-1	**0.9907**	0.9829	0.9797	0.9605	0.9820	0.9826	<u>0.9902</u>	0.9835
Urban-2	0.9946	0.9836	0.9628	0.8539	<u>0.9973</u>	0.9970	0.9941	**0.9980**
Urban-3	0.9513	0.9636	0.9415	0.9385	0.9394	0.9578	**0.9833**	<u>0.9812</u>
Urban-4	0.9887	0.9809	0.9575	0.9205	0.9868	0.9907	<u>0.9869</u>	**0.9966**

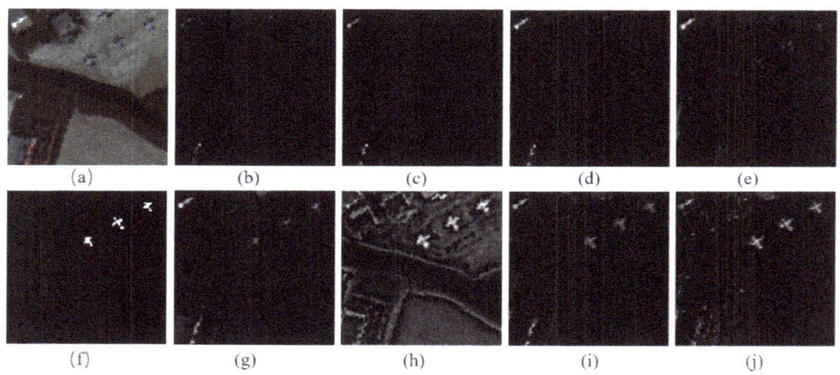

Figure 2. Detection maps obtained by all compared methods on San Diego dataset. (**a**) HSI. (**b**) RX. (**c**) LSMAD. (**d**) LRASR. (**e**) GTVLRR. (**f**) Ground-truth. (**g**) DeCNN-AD. (**h**) PTA. (**i**) PCA-TLRSR. (**j**) Ours.

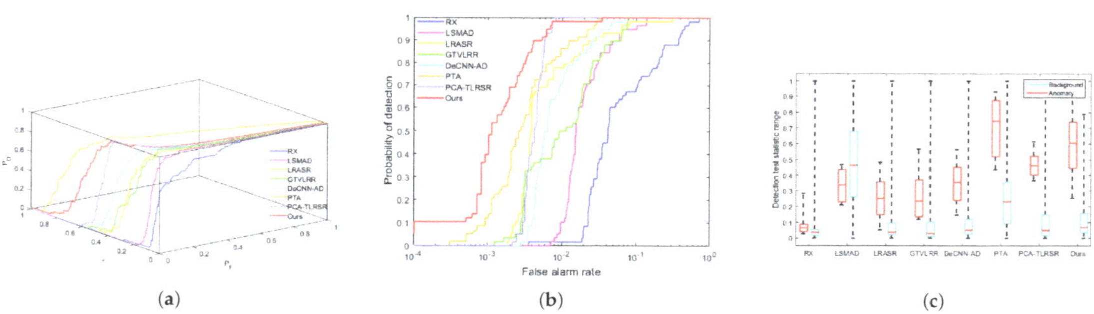

Figure 3. Anomaly detection evaluation metrics obtained by different methods on the San Diego dataset. (**a**) Three-dimensional (3D) ROC curves, (**b**) 2D ROC curves, (**c**) separability map.

4.3.3. Airport 1–4

The AUC values of four airport datasets are provided in Table 2. Our method has achieved the highest values. The false-color images, ground-truth maps, and detection maps of four airport dates are demonstrated in Figure 5. In the detection maps of Airport-1, RX, LSMAD, LRASR, and DeCNN-AD datasets, it is difficult to distinguish the aircrafts.

GTVLRR, PTA, PCA-TLRSR, and ours can distinguish the aircraft in the middle, but they contain a lot of roof information and the aircraft in the upper left corner is not visible. In the detection maps of Airport-2, our method can clearly observe the middle plane of the figure with less background information compared to other methods, but does not fully preserve its edge information due to the effect of some mixed pixels. Compared to the ground truth of Airport-3, the detection maps of all comparison methods can barely detect the outline of an aircraft. This indicates that the existing methods are not sensitive to detecting dense small targets and are easily contaminated by background information. In the detection maps of the Airport-4 dataset, our detection result shows a clear outline of the aircraft compared to other methods. There is no interference from road information compared to LSMAD, LRASR, DeCNN-AD, GTVLRR, PTA, and PCA-TLRSR. The first row of Figures 6 and 7 show the 2D and 3D ROC curves of different anomaly detectors for Airport 1–4. They demonstrate that our method produces detection maps with relatively little interference from background information. The separability maps on the Airport dataset are shown in the first row of Figure 8. The compared methods fail to effectively separate the background boxes and anomaly boxes, while the proposed MDLR achieves a bigger gap.

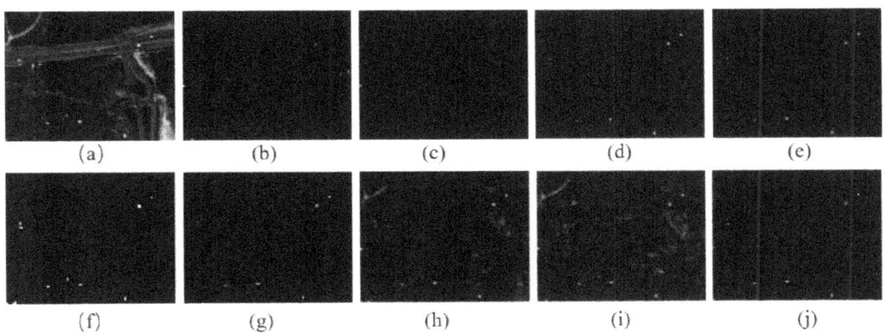

Figure 4. Detection maps on HYDICE-Urban dataset obtained by all compared methods. (**a**) HSI. (**b**) RX. (**c**) LSMAD. (**d**) LRASR. (**e**) GTVLRR. (**f**) Ground-truth. (**g**) DeCNN-AD. (**h**) PTA. (**i**) PCA-TLRSR. (**j**) Ours.

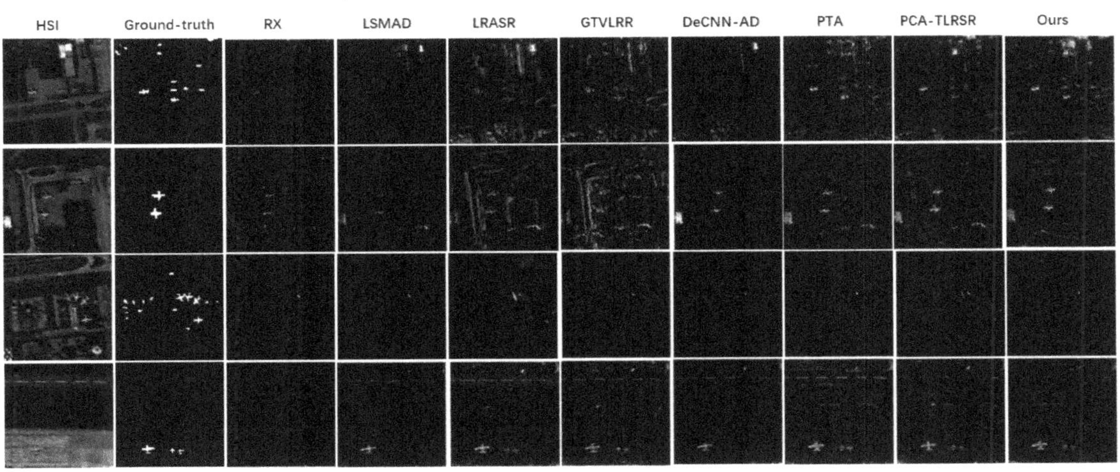

Figure 5. Detection maps obtained by all compared methods on Airport-1 (first line), Airport-2 (second line), Airport-3 (third line), and Airport-4 (forth line) datasets.

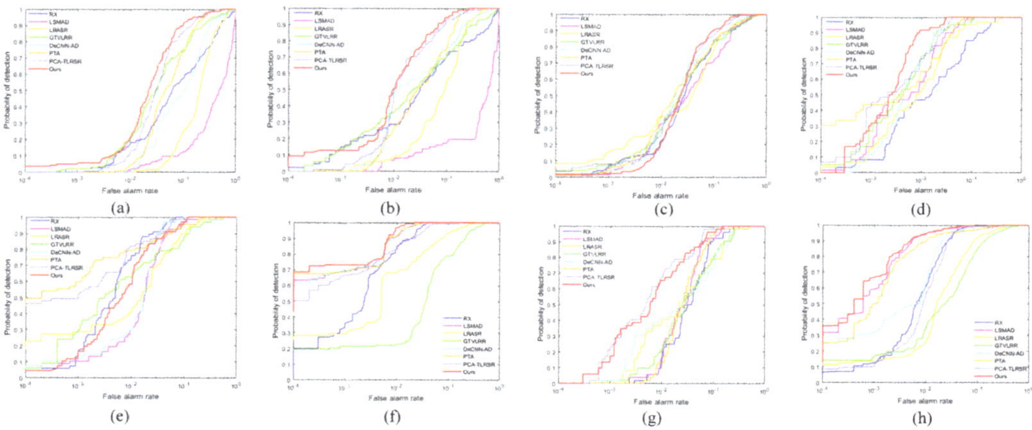

Figure 6. Two-dimensional (2D) ROC curves obtained by all compared methods. (**a**) Airport-1, (**b**) Airport-2, (**c**) Airport-3, (**d**) Airport-4, (**e**) Urban-1, (**f**) Urban-2, (**g**) Urban-3, (**h**) Urban-4.

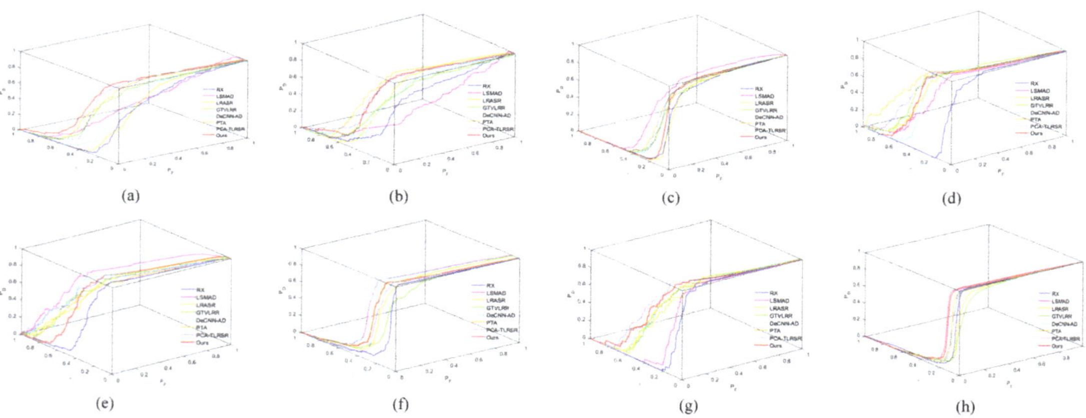

Figure 7. Three-dimensional (3D) ROC curves obtained by all compared methods. (**a**) Airport-1, (**b**) Airport-2, (**c**) Airport-3, (**d**) Airport-4, (**e**) Urban-1, (**f**) Urban-2, (**g**) Urban-3, (**h**) Urban-4.

4.3.4. Urban 1–4

The AUC values in Table 2 are optimal for all datasets except Urban-1 and Urban-3. For the Urban-1 dataset, from the detection maps shown in Figure 9 we can observe clear lines running through the maps in LSMAD, LRASR, GTVLRR, DeCNN-AD, PTA, and our method. However, the detection map of PCA-TLRSR is difficult to interpret. PCA-TLRSR uses PCA for dimensionality reduction, which aims to reduce noise in the image. The presence of noise in a particular image may have hindered the achievement of optimal results. In the false-color image of Urban-3, there are many large and obvious targets in the background. In the third row of Figure 9, the detection maps of RX and LSMAD barely show the anomaly targets. Other detection algorithms can detect the anomaly targets but retain most of the background contour information. In the detection maps of Urban-2 and Urban-4, compared to the detection maps obtained by other algorithms and the ROC curves in the second row of Figures 6 and 7 obtained by other methods, our method obtains a clearer and more accurate observation of the anomalies. In the separability maps of the Urban dataset in Figure 8, it can be seen that the background boxes and the anomaly boxes of proposed MDLR are obviously separated, which also proves that our method can achieve effective separation of background and anomaly.

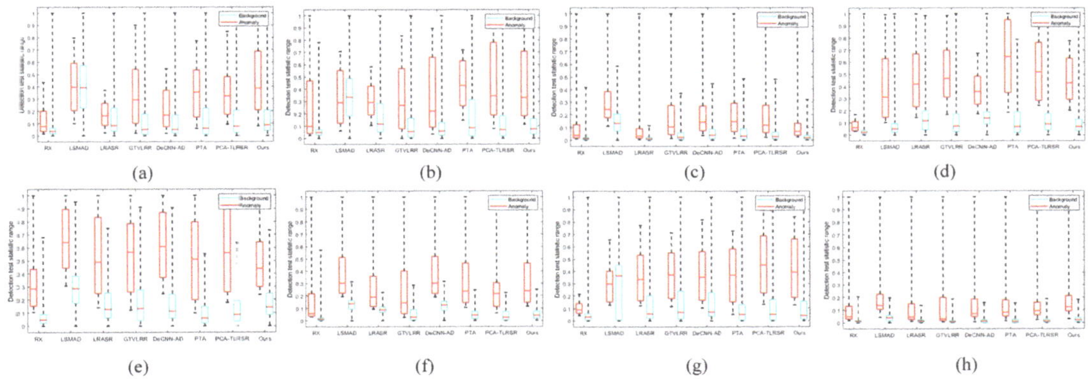

Figure 8. Separability maps obtained by all compared methods. (**a**) Airport-1, (**b**) Airport-2, (**c**) Airport-3, (**d**) Airport-4, (**e**) Urban-1, (**f**) Urban-2, (**g**) Urban-3, (**h**) Urban-4.

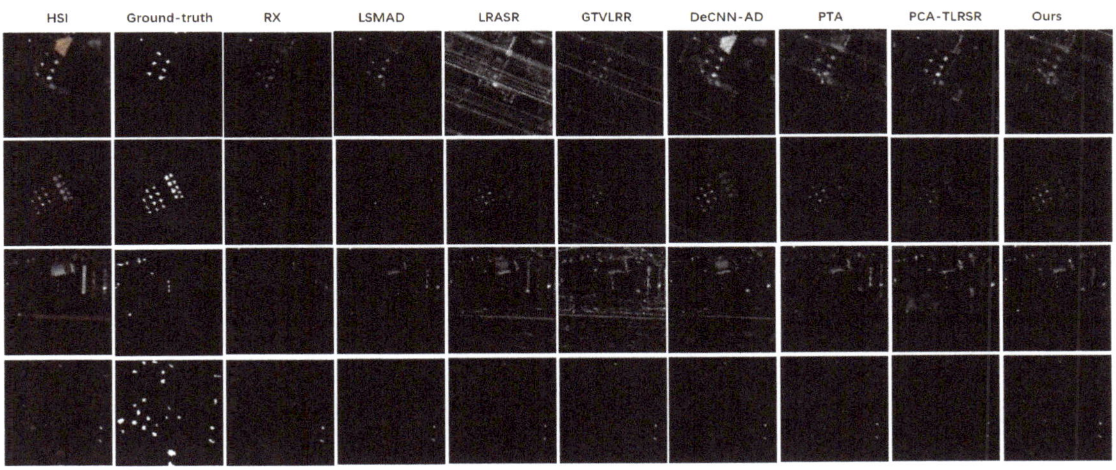

Figure 9. Detection maps obtained by all compared methods on Urban-1 (first line), Urban-2 (second line), Urban-3 (third line), and Urban-4 (forth line) datasets.

4.4. Discussion of Multi-Dimensional Low-Rank

In this section, we analyze the necessity of reconstructing the background with multi-dimensional low-rank.

The discussion on single-dimensional and three-dimensional low-rank: The results presented in Table 3 show that reconstructing the background along a multi-dimensional (w, h, d dimension) gives significantly higher AUC values compared to reconstructing along a single-dimensional (d dimension). This improvement is evident across all datasets, with notable increases in AUC values for the HYDICE-Urban and Airport-1 HSI datasets. Specifically, the AUC values for anomaly detection increased by 4 percent and 6 percent for HYDICE-Urban and Airport-1 datasets, respectively, compared to the one-dimensional reconstruction. These improvements demonstrate the benefit of using multi-dimensional information to effectively separate the background from anomalies in the HSI data. By considering the data from multiple dimensions simultaneously, the proposed method is able to capture more comprehensive and discriminative information about the background, leading to an improved detection performance. This highlights that multi-dimensional has an advantage over single-dimensional in separating background and anomaly.

Table 3. AUC values of single-dimensional and multi-dimensional low-rank.

HSI dataset	San Diego	Airport-1	Airport-2	Airport-3	Airport-4
S-dimensional	0.9966	0.8957	0.9655	0.9345	0.9921
M-dimensional	**0.9976**	**0.9538**	**0.9738**	**0.9590**	**0.9953**
HSI dataset	HYDIE-Urban	Urban-1	Urban-2	Urban-3	Urban-4
S-dimensional	0.9546	0.9619	0.9928	0.9527	**0.9966**
M-dimensional	**0.9975**	**0.9835**	**0.9980**	**0.9812**	**0.9966**

The discussion on two-dimensional and three-dimensional low-rank: From the results shown in Figure 10, it can be seen that the AUC values obtained by reconstructing the background with different combinations of two dimensions are different. In the case of Airport-1 in Figure 10a,b, reconstructing the background with $\mathcal{X}_d$ and $\mathcal{X}_w$ achieves higher AUC values compared to reconstructing with $\mathcal{X}_d$ and $\mathcal{X}_h$. On the other hand, for Airport-3 in Figure 10c,d, reconstructing the background with $\mathcal{X}_h$ and $\mathcal{X}_w$ yields higher AUC values compared to reconstructing with $\mathcal{X}_d$ and $\mathcal{X}_h$. These results indicate that there is no fixed combination of two dimensions that consistently gives the best performance for background tensor reconstruction. The optimal combination may vary depending on the specific dataset and the characteristics of the HSI data. Therefore, it is important to explore and analyze the relationship between different dimensional background tensors to achieve the best reconstruction results.

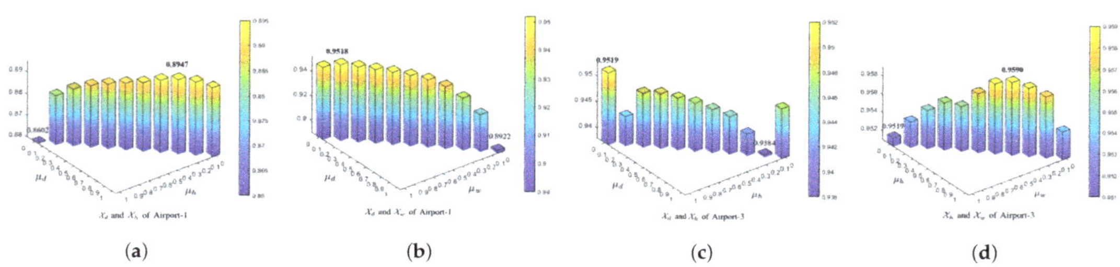

Figure 10. AUC value bars obtained from coefficients μ via the reconstruction of the background with two different dimensions. (**a**) Airport-1: μ_w and μ_h, (**b**) Airport-1: μ_w and μ_d, (**c**) Airport-3: μ_w and μ_h, (**d**) Airport-3: μ_h and μ_d.

In the analysis of the reconstruction of the background along two different dimensions of the HSI, the focus is on investigating the relationship between these dimensions. For this purpose, the experimental datasets Airport-1 and Airport-3 are selected. Two dimensions of the data are chosen to reconstruct the background, resulting in four different comparison experiments: (a) Reconstruction using $\mathcal{X}_d$ and $\mathcal{X}_h$ from Airport-1; (b) Reconstruction using $\mathcal{X}_d$ and $\mathcal{X}_w$ from Airport-1; (c) Reconstruction using $\mathcal{X}_d$ and $\mathcal{X}_h$ from Airport-3; (d) Reconstruction using $\mathcal{X}_h$ and $\mathcal{X}_w$ from Airport-3. The aim of these experiments is to investigate the performance and effectiveness of background reconstruction when using different combinations of two dimensions.

Effects of coefficient μ in two reconstructed background tensors: This study investigates the relationship between the reconstructed background tensors from different dimensions, which aims to better understand their impact on anomaly detection performance. The coefficients between two dimensions are varied in the range [0:0. 1:1] so that the results can be observed when each dimension acts alone and when two dimensions work together. The AUC values of the comparison experiments are visualized in Figure 10. In Figure 10a, the individual reconstruction of $\mathcal{X}_d$ in Airport-1 achieves an accuracy of 0.8922, while the individual reconstruction of $\mathcal{X}_h$ only achieves an accuracy of 0.8602. However, when $\mathcal{X}_d$ and $\mathcal{X}_h$ are combined in the reconstruction, they achieve a maximum AUC value of 0.8947. Similarly, in Figure 10b, the individual reconstruction of $\mathcal{X}_d$ in Airport-1 gives an AUC of 0.8922, while the individual reconstruction of $\mathcal{X}_w$ gives an AUC of 0.9487. However, their combination leads to a maximum AUC value of 0.9518. The same trend can be seen in Figure 10c,d for Airport-3. From Figure 10a,c, it is clear that both Airport-1 and Airport-3 benefit from the background reconstruction using $\mathcal{X}_d$ and $\mathcal{X}_h$. Interestingly, the coefficients of the reconstructed background from the same dimensions differ between the two datasets, indicating that the relationship between the reconstructed background tensors can vary for different datasets. The final AUC values in Table 2 also support the effectiveness of reconstructing the background along multiple dimensions. Although the AUC values of Airport-1 and Airport-3 in Figure 10 are slightly lower, they still demonstrate the validity of the multi-dimensional reconstruction approach in improving the anomaly detection performance.

4.5. Parameter Tuning

In this section, we focus on analyzing the effect of the values of λ and p on the AUC results.

(1) **Effects of parameter λ:** The influence of parameter λ on model performance was analysed on four HSI datasets. The parameter λ was selected from the set [0.001, 0.005, 0.01, 0.05, 0.1, 1, 2] while keeping the other parameters fixed. AUC value curves with respect to λ on four datasets are shown in Figure 11a. The AUC values of the San Diego, HYIDE-Urban and Airport-4 datasets reach their maximum value when λ is equal to 1. Airport-1 and Airport-2 datasets both reach their maximum value when λ is 2. Airport-3 has a downward trend when λ is 1. For the experiment as a whole, when λ is 0.1 or 1, the AUC values are relatively stable. So for San Diego, HYDICE-Urban, and Airport-4 datasets, λ is set to 1. λ is set to 0.1 on Airport 1-3.

(2) **Effects of parameter p:** The AUC value curves for the parameter p are shown in Figure 11b. The parameter p has been chosen in the range [0.1, 1]. As the value of p increases, the AUC values on different HSI datasets start to improve. The growing trend of the AUC values for San Diego, HYDICE-Urban, Airport-2, Airport-3, and Airport-4 datasets tends to level off when p reaches 0.6. However, the AUC value of Airport-1 continues to increase as p increases. Based on the AUC value curves, the following choices of p are made. For the San Diego, HYDICE-Urban, Airport-1, and Airport-4 datasets, p is set to 1. For the Airport-2 dataset, p is set to 0.9. For the Airport-3 dataset, p is set to 0.6.

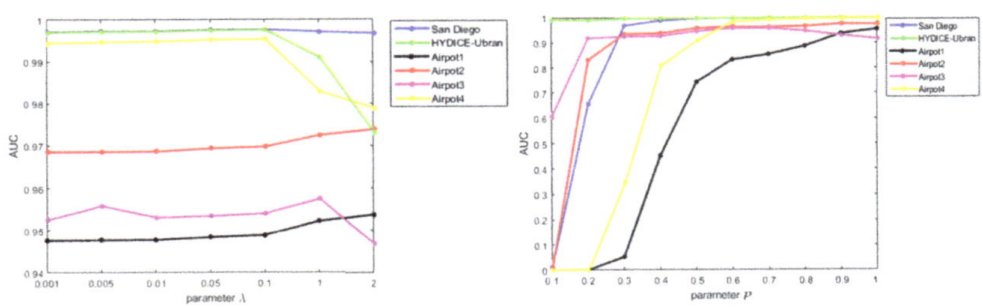

(a) (b)

Figure 11. The effect of parameter tuning on AUC values. (**a**) AUC value curves with respect to λ on four datasets; (**b**) AUC value curves with respect to p on three datasets.

5. Conclusions

In this paper, a novel multi-dimensional low-rank (MDLR) method is proposed for HSI anomaly detection. The MDLR method exploits the low-rank properties of HSI from three dimensions, namely the spatial and spectral dimensions. Multi-dimensional background tensors are reconstructed. Weighted Schatten p-norm minimization is used to enforce the low-rank constraints. In addition, the $L_{F,1}$ norm is used to penalize the anomaly tensor to promote joint spectral–spatial sparsity. The optimization problem is solved using ADMM Experimental results on real HSI datasets demonstrate its effectiveness compared with SOTA in terms of anomaly detection. However, one of the major limitations of MDLR is the computational complexity introduced by the t-SVD operation, especially when dealing with large spectral bands. In future work, we would like to try to incorporate dimensionality reduction preprocessing techniques, which is a promising direction to take in order to address this computational challenge.

Author Contributions: All authors made significant contributions to this work. Conceptualization, X.C.; methodology, Z.W., X.C. and K.W.; investigation, Z.W.; software, Z.W. and X.C. ; datacuration, X.C. and H.J.; validation, K.W.; writing original draft preparation, Z.W. and X.C. ; writing review and editing, K.W., Z.H. and Y.T. All authors have read and agreed to the published version of the manuscript.

Funding: This research was supported by the Youth Innovation Promotion Association of the Chinese Academy of Sciences under Grant 2022196 and Y202051; National Natural Science Foundation of China under Grant 61873259 and 61821005; CAS Project for Young Scientists in Basic Research under Grant YSBR-041 and the State Key Laboratory of Robotics under Grant 2023-O28.

Institutional Review Board Statement: Not applicable.

Informed Consent Statement: Not applicable.

Data Availability Statement: No new data were created or analyzed in this study. Data sharing is not applicable to this article.

Conflicts of Interest: The authors declare no conflict of interest. The funders had no role in the design of the study; in the collection, analyses, or interpretation of data; in the writing of the manuscript,or in the decision to publish the results.

References

1. Zhang, X.; Wu, H.; Sun, H.; Ying, W. Multireceiver SAS Imagery Based on Monostatic Conversion. *IEEE J. Sel. Top. Appl. Earth Obs. Remote Sens.* **2021**, *14*, 10835–10853. [CrossRef]
2. Zhu, J.; Song, Y.; Jiang, N.; Xie, Z.; Fan, C.; Huang, X. Enhanced Doppler Resolution and Sidelobe Suppression Performance for Golay Complementary Waveforms. *Remote Sens.* **2023**, *15*, 2452. [CrossRef]
3. Hong, D.; Gao, L.; Yokoya, N.; Yao, J.; Chanussot, J.; Du, Q.; Zhang, B. More Diverse Means Better: Multimodal Deep Learning Meets Remote-Sensing Imagery Classification. *IEEE Trans. Geosci. Remote Sens.* **2020**, *59*, 4340–4354. [CrossRef]
4. Liu, F.; Wang, Q. A sparse tensor-based classification method of hyperspectral image. *Signal Process.* **2020**, *168*, 107361. [CrossRef]
5. An, W.; Zhang, X.; Wu, H.; Zhang, W.; Du, Y.; Sun, J. LPIN: A Lightweight Progressive Inpainting Network for Improving the Robustness of Remote Sensing Images Scene Classification. *Remote Sens.* **2021**, *14*, 53. [CrossRef]
6. Tan, K.; Wu, F.; Du, Q.; Du, P.; Chen, Y. A Parallel Gaussian–Bernoulli Restricted Boltzmann Machine for Mining Area Classification With Hyperspectral Imagery. *IEEE J. Sel. Topics Appl. Earth Observ. Remote Sens.* **2019**, *12*, 627–636. [CrossRef]
7. Ren, Z.; Sun, L.; Zhai, Q.; Liu, X. Mineral Mapping with Hyperspectral Image Based on an Improved K-Means Clustering Algorithm. In Proceedings of the IGARSS 2019 IEEE International Geoscience and Remote Sensing Symposium, Yokohama, Japan, 28 July–2 August 2019; pp. 2989–2992.
8. Rukhovich, D.I.; Koroleva, P.V.; Rukhovich, D.D.; Rukhovich, A.D. Recognition of the Bare Soil Using Deep Machine Learning Methods to Create Maps of Arable Soil Degradation Based on the Analysis of Multi-Temporal Remote Sensing Data. *Remote Sens.* **2022**, *14*, 2224. [CrossRef]
9. Wang, Q.; Li, J.; Shen, Q.; Wu, C.; Yu, J. Retrieval of water quality from China's first satellite-based Hyperspectral Imager (HJ-1A HSI) data. In Proceedings of the 2010 IEEE International Geoscience and Remote Sensing Symposium, Honolulu, Hawaii, USA, 25–30 July 2010; pp. 371–373.
10. Zhang, X.; Han, L.; Dong, Y.; Shi, Y.; Huang, W.; Han, L.; González-Moreno, P.; Ma, H.; Ye, H.; Sobeih, T. A Deep Learning-Based Approach for Automated Yellow Rust Disease Detection from High-Resolution Hyperspectral UAV Images. *Remote Sens.* **2019**, *11*, 1554. [CrossRef]
11. Wan, Y.; Hu, X.; Zhong, Y.; Ma, A.; Wei, L.; Zhang, L. Tailings Reservoir Disaster and Environmental Monitoring Using the UAV-ground Hyperspectral Joint Observation and Processing: A Case of Study in Xinjiang, the Belt and Road. In Proceedings of the IGARSS 2019 IEEE International Geoscience and Remote Sensing Symposium, Yokohama, Japan, 28 July–2 August 2019; pp. 9713–9716.

12. Farrar, M.B.; Wallace, H.M.; Brooks, P.R.; Yule, C.M.; Tahmasbian, I.; Dunn, P.K.; Bai, S.H. A Performance Evaluation of Vis/NIR Hyperspectral Imaging to Predict Curcumin Concentration in Fresh Turmeric Rhizomes. *Remote Sens.* **2021**, *13*, 1807. [CrossRef]
13. Légaré, B.; Bélanger, S.; Singh, R.K.; Bernatchez, P.; Cusson, M. Remote Sensing of Coastal Vegetation Phenology in a Cold Temperate Intertidal System: Implications for Classification of Coastal Habitats. *Remote Sens.* **2022**, *14*, 3000. [CrossRef]
14. Cen, Y.; Huang, Y.H.; Hu, S.; Zhang, L.; Zhang, J. Early Detection of Bacterial Wilt in Tomato with Portable Hyperspectral Spectrometer. *Remote Sens.* **2022**, *14*, 2882. [CrossRef]
15. Yin, C.; Lv, X.; Zhang, L.; Ma, L.; Wang, H.; Zhang, L.; Zhang, Z. Hyperspectral UAV Images at Different Altitudes for Monitoring the Leaf Nitrogen Content in Cotton Crops. *Remote Sens.* **2022**, *14*, 2576. [CrossRef]
16. Thornley, R.H.; Verhoef, A.; Gerard, F.F.; White, K. The Feasibility of Leaf Reflectance-Based Taxonomic Inventories and Diversity Assessments of Species-Rich Grasslands: A Cross-Seasonal Evaluation Using Waveband Selection. *Remote. Sens.* **2022**, *14*, 2310. [CrossRef]
17. Pang, D.; Ma, P.; Shan, T.; Li, W.; Tao, R.; Ma, Y.; Wang, T. STTM-SFR: Spatial–Temporal Tensor Modeling With Saliency Filter Regularization for Infrared Small Target Detection. *IEEE Trans. Geosci. Remote Sens.* **2022**, *60*, 1–18. [CrossRef]
18. Gao, Q.; Zhang, P.; Xia, W.; Xie, D.; Gao, X.; Tao, D. Enhanced Tensor RPCA and its Application. *IEEE Trans. Pattern Anal. Mach. Intell.* **2020**, *43*, 2133–2140. [CrossRef] [PubMed]
19. Reed, I.S.; Yu, X. Adaptive multiple-band CFAR detection of an optical pattern with unknown spectral distribution. *IEEE Trans. Acoust. Speech Signal Process.* **1990**, *38*, 1760–1770. [CrossRef]
20. Manolakis, D.G.; Shaw, G.A. Detection algorithms for hyperspectral imaging applications. *IEEE Signal Process. Mag.* **2002**, *19*, 29–43. [CrossRef]
21. Molero, J.M.; Garzón, E.M.; García, I.; Plaza, A.J. Analysis and Optimizations of Global and Local Versions of the RX Algorithm for Anomaly Detection in Hyperspectral Data. *IEEE J. Sel. Topics Appl. Earth Observ. Remote Sens.* **2013**, *6*, 801–814. [CrossRef]
22. Taitano, Y.P.; Geier, B.A.; Bauer, K.W. A Locally Adaptable Iterative RX Detector. *Signal Process.* **2010**, *2010*, 1–10. [CrossRef]
23. Sun, W.; Liu, C.; Li, J.; Lai, Y.M.; Li, W. low-rank and sparse matrix decomposition-based anomaly detection for hyperspectral imagery. *J. Appl. Remote Sens.* **2014**, *8*, 15823048. [CrossRef]
24. Farrell, M.D.; Mersereau, R.M. On the impact of covariance contamination for adaptive detection in hyperspectral imaging. *IEEE Signal Process. Lett.* **2005**, *12*, 649–652. [CrossRef]
25. Billor, N.; Hadi, A.S.; Velleman, P.F. BACON: blocked adaptive computationally efficient outlier nominators. *Comput. Stat. Data Anal.* **2000**, *34*, 279–298. [CrossRef]
26. Sun, W.; Tian, L.; Xu, Y.; Du, B.; Du, Q. A Randomized Subspace Learning Based Anomaly Detector for Hyperspectral Imagery. *Remote Sens.* **2018**, *10*, 417. [CrossRef]
27. Sun, W.; Yang, G.; Li, J.; Zhang, D. Randomized subspace-based robust principal component analysis for hyperspectral anomaly detection. *J. Appl. Remote Sens.* **2018**, *12*, 015015. [CrossRef]
28. Qu, Y.; Wang, W.; Guo, R.; Ayhan, B.; Kwan, C.; Vance, S.D.; Qi, H. Hyperspectral Anomaly Detection Through Spectral Unmixing and Dictionary-Based low-rank Decomposition. *IEEE Trans. Geosci. Remote Sens.* **2018**, *56*, 4391–4405. [CrossRef]
29. Zhang, Y.; Du, B.; Zhang, L.; Wang, S. A low-rank and Sparse Matrix Decomposition-Based Mahalanobis Distance Method for Hyperspectral Anomaly Detection. *IEEE Trans. Geosci. Remote Sens.* **2016**, *54*, 1376–1389. [CrossRef]
30. Xu, Y.; Du, B.; Zhang, L.; Chang, S. A low-rank and Sparse Matrix Decomposition- Based Dictionary Reconstruction and Anomaly Extraction Framework for Hyperspectral Anomaly Detection. *IEEE Geosci. Remote Sens. Lett.* **2020**, *17*, 1248–1252. [CrossRef]
31. Li, L.; Li, W.; Du, Q.; Tao, R. low-rank and Sparse Decomposition With Mixture of Gaussian for Hyperspectral Anomaly Detection. *IEEE Trans. Cybern.* **2020**, *51*, 4363–4372. [CrossRef]
32. Liu, G.; Lin, Z.; Yu, Y. Robust Subspace Segmentation by low-rank Representation. In Proceedings of the International Conference on Machine Learning, Haifa, Israel, 21–24 June 2010.
33. Liu, G.; Lin, Z.; Yan, S.; Sun, J.; Yu, Y.; Ma, Y. Robust Recovery of Subspace Structures by low-rank Representation. *IEEE Trans. Pattern Anal. Mach. Intell.* **2010**, *35*, 171–184. [CrossRef]
34. Xu, Y.; Wu, Z.; Li, J.; Plaza, A.J.; Wei, Z. Anomaly Detection in Hyperspectral Images Based on low-rank and Sparse Representation. *IEEE Trans. Geosci. Remote Sens.* **2016**, *54*, 1990–2000. [CrossRef]
35. Cheng, T.; Wang, B. Graph and Total Variation Regularized low-rank Representation for Hyperspectral Anomaly Detection. *IEEE Trans. Geosci. Remote Sens.* **2020**, *58*, 391–406. [CrossRef]
36. Fu, X.; Jia, S.; Zhuang, L.; Xu, M.; Zhou, J.; Li, Q. Hyperspectral Anomaly Detection via Deep Plug-and-Play Denoising CNN Regularization. *IEEE Trans. Geosci. Remote Sens.* **2021**, *59*, 9553–9568. [CrossRef]
37. Zhang, K.; Zuo, W.; Zhang, L. FFDNet: Toward a Fast and Flexible Solution for CNN-Based Image Denoising. *IEEE Trans. Image Process.* **2017**, *27*, 4608–4622. [CrossRef] [PubMed]
38. Guo, S.; Chen, X.; Jia, H.; Han, Z.; Duan, Z.; Tang, Y. Fusing Hyperspectral and Multispectral Images via low-rank Hankel Tensor Representation. *Remote Sens.* **2022**, *14*, 4470. [CrossRef]
39. Zhang, Z.; Ding, C.; Gao, Z.; Xie, C. ANLPT: Self-Adaptive and Non-Local Patch-Tensor Model for Infrared Small Target Detection. *Remote Sens.* **2023**, *15*, 1021. [CrossRef]
40. Li, L.; Li, W.; Qu, Y.; Zhao, C.; Tao, R.; Du, Q. Prior-Based Tensor Approximation for Anomaly Detection in Hyperspectral Imagery. *IEEE Trans. Neural Netw. Learn. Syst.* **2020**, *33*, 1037–1050. [CrossRef] [PubMed]

41. Song, S.; Zhou, H.; Gu, L.; Yang, Y.; Yang, Y. Hyperspectral Anomaly Detection via Tensor- Based Endmember Extraction and low-rank Decomposition. *IEEE Geosci. Remote. Sens. Lett.* **2020**, *17*, 1772–1776. [CrossRef]
42. Shang, W.; Peng, J.; Wu, Z.; Xu, Y.; Jouni, M.; Mura, M.D.; Wei, Z. Hyperspectral Anomaly Detection via Sparsity of Core Tensor Under Gradient Domain. *IEEE Trans. Geosci. Remote Sens.* **2023**, *61*, 1–16. [CrossRef]
43. Wang, M.; Wang, Q.; Hong, D.; Roy, S.K.; Chanussot, J. Learning Tensor low-rank Representation for Hyperspectral Anomaly Detection. *IEEE Trans. Cybern.* **2022**, *53*, 679–691. [CrossRef]
44. Sun, S.; Liu, J.; Zhang, Z.; Li, W. Hyperspectral Anomaly Detection Based on Adaptive low-rank Transformed Tensor. *IEEE Trans. Neural Netw. Learn. Syst.* **2023**, 1–13. [CrossRef]
45. Lu, C.; Feng, J.; Chen, Y.; Liu, W.; Lin, Z.; Yan, S. Tensor Robust Principal Component Analysis: Exact Recovery of Corrupted low-rank Tensors via Convex Optimization. In Proceedings of the 2016 IEEE Conference on Computer Vision and Pattern Recognition (CVPR), Las Vegas, NV, USA, 26 June–1 July 2016; pp. 5249–5257.
46. Zhang, D.; Hu, Y.; Ye, J.; Li, X.; He, X. Matrix completion by Truncated Nuclear Norm Regularization. In Proceedings of the 2012 IEEE Conference on Computer Vision and Pattern Recognition, Providence, RI, USA, 16–21 June 2012; pp. 2192–2199.
47. Oh, T.H.; Kim, H.; Tai, Y.W.; Bazin, J.C.; Kweon, I.S. Partial Sum Minimization of Singular Values in RPCA for Low-Level Vision. In Proceedings of the 2013 IEEE International Conference on Computer Vision, Sydney, Australia, 1–8 December 2013; pp. 145–152.
48. Gu, S.; Zhang, L.; Zuo, W.; Feng, X. Weighted Nuclear Norm Minimization with Application to Image Denoising. In Proceedings of the 2014 IEEE Conference on Computer Vision and Pattern Recognition, Columbus, OH, USA, 23–28 June 2014; pp. 2862–2869.
49. Lu, C.; Feng, J.; Chen, Y.; Liu, W.; Lin, Z.; Yan, S. Tensor Robust Principal Component Analysis with a New Tensor Nuclear Norm. *IEEE Trans. Pattern Anal. Mach. Intell.* **2018**, *42*, 925–938. [CrossRef] [PubMed]
50. Liu, G.; Yan, S. Latent low-rank Representation for subspace segmentation and feature extraction. In Proceedings of the 2011 International Conference on Computer Vision, Barcelona, Spain, 6–13 November 2011; pp. 1615–1622.
51. Xie, Y.; Gu, S.; Liu, Y.; Zuo, W.; Zhang, W.; Zhang, L. Weighted Schatten p-Norm Minimization for Image Denoising and Background Subtraction. *IEEE Trans. Image Process.* **2015**, *25*, 4842–4857. [CrossRef]
52. Kerekes, J.P. Receiver Operating Characteristic Curve Confidence Intervals and Regions. *IEEE Geosci.* **2008**, *5*, 251–255. [CrossRef]
53. Khazai, S.; Homayouni, S.; Safari, A.; Mojaradi, B. Anomaly Detection in Hyperspectral Images Based on an Adaptive Support Vector Method. *IEEE Geosci.* **2011**, *8*, 646–650. [CrossRef]
54. Chang, C.I. Comprehensive Analysis of Receiver Operating Characteristic (ROC) Curves for Hyperspectral Anomaly Detection. *IEEE Trans. Geosci. Remote Sens.* **2022**, *60*, 1–24. [CrossRef]
55. Li, W.; Du, Q. A survey on representation-based classification and detection in hyperspectral remote sensing imagery. *Pattern Recognit. Lett.* **2016**, *83*, 115–123. [CrossRef]
56. Ma, L.; Crawford, M.M.; Tian, J. Local Manifold Learning-Based k-Nearest-Neighbor for Hyperspectral Image Classification. *IEEE Trans. Geosci. Remote Sens.* **2010**, *48*, 4099–4109. [CrossRef]
57. Kang, X.; Zhang, X.; Li, S.; Li, K.; Li, J.Y.; Benediktsson, J.A. Hyperspectral Anomaly Detection with Attribute and Edge-Preserving Filters. *IEEE Trans. Geosci. Remote Sens.* **2017**, *55*, 5600–5611. [CrossRef]

remote sensing

MDPI

Article

Tracking-by-Detection Algorithm for Underwater Target Based on Improved Multi-Kernel Correlation Filter

Wenrong Yue [1,2], Feng Xu [1] and Juan Yang [1,*]

[1] Ocean Acoustic Technology Laboratory, Institute of Acoustics, Chinese Academy of Sciences, Beijing 100190, China; xf@mail.ioa.ac.cn (F.X.)
[2] University of Chinese Academy of Sciences, Beijing 100049, China
[*] Correspondence: yangjuan@mail.ioa.ac.cn; Tel.: +86-15910836396

Abstract: Joint detection and tracking of weak underwater targets are challenging problems whose complexity is intensified when the target is disturbed by reverberation. In the low signal-to-reverberation ratio (SRR) environment, the traditional detection and tracking methods perform poorly in tracking robustness because they only consider the target motion characteristics. Recently, the kernel correlation filter (KCF) based on target features has received lots of attention and gained great success in visual tracking. We propose an improved multi-kernel correlation filter (IMKCF) tracking-by-detection algorithm by introducing the KCF into the field of underwater weak target detection and tracking. It is composed of the tracking-by-detection, the adaptive reliability check, and the re-detection modules. Specifically, the tracking-by-detection part is built on the multi-kernel correlation filter (MKCF), and it uses multi-frame data weighted averaging to update. The reliability check helps keep the tracker from corruption. The re-detection module, integrated with a Kalman filter, identifies target positions when the tracking is unreliable. Finally, the experimental data processing and analysis show that the proposed method outperforms the single-kernel methods and some traditional tracking methods.

Keywords: active sonar; low signal-to-reverberation ratio; underwater target; tracking by detection; kernel correlation filter

Citation: Yue, W.; Xu, F.; Yang, J. Tracking-by-Detection Algorithm for Underwater Target Based on Improved Multi-Kernel Correlation Filter. *Remote Sens.* **2024**, *16*, 323. https://doi.org/10.3390/rs16020323

Academic Editor: Paolo Tripicchio

Received: 30 October 2023
Revised: 22 December 2023
Accepted: 30 December 2023
Published: 12 January 2024

1. Introduction

Underwater target detection and tracking in active sonar systems has always been a hot topic in underwater applications. The conventional approach to detect and track underwater targets involves threshold detection, followed by data association and filtering tracking [1–4]. However, practical sonar systems often encounter strong reverberation interference. In this low signal-to-reverberation ratio (SRR) environment, only setting a lower threshold can ensure that the target is not missed, but it also causes a lot of false alarms [5,6]. The higher false alarm rate adversely affects target associations, thereby increasing the risk of tracker drift during the tracking process.

In order to solve the problem of weak target detection and tracking in low SRR, the methods commonly used at present can be categorized into three groups. The first group involves traditional data association methods, such as joint probabilistic data association (JPDA) [7,8] and multiple hypothesis tracking (MHT) [9,10]. However, these approaches suffer from high computational costs when confronted with a high false alarm rate. The second group focuses on methods based on random finite set (RFS) [11,12], which eliminate the need for data association. These methods employ filtering techniques based on the motion characteristics of the target. A crucial aspect of accurate tracking filtering is establishing an appropriate motion model that aligns with the target's motion type. The filtering algorithms with multiple models (MMs) and the jump Markov system (JMS) have been shown to be effective approaches for maneuvering target tracking [13–15]. In addition,

Yue et al. established a multi-directional motion model set according to the motion characteristics of the diver target [16]. However, due to the diverse underwater target motion types, it is challenging to apply a single modeling method to all underwater targets of interest. Furthermore, in environments with reverberation interference, the aforementioned tracking methods face difficulties in accurately determining the target's position solely based on trajectory information. The third group involves deep learning methods, such as Convolutional Neural Network (CNN) [17], Recurrent Neural Network (RNN) [18], and Siamese Network [19]. These approaches typically require a substantial number of samples for model training. However, in practical applications, it may not be feasible to collect a sufficient amount of sample data.

Nevertheless, the kernel correlation filter (KCF) algorithm, proposed by Joao F. Henriques et al., presents a promising solution for tracking multiple target types without the need for predefined target motion models [20,21]. Presently, the KCF algorithm is primarily used in the visual tracking field [22], and there has been no prior instance of its application in underwater target tracking domestically or internationally. Hence, the primary contribution of this study lies in the application of the KCF algorithm to resolve the detection and tracking challenges associated with underwater weak targets.

The effectiveness of single-feature-based tracking is limited due to the absence of prior knowledge about the target in the model-free kernelized correlation filter (KCF) algorithm [23–26]. To enhance robustness in tracking, researchers have explored multi-feature fusion tracking methods [27–30], which leverage a shared kernel function with multiple complementary features. However, these methods face challenges in achieving the optimal solution because different features may require distinct kernel functions. To adaptively use multiple complementary features, Tang et al. introduced multi-kernel learning (MKL) into the correlation filtering algorithm to dynamically update multiple nonlinear kernels, namely the multi-kernel correlation filter (MKCF) [31]. However, the MKCF algorithm only utilizes adjacent frame information for filter updates, which could lead to model update errors in the presence of reverberation occlusion. The second contribution of this paper is to utilize weighted information from historical samples to adaptively solve the parameters of multiple nonlinear kernels and make full use of multiple complementary features to enhance the robustness and tracking accuracy of the long-term tracking process.

In scenarios with low SRR, the involvement of non-target information in training frames may result in error propagation during the model update phase, increasing the risk of drift. Thus, it becomes crucial to assess the reliability of tracking results and identify a more dependable result when the tracking outcome is unreliable.

In terms of assessing the reliability of tracking results, Bolme [32] computed the peak-to-sidelobe ratio (PSR) score of the relevant response, comparing it with a fixed threshold to determine reliability. However, this method exhibited limited effectiveness in complex environments. Wang et al. (their tracker is abbreviated as RRLT) [33] proposed a more effective reliability criterion for evaluating the confidence of the current tracking result. This criterion adaptively updates the mean value of multi-frame PSR scores as a threshold, thereby improving the accuracy of evaluation results in complex environments. Regarding tracking methods, the long-term correlation tracker (LCT) employs an online random fern classifier to generate potential target locations [34], while Wang [33] utilizes a particle filter to generate numerous candidate target positions around the previous frame's target position. Nonetheless, neither LCT nor the random RRLT tracker can successfully re-detect a target that has been obscured for an extended period [35]. The third contribution of this paper entails a real-time assessment of target tracking result reliability and proposes an effective re-detection module. The reliability check module adopts the approach outlined by Wang et al. [33] to evaluate the reliability of both detection and tracking results obtained using the MKCF. When the tracking result is deemed unreliable, the re-detection module utilizes the historical reliability tracking result to drive the Kalman filter, predicting the target candidate position. Subsequently, several candidate positions are generated around this predicted position, following a Gaussian distribution. Finally, a stringent replacement criterion is applied to determine the final tracking result.

In summary, this paper presents an improved multi-kernel correlation filter (IMKCF) algorithm for robust detection and tracking of weak underwater targets. A novel adaptation of the KCF algorithm from visual tracking to the domain of underwater multi-motion weak target detection and tracking is proposed. To address the issue of limited robustness in single-feature tracking, the weighted information from historical samples is utilized to adaptively resolve the coefficients of multiple nonlinear kernels. The MKCF algorithm is analyzed from a maximum likelihood perspective to determine the target position based on the maximum likelihood criterion. Real-time estimation of the target tracking result reliability is performed, and an effective re-detection module is introduced. The efficacy of the algorithm is validated through the analysis of sea trial data.

The rest of this article is organized as follows: Section 2 introduces the target model and sonar measurement model. Section 3 reviews the KCF algorithm. Section 4 introduces the framework of the IMKCF tracking-by-detection algorithm and introduces the components of each module in detail. Section 5 analyzes the performance of the algorithm by processing experimental data. Section 6 summarizes the work of this paper.

2. Model Establishment

2.1. Target Model

Given the position of the target x^k at time k denoted by x_{p_k} and y_{p_k}, and the corresponding velocities v_{x_k} and v_{y_k}, the state of the target can be represented by $x^k = \left[x_{p_k}, y_{p_k}, v_{x_k}, v_{y_k}\right]$. The evolution of x^k is formulated as a first-order Markov process,

$$x^{k|k-1} = p\left(x^k \middle| x^{k-1}\right) \tag{1}$$

where the p is the a priori probability density function. The specific form of p is determined by the target model.

2.2. Measurement Model

The algorithm employs raw sonar data measurements in range-azimuth format. When the influence of noise is disregarded, the correlation between the range, angle of the sonar, and position of the target is established.

$$\begin{cases} r_k = \sqrt{x_{p_k}^2 + y_{p_k}^2} \\ \theta_k = \arctan\left(\frac{x_{p_k}}{y_{p_k}}\right). \end{cases} \tag{2}$$

At time k, the resolution of the measurement area is $N_r \times N_b$. With the sonar position serving as the reference point or origin, the measurement area is characterized by a distance range $[R_{min}, R_{max}]$, which is discretized into N_r distance units, and an azimuth range $[\theta_{min}, \theta_{max}]$, which is discretized into N_b azimuth units.

$$N_r = \frac{2(R_{max} - R_{min})}{c} \times F_s, \tag{3}$$

where F_s is the sampling frequency, and N_b can be determined by the azimuth resolution unit $\Delta\theta$. The element $z_k(x,y)$ of the xth azimuth and yth distance cell in the measurement z_k is the signal echo intensity. $z_k(x,y)$ can be modeled as

$$z_k(x,y) = \begin{cases} a_k h(x,y;x_k) + w_k & \text{if target is in } (x,y) \\ w_k & \text{otherwise} \end{cases}, \tag{4}$$

where a_k is the peak amplitude of the target, h is the point spread function, and w_k is the measured noise and reverberation of the sonar system at moment k.

3. Preliminaries of the KCF Algorithm

The KCF algorithm involves three steps: training, detection, and updating. In the training step, it aims to optimize the correlation filter parameter using the training feature–label pairs $\{x_i, y_i\}_{i=1}^{m}$. It maps the input x to a new space $\varphi(x)$ with higher dimensions and puts the $\varphi(x)$ into the optimization process. The kernel function κ and the objective of optimization are as follows.

$$\kappa(x_i, x_j) = \langle \varphi(x_i), \varphi(x_j) \rangle. \tag{5}$$

$$\min_{w} \sum_{i=1}^{m} \left(y_i - \sum_{j=1}^{m} a_j \kappa(x_i, x_j) \right)^2 + \lambda \|w\|^2, \tag{6}$$

where w is defined as

$$w = \sum_{i=1}^{m} a_i \varphi(x_i). \tag{7}$$

The solution of (6) is given by employing the circulant structure for fast training and testing,

$$\alpha = F^{-1}\left(\frac{F(\mathbf{Y})}{F^*(\mathbf{K}_{xx}) + \lambda} \right), \tag{8}$$

where the * indicates the complex conjugate, and F and F^{-1} denote the Fourier transform and the inverse, respectively. $\mathbf{K}_{xx}$ denotes kernel matrix.

In the detection step, we can compute the probability of a new input z being from the target feature.

$$\mathbf{Y}' = F^{-1}(F * (\mathbf{K}_{xz})F(\alpha)). \tag{9}$$

In the updating step, the reference feature x and the correlation filter parameters are calculated as follows

$$x_t = z_t \times \eta + (1 - \eta) \times x_{t-1}, \tag{10}$$

$$\alpha_t = \alpha_{z_t} \times \eta + (1 - \eta) \times \alpha_{t-1}, \tag{11}$$

where η is the learning rate.

As above, the KCF can be performed on general machines due to its high computational efficiency. However, the performance of KCF is related to the features extracted. Making full use of multiple complementary features can improve tracking accuracy and robustness. Therefore, in our proposed algorithm, to avoid the interference of different features in the single kernel, we use MKCF to assign a kernel for each feature.

4. Improved MKCF Tracking-by-Detection

This study aims to tackle the challenge of detecting and tracking weak targets with various motion types in shallow sea environments. We propose an IMKCF tracking-by-detection algorithm, which consists of three components: the MKCF tracking-by-detection, the adaptive reliability check, and the re-detection modules. The overall framework of the proposed approach is depicted in Figure 1.

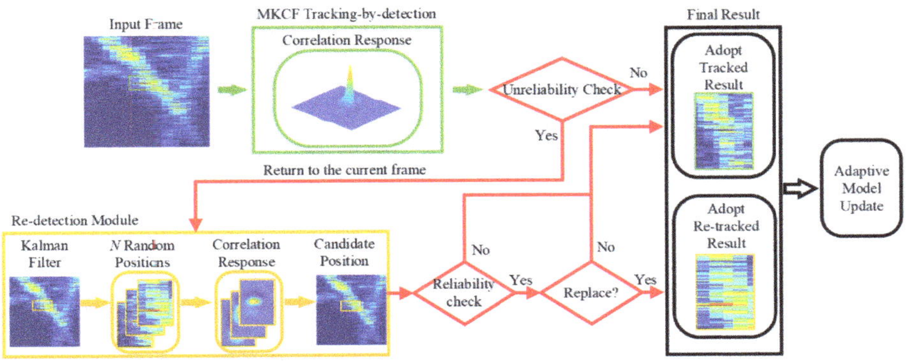

Figure 1. The IMKCF tracking-by-detection algorithm framework diagram.

4.1. The MKCF Tracking-by-Detection

Studies have shown that the incorporation of multiple kernels can improve the discriminatory ability of classifiers in comparison to a single-kernel approach [36]. A prevalent approach is to employ a basis kernel, $k_m(m=1,\cdots,M)$, and then view $k(\mathbf{x}_i,\mathbf{x}_j)=\mathbf{b}^{\mathrm{T}}\mathbf{k}(\mathbf{x}_i,\mathbf{x}_j)$ as a composition of the basis kernels, where $\mathbf{k}(\mathbf{x}_i,\mathbf{x}_j)=(k_1(\mathbf{x}_i,\mathbf{x}_j),\cdots,k_M(\mathbf{x}_i,\mathbf{x}_j))^{\mathrm{T}}$, $\mathbf{b}=(b_1,\cdots,b_m)^{\mathrm{T}}$, $\sum_{m=1}^{M}b_m=1$, and $b_m\geq 0$. Therefore $\mathbf{K}=\sum_{m=1}^{M}b_m\mathbf{K}_m$, $\mathbf{K}_m$ is the base core of group m, whose elements are $k_{ij}^{m}=k_m(\mathbf{x}_i,\mathbf{x}_j)$. The optimization problem is used to minimize the loss,

$$\begin{gathered}\min_{\boldsymbol{\alpha},\mathbf{b}}\frac{1}{2}\left\|\mathbf{y}-\sum_{m=1}^{M}b_m\mathbf{K}_m\boldsymbol{\alpha}\right\|_2^2+\frac{\lambda}{2}\boldsymbol{\alpha}^{\mathrm{T}}\sum_{m=1}^{M}b_m\mathbf{K}_m\boldsymbol{\alpha}=\min_{\boldsymbol{\alpha},\mathbf{b}}F(\boldsymbol{\alpha},\mathbf{b})\\ \text{s.t.}\sum_{m=1}^{M}b_m=1,b_m\geq 0,m=1,\cdots,M\end{gathered}\quad(12)$$

The optimal solution can be expressed as Equation (13), and the * indicates the optimal.

$$f^*(\mathbf{x})=\sum_{i=0}^{l-1}\alpha_i\mathbf{b}^{\mathrm{T}}k(\mathbf{x}_i,\mathbf{x}).\quad(13)$$

The diagram of the MKCF tracking-by-detection algorithm is presented in Figure 2.

In order to achieve localization robustness, the MKCF tracking-by-detection algorithm uses the weighted average of historical samples to update the training coefficients $\boldsymbol{\alpha}$ and $\boldsymbol{b}$. The optimization function is represented by

$$F_p(\boldsymbol{\alpha}_p,\mathbf{b}_p)\equiv\frac{1}{2}\sum_{j=1}^{p}\sum_{m=1}^{M}\beta_m^j u_{F(\boldsymbol{\alpha},\mathbf{b})}^{j,m},\quad(14)$$

where β^j is the weight of the sample optimization function of the jth frame, $u_{F(\boldsymbol{\alpha},\mathbf{b})}^{j,m}=\left\|\mathbf{y}_c-b_{m,p}\mathbf{K}_m^j\boldsymbol{\alpha}_p\right\|_2^2+\lambda b_{m,p}\boldsymbol{\alpha}_p^{\mathrm{T}}\mathbf{K}_m^j\boldsymbol{\alpha}_p, j=2,\cdots,p, \beta_m^1=(1-\gamma_m)^{p-1}, \beta_m^j=\gamma_m(1-\gamma_m)^{p-j}$, $\boldsymbol{\alpha}_p=\left(\alpha_{0,p},\cdots,\alpha_{l-1,p}\right)^T$, $\mathbf{b}_p=(b_{1,p},\cdots,b_{M,p})^T$, $\sum_{m=1}^{M}b_{m,p}=1$, p is the number of historical frames, $\gamma_m\in(0,1)$ is the learning rate, and $\mathbf{K}_m^j$ is the Gram matrix of kernel m. The new optimization problem is

$$\begin{gathered}\min_{\boldsymbol{\alpha}_p,b_p}F_p(\alpha_p,\mathbf{b}_p)\\ \text{s.t.}\sum_{m=1}^{M}d_{m,p}=1\\ d_{m,p}\geq 0,m=1,\cdots,M.\end{gathered}\quad(15)$$

First, given $\mathbf{b}_p$ to solve $\boldsymbol{\alpha}_p$, the above optimization problem becomes an unconstrained optimization problem. Let $\nabla_{\boldsymbol{\alpha}_p} F_p(\boldsymbol{\alpha}_p, \mathbf{b}_p)$, then,

$$\boldsymbol{\alpha}_p = \left(\sum_{j=1}^{p}\sum_{m=1}^{M} \beta_m^j \left(\left(b_{m,p}\mathbf{K}_m^j\right)^2 + \lambda b_{m,p}\mathbf{K}_m^j\right)\right)^{-1} \cdot \sum_{j=1}^{p}\sum_{m=1}^{M} \beta_m^j b_{m,p}\mathbf{K}_m^j \mathbf{y}_c. \tag{16}$$

The efficient evaluation can be achieved through the utilization of FFT.

$$A_p \equiv F(\boldsymbol{\alpha}_p) = \frac{\sum_{j=1}^{p}\sum_{m=1}^{M} \beta_m^j F\left(b_{m,p}\mathbf{k}_m^j\right) \odot F(\mathbf{y}_c)}{\sum_{j=1}^{p}\sum_{m=1}^{M} \beta_m^j F\left(b_{m,p}\mathbf{k}_m^j\right) \odot \left(F\left(b_{m,p}\mathbf{k}_m^j\right) + \lambda\right)}. \tag{17}$$

Set

$$A_p = \frac{A_p^N}{A_p^D} = \frac{\sum_{m=1}^{M} A_{m,p}^N}{\sum_{m=1}^{M} A_{m,p}^D}. \tag{18}$$

when $p = 1$,

$$\begin{aligned} A_{m,1}^N &= F\left(b_{m,1}\mathbf{k}_m^1\right) \odot F(\mathbf{y}_c) \\ A_{m,1}^D &= F\left(b_{m,1}\mathbf{k}_m^1\right) \odot \left(F\left(b_{m,1}\mathbf{k}_m^1\right) + \lambda\right). \end{aligned} \tag{19}$$

when $p > 1$,

$$\begin{aligned} A_{m,p}^N &= (1-\gamma_m)A_{m,p-1}^N + \gamma_m F\left(b_{m,p}\mathbf{k}_m^p\right) \odot F(\mathbf{y}_c) \\ A_{m,p}^D &= (1-\gamma_m)A_{m,p-1}^D + \gamma_m F\left(b_{m,p}\mathbf{k}_m^p\right) \odot \left(F\left(b_{m,p}\mathbf{k}_m^p\right) + \lambda\right). \end{aligned} \tag{20}$$

The optimal solution α_p^* can be attained by means of the aforementioned iteration. Subsequently, when presented with the task of solving $\mathbf{b}_p$ given $\boldsymbol{\alpha}_p$, the optimization problem outlined earlier transforms into a constrained optimization problem. To address this issue, we initially posit it as an unconstrained optimization problem and subsequently demonstrate that the resulting solution $\mathbf{b}_p$ conforms to the prescribed constraint conditions. Let $\nabla_{\mathbf{b}_p} F_p(\boldsymbol{\alpha}_p, \mathbf{b}_p)$, then,

$$b_{m,p} = \frac{\sum_{j=1}^{p} \beta_m^j \left(\mathbf{K}_m^j \boldsymbol{\alpha}_p\right)^{\mathrm{T}} (2\mathbf{y}_c - \lambda\boldsymbol{\alpha}_p)}{2\sum_{j=1}^{p} \beta_m^j \left(\mathbf{K}_m^j \boldsymbol{\alpha}_p\right)^{\mathrm{T}} \left(\mathbf{K}_m^j \boldsymbol{\alpha}_p\right)}, \tag{21}$$

where $m = 1, \cdots, M$, let

$$b_{m,p} = \frac{b_{m,p}^N}{b_{m,p}^D}. \tag{22}$$

when $p = 1$,

$$\begin{aligned} b_{m,p}^N &= (1-\gamma_m)b_{m,p-1}^N + \gamma_m \left(\mathbf{K}_m^p \boldsymbol{\alpha}_p\right)^{\mathrm{T}} (2\mathbf{y}_c - \lambda\boldsymbol{\alpha}_p) \\ b_{m,p}^D &= (1-\gamma_m)b_{m,p-1}^D + 2\gamma_m \left(\mathbf{K}_m^p \boldsymbol{\alpha}_p\right)^{\mathrm{T}} \left(\mathbf{K}_m^p \boldsymbol{\alpha}_p\right) \end{aligned}. \tag{23}$$

when $p > 1$,

$$\begin{aligned} b_{m,1}^N &= \left(\mathbf{K}_m^1 \boldsymbol{\alpha}_1\right)^{\mathrm{T}} (2\mathbf{y}_c - \lambda\boldsymbol{\alpha}_1) \\ b_{m,1}^D &= 2\left(\mathbf{K}_m^1 \boldsymbol{\alpha}_1\right)^{\mathrm{T}} \left(\mathbf{K}_m^1 \boldsymbol{\alpha}_1\right) \end{aligned}. \tag{24}$$

The FFT method enables the rapid calculation of $\mathbf{K}_m^p \boldsymbol{\alpha}_p$ as $F^{-1}\left(F^*\left(\mathbf{k}_m^p\right) \odot F(\alpha_p)\right) = F^{-1}\left(F^*\left(\mathbf{k}_m^p\right) \odot A_p\right)$. Subsequently, the optimal solution can be attained through the aforementioned iteration. Finally, the solution is verified to conform to the prescribed constraint conditions.

The kernel function employed in this algorithm utilizes a Gaussian kernel, and the elements within the kernel matrix can be computed using the following formula,

$$k_m(\mathbf{x}_i, \mathbf{x}_j) = \exp\left(-\frac{1}{\sigma_k^2}\left(\|\mathbf{x}_i - \mathbf{x}_j\|^2\right)\right). \tag{25}$$

The exponent within the exponential function $\exp(\cdot)$ used in this study is determined by the negative normalized Euclidean distance between $\mathbf{x}_i$ and $\mathbf{x}_j$. The higher the similarity between $\mathbf{x}_i$ and $\mathbf{x}_j$ in Euclidean space, the greater the value of $k_m(\mathbf{x}_i, \mathbf{x}_j)$. This function is commonly known as the likelihood function in filter-based tracking [37]. It enables the calculation of a value that represents the likelihood of the measured real target given any measurement. Based on the maximum likelihood estimation criterion, the target position is estimated by determining the peak position y_p of the correlation response $\mathbf{y}$.

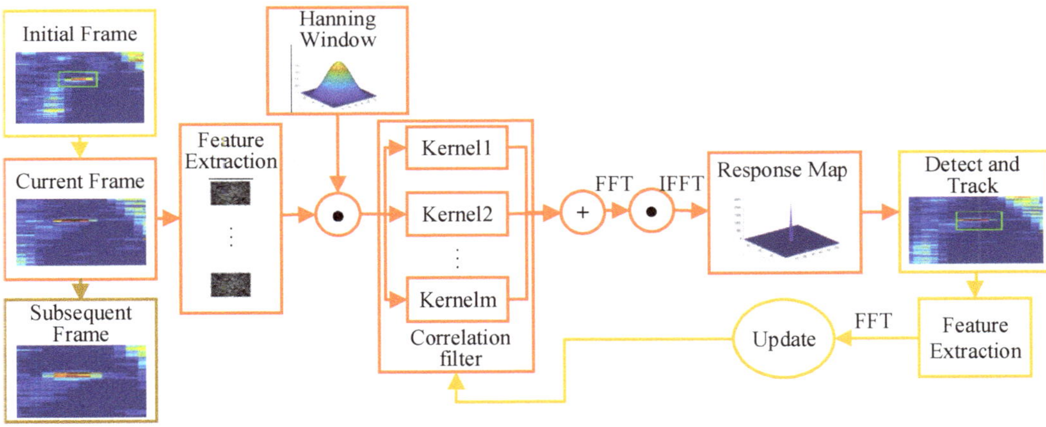

Figure 2. The diagram of the MKCF tracking-by-detection algorithm.

4.2. The Adaptive Reliability Check

In the correlation filter response map, a single peak is observed, and the sharpness of the peak corresponds to the reliability of the tracking result. The work conducted by Bolme [34] has proposed the idea that the peak-to-sidelobe ratio (PSR) possesses the potential to serve as an indicator of the sharpness of the response peak. The PSR is defined as

$$S_p = \frac{\max(R_p) - \mu_p}{\sigma_p}, \tag{26}$$

where R_p represents the response map calculated by the correlation filter at frame p, and μ_p and σ_p are the mean and standard deviation of R_p, respectively. In cases where the tracking result is unreliable, as exemplified in Figure 3, the response map may exhibit multiple peaks with low values, resulting in a significant decrease in the PSR. Therefore, the PSR can serve as an indicator of tracking result quality to a certain degree.

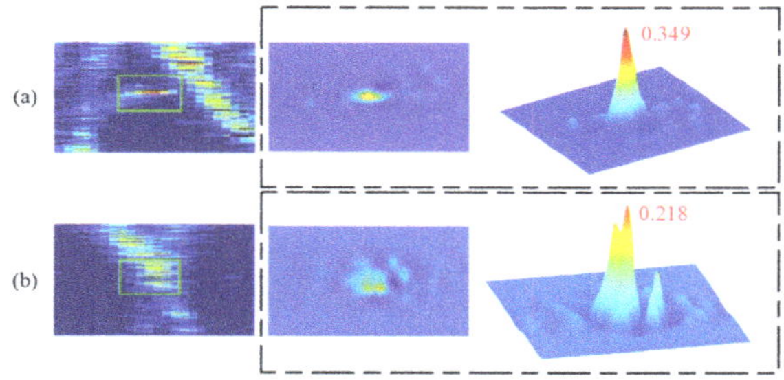

Figure 3. From left to right: input image, correlation response map (with a 2D visualization), and correlation response map (with a 3D visualization). (**a**) The target is in the target box. (**b**) The non-target is in the target box.

It is not suitable to pre-define a constant threshold to judge the reliability of the current tracking because the PSR fluctuates between different values due to the uncertainty factor in different scenarios. In order to mitigate the impact of fluctuations in the PSR, we utilize the historical frames to compute the average score to determine the reliability of the tracking results. We combine the PSR values of the historical frames to $C = \{S_2, \cdots, S_{p-1}\}$ with the mean of M. Furthermore, we introduce a small coefficient τ_1, whereby the PSR S_i of the ith frame is stored in C if $S_i < \tau_1 \cdot M$ and discarded otherwise. Finally, the evaluation criteria of the MKCF algorithm adaptively changes from frame to frame as the average of the multi-frame PSR is computed.

We check the reliability of the tracking result in each frame. The tracking result is deemed unreliable if $S_i < \tau_1 \cdot M$ ("Unreliability Check" in Figure 1). On the other hand, the tracking result is likely to be reliable if it satisfies $S_i > \tau_2 \cdot M$ ("Reliability Check" in Figure 1) and the coefficient τ_2 is higher than τ_1. Once the initial tracking result is determined to be unreliable, the re-detection module is initiated.

4.3. The Re-Detection Module

This section provides an introduction to the re-detection module, which plays a crucial role in generating candidate target locations and determining whether to substitute the initial tracking result with the optimal candidate target location. A key component of this module is the implementation of the Kalman filter, which utilizes reliable tracking results from the current frame for filter updates. Assuming that the motion between adjacent frames adheres to a linear Gaussian distribution as prior knowledge, the motion model in the Kalman filter can be established as a uniform linear motion model.

In cases where the track result is deemed unreliable, the target positions from the last reliable tracking results are saved and used to drive a Kalman filter, providing an estimated position $position_p$ for the current target position with the variance in $\psi\left(position_p\right)$. Subsequently, N random locations $g_j (j = 1, \cdots, N)$ are generated around the location $position_p$, following a Gaussian distribution with the mean $position_p$ and the variance $\psi\left(position_p\right)$. With g_j as the center, N target candidate bounding boxes B_j are generated. For the obtained B_j, their response maps R_p^j are generated by Equation (6) and the maximum values of these maps, $q_j = \max_u R_p^j(u)$, are also computed. Suppose q_j^* is the maximum among $\{q_1, \ldots, q_N\}$. The best candidate bounding box of the target B_j^* and the best candidate location of the target g_j^* are determined accordingly. Finally, the decision of whether the best candidate location g_j^* replaces the initial tracking result is determined through the following two steps.

(1) If it does not meet the reliability check $S_i > \tau_2 \cdot M$, the initial tracking result is used;

(2) If it meets the reliability check $S_i > \tau_2 \cdot M$, we compute the correlation response at the initial tracking location, and the highest response value is recorded as k_p. If (27) is met, the initial tracking result is replaced with g_j^*.

$$q_j^* \geq \gamma \times k_p, \tag{27}$$

where γ is the penalty parameter, and the * indicates the optimal. If the above equation is not satisfied, the initial tracking result is not replaced.

5. Experimental Results

In this section, we apply the proposed method on two test scenarios and compare it with the traditional tracking methods and the original KCF algorithms. All the algorithms are implemented in MATLAB 2018b, utilizing an Intel i5-6200U CPU with a main frequency of 2.3 GHz and 8 GB of memory. We make a Table 1 to summarize the abbreviations of various algorithms.

Table 1. The abbreviations of various algorithms.

Algorithm	Abbreviation
Multiple Hypothesis Tracking	MHT
Joint Probabilistic Data Association	JPDA
Probability Hypothesis Density	PHD
Multi-Feature Kernel Correlation Filter	MF-KCF
Multi-Kernel Correlation Filter	MKCF
Improved Multi-Feature Kernel Correlation Filter	IMF-KCF
Improved Multi-Kernel Correlation Filter	IMKCF

5.1. Evaluation Metrics

Evaluation metrics of performance are discussed below.

(1) Root-mean-square error (RMSE) and precision: the average distance error between the estimated position and the actual position, defined as (28). Given a threshold M, the centroid position is properly estimated if its RMSE is less than M. The RMSE accuracy is defined as the percentage of the total number of frames for which the location is correctly estimated.

$$\text{RMSE} = \sqrt{\sum_{i=1}^{m}\left[\left(\hat{x}_k^i - x_k\right)^2 + \left(\hat{y}_k^i - y_k\right)^2\right]/m}, \tag{28}$$

where m is the number of Monte Carlo experiments.

(2) Intersection over union (IOU) and precision: IOU is defined as (29). A larger IOU value indicates a more accurate estimation of the target. Given a threshold N, the target box is considered correctly estimated if its IOU is greater than N. The IOU accuracy is defined as the percentage of the total number of frames for which the target box is correctly estimated.

$$\text{IOU} = \frac{E \cap G}{E \cup G}, \tag{29}$$

where E denotes the area of the estimated target and G denotes the area of the real target. Operator $\cap$ represents intersection and $\cup$ means the union.

(3) Frames Per Second (FPS): The number of frames processed per second, the greater the FPS; the higher the efficiency of the algorithm.

5.2. Test Scenarios and Parameter Settings

5.2.1. Test Scenarios

In order to verify the effect of the proposed algorithm on tracking targets exhibiting different types of motion, we have designed two scenarios. Notably, both the ball and the diver have equivalent target strength, with the difference lying in their respective motion types. The experiment used a GPS device to record the actual motion trajectory of the target. The GPS device is soft-connected to the target and floats on the ocean directly above the target.

(a) Maneuvering target: As shown in Figure 4a, the small boat drags the ball to perform a turning motion. Figure 5a is the sound speed profile of the experiment. The actual trajectory of the target is plotted in Figure 5b, represented by the red line. Subsequent to the 70th frame, the SRR is lower than 0 dB for the majority of frames within this specific scene.

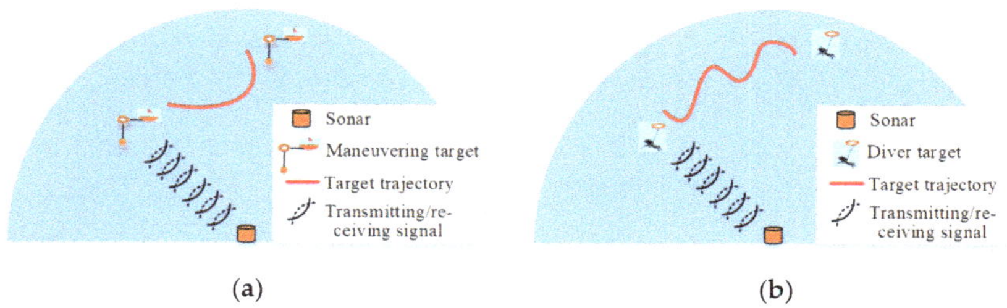

Figure 4. The test scenarios. (**a**) Maneuvering target. (**b**) Diver target.

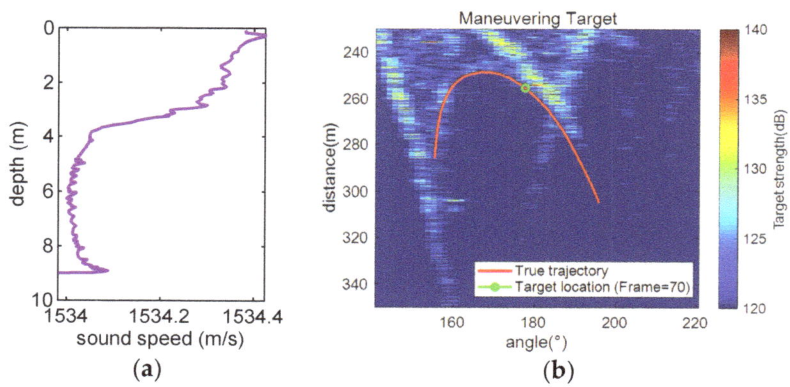

Figure 5. (**a**) Sound speed profile. (**b**) The actual trajectory of maneuvering target in acoustic image.

(b) Diver target: As represented in Figure 4b, the closed diver moves in a Z-shaped manner. The sound speed profile is shown in Figure 6a. The actual trajectory of the target is plotted in Figure 6b, represented by the red line. In this scene, the SRR surpasses 0 dB for the majority of frames, indicating a higher quality data set.

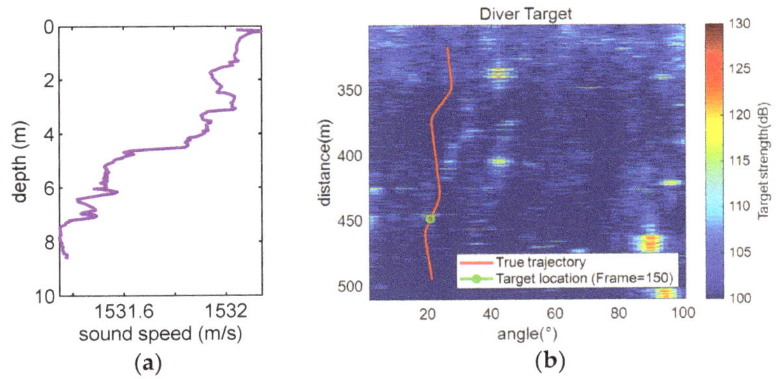

Figure 6. (**a**) Sound speed profile. (**b**) The actual trajectory of diver target in acoustic image.

5.2.2. Parameter Settings

In this paper, we use the histogram of oriented gradient (HOG) and invariant moment features. The HOG feature has nine gradient orientations, and the cell size is 4×4. In the KCF, it uses a single Gaussian kernel with parameter $\sigma = 0.5$ and the learning factor $\eta = 0.01$. It is worth noting that the multi-feature KCF (MF-KCF) trains the tracker based on the above features' fusion. The MKCF uses two Gaussian kernels with parameters $\sigma_1 = 0.3$ and $\sigma_2 = 0.3$ and learning factors $\eta = 0.0175$ and $\eta = 0.018$, respectively. Both methods employ a regularization parameter of $\lambda = 10^{-4}$ [38]. In the adaptive reliability check module, the coefficients τ_1 and τ_2 are 0.75 and 0.9, respectively.

To address the issue of high-frequency noise caused by abrupt edges in samples after the cyclic shift, this study uses the Hanning window when processing sample features. The application of the Hanning window aids in smoothing boundaries and minimizing interference from background information, consequently enhancing overall tracking accuracy. Furthermore, due to the potential decline in detection and tracking performance associated with an excessive search area in the KCF, it becomes necessary to restrict the size of the search area. Hence, the search area is limited to 2.5 times the size of the target box [39].

5.3. Data Processing and Analysis

5.3.1. Comparison with Traditional Tracking Algorithms

In this section, a comparison is made between the proposed algorithm and three commonly used algorithms in underwater target detection and tracking: MHT, JPDA, and PHD. To minimize target loss during thresholding, a relatively low threshold is adopted in the data preprocessing stage. It should be noted that this approach can lead to an increased probability of false alarms, resulting in an increased number of false targets, thereby adding difficulties of data association and computational costs.

The outcomes of data processing for maneuvering and diver targets are depicted in Figures 7 and 8. Analysis of these figures reveals that the proposed IMKCF algorithm consistently maintains a relatively low RMSE across the majority of frames in both maneuvering and diver target tracking processes, when compared to the other three algorithms. This superior performance can mainly be attributed to two factors. Firstly, our response map, which is based on target features, effectively identifies target areas. Secondly, the re-detection module with a Kalman filter estimates a reliable target location when the tracking result is deemed unreliable.

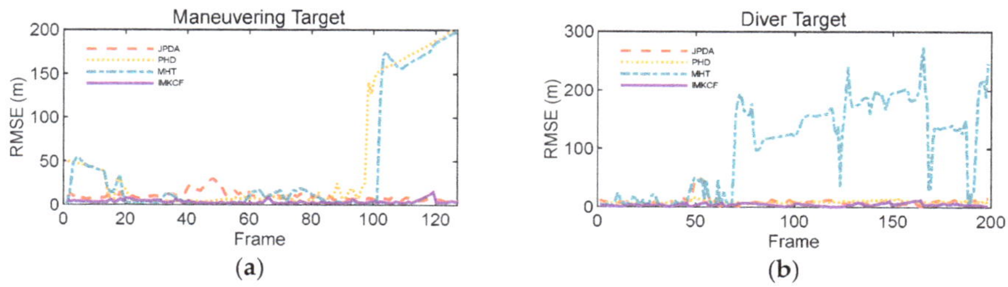

Figure 7. (**a**) RMSE with maneuvering target. (**b**) RMSE with diver target.

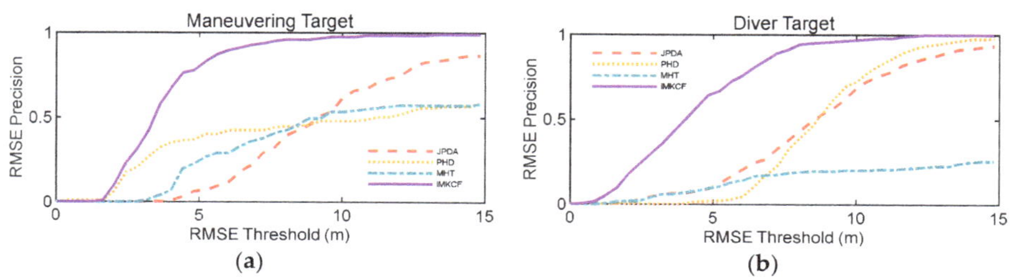

Figure 8. (**a**) RMSE precision with maneuvering target. (**b**) RMSE precision with diver target.

As shown in Figure 7, in the tracking of maneuvering targets, the three traditional tracking algorithms perform poorly, particularly the PHD and MHT algorithms. Both exhibit significant tracking drift at approximately 100 frames. The JPDA and PHD algorithms demonstrate satisfactory performance in tracking the diver target, while the MHT algorithm experiences tracking drift at about 70 frames. The JPDA algorithm excels in cluttered environments as it does not require prior information about the target and clutter, allowing for successful target tracking. Nevertheless, because this study focuses on a single-target tracking scenario, the JPDA algorithm performs relatively well. In contrast, the MHT algorithm necessitates prior information about the target and clutter, and the computational complexity increases exponentially with the clutter density. To enhance computational efficiency, we set a smaller value for N-scan pruning, but this compromises the tracking performance of the algorithm. The PHD algorithm, which is based on RFS theory, avoids the intricate correlation process associated with traditional methods and exhibits high computational efficiency. In the data processing of the two scenarios, we assume a uniform linear motion model for both the MHT and PHD algorithms. However, this motion constraint is not robust in low SRR environments, leading to the failure of the PHD algorithm in tracking maneuvering targets. Our proposed method surpasses these algorithms by utilizing multiple features and incorporating reliability estimation to identify a reliable re-detected target for self-correction.

Figure 8a,b display the percentage of frames within a given RMSE threshold for the error between the estimated position and the true position of the maneuvering target and the diver target, respectively. Analysis of these figures demonstrates that the IMKCF algorithm outperforms the other three tested algorithms in terms of RMSE precision. Specifically, when the RMSE threshold is set at 5, the IMKCF algorithm achieves an accuracy rate close to 80%, while the other algorithms only have approximately 50% accuracy when the RMSE threshold is 10.

5.3.2. Comparison with Original KCF Algorithms

In this section, a comparative analysis is conducted between the proposed algorithm and three original KCF algorithms, namely, MF-KCF, improved MF-KCF (IMF-KCF), and MKCF. Figure 9 illustrates the target position in the final frame and tracking outcomes for all four algorithms. It can be observed from the figure that both the MF-KCF and MKCF algorithms fail to track the maneuvering target. Conversely, the IMF-KCF and IMKCF algorithms, incorporating adaptive reliability checks and a re-detection module, successfully track the target. At approximately the 70th frame, target tracking is interfered with by reverberation, and the training samples are contaminated, resulting in error propagation during model training and subsequent tracker drift. The IMKCF algorithm checks real-time reliability on target tracking results and re-detects the target position when deemed unreliable, preventing frame drift. The versatility of the re-detection module is demonstrated by its successful implementation in both algorithms.

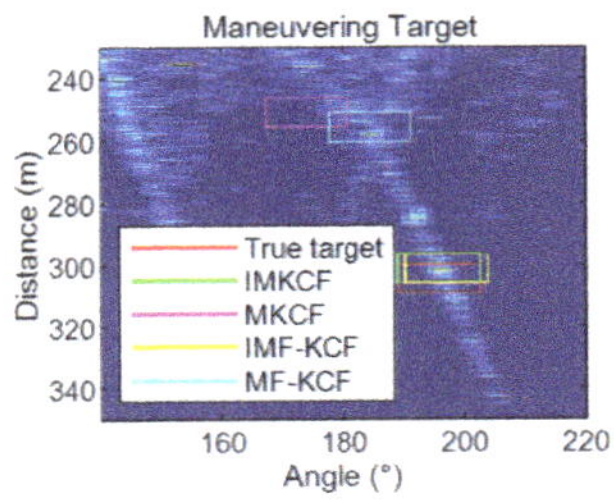

Figure 9. Track results (frame = 127).

To further elucidate tracking performance under low-SRR scenarios, Figure 10 presents the frame-by-frame RMSE and IOU of maneuvering target tracking. Higher IOU values and lower RMSE values signify more accurate tracking outcomes. As shown in Figure 10, both the MF-KCF and MKCF algorithms lose track of the targets around the 70th frame. The IMKCF algorithm outperforms the IMF-KCF algorithm, exhibiting relatively low RMSE and high IOU values across most frames, indicating superior accuracy. This improvement can be attributed to the MKCF tracking-by-detection module, which enables the algorithm to make full use of the complementary features and improve tracking accuracy.

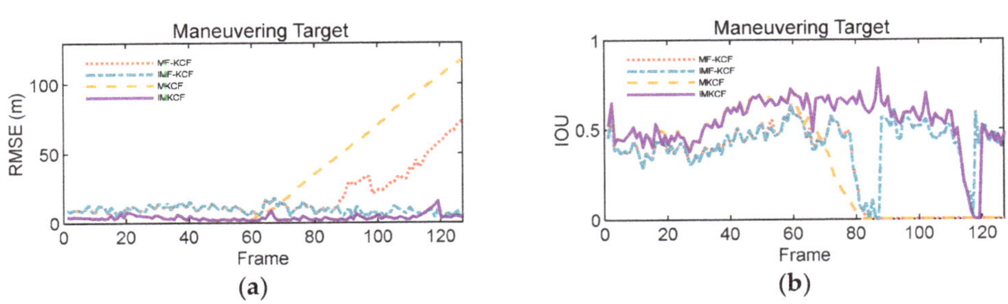

Figure 10. (**a**) RMSE. (**b**) IOU.

Figure 11a depicts the RMSE precision. The figure reveals that when the RMSE threshold is set at 10, the IMKCF algorithm achieves nearly 100% precision, while the other three algorithms fall below 60%. Additionally, Figure 11b demonstrates the IOU precision. Notably, when the IOU threshold is set at 0.5, the IMKCF algorithm attains nearly 100% precision, whereas the other three algorithms exhibit less than 60%. The results show that, compared with the single-kernel correlation filter, the correlation filter based on multiple kernels has greater advantages in tracking accuracy.

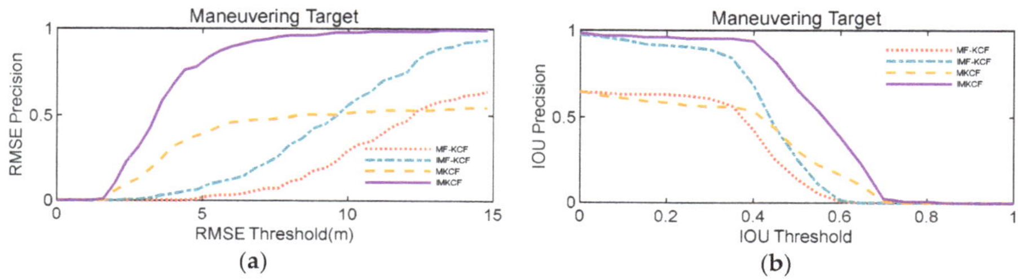

Figure 11. (**a**) RMSE precision. (**b**) IOU precision.

Figure 12 illustrates the PSR curve during the tracking process of the MKCF and IMKCF algorithms. It can be observed that the PSR of the MKCF algorithm experiences a significant decline around the 70th frame, whereas the PSR of the IMKCF algorithm fluctuates steadily throughout the tracking process. These findings indicate that the PSR score serves as an indicator of tracking result reliability, and the adaptive reliability check and re-detection modules within the IMKCF algorithm play a vital role in enhancing tracking robustness.

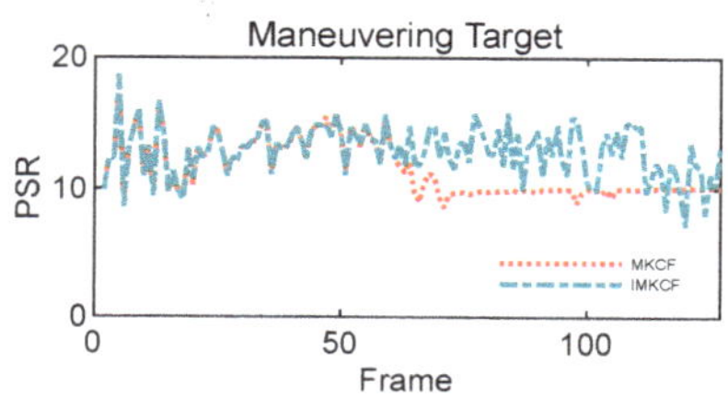

Figure 12. The comparison of PSR between MKCF and IMKCF algorithms.

5.3.3. Algorithm Efficiency

Table 2 displays the PFS and average RMSE of the above algorithms. The results indicate that the PHD algorithm has the highest FPS and computational efficiency but exhibits the poorest tracking accuracy. In contrast, the KCF algorithms exhibit relatively lower computational efficiency but has higher tracking accuracy compared to the classical filtering algorithm. The analysis suggests that MKCF exhibits a slight improvement in computational efficiency when compared MF-KCF. This implies that the incorporation of an additional kernel in kernel correlation filtering does not result in a significant increment in computational cost. Moreover, the results demonstrate that IMKCF achieves superior tracking accuracy in comparison to both IMF-KCF and MF-KCF. Notably, Table 2 reveals that the inclusion of a re-detection module can lead to an increase in computational cost, as it carries out target re-detection when the tracking results are deemed unreliable.

Table 2. FPS and the average of RMSE.

	MHT	JPDA	PHD	MF-KCF	MKCF	IMF-KCF	IMKCF
FPS	7.2	107.1	122.3	22.3	27.2	11.3	14.2
RMSEaver	44.82	10.26	50.83	19.94	33.52	9.6	3.86

6. Conclusions

In this paper, we propose an IMKCF algorithm to solve the challenging problem of detecting and tracking weak targets with varying movements in a complex marine environment. The IMKCF algorithm consists of three modules: the MKCF tracking-by-detection, the adaptive reliability check, and the re-detection modules. The MKCF tracking-by-detection module employs a multi-frame data weighted average technique to adaptively update the coefficients of multiple kernels, thereby enhancing tracking accuracy. We conduct a comprehensive analysis of the MKCF algorithm using a maximum likelihood perspective and prove that the target location can be precisely determined based on the location of the maximum value of the correlation response. The remaining two modules work collaboratively to improve the robustness of target tracking. In particular, the previous reliable tracking results are utilized to drive a Kalman filter, generating a position estimate when the tracking result is considered unreliable. A decision is then made about whether to replace the original target position with the estimated one.

In data processing, we extracted HOG features and invariant moment features to train the proposed IMKCF algorithm, which has been compared with traditional tracking algorithms and original KCF algorithms. The experimental results demonstrate that our proposed algorithm not only exhibits the capability of effectively tracking underwater targets with diverse motion types but also achieves long-term robust tracking in low-SRR environments. Moreover, the tracking accuracy of our algorithm surpasses that of the single-core correlation filter. Currently, the method proposed in this paper is only suitable for single-target tracking. In future research, we plan to delve deeper into the challenges of multiple weak underwater target tracking. Additionally, we aim to mine more target feature information to enhance the algorithm's robustness.

Author Contributions: W.Y. conceived the main idea, designed the main algorithm, and wrote the manuscript. The experimental results were analyzed by W.Y., F.X. and J.Y. provided suggestions for the proposed algorithm. All authors have read and agreed to the published version of the manuscript.

Funding: This work was supported in part by the Science Foundation of the Chinese Academy of Sciences under Grant 8091A120105.

Data Availability Statement: The data presented in this study are available on request from the corresponding author. The data are not publicly available due to our laboratory policy.

Conflicts of Interest: The authors declare no conflicts of interest.

References

1. Jeong, T.T. Particle PHD Filter Multiple Target Tracking in Sonar Image. *IEEE Trans. Aerosp. Electron. Syst.* **2007**, *43*, 409–416. [CrossRef]
2. Rodningsby, A.; Bar-Shalom, Y. Tracking of Divers using a Probabilistic Data Association Filter with a Bubble Model. *IEEE Trans. Aerosp. Electron. Syst.* **2009**, *45*, 1181–1193. [CrossRef]
3. Zhang, H.; Tian, M.; Ouyang, Q.; Liu, J.; Shao, G.; Cheng, J. Track Detection of Underwater Moving Targets Based on CFAR. *J. Phys. Conf. Ser.* **2023**, *2486*, 012076. [CrossRef]
4. Zhu, J.; Song, Y.; Jiang, N.; Xie, Z.; Fan, C.; Huang, X. Enhanced Doppler Resolution and Sidelobe Suppression Performance for Golay Complementary Waveforms. *Remote Sens.* **2023**, *15*, 2452. [CrossRef]
5. Zhang, D.; Gao, L.; Sun, D.; Teng, T. Soft-decision Detection of Weak Tonals for Passive Sonar using Track-before-detect Method. *Appl. Acoust.* **2022**, *188*, 108549. [CrossRef]
6. Yi, W.; Fu, L.; García-Fernández, Á.F.; Xu, L.; Kong, L. Particle Filtering based Track-before-detect Method for Passive Array Sonar Systems. *Signal Process.* **2019**, *165*, 303–314. [CrossRef]
7. Vivone, G.; Braca, P. Joint Probabilistic Data Association Tracker for Extended Target Tracking Applied to X-Band Marine Radar Data. *IEEE J. Ocean. Eng.* **2016**, *41*, 1007–1019. [CrossRef]
8. Yang, S.; Thormann, K.; Baum, M. Linear-Time Joint Probabilistic Data Association for Multiple Extended Object Tracking. In Proceedings of the 2018 IEEE 10th Sensor Array and Multichannel Signal Processing Workshop (SAM), Sheffield, UK, 8–11 July 2018; pp. 6–10.
9. Blackman, S.S. Multiple Hypothesis Tracking for Multiple Target Tracking. *IEEE Aerosp. Electron. Syst. Mag.* **2004**, *19*, 5–18. [CrossRef]
10. Li, X.; Zhao, C.; Lu, X.; Wei, W. Underwater Bearings-Only Multitarget Tracking Based on Modified PMHT in Dense-Cluttered Environment. *IEEE Access* **2019**, *7*, 93678–93689. [CrossRef]
11. Zhou, T.; Wang, Y.; Chen, B.; Zhu, J.; Yu, X. Underwater Multitarget Tracking with Sonar Images Using Thresholded Sequential Monte Carlo Probability Hypothesis Density Algorithm. *IEEE Geosci. Remote Sens. Lett.* **2022**, *19*, 1506305. [CrossRef]
12. Williams, J.L. Marginal Multi-bernoulli Filters: RFS Derivation of MHT, JIPDA, and Association-based Member. *IEEE Trans. Aerosp. Electron. Syst.* **2015**, *51*, 1664–1687. [CrossRef]
13. Chai, L.; Kong, L.; Li, S.; Yi, W. The Multiple Model Multi-Bernoulli Filter based Track-before-detect using a Likelihood based Adaptive Birth Distribution. *Signal Process.* **2020**, *171*, 107501. [CrossRef]

14. Li, S.; Yi, W.; Kong, L.; Wang, B. Multi-bernoulli Filter based Track-before-detect for Jump Markov Models. In Proceedings of the 2014 IEEE Radar Conference, Cincinnati, OH, USA, 19–23 May 2014; pp. 1257–1261.
15. Liu, Z.-X.; Zhang, Q.-Q.; Li, L.-Q.; Xie, W.-X. Tracking Multiple Maneuvering Targets using a Sequential Multiple Target Bayes Filter with Jump Markov System Models. *Neurocomputing* **2016**, *216*, 183–191. [CrossRef]
16. Yue, W.; Xu, F.; Xiao, X.; Yang, J. Track-before-Detect Algorithm for Underwater Diver Based on Knowledge-Aided Particle Filter. *Sensors* **2022**, *22*, 9649. [CrossRef]
17. Nam, H.; Han, B. Learning multi-domain convolutional neural networks for visual tracking. In Proceedings of the 2016 IEEE Conference on Computer Vision and Pattern Recognition (CVPR), Las Vegas, NV, USA, 27–30 June 2016; pp. 4293–4302.
18. Fan, H.; Ling, H. SANet: Structure-aware network for visual tracking. In Proceedings of the 2017 IEEE Conference on Computer Vision and Pattern Recognition Workshops (CVPRW), Honolulu, HI, USA, 21–26 July 2017; pp. 2217–2224.
19. Bertinetto, L.; Valmadre, J.; Henriques, J.F.; Vedaldi, A.; Torr, P.H.S. Fully-convolutional Siamese networks for object tracking. In Proceedings of the European Conference on Computer Vision (ECCV), Amsterdam, The Netherlands, 8–10 October 2016; pp. 850–865.
20. Henriques, J.F.; Caseiro, R.; Martins, P.; Batista, J. Exploiting the circulant structure of tracking-by-detection with kernels. In Proceedings of the 12th European Conference on Computer Vision (ECCV), Florence, Italy, 7–13 October 2012; pp. 702–715.
21. Henriques, J.F.; Caseiro, R.; Martins, P.; Batista, J. High-speed Ttracking with Kernelized Correlation Filters. *IEEE Trans. Pattern Anal. Mach. Intell.* **2015**, *37*, 583–596. [CrossRef] [PubMed]
22. Shin, J.; Kim, H.; Kim, D.; Paik, J. Fast and Robust Object Tracking Using Tracking Failure Detection in Kernelized Correlation Filter. *Appl. Sci.* **2020**, *10*, 713. [CrossRef]
23. Zhang, L.; Suganthan, P.N. Robust Visual Tracking via Co-trained Kernelized Correlation Filters. *Pattern Recognit.* **2017**, *69*, 82–93. [CrossRef]
24. Bertinetto, L.; Valmadre, J.; Golodetz, S.; Miksik, O.; Torr, P.H.S. Staple: Complementary Learners for Real-Time Tracking. In Proceedings of the 2016 IEEE Conference on Computer Vision and Pattern Recognition (CVPR), Las Vegas, NV, USA, 27–30 June 2016; pp. 1401–1409.
25. Hao, Z.; Liu, G.; Gao, J.; Zhang, H. Robust Visual Tracking Using Structural Patch Response Map Fusion Based on Complementary Correlation Filter and Color Histogram. *Sensors* **2019**, *19*, 4178. [CrossRef]
26. Sun, X.; Cheung, N.-M.; Yao, H.; Guo, Y. Non-rigid Object Tracking via Deformable Patches using Shape-Preserved KCF and Level Sets. In Proceedings of the 2017 IEEE International Conference on Computer Vision (ICCV), Venice, Italy, 22–29 October 2017; pp. 5496–5504.
27. Zhou, Y.; Su, H.; Tian, S.; Liu, X.; Suo, J. Multiple Kernelized Correlation Filters based Track-Before-Detect Algorithm for Tracking Weak and Extended Target in Marine Radar Systems. *IEEE Trans. Aerosp. Electron. Syst.* **2022**, *58*, 3411–3426. [CrossRef]
28. Zhou, Y.; Wang, T.; Hu, R.; Su, H.; Liu, Y.; Liu, X.; Suo, J.; Snoussi, H. Multiple Kernelized Correlation Filters (MKCF) for Extended Object Tracking Using X-Band Marine Radar Data. *IEEE Trans. Signal Process.* **2019**, *67*, 3676–3688. [CrossRef]
29. Zeng, X.; Xu, L.; Cen, Y.; Zhao, R.; Hu, S.; Xiao, G. Visual Tracking Based on Multi-Feature and Fast Scale Adaptive Kernelized Correlation Filter. *IEEE Access* **2019**, *7*, 83209–83228. [CrossRef]
30. Ren, H.; Qiao, J.; Shi, T. Multifeature Fusion Tracking Algorithm Based on Self-Associative Memory Learning Mechanism. *IEEE Access* **2022**, *10*, 100605–100614. [CrossRef]
31. Tang, M.; Feng, J. Multi-kernel Correlation Filter for Visual Tracking. In Proceedings of the 2015 IEEE International Conference on Computer Vision (ICCV), Santiago, Chile, 7–13 December 2015; pp. 3038–3046.
32. Bolme, D.S.; Beveridge, J.R.; Draper, B.A.; Lui, Y.M. Visual Object Tracking using Adaptive Correlation Filters. In Proceedings of the 2010 IEEE Computer Society Conference on Computer Vision and Pattern Recognition, San Francisco, CA, USA, 13–18 June 2010; pp. 2544–2550.
33. Wang, N.; Zhou, W.; Li, H. Reliable Re-Detection for Long-Term Tracking. *IEEE Trans. Circuits Syst. Video Technol.* **2019**, *29*, 730–743. [CrossRef]
34. Ma, C.; Yang, X.; Zhang, C.; Yang, M.-H. Long-term Correlation Tracking. In Proceedings of the 2015 IEEE Conference on Computer Vision and Pattern Recognition (CVPR), Boston, MA, USA, 7–12 June 2015; pp. 5388–5396.
35. Tang, F.; Ling, Q. Contour-Aware Long-Term Tracking with Reliable Re-Detection. *IEEE Trans. Circuits Syst. Video Technol.* **2020**, *30*, 4739–4754. [CrossRef]
36. Varma, M.; Ray, D. Learning the Discriminative Power-Invariance Trade-Off. In Proceedings of the 2007 IEEE 11th International Conference on Computer Vision, Rio de Janeiro, Brazil, 14–21 October 2007; pp. 1–8.
37. Esmzad, R.; Mahboobi Esfanjani, R. Modified likelihood probabilistic data association filter for tracking systems with delayed and lost measurements. *Digit. Signal Process.* **2018**, *76*, 66–74. [CrossRef]
38. Tang, M.; Yu, B.; Zhang, F.; Wang, J. High-Speed Tracking with Multi-kernel Correlation Filters. In Proceedings of the 2018 IEEE Conference on Computer Vision and Pattern Recognition (CVPR), Salt Lake City, CA, USA, 18–23 June 2018; pp. 4874–4883.
39. Danelljan, M.; Khan, F.S.; Felsberg, M.; van de Weijer, J. Adaptive Color Attributes for Real-Time Visual Tracking. In Proceedings of the 2014 IEEE Conference on Computer Vision and Pattern Recognition (CVPR), Columbus, OH, USA, 23–28 June 2014; pp. 1090–1097.

remote sensing

MDPI

Article

Infrared Dim Star Background Suppression Method Based on Recursive Moving Target Indication

Lei Zhang [1], Peng Rao [2,3,*], Yang Hong [2,3,4], Xin Chen [2,3] and Liangjie Jia [2,3,4]

1 College of Astronautics, Nanjing University of Aeronautics and Astronautics, Nanjing 210016, China
2 Shanghai Institute of Technical Physics, Chinese Academy of Sciences, Shanghai 200083, China; hongyang@mail.sitp.ac.cn (Y.H.); chenxin@mail.sitp.ac.cn (X.C.); jialiangjie@mail.sitp.ac.cn (L.J.)
3 Key Laboratory of Intelligent Infrared Perception, Chinese Academy of Sciences, Shanghai 200083, China
4 University of Chinese Academy of Sciences, Beijing 100049, China
* Correspondence: peng_rao@mail.sitp.ac.cn

Abstract: Space-based infrared target detection can provide full-time and full-weather observation of targets, thus it is of significance in space security. However, the presence of stars in the background can severely affect the accuracy and real-time performance of infrared dim and small target detection, making star suppression a key technology and hot spot in the field of space target detection. The existing star suppression algorithms are all oriented towards the detection before track method and rely on the single image properties of the stars. They can only effectively suppress bright stars with a high signal-to-noise ratio (SNR). To address this problem, we propose a new method for infrared dim star background suppression based on recursive moving target indication (RMTI). Our proposed method is based on a more direct analysis of the image sequence itself, which will lead to more robust and accurate background suppression. The method first obtains the motion information of stars through satellite motion or key star registration. Then, the advanced RMTI algorithm is used to enhance the stars in the image. Finally, the mask of suppressing stars is generated by an accumulation frame adaptive threshold. The experimental results show that the algorithm has a less than 8.73% leakage suppression rate for stars with an SNR ≤ 2 and a false suppression rate of less than 2.3%. The validity of the proposed method is verified in real data. Compared with the existing methods, the method proposed in this paper can stably suppress stars with a lower SNR.

Keywords: star background suppression; recursive moving target indication; dim space target detection

Citation: Zhang, L.; Rao, P.; Hong, Y.; Chen, X.; Jia, L. Infrared Dim Star Background Suppression Method Based on Recursive Moving Target Indication. *Remote Sens.* **2023**, *15*, 4152. https://doi.org/10.3390/rs15174152

Academic Editor: Paolo Tripicchio

Received: 24 July 2023
Revised: 21 August 2023
Accepted: 23 August 2023
Published: 24 August 2023

1. Introduction

The utilization of space resources has led to an increase in space debris, asteroids, and failed satellites, which pose a serious threat to working satellites [1,2]. Ensuring space security is a vital mission, and the ability to surveil these space targets is essential in achieving this goal [3,4]. Space infrared remote sensing provides full-time and full-weather observation of objects, making it the main tool for surveilling space targets. However, due to the long detection range, limited resolution, low radiant energy, and small target size, these space targets appear as dim point targets in the focal plane [5–9]. These low signal-to-noise ratio (SNR) point targets are inherently difficult to detect, especially with complex backgrounds and noise [10–14]. Therefore, background suppression is crucial for space target observation.

Stars are a crucial component of the deep space background and, like other space targets, appear as point sources in images [15,16]. However, stars are a significant source of false alarms in space target detection [17,18]. If the star background cannot be effectively suppressed, it can negatively impact the accuracy and real-time performance of the detection process [19,20]. While most algorithms proposed in the past few decades have focused on clearing bright stars as part of high-SNR target detection, few have addressed the suppression of dim stars. A few algorithms that can suppress dim stars are limited by specific application conditions. Nevertheless, the presence of dim stars in the background cannot be ignored. Therefore, this paper aims to propose a universal dim star background suppression method that can ensure the accurate and real-time detection of low-SNR targets.

1.1. Research Status

Detecting stars in space images can be challenging due to their similar characteristics to other space targets. Existing algorithms can be generally classified into two kinds, which are based on star catalogue and image. The algorithms based on images often rely on multi-frame images to introduce kinematic characteristics. Stars, being stationary in celestial coordinates for short periods, can be distinguished from other moving space targets [21]. Spatial–temporal correlation information is used to classify detection methods into two categories: space before time (SBT) and time before space (TBS). Previous research has explored both approaches. SBT methods prioritize spatial information before temporal [22–26], while TBS methods prioritize temporal information before spatial [27–31].

SBT utilizes the detection before track (DBT) method to detect potential targets, which may include stars. Then, stars are stationary in celestial coordinates and have fixed positions relative to each other in time, while real space targets and stars constantly change positions, as shown in Figure 1. The blue stars present stars and the red point presents space targets Therefore, SBT can suppress stars from other space targets by extracting the position information of potential targets from a single frame. This approach allows for more accurate tracking of real space targets.

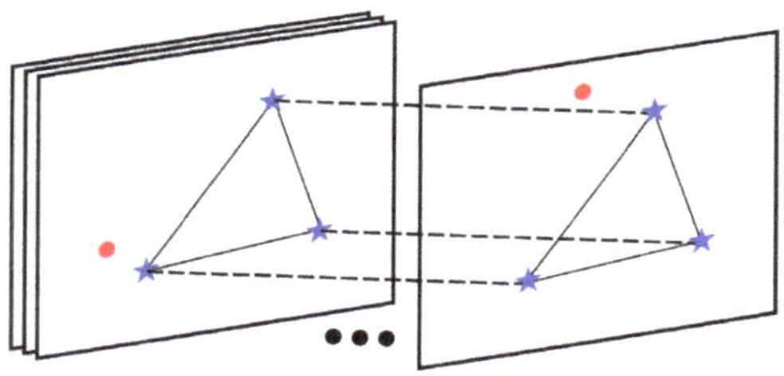

Figure 1. The schematic of stars suppression of SBT.

For instance, Hong Zhang et al. [22] define a feature space of distance (FSD) between stars to describe the invariance of distance among the stars. This method is used for feature matching and image transformation to achieve image registration. After image registration, the star background can be easily suppressed by the image difference. Other SBT methods based on star image registration adopt different matching features and registration algorithms. Yu Zhu et al. [23] propose the longest common sub-sequence (LCS) to find the isomorphism sub-graph which represents the matched feature pairs. Qingqing Luo et al. [24] introduce an iterative closest point algorithm which is a widely used point cloud registration algorithm. They also proposed a Gaussian mixture probability hypothesis density filter to avoid the target being mistakenly associated with stars. Feng Liu et al. [25] apply a mature and effective triangle algorithm to register stars. Meanwhile, target motion track detection is also considered to further suppress noise. Recently, the interior angle matrix has been used to describe the topological invariance of stars [26]. Then, stars are suppressed by sequence frame offset statistics histogram.

SBT methods rely on detecting stars using a single frame, which severely limits the SNR of suppressed stars. As a result, these methods are not suitable for aiding the detection of very-low-SNR targets.

Figure 2 illustrates the TBS method, which utilizes the position invariant theory to identify potential targets. The blue stars present stars and the red point presents space targets. The black line denotes the movement of the target and the star. Unlike SBT, TBS accumulates the energy of potential targets from the time dimension. This allows all potential targets to gain their corresponding route. The route of space targets is markedly different from the route of stars, whether the field of the camera is moving or still. By identifying the routes of space targets, TBS can suppress the stars and effectively identify potential targets.

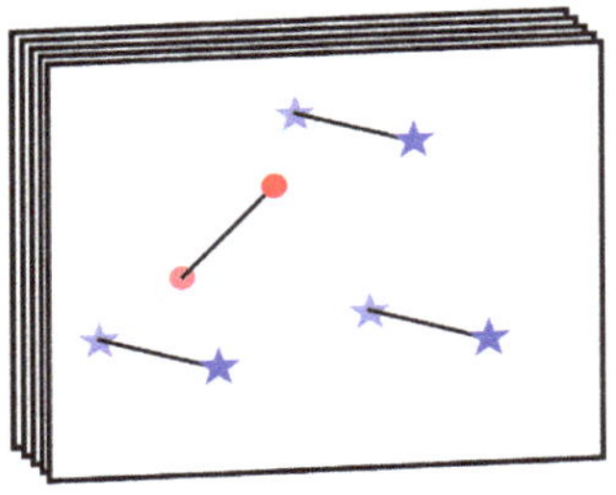

Figure 2. The schematic of stars suppression of SBT.

The key to TBS is to identify the route difference to suppress stars. Wang Hou et al. [27] propose a main directional suppression high pass filter for star line suppression, which considers the phase of the spectrum as the velocity of the target. Similarly, an adaptive linear filtering method is proposed to decrease the influence region of stars filtering uniformly [28]. Furthermore, the moving target indicator (MTI) algorithm is an effective trajectory detection method that achieves target detection and star suppression through route energy accumulation and direction judgment [29]. Another set of TBS methods focuses on the still field of the camera. Interval frame subtraction is applied to suppress stars, and back neighborhood frame correlation is proposed to protect the targets covered by stars [30]. Additionally, a star subtraction mask is obtained by introducing the maximum frame and medium frame, which can suppress the still star residues [31].

The preceding methods are highly dependent on their corresponding target detection methods, making them challenging to apply to alternative methods. Furthermore, the use of multiple frames superimposition is utilized to derive the star route, but this approach has limited energy enhancement and is unable to effectively suppress stars with a very low SNR.

The algorithms based on star catalogue are theoretically not limited by the SNR of stars. According to the reference method of the star catalogue, it can be divided into two categories. The first is star identification. Star identification determines the correspondence between stars based on the feature matching of the star catalogue consisting of the observed stars in the field of view. Star identification is the key technic of satellite position confirmation from the star sensor. To pursue a faster and more robust star identification performance, neural networks [32], color ratio information [33], rotation-invariant additive vector sequence [34], and so on are introduced and have obtained good results. In the application of star suppression, star identification matches the stars in the image for suppression. These algorithms can suppress all the stars recorded by the star catalogue regardless of the SNR. The second is the star map mask. These algorithms generate a star mask from the observation direction and star catalogue [35]. Like star identification, these algorithms are not limited by star SNR because they rely on star catalogue.

In addition to the previously mentioned methods, there exist various algorithms for star observation. Some of these algorithms aim to suppress stars or enhance them. For example, one approach involves using two spectral band sensors to estimate the temperature of targets and differentiate stars based on temperature differences [19]. However, this method requires high-quality detection hardware. In cases where the target is at a finite distance and its scale is larger than 3×3 pixels [36], connected components analysis can be used to cluster stars. This method cannot be generalized to other applications.

Currently, neural networks are widely used for the classification of remote sensing image data. An improved CBDNet network structure has been proposed for star background suppression, which is trained using real images [37]. However, it is important to note that the signal-to-noise ratio (SNR) of stars that this algorithm can handle is limited.

In conclusion, there is no universal infrared dim star background suppression. The algorithms based on images are mainly used to suppress the high-SNR star background. Although the algorithms based on star catalogue can suppress low-SNR stars, it is not suitable for the preprocessing of space target detection and tracking. These algorithms may suppress some stars that are not in the image, causing information loss and consuming more computing resources. Other algorithms require specific application conditions that are not universally applicable. If the large amounts of dim infrared stars cannot be suppressed by preprocessing, the real-time on-board intelligent information processing would be catastrophic. Therefore, an effective and less consuming infrared dim star background suppression method is vitally important in practical application.

1.2. Motivation

Previous research has largely overlooked the impact of very-low-SNR stars that cannot be detected using a single frame. While these stars may not significantly affect the performance of DBT methods, they can still hurt real-time performance and accuracy in TBD methods. Existing TBD methods, such as particle filter [38,39], dynamic programming [40,41], and Hough transform [42,43], have primarily focused on digging targets that are covered by heavy noise. Usually, these TBD methods need extra processes to identify real targets and stars, which can negatively impact real-time performance and accuracy. For instance, the particle filter may cancel the route of stars, but the small number of stars as potential targets requires a large number of additional particles to track, which can waste computing resources and negatively affect real-time performance.

The main contributions of this paper are as follows: (1) Recursive moving target indication (RMTI) is improved in a motion vector to enhance dim stars efficiently and accurately. (2) An adaptive multi-frame accumulation threshold segmentation is proposed, which can create an accurate star mask. Dim stars can be suppressed in real-time. (3) The set value of key parameters is provided by analyzing the experiment. Meanwhile, a simulation experiment was designed to verify the feasibility and robustness of this method. The proposed algorithm fills the low-SNR star background suppression gap in space target detection and tracking. It can be used as an efficient preprocessing step for most target detection and tracking methods and has great practical value.

The remainder of this paper is structured as follows. Section 2 provides a detailed explanation of our method. In Section 3, we present our experimental approach, including the set values of main parameters and the resulting experimental results. Section 4 discusses the performance of our proposed methods. Finally, in Section 5, we present our conclusions.

2. Methodology

Figure 3 displays the block diagram of the proposed infrared dim star background suppression method based on recursive moving target indication. Firstly, the star motion is extracted from the high-SNR star map or is deduced from the satellite attitude and orbit data. Then, the RMTI is used to enhance the dim star and generate the frame mask of multi-frame accumulation (FMMA). The FMMA carries the number of frames that have been accumulated for each pixel. Therefore, the adaptive threshold for each pixel is derived by FMMA. Finally, the star mask can be extracted from a multi-frame enhanced star image using adaptive threshold segmentation. The dim stars can be suppressed by the star mask.

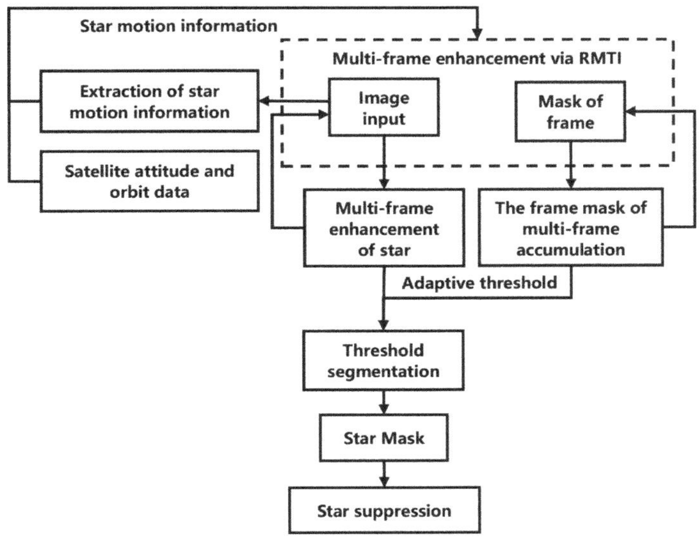

Figure 3. Block diagram of the proposed infrared dim star background suppression method.

2.1. The Multi-Frame Enhancement of Advanced RMTI

RMTI can produce significant SNR gains when the target motion state is known [44,45]. Unlike noncooperative targets, such as space debris, stars remain stationary relative to the Earth. As a result, estimating the motion information of stars in an image sequence is a relatively simple task. The phase matrix of the motion spectrum for each frame can be calculated based on the star motion information, and the spectrum of the current enhanced frame can be obtained by multiplying the phase matrix of the motion spectrum for the current frame and the spectrum of the previous enhanced frames. In the space domain, RMTI enhances stars by registration and accumulation [46]. RMTI processes each frame and stores the result for further processing. The output of the previous frame is used as the input for the next frame, allowing for easy adaptation to digital processing.

To continuously, accurately, and efficiently enhance the star, we improve the RMTI in the motion vector. The multi-frame enhancement of RMTI for stars proceeds is described below. The signal intensity of the star in the focal plane is denoted by $s(\boldsymbol{r}, nt_0)$, where $\boldsymbol{r}$ represents the coordinate of stars, and t_0 is the image sampling period. The star state transition function is defined as

$$s(\boldsymbol{r}, nt_0) = s(\boldsymbol{r} - \boldsymbol{v_n} t_0, (n-1)t_0) \tag{1}$$

where $\boldsymbol{v_n}$ is the velocity of the stars in nt_0. The image $y(\boldsymbol{r}, nt_0)$ can be described as:

$$y(\boldsymbol{r}, nt_0) = s(\boldsymbol{r}, nt_0) + n(\boldsymbol{r}, nt_0) \tag{2}$$

where $n(\boldsymbol{r}, nt_0)$ represents the noise of the image. The two-dimensional spatial Fourier transform of the image is:

$$Y(\boldsymbol{k}, nt_0) = S_{n-1}(\boldsymbol{k}) exp\{-i\boldsymbol{k} \cdot \boldsymbol{v_n} t_0\} + N(\boldsymbol{k}, nt_0) \tag{3}$$

where $\boldsymbol{k}$ denotes a two-dimensional spatial wavenumber vector. $S_{n-1}(\boldsymbol{k})$ denotes the two-dimensional spatial Fourier transform of the star signal of the previous frame, and $N(\boldsymbol{k}, nt_0)$ is the two-dimensional spatial Fourier transform of noise. The star registration of adjacent frames in the space domain can be achieved by multiplying $exp\{-i\boldsymbol{k} \cdot \boldsymbol{v_n} t_0\}$ in the frequency domain. For convenience, let $\alpha_n = exp\{-i\boldsymbol{k} \cdot \boldsymbol{v_n} t_0\}$. α_n is the phase matrix of the motion spectrum for nt_0. When $n = 0$, the two-dimensional spatial Fourier transform of the image is:

$$Y(\boldsymbol{k}, 0) = S_0(\boldsymbol{k}) + N(\boldsymbol{k}, 0) \tag{4}$$

Let $X_0(k) = Y(\boldsymbol{k}, 0)$, where $X_n(k)$ denotes the enhanced frequency spectrum of stars When $n = 1$, the two-dimensional spatial Fourier transform of the image is:

$$Y(\boldsymbol{k}, 1) = S_0(\boldsymbol{k})\alpha_1 + N(\boldsymbol{k}, 1) \tag{5}$$

Since the noise in different positions and times is mutually uncorrelated, we can describe $X_1(k)$ as:

$$X_1(k) = Y(\boldsymbol{k}, 1) + X_0(k)\alpha_1 \tag{6}$$

Similarly, when $n = 2$, $X_2(k)$ is defined as:

$$X_2(k) = Y(\boldsymbol{k}, 2) + X_1(k)\alpha_2 \tag{7}$$

Therefore, the enhanced frequency spectrum of stars in n can be represented as:

$$X_n(k) = Y(\boldsymbol{k}, n) + X_{n-1}(k)\alpha_n \tag{8}$$

To obtain the frequency spectrum of all the superposed frames, we can add the two-dimensional spatial Fourier transform of the current frame to the frequency spectrum of all the previous superposed frames, as shown in Equation (8). The result of the current frame will be superposed by the next frame, and this iteration can output the frequency spectrum of the enhanced stars of every frame easily.

In practical applications, digital images are represented by integer coordinates, and the velocity of stars needs to be converted to an integer to avoid artifacts. Let $I(x, y, f)$ be the input image with pixel coordinates x, y and frame number f. The image size is $M \times N$, and $x = 1, \ldots, M, y = 1, \ldots, N$. The two-dimensional spatial Fourier transform of the input image is given by:

$$FI(u, v, f) = \sum_{x=0}^{M-1} \sum_{y=0}^{N-1} I(x, y, f) e^{-i2\pi(\frac{ux}{M} + \frac{vy}{N})} \tag{9}$$

where (u, v) denotes the two-dimensional spatial wavenumber vector.

However, the velocity of stars in the f-th frame, denoted by $V_x(f)$ and $V_y(f)$, is typically non-integer. To address this, we introduce an offset of velocity ($D_{x-offset}(f)$ and $D_{y-offset}(f)$) to compensate for the error of converting velocity to an integer. This allows us to obtain a more accurate representation of the image without introducing artifacts. To determine the motion vector of the current frame ($D_x(f)$ and $D_y(f)$), we need to consider the offset of the velocity of the previous frame and the velocity of the current frame together. This ensures that the motion vector accurately reflects the movement of the stars in the image. The offset of the velocity of the previous frame and the velocity of the current frame is given as follows:

$$\begin{cases} D_x(f) = round\left[V_x(f)t_0 + V_{x-offset}(f-1)t_0\right] \\ D_y(f) = round\left[V_y(f)t_0 + V_{y-offset}(f-1)t_0\right] \end{cases} \tag{10}$$

$$\begin{cases} D_{x-offset}(f) = D_x(f) - V_x(f)t_0 \\ D_{y-offset}(f) = D_y(f) - V_y(f)t_0 \end{cases} \tag{11}$$

where $round(\cdot)$ means the process of rounding off. When $f = 1$, $V_{x-offset}$ and $V_{y-offset}$ are equal to 0. Now, we can deduce the phase matrix of motion spectrum as follows:

$$\alpha = exp\{-i(D_x u + vD_y)\} \tag{12}$$

Hence, according to Equations (9) and (12), Equation (8) can transform as:

$$FEI(u,v,f) = FI(u,v,f) + FEI(u,v,f-1) \cdot \alpha \tag{13}$$

Notably, the enhanced spectrum is equivalent to the original spectrum for the first frame component ($FEI(u,v,1) = FI(u,v,1)$). Furthermore, in actual observations, the visual field undergoes slow movement, with a maximum motion of two pixels per frame [47]. To optimize computing resources, a lookup table is introduced to determine the phase matrix of the motion spectrum, as shown in Figure 4. This approach allows for efficient processing as each iteration only requires the operation of Equations (9) and (13) and a single lookup.

$\alpha_{-2,2}$	$\alpha_{-1,2}$	$\alpha_{0,2}$	$\alpha_{1,2}$	$\alpha_{2,2}$
$\alpha_{-2,1}$	$\alpha_{-1,1}$	$\alpha_{0,1}$	$\alpha_{1,1}$	$\alpha_{2,1}$
$\alpha_{-2,0}$	$\alpha_{-1,0}$	$\alpha_{0,0}$	$\alpha_{1,0}$	$\alpha_{2,0}$
$\alpha_{-2,-1}$	$\alpha_{-1,-1}$	$\alpha_{0,-1}$	$\alpha_{1,-1}$	$\alpha_{2,-1}$
$\alpha_{-2,-2}$	$\alpha_{-1,-2}$	$\alpha_{0,-2}$	$\alpha_{1,-2}$	$\alpha_{2,-2}$

Figure 4. The lookup table of motion vector.

Furthermore, the stars that have just entered the field of view have fewer superposed frames than those almost leaving the field of view. Stars that have more superposed frames are enhanced to a greater degree. Therefore, to achieve adaptive threshold segments for different pixels, a mask of superposed frames is introduced. This mask helps to differentiate between pixels that have a high degree of superposition and those that do not, resulting in a more accurate representation of the image. The generation of this mask is similar to the star enhancement and can be described as follows:

$$FM(u,v,f) = E(u,v) + FM(u,v,f-1) \cdot \alpha \tag{14}$$

Here, $FM(u,v,f)$ is the spectrum of the mask of superposed frames, $E(u,v)$ denotes the two-dimensional spatial Fourier transform of the unit image, and alpha is a constant. In particular, $FM(u,v,1) = E(u,v)$. Figure 5 shows the generation of this mask. Owing to the invariability of the unit image of each frame, the procedure is easy to compute and implement.

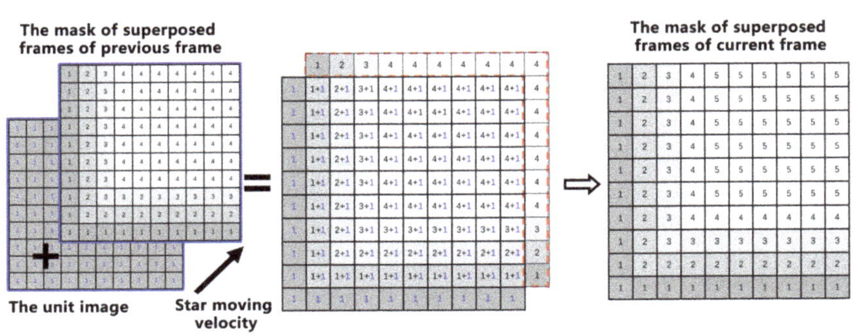

Figure 5. The implementation of the mask of superposed frames.

2.2. Adaptive Star Map

In order to improve the accuracy of star detection, an adaptive threshold segmentation approach is employed. This approach adjusts the threshold based on the number of superimposed frames for each pixel. Pixels with fewer superimposed frames are assigned a higher threshold to avoid false detection, while pixels with more superimposed frames are assigned a lower threshold to improve the detection rate. By using this approach, the algorithm can effectively detect stars while minimizing the impact of noise.

To determine the threshold for star detection, it is important to understand the probability distribution of stars, background, and noise. According to engineering practice, the noise in infrared images is distributed nearly normally [48]. Additionally, the temperature of deep space is less than 4 K [49], which has a negligible effect on stars. Therefore, in this paper, the background and noise are estimated using the normal distribution function. Suppose the distribution of background and noise is $\mathcal{N}(\mu_{noise}, {\sigma_{noise}}^2)$. Then, the stars can be expressed as $\mathcal{N}(\mu_{noise} + I_{star}, {\sigma_{noise}}^2)$, where μ_{noise} and σ_{noise} represent the mean and standard deviation of the noise, and I_{star} denotes the responsive intensity of the stars. Since the background noise is independent between different pixels and different frames, after n frames accumulation the distribution of the star and noise is still Gaussian, with the mean amplified by n times and the standard deviation amplified by $\sqrt{n}$ times. The noise and stars distribution will change to $\mathcal{N}(n\mu_{noise}, n{\sigma_{noise}}^2)$ and $\mathcal{N}(n\mu_{noise} + nI_{star}, n{\sigma_{noise}}^2)$, respectively. Ideally, the SNR of the stars will increase by a factor of $\sqrt{n}$. This superimposed process makes star detection easier, as shown in Figure 6.

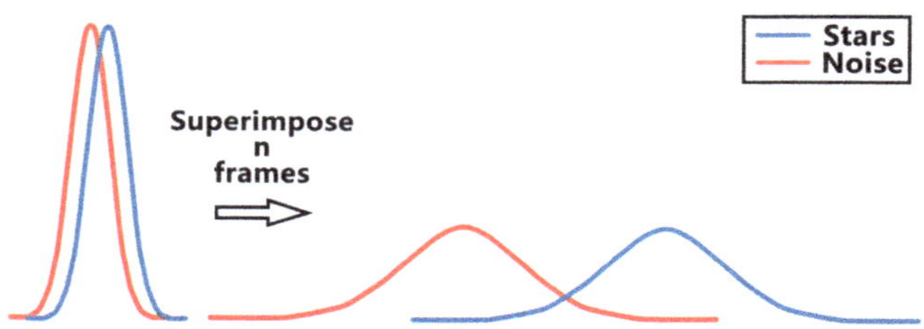

Figure 6. The probability distribution of stars and noise.

The calculation of the adaptive threshold is introduced in detail below. Firstly, the enhanced star image $EI(x,y,f)$ and the mask of superposed frames $M(x,y,f)$ is obtained by applying the inverse transformation of two-dimensional spatial Fourier as follows:

$$EI(x,y,f) = \frac{1}{MN}\sum_{u=0}^{U-1}\sum_{v=0}^{V-1} FEI(u,v,f)e^{i2\pi(\frac{ux}{M}+\frac{vy}{N})} \tag{15}$$

$$M(x,y,f) = \frac{1}{MN}\sum_{u=0}^{M-1}\sum_{v=0}^{N-1} FM(u,v,f)e^{i2\pi(\frac{ux}{M}+\frac{vy}{N})} \tag{16}$$

Then, we apply median filtering to eliminate stars and other targets. Once the background is relatively clean, we can estimate the mean (I_{mean}) and standard deviation (I_{std}) of the background. These values are used to calculate the threshold for eliminating noise and stars. We use the three sigma criteria along with an analysis of the probability distribution of stars and noise. By applying these criteria, we obtain the threshold for eliminating noise (T_{EN}) and the threshold for detecting stars (T_{DS}) as follows:

$$T_{EN}(x,y,f) = I_{mean}(f) \cdot M(x,y,f) + C_{\sigma N} \cdot I_{std}(f) \cdot \sqrt{M(x,y,f)} \tag{17}$$

$$T_{DS}(x,y,f) = [I_{mean}(f) + Psnr \cdot I_{std}(f)]M(x,y,f) + C_{\sigma S} \cdot I_{std}(f) \cdot \sqrt{M(x,y,f)} \tag{18}$$

where $Psnr$ denotes the lowest signal-to-noise ratio of the stars that plan to suppress. This parameter can limit the lower limit of the star SNR that needs to be suppressed and enhance the robustness of the proposed method for blurred images. Then, $C_{\sigma N}$ and $C_{\sigma S}$ are the coefficients of sigma for eliminating noise and the coefficients of sigma for detecting stars, respectively. The effectiveness of noise suppression in the image processing algorithm is directly proportional to the value of $C_{\sigma N}$. However, it is important to note that an excessively high value of $C_{\sigma N}$ may result in the erroneous detection of stars. Similarly, the selection of $C_{\sigma S}$ should aim to balance noise suppression and star detection. Typically, a value of $C_{\sigma N}$ greater than 4.5 and a value of $C_{\sigma S}$ greater than 3 are recommended. The specific parameter selection will be explained in Section 3.2. The optimal values of $C_{\sigma N}$ and $C_{\sigma S}$ can be increased with a higher $Psnr$. In practical applications, these parameters can be adjusted based on the acceptable level of false positives and missed detections. To obtain the final threshold (T_{star}), Equations (17) and (18) are used to calculate the number of frames that make T_{EN} and T_{DS} equal. The final threshold is obtained by fusing the two thresholds as in the below equation:

$$T_{star}(x,y,f) = \begin{cases} \frac{T_{EN}(x,y,f)+T_{DS}(x,y,f)}{2} & M(x,y,f) > \left(\frac{C_{\sigma 1}+C_{\sigma 2}}{Psnr}\right)^2 \\ T_{EN}(x,y,f) & M(x,y,f) \leq \left(\frac{C_{\sigma 1}+C_{\sigma 2}}{Psnr}\right)^2 \end{cases} \quad (19)$$

It is crucial to note that the selection of the final threshold depends on the number of available frames. Figure 7 depicts the probability distribution of star superposition with different frame numbers and the selection principle of the final threshold. In Figure 7a, the noise and the star have a lot of overlap, and it is not good to separate the two. Our primary objective is to eliminate noise and prevent false detection when the number of frames is less. Owing to the lack of enhanced frames, the stars are mixed with noise, thereby preventing false detection as the main object. Conversely, if we have sufficient frames, we have the conditions to distinguish the stars from the noise. Then, we will select the middle value of the threshold for eliminating noise and detecting stars as the final threshold, as shown in Figure 7b.

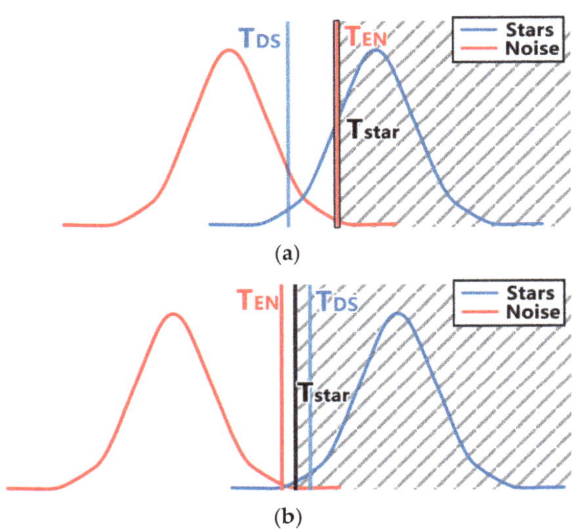

Figure 7. The selection principle of the final threshold: (**a**) noise elimination priority; (**b**) star detection and noise elimination are considered comprehensively.

After the above steps, the adaptive threshold is used to segment the mask of stars from the enhanced star image. The detection of stars can be suppressed to avoid their influence on the detection of other space objects. In some cases, it may be beneficial to further improve the suppression of the star background by applying a morphology dilation operation to the mask. This process should be chosen based on the specific optical system under consideration.

3. Experiment and Parameter Setting

In this section, we provide a detailed explanation of the experimental design. Meanwhile, the set value of the main parameters is discussed in order to obtain a good performance. Finally, the experimental results compared with other new methods are presented

3.1. Experimental Setup

In this paper, three satellite real star data were used to verify the robustness and effectiveness of the proposed method. The parameters of the infrared camera used in this satellite are listed in Table 1.

Table 1. The parameters of infrared camera used in the satellite.

Parameters	Value
Format	512 × 512
The angle resolution of pixel	0.02464°
The angle of field of view	12.6° × 12.6°
Framerate	20 Hz
Bits per pixel	14 bits
Spectrum	2.1~3.3 μm
Field direction	Seq.1: De = 340.668 Ra = −46.885 Seq.2: De = 298.808 Ra = −59.196 Seq.3: De = 252.166 Ra = −69.028

To evaluate the effectiveness of the proposed method quantificationally in this paper, we made use of a star table obtained from NASA's Wide-field Infrared Survey Explorer (WISE) [50]. Figure 8 shows the procedure for simulating star images. Each color in the picture corresponds to a star. Arrows indicate camera acquisition direction. The main method was referenced from star identification described by Zhang Guangjun [51]. The simulation procedure involved the establishment of a celestial, satellite camera, and image coordinates. Next, rotation transformation and perspective projection transformation were employed to convert the stars in the star table to image coordinates. The response intensities of the stars in the resultant image were then calculated based on factors such as the minimum detectable star magnitude, corresponding SNR, as well as the star's magnitude in the star list. Subsequently, each star was simulated using the point diffusion function and its response intensity, as well as its sub-pixel position concerning the camera. It is worth noting that the point diffusion function was approximated by the circular symmetric two-dimensional Gaussian distribution. In addition to the aforementioned techniques, a nearly constant velocity model for camera motion was used following the method in [52]. A randomized approach was adopted to generate the initial right ascension and declination of the camera, as well as its initial moving speed within the range of 1~3 pixels/frame. Other simulation parameters, as well as their relevant values, are listed in Table 2. Overall, these simulation procedures accurately reflected the expected behavior of the proposed method in different scenarios, which validates the effectiveness of the proposed solution.

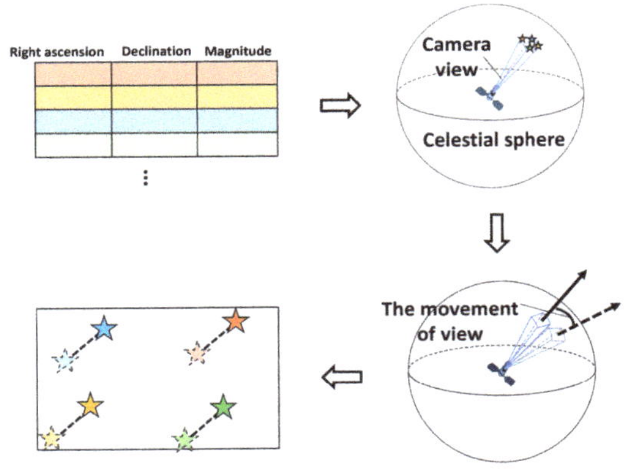

Figure 8. The procedure of simulating star images.

Table 2. The parameters of star map simulation.

Parameters	Value
Format	320 × 256
The angle resolution of pixel	0.01784°
The angle of field of view	4.568° × 5.710°
Framerate	30 Hz
Bits per pixel	14 bits
Spectrum	3 μm
Minimum detectable magnitude (corresponding SNR = 1)	9.56

To quantitatively analyze the effectiveness of the proposed method, three testing metrics are introduced: the accuracy of star suppression R_{ts}, the ratio of star suppression R_{ss}, and the average running time per frame T_{pf}. Suppose the number of stars suppressed by the method is N_s, and the number of stars correctly suppressed by the method is N_{true}. Furthermore, the number of stars whose SNR is larger than the lowest SNR of the stars that plan to suppress is N_{total}. Then, these metrics can be defined as follows:

$$R_{ts} = \frac{N_{true}}{N_s} \tag{20}$$

$$R_{ss} = \frac{N_{true}}{N_{total}} \tag{21}$$

3.2. Parameter Setting

To effectively suppress stars, it is crucial to determine the appropriate coefficients of sigma for both eliminating noise and detecting stars. To test the effectiveness of different coefficients, we conducted simulations and analyzed the results. Based on our findings, we will recommend the coefficients for optimal star suppression.

Where $N_{fsf}(n)$ refers to the frame amount that falsely suppressed the *nth* flickering pixel. As mentioned above, the evaluation of flickering pixel suppression uses simulation data. Therefore, the $N_{rf}(n)$, $N_{fsf}(n)$, and $N_{msf}(n)$ in Equations (13) and (14) can be recorded while the simulation is underway.

We introduce $R_{ts} \times R_{ss}$ to evaluate the performance of the proposed method. This index expresses the suppression precision and the suppressing rate, with a range of 0 to 1. A large value of $R_{ts} \times R_{ss}$ can only be obtained when the suppressed stars are many and accurate. We simulate hundreds of star image sequences, each using different values of $C_{\sigma N}$, $C_{\sigma S}$, and SNR to suppress stars. The results of these tests are shown in Figures 9–11, with SNR values of 1, 1.5, and 2, respectively. Figures 9–11 (a) show the 3D map of $R_{ts} \times R_{ss}$ with different values of $C_{\sigma N}$ and $C_{\sigma S}$, while Figures 9–11 (b) show the distribution of $R_{ts} \times R_{ss}$ for different values of $C_{\sigma N}$ under maximum $C_{\sigma S}$, and Figures 9–11 (c) show the distribution of $R_{ts} \times R_{ss}$ for different values of $C_{\sigma S}$ under maximum $C_{\sigma N}$. It is clear that the suppression performance is better when $C_{\sigma N}$ is between 4 and 5 and when $C_{\sigma S}$ approaches 6. Table 3 provides a list of typical parameter selections.

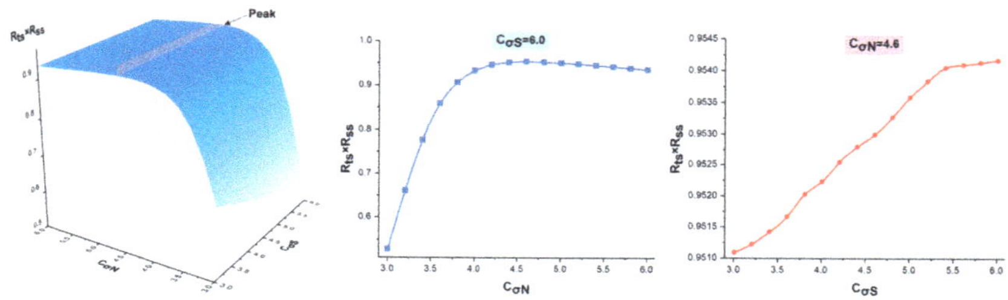

(a) (b) (c)

Figure 9. The distribution of $R_{ts} \times R_{ss}$ in Psnr = 1. (**a**) The 3D map of $R_{ts} \times R_{ss}$ with different values of $C_{\sigma N}$ and $C_{\sigma S}$; (**b**) the distribution of $R_{ts} \times R_{ss}$ for different values of $C_{\sigma N}$ under maximum $C_{\sigma S}$; (**c**) the distribution of $R_{ts} \times R_{ss}$ for different values of $C_{\sigma S}$ under maximum $C_{\sigma N}$.

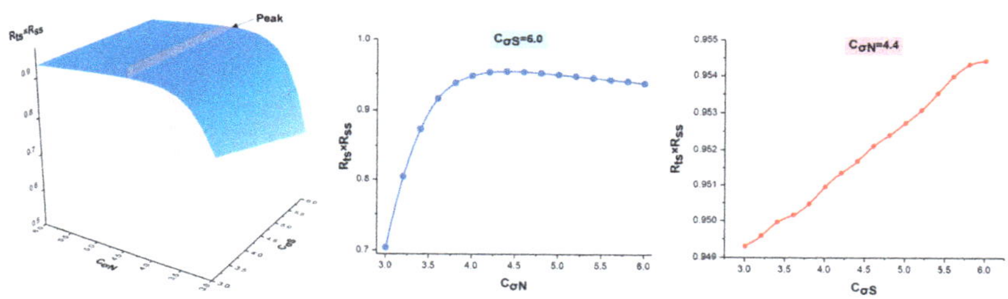

(a) (b) (c)

Figure 10. The distribution of $R_{ts} \times R_{ss}$ in Psnr = 1.5. (**a**) The 3-D map of $R_{ts} \times R_{ss}$ with different values of $C_{\sigma N}$ and $C_{\sigma S}$; (**b**) the distribution of $R_{ts} \times R_{ss}$ for different values of $C_{\sigma N}$ under maximum $C_{\sigma S}$; (**c**) the distribution of $R_{ts} \times R_{ss}$ for different values of $C_{\sigma S}$ under maximum $C_{\sigma N}$.

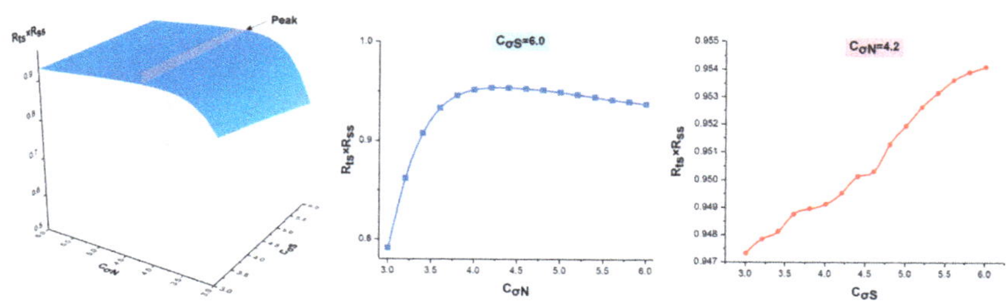

(a) (b) (c)

Figure 11. The distribution of $R_{ts} \times R_{ss}$ in Psnr = 2. (**a**) The 3-D map of $R_{ts} \times R_{ss}$ with different values of $C_{\sigma N}$ and $C_{\sigma S}$; (**b**) the distribution of $R_{ts} \times R_{ss}$ for different values of $C_{\sigma N}$ under maximum $C_{\sigma S}$; (**c**) the distribution of $R_{ts} \times R_{ss}$ for different values of $C_{\sigma S}$ under maximum $C_{\sigma N}$.

Table 3. The typical selection of parameters.

Psnr	$C_{\sigma N}$	$C_{\sigma S}$
1	4.2~4.8	5.5~6.0
1.5	4.0~4.6	5.6~6.0
2	4.0~4.4	5.8~6.0

3.3. Experimental Result

With the above experimental data and evaluation criterion, the proposed method is compared with star map registration via topology invariance (SMRTI) [26] and an enhanced moving target indicator (EMTI) [29]. The SMRTI and EMTI are the most recently proposed SBT and TBS, respectively. The proposed method uses the following parameter setting: $C_{\sigma N} = 4.6$, $C_{\sigma S} = 6$, $Psnr = 1$. The proposed method and other methods are implemented under MATLAB R2018a with an Intel Core 2.80 GHz processor and 8 GB of physical memory.

Figures 12–14 show the experimental results of real data Seq.1, Seq.2, and Seq.3, where (a), (b), and (c) are the result of the proposed method, SMRTI, and EMTI, respectively. The green star represents the real stars in the image. The red circle represents the stars suppressed by the proposed method in this paper. The blue box represents the stars suppressed by SMRTI, and the yellow triangle represents the stars suppressed by EMTI.

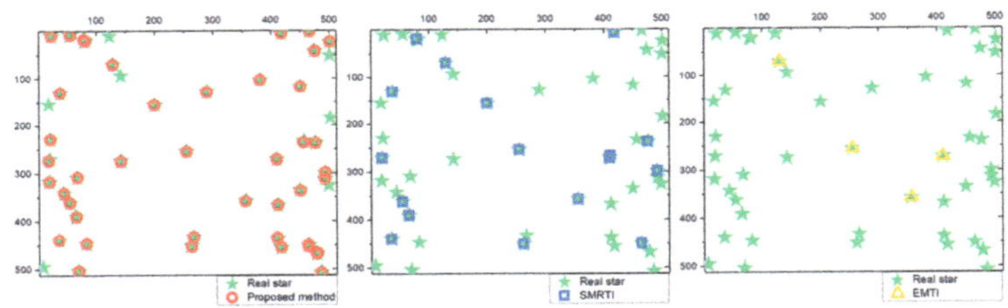

(a) (b) (c)

Figure 12. The experimental results with Seq.1 real data: (**a**) proposed method; (**b**) SMRTI; (**c**) EMTI.

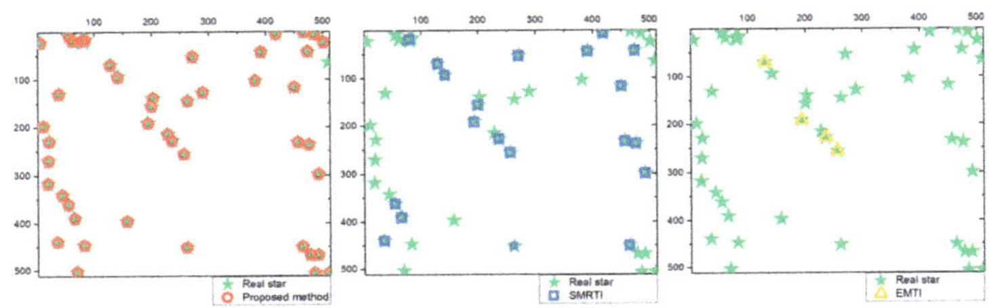

(a) (b) (c)

Figure 13. The experimental results with Seq.2 real data: (**a**) proposed method; (**b**) SMRTI; (**c**) EMTI.

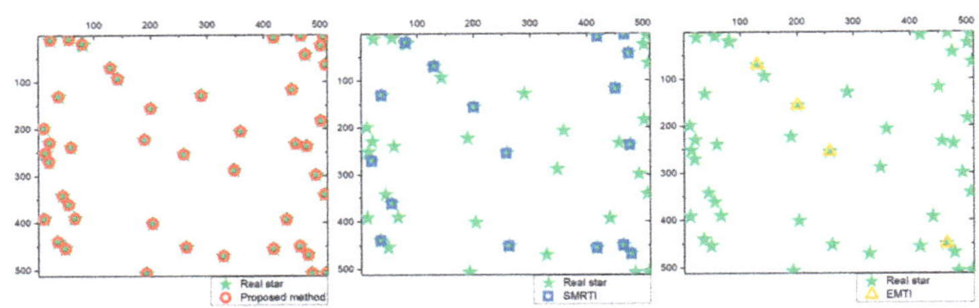

(a) (b) (c)

Figure 14. The experimental results with Seq.3 real data: (**a**) proposed method; (**b**) SMRTI; (**c**) EMTI.

Figure 15 presents the experimental results of the proposed method. In Figure 15a the key frames of the simulated star image are shown, where most stars are obscured by noise. Figure 15b displays the enhanced star images using the proposed method, revealing many low-SNR stars that were previously hidden. Figure 15c shows the stars detected by the proposed method, with the blue block indicating the detected stars and the red stars representing the actual stars present in the images. While there were a few mistaken detections, the proposed method successfully identified most of the stars, with only a small quantity of stars being missed.

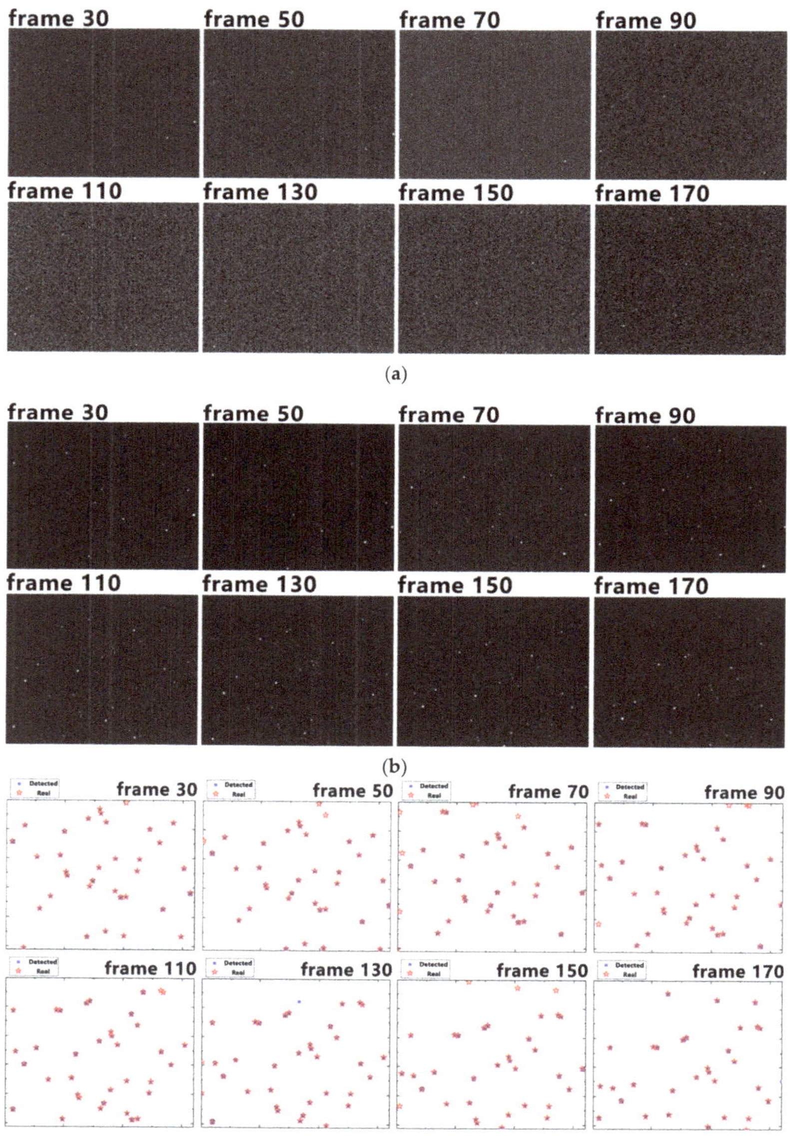

Figure 15. The experimental result of the proposed method. (**a**) Simulated star image sequence; (**b**) star image sequence enhanced advanced RMTI; (**c**) the stars detected by the proposed method.

We evaluated the performance of the three methods by recording the accuracy of star suppression (R_{ts}), the ratio of star suppression (R_{ss}), and the average running time per frame (T_{pf}) using a hundred simulation sequences. The results, presented in Table 4, demonstrate that the proposed method outperforms the other methods in terms of both speed and accuracy. Specifically, the proposed method achieves the task quickly and effectively, as evidenced by its high R_{ts} and low R_{ss} values.

Table 4. The experimental result of methods.

Method	R_{ts}	R_{ss}	T_{pf}
Proposed method	98.72%	98.82%	0.0031 s
SMRTI	98.19%	69.88%	0.2490 s
EMTI	98.59%	73.28%	0.0640 s

To offer a more intuitionistic star suppression result, the ratio of star suppression on a different SNR partition is recorded and mapped as follows.

4. Discussion

In the experiment with real data, the proposed method has an obvious advantage over other methods, as shown in Figures 12–14. The proposed method only misses a few stars. The effectiveness and robustness of the proposed method are demonstrated.

The working process shown in Figure 15 explains why the proposed method can suppress a dim star background. The proposed method for suppressing a dim star background is based on the core idea of treating stars as targets. By enhancing the SNR of stars through multi-frame accumulation, as shown in Figure 15b, the enhanced stars can be easily detected through threshold segmentation. Finally, the detected stars in Figure 15c are suppressed. The proposed method is capable of suppressing the majority of stars in the field of view, as seen in Figure 15c. However, a small minority of stars may be missed or wrongly suppressed in the top right corner of the subgraph of Figure 15c due to the lack of superimposed frames. If we do not accept the suppression results of this region, the accuracy of star suppression and the ratio of star suppression will be further improved. However, this comes at the expense of the field of view. Therefore, this promotion scheme should be considered according to the actual application situation. It is worth noting that the results compared with SMRTI and EMTI in Table 4 and Figure 16 are evaluated from the entire field of view.

Space target detection under a star background has been extensively researched, with a recent focus on high-SNR star suppression and corresponding DBT target detection methods. As shown in Table 4, the proposed star suppression method has a significant advantage in the ratio of star suppression. This advantage is mainly due to the effective suppression of low-SNR stars, as demonstrated in Figure 16. When the SNR is larger than five, these three methods are evenly matched. When the SNR ratio is smaller, the advantages of the proposed method are more obvious. The SMRTI and EMTI methods can suppress a few stars when the SNR is lower than two, and EMTI can suppress more stars than SMRTI when the SNR is between two and five. This is because EMTI adopts limited energy accumulation first. However, the proposed method enhances low-SNR stars through RMTI, resulting in stable and efficient star suppression when the SNR is lower than five. Although low-SNR stars do not interfere with SMRTI and EMTI, the proposed method's ability to suppress them is vital to TBD methods. This is the most significant contribution of our proposed method. The suppression of a low-SNR star background is an urgent issue, and the proposed method is on par with EMTI and SMRTI in terms of the accuracy of star suppression, with all three methods achieving over 98% accuracy. Additionally, the proposed method has a certain advantage in running time. These results demonstrate that the proposed method can be widely used in preprocessing for low-SNR target detection and tracking.

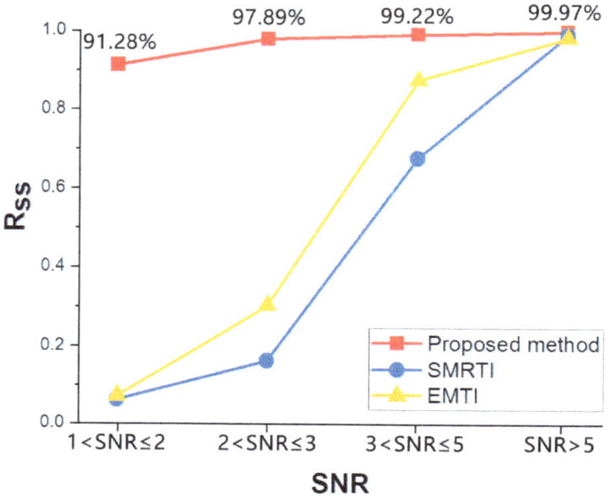

Figure 16. The ratio of star suppression on different SNR partitions.

5. Conclusions

To address the limitation of existing star suppression algorithms in effectively suppressing very-low-SNR star backgrounds, a dim star background suppression algorithm via RMTI is proposed in this paper. The proposed method involves enhancing the dim stars using advanced RMTI, followed by an adaptive threshold segmentation to filter out stars precisely. The experimental results using simulated star images demonstrate that the proposed method can stably and reliably suppress stars with an SNR of less than 2, with a star suppression rate of over 91%, and an overall star suppression accuracy of over 98.7%. Compared to the existing star suppression algorithms, the proposed method exhibits significant improvements in real-time performance and low-SNR star suppression ability. For real image processing, this method still maintains a good performance. As a preprocessing step for many TBD methods, the proposed method can effectively reduce the false detection rate of infrared dim small target detection and tracking and improve the real-time performance.

Author Contributions: All the authors contributed to this study. Conceptualization, L.Z. and P.R.; Investigation, L.Z. and Y.H.; Methodology, P.R. and X.C.; Resources, Y.H. and X.C.; Software, L.Z. and X.C.; Data curation, L.Z.; Funding acquisition, P.R. and L.J.; Project administration, Y.H. and X.C.; Supervision, L.J. Writing—Original draft preparation, L.Z. and P.R.; Writing—review and editing, P.R. and L.J.; Validation, L.J. All authors have read and agreed to the published version of the manuscript.

Funding: This research was funded by National Natural Science Foundation of China, grant number 62175251.

Data Availability Statement: Not applicable.

Conflicts of Interest: The authors declare no conflict of interest.

References

1. Wei, B.; Nener, B.D. Multi-Sensor Space Debris Tracking for Space Situational Awareness With Labeled Random Finite Sets. *IEEE Access* **2019**, *7*, 36991–37003. [CrossRef]
2. Xie, Z.; Chen, X.; Ren, Y.; Zhao, Y. Design and Analysis of Preload Control for Space Debris Impact Adhesion Capture Method. *IEEE Access* **2020**, *8*, 203845–203853. [CrossRef]
3. Guo, X.; Chen, T.; Liu, J.; Liu, Y.; An, Q. Dim Space Target Detection via Convolutional Neural Network in Single Optical Image. *IEEE Access* **2022**, *10*, 52306–52318. [CrossRef]
4. Liu, D.; Wang, X.; Xu, Z.; Li, Y.; Liu, W. Space target extraction and detection for wide-field surveillance. *Astron. Comput.* **2020**, *32*, 100408. [CrossRef]

5. Kwan, C.; Budavari, B. Enhancing small moving target detection performance in low-quality and long-range infrared videos using optical flow techniques. *Remote Sens.* **2020**, *12*, 4024. [CrossRef]
6. Rawat, S.S.; Verma, S.K.; Kumar, Y. Review on recent development in infrared small target detection algorithms. *Procedia Comput. Sci.* **2020**, *167*, 2496–2505. [CrossRef]
7. Zou, Y.; Zhao, J.; Wu, Y.; Wang, B.; Dong, L. Reverse Procedure Detection of Space Target Streaks Based on Motion Parameter Estimation. *IEEE Access* **2021**, *9*, 21823–21831. [CrossRef]
8. Zhao, F.; Wang, T.; Shao, S.; Zhang, E.; Lin, G. Infrared moving small-target detection via spatiotemporal consistency of trajectory points. *IEEE Geosci. Remote Sens. Lett.* **2020**, *17*, 122–126. [CrossRef]
9. Cao, Y.; Wang, G.; Yan, D.; Zhao, Z. Two Algorithms for the Detection and Tracking of Moving Vehicle Targets in Aerial Infrared Image Sequences. *Remote Sens.* **2016**, *8*, 28. [CrossRef]
10. Chen, S.T.; Jin, M.; Zhang, Y.Y.; Zhang, C. Infrared blind-pixel compensation algorithm based on generative adversarial networks and Poisson image blending. *Signal Image Video Process* **2020**, *14*, 77–85. [CrossRef]
11. Tchendjou, G.T.; Simeu, E. Detection, location and concealment of defective pixels in image sensors. *IEEE Trans. Emerg. Top. Comput.* **2021**, *9*, 664–679. [CrossRef]
12. Wan, M.; Ye, X.; Zhang, X.; Xu, Y.; Gu, G.; Chen, Q. Infrared small target tracking via gaussian curvature-based compressive convolution feature extraction. *IEEE Geosci. Remote Sens. Lett.* **2021**, *19*, 7000905. [CrossRef]
13. Wan, M.J.; Gu, G.H.; Cao, E.C.; Hu, X.B.; Qian, W.X.; Ren, K. In-frame and inter-frame information based infrared moving small target detection under complex cloud backgrounds. *Infrared Phys. Technol.* **2016**, *76*, 455–467. [CrossRef]
14. Li, M.; Peng, L.; Chen, Y.; Huang, S.; Qin, F.; Peng, Z. Mask Sparse Representation Based on Semantic Features for Thermal Infrared Target Tracking. *Remote Sens.* **2019**, *11*, 1967. [CrossRef]
15. Cao, L.; Wan, C.; Zhang, Y.; Li, N. Infrared radiation characteristic measure method of point target. *J. Infrared Millim. Waves* **2015**, *34*, 5. [CrossRef]
16. Jia, L.; Rao, P.; Chen, X.; Qiu, S. On-Board Flickering Pixel Dynamic Suppression Method Based on Multi-Feature Fusion. *Appl. Sci.* **2022**, *12*, 198. [CrossRef]
17. Zhou, D.; Wang, X. Stray Light Suppression of Wide-Field Surveillance in Complicated Situations. *IEEE Access* **2023**, *11*, 2424–2432. [CrossRef]
18. Xu, Z.; Liu, D.; Yan, C.; Hu, C. Stray light nonuniform background correction for a wide-field surveillance system. *Appl. Opt.* **2020**, *59*, 10719–10728. [CrossRef]
19. Johnson, C.R.; Sentovich, M.F.; Ho, C.q. Star Background Cancellation for Deep Space Surveillance. *IEEE Trans. Aerosp. Electron. Syst.* **1981**, *AES-17*, 314–319. [CrossRef]
20. Xue, D.; Sun, J.; Hu, Y.; Zheng, Y.; Zhu, Y.; Zhang, Y. Dim small target detection based on convolutinal neural network in star image. *Multimed. Tools Appl.* **2020**, *79*, 4681–4698. [CrossRef]
21. Jun, Z.; Hongjian, Z.; Dakai, S.; Li, W.; Yanpeng, W.; Chunyan, L. High sensitive automatic detection technique for space objects. *Infrared Laser Eng.* **2020**, *49*, 88–94.
22. Zhang, H.; Bai, Y.; Li, J. An algorithm of small and dim target detection in deep space background. In Proceedings of the 2009 International Conference on Information and Automation, Zhuhai, China, 22–24 June 2009; pp. 985–989.
23. Zhu, Y.; Hu, W.; Zhou, J.; Duan, F.; Sun, J.; Jiang, L. A new starry images matching method in dim and small space target detection. In Proceedings of the 2009 Fifth International Conference on Image and Graphics, Xi'an, China, 20–23 September 2009; pp. 447–450.
24. Luo, Q.; Gao, Z.; Xie, C. Improved GM-PHD filter based on threshold separation clusterer for space-based starry-sky background weak point target tracking. *Digit. Signal Process.* **2020**, *103*, 102766. [CrossRef]
25. Feng, L.; Xiaoliang, X.; Tongsheng, S. Space small targets detection based on maximum projection and quick registration. *Infrared Laser Eng.* **2016**, *45*, 145–150.
26. Jiang, F.; Yuan, J.; Qi, Y.; Liu, Z.; Cai, L. Space target detection based on the invariance of inter-satellite topology. In Proceedings of the 2022 IEEE 10th Joint International Information Technology and Artificial Intelligence Conference (ITAIC), Chongqing, China, 17–19 June 2022; pp. 2151–2155.
27. Hou, W.; Lei, Z.; Yu, Q.; Liu, X. Small target detection using main directional suppression high pass filter. *Optik* **2014**, *125*, 3017–3022. [CrossRef]
28. Jianlin, L.; Xijian, P.; Debao, M. A novel method of drift-scanning stars suppression based on the standardized linear filter. In Proceedings of the 2011 International Conference on Optical Instruments and Technology: Optoelectronic Imaging and Processing Technology, Beijing, China, 28 November 2011.
29. Zhang, Y.; Rao, P.; Jia, L.; Chen, X. Dim moving infrared target enhancement based on precise trajectory extraction. *Infrared Phys. Technol.* **2022**, *128*, 104374. [CrossRef]
30. Wenkang, D.; Zongxi, S. Detection and tracking of multi-space junks in star images. In Proceedings of the Eighth International Conference on Digital Image Processing (ICDIP 2016), Chengu, China, 20–22 May 2016; p. 100330N.
31. Dong, W.; Yan, W.; Zhao, L. Moving space target detection algorithm based on trajectory similarity. In Proceedings of the SPIE/COS Photonics Asia, Beijing, China, 11 October 2018; p. 108161B.

32. Chen, B.; Qin, S.; Dai, D. A Star Identification Algorithm based on Radial Basis Neural Network. In Proceedings of the 2022 4th International Academic Exchange Conference on Science and Technology Innovation (IAECST), Guangzhou, China, 9–11 December 2022; pp. 1274–1278.
33. Niu, Y.; Wei, X.; Li, J. Fast and Robust Star Identification Using Color Ratio Information. *IEEE Sens. J.* **2022**, *22*, 20401–20412 [CrossRef]
34. Mehta, D.S.; Chen, S.; Low, K.S. A Rotation-Invariant Additive Vector Sequence Based Star Pattern Recognition. *IEEE Trans Aerosp. Electron. Syst.* **2019**, *55*, 689–705. [CrossRef]
35. Coupon, J.; Czakon, N.; Bosch, J.; Komiyama, Y.; Medezinski, E.; Miyazaki, S.; Oguri, M. The bright-star masks for the HSC-SSP survey. *Publ. Astron. Soc. Jpn.* **2018**, *70*, S7. [CrossRef]
36. Han, K.; Pei, H.; Huang, Z.; Huang, T.; Qin, S. Non-cooperative Space Target High-Speed Tracking Measuring Method Based on FPGA. In Proceedings of the 2022 7th International Conference on Image, Vision and Computing (ICIVC), Xi'an, China, 26–28 July 2022; pp. 222–231.
37. Li, Y.; Niu, Z.; Sun, Q.; Xiao, H. Background Suppression Method of Star Image Based on Improved CBDNet. In Proceedings of the 2022 3rd International Conference on Computer Vision, Image and Deep Learning & International Conference on Computer Engineering and Applications (CVIDL & ICCEA), Changchun, China, 20–22 May 2022; pp. 671–674.
38. Hu, Z.; Su, Y. Infrared target tracking based on improved particle filtering. *Int. J. Pattern Recognit. Artif. Intell.* **2021**, *35*, 2154015 [CrossRef]
39. Jia, L.; Rao, P.; Zhang, Y.; Su, Y.; Chen, X. Low-SNR Infrared Point Target Detection and Tracking via Saliency-Guided Double-Stage Particle Filter. *Sensors* **2022**, *22*, 2791. [CrossRef]
40. Barniv, Y. Dynamic Programming Solution for Detecting Dim Moving Targets. *IEEE Trans. Aerosp. Electron. Syst.* **1985**, *AES-21*, 144–156. [CrossRef]
41. Sun, X.; Liu, X.; Tang, Z.; Long, G.; Yu, Q. Real-time visual enhancement for infrared small dim targets in video. *Infrared Phys. Technol.* **2017**, *83*, 217–226. [CrossRef]
42. Liu, H.; Rosenfeld, A.; Bhattacharya, P. Hough-transform detection of lines in 3-D space. *Pattern Recognit. Lett.* **2000**, *21*, 843–849
43. Kultanen, P.; Xu, L.; Oja, E. Randomized Hough transform (RHT). In Proceedings of the 10th International Conference on Pattern Recognition, Atlantic City, NJ, USA, 16–21 June 1990; Volume 631, pp. 631–635.
44. Reed, I.S.; Gagliardi, R.M.; Stotts, L.B. Optical moving target detection with 3-D matched filtering. *IEEE Trans. Aerosp. Electron. Syst.* **1988**, *24*, 327–336. [CrossRef]
45. Reed, I.S.; Gagliardi, R.M.; Stotts, L.B. A recursive moving-target-indication algorithm for optical image sequences. *IEEE Trans. Aerosp. Electron. Syst.* **1990**, *26*, 434–440. [CrossRef]
46. Hou, W.; Yu, Q.F.; Lei, Z.H.; Liu, X.C. A block-based improved recursive moving-target-indication algorithm. *Acta Phys. Sin.* **2014**, *63*, 13. [CrossRef]
47. Zongfu, H.; Jinzhen, W.; Zengping, C. Motion characteristics analysis of space target and stellar target in opto-electronic observation. *Opto-Electron. Eng.* **2012**, *39*, 67–72.
48. Ibarra-Castanedo, C.; González, D.; Klein, M.; Pilla, M.; Vallerand, S.; Maldague, X. Infrared image processing and data analysis. *Infrared Phys. Technol.* **2004**, *46*, 75–83. [CrossRef]
49. Hong, S.H.; Choi, G.B.; Baek, R.H.; Kang, H.S.; Jung, S.W.; Jeong, Y.H. Low-Temperature Performance of Nanoscale MOSFET for Deep-Space RF Applications. *IEEE Electron Device Lett.* **2008**, *29*, 775–777. [CrossRef]
50. Wright, E.L.; Eisenhardt, P.R.M.; Mainzer, A.K.; Ressler, M.E.; Cutri, R.M.; Jarrett, T.; Kirkpatrick, J.D.; Padgett, D.; McMillan, R.S.; Skrutskie, M.; et al. The Wide-field Infrared Survey Explorer (WISE): Mission Description and Initial On-orbit Performance. *Astron. J.* **2010**, *140*, 1868–1881. [CrossRef]
51. Zhang, G. *Star Identification*; Nation Defense Industry Press: Beijing, China, 2011.
52. Ristic, B.; Arulampalam, S.; Gordon, N. Detection and tracking of stealthy targets. In *Beyond the Kalman Filter Particle Filters for Tracking Applications*; Barton, D.K., Ed.; Artech House: Boston, MA, USA; London, UK, 2004; pp. 240–251.

remote sensing

MDPI

Article

Implementation of Real-Time Space Target Detection and Tracking Algorithm for Space-Based Surveillance

Yueqi Su [1,2,3], Xin Chen [1,2], Gaorui Liu [1,2], Chen Cang [1,2] and Peng Rao [1,2,*]

1 Shanghai Institute of Technical Physics, Chinese Academy of Sciences, Shanghai 200083, China; suyueqi@mail.sitp.ac.cn (Y.S.); chenxin@mail.sitp.ac.cn (X.C.); liugaorui@mail.sitp.ac.cn (G.L.); cangchen@mail.sitp.ac.cn (C.C.)
2 Key Laboratory of Intelligent Infrared Perception, Chinese Academy of Sciences, Shanghai 200083, China
3 University of Chinese Academy of Sciences, Beijing 100049, China
* Correspondence: peng_rao@mail.sitp.ac.cn

Abstract: Space-based target surveillance is important for aerospace safety. However, with the increasing complexity of the space environment, the stellar target and strong noise interference pose difficulties for space target detection. Simultaneously, it is hard to balance real-time processing with computational performance for the onboard processing platform owing to resource limitations. The heterogeneous multi-core architecture has corresponding processing capabilities, providing a hardware implementation platform with real-time and computational performance for space-based applications. This paper first developed a multi-stage joint detection and tracking model (MJDTM) for space targets in optical image sequences. This model combined an improved local contrast method and the Kalman filter to detect and track the potential targets and use differences in movement status to suppress the stellar targets. Then, a heterogeneous multi-core processing system based on a field-programmable gate array (FPGA) and digital signal processor (DSP) was established as the space-based image processing system. Finally, MJDTM was optimized and implemented on the above image processing system. The experiments conducted with simulated and actual image sequences examine the accuracy and efficiency of the MJDTM, which has a 95% detection probability while the false alarm rate is 10^{-4}. According to the experimental results, the algorithm hardware implementation can detect targets in an image with 1024×1024 pixels in just 22.064 ms, which satisfies the real-time requirements of space-based surveillance.

Keywords: space target; heterogeneous multi-core system; detection and tracking; MJDTM; FPGA; DSP

Citation: Su, Y.; Chen, X.; Liu, G.; Cang, C.; Rao, P. Implementation of Real-Time Space Target Detection and Tracking Algorithm for Space-Based Surveillance. *Remote Sens.* **2023**, *15*, 3156. https://doi.org/10.3390/rs15123156

Academic Editor: Paolo Tripicchio

Received: 10 May 2023
Revised: 3 June 2023
Accepted: 14 June 2023
Published: 16 June 2023

1. Introduction

The term space target refers to all outer space objects, including nonfunctional spacecraft, spent upper stages, and space debris [1]. With the development of human activities, the amount of space debris is multiplying. The collision between different space targets such as space debris and spacecraft may lead to equipment damage and mission failure and even produce more space debris, which poses a significant threat to aerospace safety. Therefore, space target detection and tracking are essential to avoid space collision and ensure the operation safety of on-orbit spacecraft. Space situational awareness technology is important to guarantee on-orbit safety, monitoring space targets, evaluating space events and providing space situational information for on-orbit spacecraft using the space-based or ground-based detection equipment. Its primary mission is to accurately detect and track space targets and calculate important characteristic parameters such as the size and shape of space targets that may pose a threat to the on-orbit spacecraft [2].

Compared to ground-based observation, a space-based photoelectric detection system has the advantages of high maturity, high precision, and low energy consumption, making it possible to realize all-weather space target detection and on-orbit spacecraft protection [3]. The subject of the detection and tracking of space targets using optical detection equipment consists of a set of problems that are central to the disciplines of space-based space target awareness. However, as space-based optical detection technology is upgraded, the detection field of view is gradually expanding, and space-based images are including increasingly complex information about the space environment. The existing methods have limited ability to suppress background noise in space-based images and are insufficient for space target perception, which generally concentrates on a single task such as detection or tracking. Therefore, developing a target detection algorithm with improved detection precision and a low rate of false alarms to separate space targets from the background is the critical problem of space target detection and tracking algorithms. Meanwhile, there are few space-based image processing system solutions presented by the current researchers, which have not been able to solve the problems of small space target detection and stellar target suppression, and the detection performance of the systems cannot keep up with the demand for the real-time processing of high-resolution space target images. The development of miniaturized, dedicated, high-speed processing systems for real-time space target detection with constrained on-orbit hardware resources remains a difficult task.

To achieve real-time space target surveillance, we proposed a high-precision detection and tracking architecture for space targets and developed a high-speed image processing platform to fulfill the algorithm implementation while maintaining real-time processing requirements. The main contributions of our paper are as follows. First, inspired by the human visual contrast mechanism, we improved the local feature contrast and energy concentration degree method to extract the potential small space targets in the optical image sequences. The local subtraction of the target detection algorithm suppresses background noise, and the accumulation of the target area boosts the target to achieve high-precision detection for space targets. Second, to eliminate false alarms of stellar targets with similar imaging features to the real space target, we proposed a stellar target suppression method that uses differences in motion relative to the Earth and real-time satellite attitude data to distinguish between the space and stellar targets. The algorithm is based on the historical coordinate data of the tracking trajectory and uses the platform parameters to determine the target type accurately. Finally, a comprehensive and lightweight space target perception architecture, called the multi-stage joint detection and tracking model (MJDTM), is given. It combines the space target detection method based on the LFC, the Kalman filter algorithm, and the proposed stellar target suppression method to accurately detect and track space targets. The architecture is implemented on a specialized heterogeneous multi-core processing platform based on FPGA and DSP. Additionally, the performance measures of the architecture and its implementation are evaluated using the simulated and real image sequences including computation time, resource usage, and detection capability.

The remainder of this paper is divided into the following sections. Section 2 summarizes related works on space target detection and tracking algorithms and existing hardware implementation schemes. In Section 3, the proposed multi-stage joint detection and tracking model is elaborated. In Section 4, we present the proposed hardware architecture and the algorithm implementation. Section 5 validates the performance and effectiveness of the proposed implementation. Section 6 gives the discussion of this paper. The conclusions are provided in Section 7.

2. Related Work

Various algorithms and relevant hardware implementations have been exploited for faint and tiny moving space target detection and tracking in space-based optical image sequences. Track-before-detect (TBD) and detect-before-track (DBT) are the dominating solutions to the difficulty of moving target detection and tracking. A dynamic programming approach was developed by BARNIV [4,5] that utilizes the velocity and shape information to detect linear moving objects with a low signal to noise ratio (SNR). The particle filter method [6] is a nonlinear dynamic filter based on the Monte Carlo method. A TBD algorithm has been realized using a Bayesian particle filter to approximate the posterior probability distribution of the target state [7]. Reed et al. [8] established a dim and small target detection method based on three-dimensional matching filtering that matched and filtered the feature information of moving targets in the Fourier domain. These three methods described above could be defined as the TBD method. In actual scenarios, since the energy distribution and pattern of stellar and space targets are similar, it is challenging for the TBD approach to distinguish between them. Moreover, the variety of the target motion state will enhance the computational burden of the algorithm, making it hard for the TBD method to satisfy real-time application requirements. Accordingly, the DBT method is more suitable for space target surveillance in the space-based scenario.

The detection stage of the DBT method needs to extract the possible target and acquire the target region. The star map registration algorithm [9–12] is a common method for space target detection. In contrast, the satellite platform attitude variation increases the image background uncertainty and complexity, which makes the star map registration method unsuitable for space-based scenarios. Some threshold segmentation methods based on target enhancement have been studied, including the wavelet filtering method [13,14], local contrast method [15,16], and morphology filtering. Boccignone et al. [13] presented a small target detection method using wavelets. Jiang et al. [14] improved this method and developed an automatic space debris extraction algorithm. It utilized wavelet transform and variational hybrid filtering algorithms to suppress noise and detected candidate debris targets using the Hough transform. Mathematical morphology-based algorithms usually use image filters to eliminate background noise and enhance small targets, such as median filters [17], max-mean and max-median filters [18], and top-hat [19]. The local contrast method [15] is a powerful small target detection algorithm that was inspired by the human visual system contrast mechanism. It can enhance the target by calculating the local contrast map of the infrared image. Chen et al. [16] combined the local contrast method with energy concentration degree and proposed an infrared dim and small target detection algorithm. Lv et al. [20] developed a novel algorithm called neighborhood saliency map (NSM) based on the contrast mechanism of the human visual system. Han et al. [21] improved the local contrast method (LCM) and designed a detection architecture named multiscale tri-layer local contrast measure (TLLCM). The image filter algorithm based on the LCM method, which has been applied to the problem of infrared (IR) small target detection, could effectively boost the dim target and increase the detection accuracy. In recent years, researchers have also proposed deep learning-based solutions to the problem of dim and small target detection [22]. However, the network structure of these algorithms is frequently complex, and they frequently require a large quantity of experimental data to learn, making their implementation and application challenging.

Once the target has been extracted, it requires a tracker to predict the target motion state and update the trajectories in the subsequent frames. Fan Shi et al. [23] tracked a moving target using a primary scale invariant feature transform (P-SIFT) keypoint matching algorithm. In this way, the deviation of feature extraction will also have an impact on tracking. K. Fujita et al. [24] described a computer vision technique called an optical flow algorithm to detect and track GEO debris. However, its computational complexity makes it challenging to meet the requirements of real-time applications. The Kalman filter [25] is a classic target tracking algorithm used in dynamic procedures where the measured process is linear and Gaussian. Scala and Bitmeand [26] proposed the extended Kalman filter for solving the tracking problem when both the dynamic and measurement processes are nonlinear. Tao et al. [27] presented a space target surveillance algorithm that contains a variance detector and uses a Markov-based dynamic model to forecast the potential target position.

In addition to noise interference, hundreds of thousands of stars are the primary interference sources for space target detection in space-based detection scenarios. The image difference method [28] directly differentiates adjacent image frames, but when the platform moves, the imaging position of the stars changes, which will cause a false alarm for detection. The star mask frame method [29] employs multi-frame image accumulation to calculate the position of a star and generates a star mask frame to filter out the stars in the image. However, detection fails when the target is near the imaging distance of the star. The star image recognition method [30] matches the image with the star map to extract the matching star point, but it is hard to implement it in hardware due to the heavy calculation burden. In this paper, a stellar target suppression method that uses differences in motion and real-time satellite attitude data is provided to distinguish between the space and stellar targets.

The target only takes up a small portion of the image pixels due to the large separation between space targets and detector, and the contrast between targets and background may not be strong enough for the detection method to utilize the texture feature data effectively. The space target detection method mentioned above can only solve the problem of target detection in some specific scenes, and it is challenging to overcome the problem of background stellar false alarms and strong noise in space-based scenes. In addition, these works have not been implemented by the hardware platform, and its real-time processing capability needs to be evaluated.

Moreover, the research community has designed some image processing systems using the limited hardware system resources that could implement the related target detection and tracking algorithms in the space-based scenario. A high-performance embedded processing platform based on a graphics processing unit (GPU), DSP, and FPGA has become the potential solution for onboard image processing [31–34]. As the specialized image processor, the embedded GPU processing platform [35] has been widely used in unmanned driving technology, AI computation, and video image processing. Its parallel processing capability supports it in handling complex data and geometry computing [36]. However, the disadvantages of poor independence and high power consumption hinder the application of GPUs in onboard applications. In the meantime, DSP and ARM processors with computing capability, flexibility, and large-scale integration have been adopted to implement the vision and image processing algorithms [37]. Sun et al. [38] described an onboard space debris detection approach on a multi-core DSP platform that can process a 2048×2048 image in 600 ms. The parallelism of this system constrains the throughput of processing data streams, making it challenging to process intensive computing with large data volumes. Over other embedded systems, the use of FPGAs in high-speed parallel data processing has become more prevalent due to their parallel processing capability. The FPGA platform is suitable for onboard image processing because of its flexibility, reconfigurability, and high energy efficiency [39]. Han et al. [40] proposed a high-speed tracking and measurement method for non-cooperative space targets and applied it to an FPGA-based space-embedded system. However, their scheme does not consider the situation of small space targets and establishes an overly ideal stellar interference model that may malfunction in practical space-based scenarios. Yang et al. [41] implemented the ATGP algorithm on FPGAs to achieve real-time target and anomaly detection in hyperspectral image sequences. Nevertheless, it is not feasible for FPGAs to implement high-precision data operations, and their programming development is complex. A heterogeneous processing platform based on FPGA and DSP is one of the most commonly used embedded image processing systems, and has been relatively maturely applied to the field of space-based image processing [42,43]. The high-speed parallel processing capability of FPGA has considerable advantages in large-scale image data processing. The DSP processor has the characteristics of large-scale integration and stability, which can realize high-precision digital signal processing.

For space-based surveillance, an integrated processing system with high flexibility, powerful processing performance, and low power consumption can quickly complete image processing. The current research in space target detection and tracking and other image processing performance needs to be improved, as it has failed to give a high-performance space target sensing method and its hardware implementation. However, a high-performance space target perception method and its hardware implementation have not been provided by the present research.

In this paper, we provide a complete space target perception architecture that realizes the accurate detection and tracking of small space targets. A space-based image processing system platform based on FPGA + DSP has been constructed to implement this architecture.

3. Methodology

A flow diagram of the MJDTM architecture is outlined in Figure 1. With the detection range extension of the space-based optical detector, there are more stellar targets and noise points in images, making it challenging to accurately identify space targets only occupying one to several pixels in the image plane. To ensure target detection accuracy and reduce the false alarm rate, the interference of the stellar targets and background noise points needs to be suppressed. The proposed space target perception architecture contains three main parts: space target detection and tracking, stellar target suppression, and target feature calculation. As shown in Figure 1, we first adopt an improved local contrast method to extract the potential space point target during the target detection and tracking stage. Then, the classical Kalman filter algorithm and the Hungarian matching algorithm are combined to predict the target state and correlate tracking trajectories. The sidereal targets with similar imaging properties to space targets are suppressed during the stellar target suppression stage. A schematic diagram of the image sequence after target detection tracking and stellar suppression is given in Figure 1. After that, the feature information of the space target that is confirmed as the real target is calculated. Details are as follows.

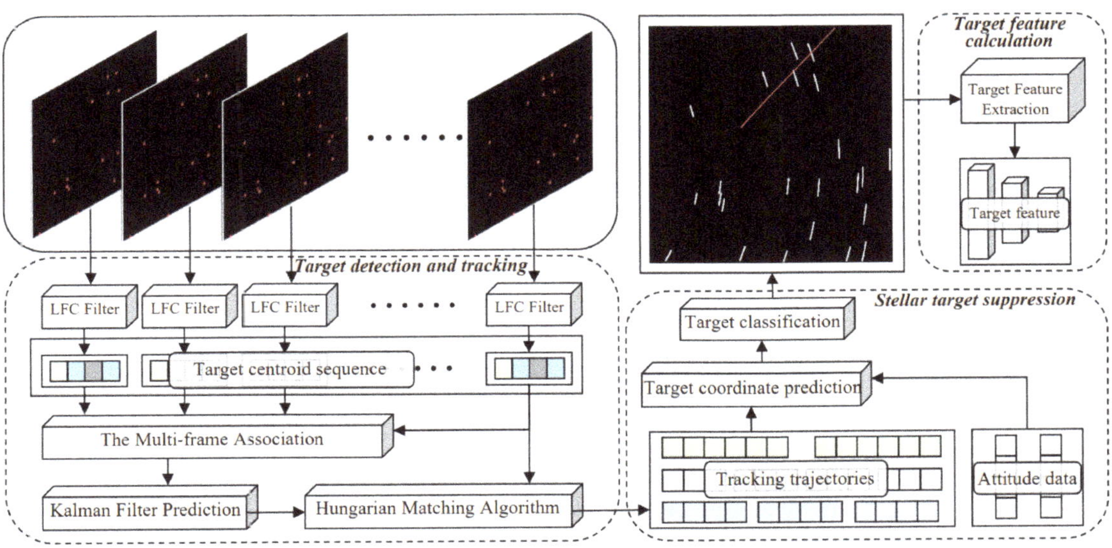

Figure 1. Workflow of the proposed architecture.

3.1. Target Detection and Tracking

3.1.1. Target Detection Algorithm

The space-based optical image of space can be modeled as follows:

$$F(i,j) = T(i,j) + S(i,j) + B(i,j) + N(i,j) \tag{1}$$

where (i,j) represents the pixel coordinates of the image and $F(i,j)$ denotes the grayscale value of the pixel coordinate (i,j) in the image. $T(i,j)$ and $S(i,j)$, respectively, denote the space target and the stellar targets, which obey the Gaussian distribution model. $B(i,j)$ represents the background of the deep space environment and $N(i,j)$ denotes the noise generated by the internal noise of the imaging system and external environment interference.

It can be seen from the above model that most of the image information obtained by the space-based space target detection equipment is from the deep space background environment. The space targets and stellar targets only account for a small part of the image, and various noise disturbances are randomly distributed throughout the image. Due to the limitation of the detection distance and the short exposure time, the space targets and stellar targets occupy only one pixel in the image and the energy of the target is weak. In order to accurately detect real space targets, the detection algorithm should enhance the target region for better target segmentation and extraction, and reduce the independent noise points on the image to lower the false alarm rate. All possible targets in the image must be segmented during the target detection phase to avoid missing actual space targets since the space target and stellar target have very similar imaging characteristics.

In this paper, the target detection algorithm uses the target energy feature and the local standard deviation feature to establish the LFC model and employs this model to realize the image filtering and the detection of the space and stellar targets. The local contrast method [15] is an image-filtering method based on the contrast mechanism of the human visual system and is commonly used to solve IR dim target detection problems. Similar to IR faint targets, space targets in space-based optical image sequences have weak energy and occupy only a few pixels without shape and texture features. Therefore, the detection rate could be guaranteed by using this technique to locate space targets in the deep space background. Chen et al. [15] proposed a local feature contrast and energy concentration degree method (LFC-ECD) that combines the LCM algorithm with the energy concentration algorithm to detect small infrared targets. This algorithm suppresses the neighboring regions of the target through local subtraction and performs energy accumulation to enhance the faint target. In the target detection stage, we remove the energy accumulation progress of the LFC-ECD and use the local feature contrast filter to extract space targets. The specific steps are as follows.

By sliding the local window on the whole image, the local feature contrast value of each point in the image is calculated. Firstly, the image slice larger than the target region is selected as the local window. Additionally, the slice of coordinate (x, y) in the image is divided into the target region S_0 and the neighboring region S, which is shown in Figure 2.

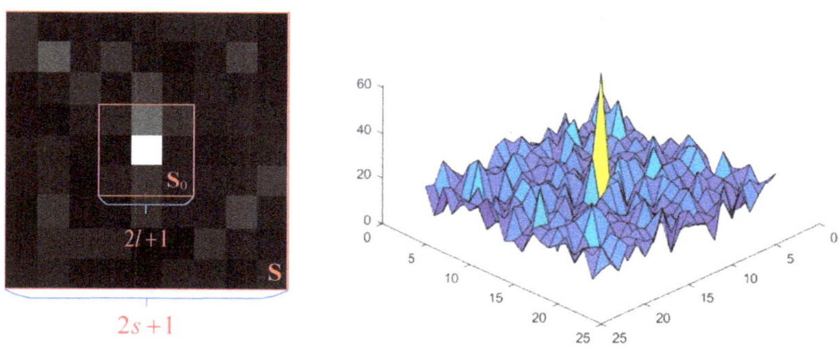

Figure 2. The target and its neighboring region.

The region of S_0 and S are represented as follows:

$$R_S = \{(p,q)|\max(|p-x|,|q-y| \leq s)\},\ s = 4,7,10,13 \quad (2)$$

$$R_{S_0} = \{(i,j)|\max(|i-x|,|j-y| \leq l)\},\ l = 1,2,3,4 \quad (3)$$

where s and l represent the radius of S_0 and S, respectively, and (i,j) and (p,q) denote the pixel coordinates of S_0 and S in the image.

The pixel grayscale average value and standard deviation in the region S are computed to represent the background and noise of the target region, and the formula is denoted as follows:

$$G_{\mathrm{m}}(x,y) = \sum\nolimits_{(p,q)\in R_S} \frac{G(p,q)}{(2s+1)^2} \tag{4}$$

$$G_{\mathrm{s}}(x,y) = \sqrt{\frac{\sum_{(p,q)\in R_s} [G(p,q) - G_{\mathrm{m}}(x,y)]^2}{(2s+1)^2}} \tag{5}$$

where $G(p,q)$ is the pixel grayscale at (p,q) in the region S and $G_{\mathrm{m}}(x,y)$ and $G_{\mathrm{s}}(x,y)$ are the pixel grayscale average value and standard deviation at (x,y) in the region S, respectively.

Then, the background should be inhibited by the regional background subtraction because of the solid local continuity, and the formula is represented as follows:

$$G_{\mathrm{t}}(x,y) = G(x,y) - G_{\mathrm{m}}(x,y), (i,j) \in \mathrm{S}_0 \tag{6}$$

where $G(x,y)$ denotes the grayscale of the pixel at (i,j) in the region S_0 and $G_{\mathrm{t}}(x,y)$ represents the grayscale of the pixel at (i,j) in the region S_0 after the background suppression.

When the target is weak, the pixel grayscale of the region S_0 will be low after the background subtraction. Therefore, the target component should be magnified by the energy accumulation to ensure the target is detected correctly. The formula of energy accumulation is denoted as follows:

$$E_{\mathrm{t}}(x,y) = \sum\nolimits_{(i,j)\in R_{S_0}} G_{\mathrm{t}}^2(i,j) \tag{7}$$

where $E_{\mathrm{t}}(x,y)$ denotes the energy accumulation value of the pixel grayscale in the region S_0 at (x,y). The sum operation helps in the rapid enhancement of targets.

Finally, the local feature contrast value of the coordinates (x,y) in the image is provided in the following formula:

$$G_{\mathrm{c}}(x,y) = E_{\mathrm{t}}(x,y)/G_{\mathrm{s}}(x,y) \tag{8}$$

$$G_{\mathrm{L}}(x,y) = G_{\mathrm{c}}(x,y) \times G_{\mathrm{t}}(x,y) \tag{9}$$

where $G_{\mathrm{c}}(x,y)$ represents the contrast factor and $G_{\mathrm{L}}(x,y)$ denotes the values of local feature contrast at the coordinates (x,y).

When obtaining the local feature contrast, the adaptive threshold segmentation will be conducted on the local feature contrast image to segment the target. The adaptive threshold T_1 is denoted as follows:

$$T_1 = m_{\mathrm{L}} + k_1 \times std_{\mathrm{L}} \tag{10}$$

where m_{L} and std_{L} represent the average value and standard deviation of local features' contrast $G_{\mathrm{L}}(x,y)$. The range of the parameter k_1 confirms that a range of 30 to 40 is efficacious. In Section 5.1, the selection of the parameter k_1 will be discussed in depth.

Then, the binary image of the LFC result is segmented by the threshold T_1. The formula is represented as follows:

$$b_1(i,j) = \begin{cases} 1, G_{\mathrm{L}}(i,j) > T_1 \\ 0, G_{\mathrm{L}}(i,j) \le T_1 \end{cases} \tag{11}$$

where $b_1(i,j)$ denotes the grayscale at the coordinates (i,j) in the segmented image and $G_{\mathrm{L}}(i,j)$ is the local feature contrast value at the coordinates (i,j) in the image.

In the ideal optical system, only one pixel of the detector is occupied by the point target. However, the targets will diffuse into several pixels due to circular aperture diffraction in practical situations. The precise pixel coordinates of the target are confirmed by the center location of the gray pixel. Since the target is susceptible to the effects of ambient background noise, the precision of target positioning might be impacted by the classic centroid method. The centroid coordinates of targets are calculated using the distance-weighted centroid method that enhances the conventional centroid with distance-weighting. Based on the conventional centroid approach, the distance-weighted centroid method adds the grayscale and distance influence factor as the weight to lessen the impact of target edge noise on target location extraction. The steps of this method are described below.

Firstly, the maximum grayscale pixel coordinate data in the target region are provided by the target detection stage result. The target region S_t is separated by extending m pixels outward from the center of the maximum pixel gray value coordinates. The size of the target area is $n = 2 \times m + 1$. Moreover, the distance $D(i,j)$ between the maximum grayscale pixel and each pixel in the target area is defined by the following formula:

$$D(i,j) = \sqrt{(i-x)^2 + (j-y)^2}(i,j \in S_t) \quad (12)$$

where (i,j) are the pixel coordinate data in the target region and (x,y) represent the maximum pixel grayscale value coordinates.

The formula of the distance weight $D'(i,j)$ is defined as:

$$D'(i,j) = 1/D(i,j) \quad (13)$$

where $D'(i,j)$ is the distance weight at the coordinates (i,j) in the target area. The distance weight at the maximum grayscale pixel is $a(3 \leq a \leq 5)$.

The following formula computes the distance-weighted centroid coordinates of the target:

$$X = \frac{\sum_{(i,j) \in R_{S_t}} G(i,j)D'(i,j)i}{\sum_{(i,j) \in R_{S_t}} G(i,j)D'(i,j)} \quad (14)$$

$$Y = \frac{\sum_{(i,j) \in R_{S_t}} G(i,j)D'(i,j)j}{\sum_{(i,j) \in R_{S_t}} G(i,j)D'(i,j)} \quad (15)$$

where the $G(i,j)$ is the grayscale value at the coordinates (i,j) in the target area and (X,Y) is the target centroid coordinates calculated through the distance-weighted centroid method.

3.1.2. Target Tracking Algorithm

The target coordinate sequence of each frame can be obtained after the target detection for the optical image sequence. We establish tracking trajectories for each potential target during the tracking stage, estimate the candidate target motion state, and predict the target position using the Kalman filter to track space targets steadily. Moreover, the Hungarian matching algorithm is adopted to correlate the tracking trajectories with the target sequence to update the target coordinate positions. The specific operation flow is shown in Figure 3.

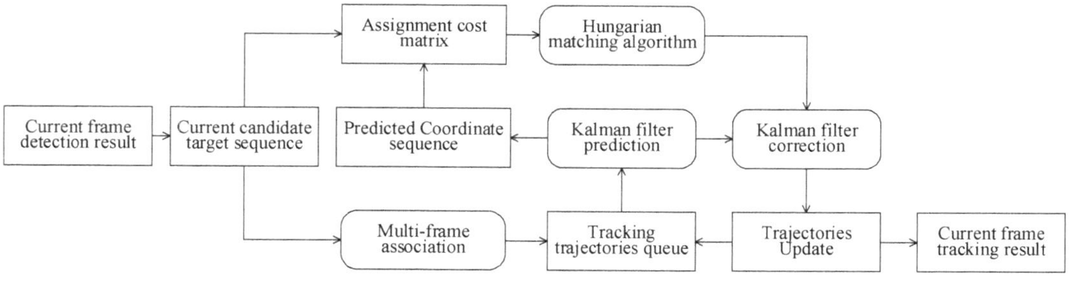

Figure 3. Procedure of the tracking stage.

Affected by the gravity of the Earth, both the space targets and the observation satellite platforms will run in a specific orbit. Thus, we can use a linear uniform model to simulate the motion of the space target relative to the platform, which is unrelated to other targets and camera motion. It can be assumed that the running path of the space target in the continuous space image sequence is connected and that the target detection results from earlier frames can be used to estimate the target motion model and forecast the target position.

Before the target motion state prediction and tracking trajectory association, the multi-frame association operation as shown in Figure 3 will confirm the current candidate target queue, which reduces the calculation amount of the subsequent tracking process while suppressing noise points. The target that satisfies the trajectory creation condition initializes the corresponding tracking trajectory after this operation. The specific processing steps are as follows.

First, based on the detection result of the first frame image, a suspicious target queue TS_s will be created for each target in the current candidate target queue TS_c.

Then, the Euclidean distance between each target in the suspicious target queue TS_s and the current candidate's target queue TS_c will be calculated after the detection of the subsequent image frames. The specific formula is as follows:

$$TD(n,m) = \sqrt{(i_n - x_m)^2 + (j_n - y_m)^2}, n = 1 \ldots N, m = 1 \ldots M. \quad (16)$$

where $TD(n,m)$ is the Euclidean distance between the target $TS_S(n)$ in the suspicious target queue TS_S and the target $TS_C(m)$ in the current candidate target queue TS_c, (i_n, j_n) is the coordinate data of the target $TS_S(n)$, (x_m, y_m) is the coordinate data of the target $TS_C(m)$, and N and M are the numbers of targets in the queue TS_S and TS_c, respectively.

For a target in the suspicious target queue TS_S, if there is a target within the predetermined distance threshold range in the candidate target queue TS_c, the number of target occurrences is determined to increase. Then, the target coordinate data and the number of target occurrences in the suspicious target queue TS_S will be updated. If the current candidate target queue does not contain a target that fulfills the criteria, the number of target disappearances will be updated.

After the multi-frame suspicious target queue update, there will be a target in the suspicious target queue TS_S that appears more than the set threshold. It can be assumed that this is a real target rather than an independent noise point. For the real target, we adopt the Kalman filter to estimate the motion state and predict the coordinate position to achieve stable tracking of the target. The Kalman filter [25] is an optimal estimation algorithm for system state that uses the linear system state equation and system input and output observation data. It has been widely applied in the fields of orbit calculation [44], target tracking, and navigation [45], such as calculations of spacecraft orbit, tracking of maneuvering targets, and positioning of GPS. The specific calculation steps of the Kalman filter are as follows.

The tracking trajectory of this target will be initialized for subsequent target tracking and trajectory update. The invalid targets that disappear more than the set threshold in the suspicious queue will be cleared. When a trajectory is created in the target trajectory queue, the motion state estimation and target prediction of the target will be performed in subsequent frames as shown in Figure 3. The state of the target is defined according to the following model:

$$x_k = [u, v, p, q]^T \quad (17)$$

where u and v represent the horizontal and vertical coordinates of the target centroid and p and q represent the velocity component of the coordinate. The Kalman filter algorithm predicts the target position in subsequent frames according to the target state and updates the target state according to the measured value associated with the current target queue using the Hungarian matching algorithm. If a target has no correlation matching, its state is simply predicted using the linear velocity model without any correction.

One of the most well-known Bayesian filter theories is the Kalman filter, a linear optimal status estimate technique [46]. The estimation process of the Kalman filter consists of the previous prediction step and the current measurement step. It includes two types of equations: status equation and observation equation. A dynamic model with the status and observation equations is given using a precise estimation that has been measured and altered. The status equation of the Kalman filter is represented as follows [47]:

$$x_k = Ax_{k-1} + Bu_k + w_k \tag{18}$$

where A is the status transition matrix, B is the control–input matrix, x_k is the status vector, u_k is the system control matrix, and w_k is the system noise vector.

The Kalman filter observation equation is defined as follows:

$$z_k = Hx_k + v_k \tag{19}$$

where H is the observation matrix, z_k is the observation vector, and v_k is the observation noise vector. w_k and v_k are assumed to be zero-mean Gaussian white noise with covariance Q and R, respectively, denoted as:

$$w \sim N(0, Q) \tag{20}$$

$$v \sim N(0, R) \tag{21}$$

When a discrete control process system satisfies the above conditions, the Kalman filter algorithm can be used to predict the system state.

The calculation procedure of the algorithm is as follows.

Firstly, the prediction equation is used to predict the next state of the system. The prediction equation is defined as follows [47]:

$$\hat{x}_k^- = A\hat{x}_{k-1} + Bu_k \tag{22}$$

where $\hat{x}_{k-1}$ represents the posterior status estimation combined with the measurements at the moment of $k-1$ and $\hat{x}_k^-$ denotes the prior status estimation derived from the status transition equation at the moment of k.

Then, the error covariance is calculated using the update equation. The update equation is as follows:

$$P_k^- = AP_{k-1}A^T + Q \tag{23}$$

where P_k^- is the prior estimation deviation covariance of the status $\hat{x}_k^-$, P_{k-1} is the posterior estimation deviation covariance of the status $\hat{x}_{k-1}$, and Q is the deviation covariance of the system noise vector.

The trajectory correlation operation follows the aforementioned prediction stage. As shown in Figure 3, the Hungarian matching algorithm is used to correlate the predicted coordinate sequence and the current candidate target sequence. The Hungarian matching algorithm [48] was proposed by two Hungarian mathematicians and is mainly used to solve some problems related to bipartite graph matching, such as data association [49], UAV task assignment [50], and multi-target tracking [51]. The core of the algorithm is to use the augmented path to find the maximum matching algorithm of the bipartite graph. The predicted coordinate sequence of the tracking trajectory and the current candidate target sequence form a bipartite graph that can be easily represented by a distance matrix.

Specifically, the Euclidean distance between the target prediction coordinates of the tracking trajectory sequence and the target coordinates in the current candidate target sequence is calculated and integrated into a distance matrix ED, as shown in Figure 4a. The formula is as follows:

$$ED = \begin{pmatrix} Ed_{1,1} & Ed_{1,2} & \cdots & Ed_{1,n} \\ Ed_{2,1} & Ed_{2,2} & \cdots & Ed_{2,n} \\ \vdots & \vdots & \ddots & \vdots \\ Ed_{m,1} & Ed_{m,2} & \cdots & Ed_{m,n} \end{pmatrix} \tag{24}$$

$$Ed_{m,n} = \sqrt{(i_n - x_m)^2 + (j_n - y_m)^2},\ n = 1 \ldots P,\ m = 1 \ldots C. \tag{25}$$

where $Ed_{m,n}$ is the Euclidean distance between the target $PC(n)$ in the predicted target coordinate sequence PC and the target $CT(m)$ in the current candidate target sequence CT, (i_n, j_n) is the coordinate data of the target $PC(n)$, (x_m, y_m) is the coordinate data of the target $CT(m)$, and P and C are the numbers of targets in the sequence PC and CT, respectively.

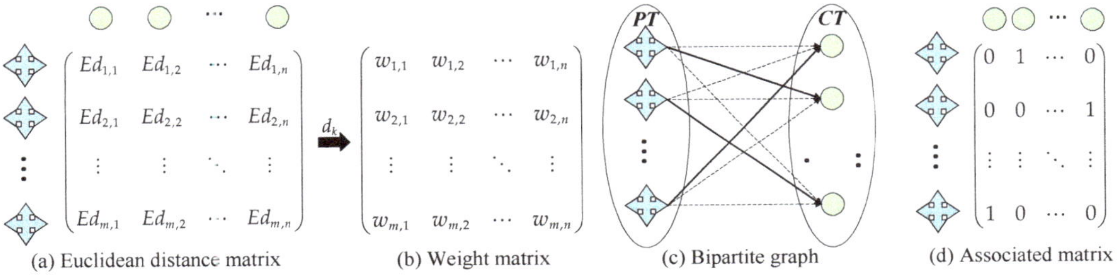

Figure 4. Diagram of the Hungarian matching algorithm.

Then, as shown in Figure 4b, the distance matrix will be transformed into a registration weight matrix WM according to the following formula:

$$WM = \begin{pmatrix} w_{1,1} & w_{1,2} & \cdots & w_{1,n} \\ w_{2,1} & w_{2,2} & \cdots & w_{2,n} \\ \vdots & \vdots & \ddots & \vdots \\ w_{m,1} & w_{m,2} & \cdots & w_{m,n} \end{pmatrix},\ w_{m,n} = \begin{Bmatrix} dk - Ed_{n,m}, & Ed_{n,m} \le dk \\ 0 & , Ed_{n,m} > dk \end{Bmatrix} \tag{26}$$

where dk is the distance threshold parameter of the weight matrix. When the distance between the coordinates is close, we hope that the corresponding correlation weight is large, so the weight matrix should be inversely proportional to the distance matrix. At the same time, we introduce the parameter dk to limit the correlation of the target coordinates far away. If the distance between the targets is greater than dk, it is considered that the possibility of a large difference between the two coordinates is low, and the corresponding weight is set to zero directly to simplify the correlation calculation. After this operation, the weight of the coordinates with smaller distances becomes larger, and the possibility of registration association is also enhanced.

Through iterative optimization, the Hungarian matching algorithm generates a maximum weight distribution matrix, as shown in Figure 4c,d, which represents the correspondence between the target in the tracking trajectory and the latest subsequent target sequence. Each tracking trajectory is assigned to a current candidate target so that the posterior state can be calculated and updated according to the associated measurement. The Euclidean distance between the target predicted coordinate of the tracking trajectory sequence and the target coordinate in the current candidate target sequence is calculated and sorted into the correlation cost matrix. The assignment problem between the tracking trajectory and the current candidate target sequence is solved optimally using the Hungarian algorithm, which can provide the best matching for the two sequences.

For the tracking trajectory with correlation detection, the correction stage combines its predicted state with the measured value to obtain the best estimation x_k. The formula is represented as follows:

$$x_k = \hat{x}_k^- + K_k(z_k - H\hat{x}_k^-) \tag{27}$$

where x_k is the optimal estimation at the moment of k and K_k is the Kalman gain matrix The formula of K_k is denoted as follows:

$$K_k = P_k^- H^T (HP_k^- H^T + R)^{-1} \tag{28}$$

where R is the deviation covariance of the observation noise vector.

The posterior estimation deviation covariance of the state x_k is calculated using the following formula to keep the Kalman filter running until the processing system is finished.

$$P_k = (I - K_k H)P_k^- \tag{29}$$

where P_k is the filter deviation matrix and I is the unit matrix.

The target parameters in the target tracking trajectory queue are updated after the tracking association operation in each frame, including the coordinate data of the target and the number of occurrences. Similar to the suspicious target queue, when the number of target disappearances in the tracking trajectory exceeds the predetermined threshold, the target is removed from the tracking trajectory. This procedure stops the unrestricted expansion of the tracker population and positioning inaccuracies brought on by excessively extended forecast durations without detector correction.

3.2. Stellar Target Suppression Algorithm

As the target detection and tracking stage run alternately, the number of target historical coordinates in the tracking trajectory queue increases cumulatively. Both the space target and the stellar target are included in the trajectory queue. This section proposes a method for classifying the stellar targets and space targets using real-time satellite attitude data and the historical coordinate data of track trajectories. We postulate that due to the remote distance between the sidereal target and the Earth, the positions of the stars relative to the Earth remain unchanged for a short time. In contrast, the coordinate of the space targets relative to the Earth will change during this time because the moving space target has a certain velocity and is closer to the Earth. Therefore, we exploit the different motion states of stellar and space targets relative to the Earth to suppress the stellar targets. The specific methods are as follows.

For the candidate trajectory formed at the time t, this stage performs the subsequent operations on the latest target coordinates of each trajectory. The latest target point coordinates of a trajectory need to be transferred to the camera coordinate using the camera's intrinsic matrix. Since the plane image can only provide two-dimensional coordinate data, it is difficult to acquire the distance data of the target. During the process of coordinate transformation, we assume the Z-axis data of the target point to be 1.

$$\begin{bmatrix} x_c(t) \\ y_c(t) \\ z_c(t) \end{bmatrix} = R_{Int}^{-1} \begin{bmatrix} x(t) \\ y(t) \\ 1 \end{bmatrix}, \; R_{Int} = \begin{bmatrix} f/dx & 0 & x_0 \\ 0 & f/dy & y_0 \\ 0 & 0 & 1 \end{bmatrix} \tag{30}$$

where R_{Int} is the intrinsic matrix of the camera, R_{Int}^{-1} is the inverse of the intrinsic matrix, $[x_c(t), y_c(t), z_c(t)]$ is the camera coordinate of the target, $[x(t), y(t)]$ is the pixel coordinate of the target at the time n, (dx, dy) is the focus of the camera on the X and Y axis, and (x_0, y_0) is the center pixel coordinate value of the camera.

The installation matrix calculates the target coordinate relative to the platform. The formula is as follows:

$$\begin{bmatrix} x_s(t) \\ y_s(t) \\ z_s(t) \\ 1 \end{bmatrix} = R_{Ins} \begin{bmatrix} x_c(t) \\ y_c(t) \\ z_c(t) \\ 1 \end{bmatrix}, \ R_{Ins} = \begin{bmatrix} R & t \\ 0 & 1 \end{bmatrix} = \begin{bmatrix} n_x & o_x & a_x & t_x \\ n_y & o_y & a_y & t_y \\ n_z & o_z & a_z & t_z \\ 0 & 0 & 0 & 1 \end{bmatrix} \quad (31)$$

where R_{Ins} is the camera's installation matrix and $[x_s(t), y_s(t), z_s(t)]$ denotes the target coordinates relative to the platform.

The target coordinates relative to the Earth are calculated using the satellite attitude matrix at the time n. The formula is as follows:

$$\begin{bmatrix} x_e \\ y_e \\ z_e \end{bmatrix} = R_{Rot}(t) \begin{bmatrix} x_s(t) \\ y_s(t) \\ z_s(t) \end{bmatrix} \quad (32)$$

where $[x_e(t), y_e(t), z_e(t)]$ represents the target coordinates relative to the Earth t and $R_{Rot}(t)$ represents the satellite platform attitude matrix at the time n. This formula converts the target coordinate relative to the camera to the platform coordinate system using the camera external parameter matrix and the fourth-dimensional coordinate data are added to facilitate the calculation. The formula for the attitude matrix R_{Rot} is as follows [52]:

$$R_{Rot} = \begin{bmatrix} q_0^2 + q_1^2 - q_2^2 - q_3^2 & 2(q_1q_2 - q_0q_3) & 2(q_1q_3 + q_0q_2) \\ 2(q_1q_2 + q_0q_3) & q_0^2 - q_1^2 + q_2^2 - q_3^2 & 2(q_2q_3 - q_0q_1) \\ 2(q_1q_3 - q_0q_2) & 2(q_2q_3 + q_0q_1) & q_0^2 - q_1^2 - q_2^2 + q_3^2 \end{bmatrix} \quad (33)$$

where (q_0, q_1, q_2, q_3) denotes the attitude quaternion matrix of the satellite.

The target coordinates relative to the Earth can be obtained by transforming the target coordinates. The target coordinate data relative to the platform at the time $t + 1$ is predicted by the satellite attitude matrix at the time $t + 1$. The formula is as follows:

$$\begin{bmatrix} x_s(t+1) \\ y_s(t+1) \\ z_s(t+1) \end{bmatrix} = R_{Rot}^{-1}(t+1) \begin{bmatrix} x_e \\ y_e \\ z_e \end{bmatrix} \quad (34)$$

where $[x_s(t+1), y_s(t+1), z_s(t+1)]$ denotes the target coordinates relative to the platform at the time $t + 1$, $[x_e, y_e, z_e]$ represents the coordinates of the target relative to the Earth, and $R_{Rot}^{-1}(t+1)$ represents the satellite attitude inverse matrix at the time $t + 1$.

The camera coordinates are predicted by the coordinates of the target relative to the platform and the installation matrix. The formula is as follows:

$$\begin{bmatrix} x_c(t+1) \\ y_c(t+1) \\ z_c(t+1) \end{bmatrix} = R_{Ins}^{-1} \begin{bmatrix} x_s(t+1) \\ y_s(t+1) \\ z_s(t+1) \end{bmatrix} \quad (35)$$

where $[x_c(t+1), y_c(t+1), z_c(t+1)]$ represents the target camera coordinate data at the time $t + 1$ and R_{Ins}^{-1} denotes the inverse of the installation matrix. The target coordinates $[x(t+1), y(t+1)]$ at the time $t + 1$ are predicted by the target camera coordinates at the time $t + 1$ and the camera internal reference matrix. The formula is as follows:

$$\begin{bmatrix} x(t+1) \\ y(t+1) \\ 1 \end{bmatrix} = R_{Int} \begin{bmatrix} x_c(t+1) \\ y_c(t+1) \\ z_c(t+1) \end{bmatrix} \quad (36)$$

where $[x(t+1), y(t+1)]$ denotes the target pixel coordinates at the time $t + 1$.

After a sequence of coordinate transformations, we can obtain the predicted target coordinates of the tracked trajectory at the time $t+1$. The actual target coordinate data in the track trajectories at the time $t+1$ are provided after the tracking association operation in the above section. The difference between the target predicted and practical coordinates can be used as a criterion to determine whether the target is a real space target or not. When the difference exceeds a set threshold, the target is judged to be a real space target; otherwise, it is a stellar target. The threshold to confirm the target type is determined based on prior experience, which is selected as 1 in this paper.

3.3. Target Angle Calculation

The stellar target suppression module described in the above section classifies the space targets and stellar targets in the tracking trajectory queue. The calculation method of space target angle information will be introduced in this section. The position of the target with respect to the optical axis of the camera determines the azimuth and pitch angle of the target, and the precise calculation procedures are as follows.

Firstly, the intrinsic matrix of the camera is used to convert the target coordinate data from the image pixel coordinate system to the image physical coordinate system, as defined in the following formula.

$$\begin{bmatrix} x \\ y \\ 1 \end{bmatrix} = R_{Int} \begin{bmatrix} u \\ v \\ 1 \end{bmatrix} = \begin{bmatrix} 1/dx & 0 & u_0 \\ 0 & 1/dy & v_0 \\ 0 & 0 & 1 \end{bmatrix} \quad (37)$$

where (u, v) denotes the coordinate value of the target in the image pixel coordinate system, (x, y) indicates the coordinate value of the target in the image physical coordinate system, and R_{Int} is the intrinsic matrix of the camera.

Due to the lens distortion of the optical system, it is necessary to correct the target coordinate to ensure the accuracy of the target angle calculation. The radial and tangential distortion [53] are the major factors affecting the imaging quality of the wide field view optical system. The radial distortion can be fitted by quadratic and higher-order polynomial functions linked to the separation between target point coordinates and the image center pixel coordinates, as shown in the following formula [54]:

$$\begin{cases} x_d = x(1 + k_1 r^2 + k_2 r^4 + k_3 r^6) \\ y_d = y(1 + k_1 r^2 + k_2 r^4 + k_3 r^6) \end{cases}, \ r^2 = x^2 + y^2 \quad (38)$$

where (x_d, y_d) indicates the point coordinates after the radial distortion, (x, y) represents the coordinate value of the target in the image physical coordinate system, r^2 is equivalent to the distance between the coordinate point and the image center, and (k_1, k_2, k_3) denotes the parameters of the radial distortion model.

The tangential distortion is similar to radial distortion, which can be fitted using two other parameters, as shown in the following formula.

$$\begin{cases} x_d = x + 2p_1 xy + p_2(r^2 + 2x^2) \\ y_d = y + p_1(r^2 + 2y^2) + 2p_2 xy \end{cases} \quad (39)$$

where (x_d, y_d) is the point coordinates after the tangential distortion, (x, y) represents the coordinate value of the target in the image's physical coordinate system, r^2 is equivalent to the distance between the coordinate point and the image center, and (p_1, p_2) denotes the parameters of the tangential distortion model.

The complete distortion model of the optical system can be determined by combining the above two types of distortion models, as shown in the following formula [54].

$$\begin{cases} x_d = x(1 + k_1 r^2 + k_2 r^4 + k_3 r^6) + 2p_1 xy + p_2(r^2 + 2x^2) \\ y_d = y(1 + k_1 r^2 + k_2 r^4 + k_3 r^6) + p_1(r^2 + 2y^2) + 2p_2 xy \end{cases} \quad (40)$$

In this paper, we use the fitting approach to correct the distortion of the coordinate data of a single target to reduce the calculation amount of the distortion correction model. The distortion correction model is shown as follows:

$$\begin{cases} x_r = x_d(1 + k'_1 r_d^2 + k'_2 r_d^4 + k'_3 r_d^6) + 2p'_1 x_d y_d + p'_2(r_d^2 + 2{x_d}^2) \\ y_r = y_d(1 + k'_1 r_d^2 + k'_2 r_d^4 + k'_3 r_d^6) + p'_1(r_d^2 + 2{x_d}^2) + 2p'_2 x_d y_d \end{cases}, \; r_d^2 = x_d^2 + y_d^2 \tag{41}$$

where (x_d, y_d) represents the target coordinates of the image's physical coordinate system after distortion correction, (x_r, y_r) denotes the target coordinates of the image's physical coordinate system after distortion correction, and $(k'_1, k'_2, k'_3, p'_1, p'_2)$ indicates the parameters of the inverse distortion model. The parameters are fitted using the measured and actual angle data of the sampling target points.

Finally, the azimuth angle θ and pitch angle φ of the target can be calculated using the following formula.

$$\begin{cases} \theta = \arctan(x_r) \\ \varphi = \arctan(y_r) \end{cases} \tag{42}$$

The calibration experiments of the camera internal and distortion parameters are conducted before the DSP implementation. We calculate the relevant parameters on the PC platform, such as the intrinsic camera matrix and distortion correction parameters. Then, the relevant calculated parameters are solidified within the DSP program to achieve a fast target angle calculation task. The relevant calculated parameters can also be changed by sending instructions from the integrated control system. The calculation of target characteristics data, particularly azimuth and pitch information, can provide comprehensive space target position and grayscale characteristics for space-based surveillance and assist in generating the decision information for spacecraft obstacle avoidance.

4. Hardware Implementation

This paper presents a hardware implementation of the MJDTM model based on an embedded image processing system composed of FPGA and DSP. The FPGA processor, which is suited for parallel computing and has a low computational complexity, has major advantages in terms of large-scale image data processing. This model is suitable for implementing the image filtering algorithm to accomplish rapid target detection. The DSP chip with high-precision digital signal processing capability can complete the algorithm with high computational resource consumption during the tracking stage such as the Kalman filter. To meet the real-time processing requirement of the optical image sequences, we assign processing tasks to the designed multi-core heterogeneous system according to the resource requirements of each processing step. In this section, the constituent modules of the algorithm implementation will be explained in detail.

4.1. Overall Hardware Design

Figure 5 depicts the hardware architecture of the designed space target detection architecture. The onboard space target surveillance system comprises the image acquisition system, the image processing system, the integrated control system, and the external storage. The space-based optical image acquisition system consists of two CMOS sensors with a 1024 × 1024 pixel resolution and a grayscale value of 12 bits that output image data in the format of LVDS (low-voltage differential signaling) data. The image acquisition system captures the space optical image at a rate of five frames per second. The proposed image processing platform consists of a DSP processor for the target tracking association algorithm and an FPGA chip for image acquisition and target detection. When the image data are received, the processing system will cache the image data and perform the operations of the space target detection and tracking. We will obtain the space target optical image and detection results with the FPGA and DSP processing the image data. A Xilinx Kintex-7 XC7K325TFFG900 FPGA device with 326,080 logic cells, 16,020 KB Block RAM, and 840 DSP slices is used in the image processing system. The TMS320C6678 DSP processor with eight C66x cores from TI, whose main frequency can reach 1GHz, has been adopted Additionally, each C66x Core-Pac contains a 512 KB secondary memory (L2), a 32 KB primary program memory (L1P), and a 32 KB data memory (L1D) and can access 4 MB multicore shared memory (MSM). The external storage module contains SDRAM and flash, which are used to cache the image and software program data. The integrated control system sends operation commands to change the working mode of the system and receives the feedback data of the running state through the RS422 interface. It also receives image data and target detection results through the LVDS interface.

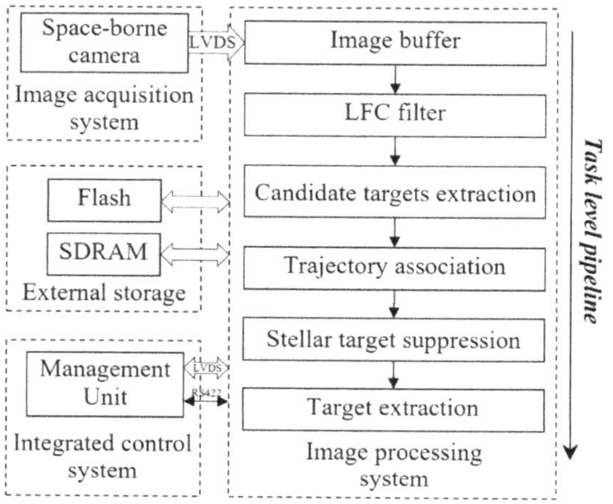

Figure 5. Overview of system conceptual schematic hardware structure.

The overall block structure of the algorithm implementation architecture is shown in Figure 6. The hardware implementation of the detection and tracking algorithm is separated into two parts. The detecting procedure, which requires pixel-by-pixel filtering of the image, is carried out on the FPGA with quick parallel processing capacity. The tracking procedure that executes predictive correlation on the detection target is carried out in the DSP with high-precision digital computing performance. The implementation architecture of the image processing platform comprises the functional modules and processing units. The input to the implementation is the LVDS digital image signals and the RS422 command signals, and the output is the detected target results. In a nutshell, the proposed architecture receives the digital image data from the LVDS interface and executes target detection and tracking operations.

The FPGA implementation comprises the data receive and analysis unit, the command analysis unit, the data cache and output buffer, the system configuration manager, the SRIO communicator, the image cache and slice extract unit, and the target detection module. The data receive and analysis unit first receives the digital image data, converts the serial LVDS data into parallel data, and reads the data packet headers to parse the data according to the communication protocol. Then, the data cache and output buffer store the image data in the external memory. The image cache and slice extract unit stores a whole frame of image data and sends it to the target detection module. The target detection module uses the local feature contrast filter to detect the space and stellar target. The command analysis unit receives various commands such as image processing parameters and configuration management commands from the integrated management unit via the RS422 interface. This unit is also responsible for forwarding instruction data to the DSP processor. Finally, the SRIO communicator packages and send the detected target coordinate and slice data and the auxiliary data to the DSP via the SRIO interface.

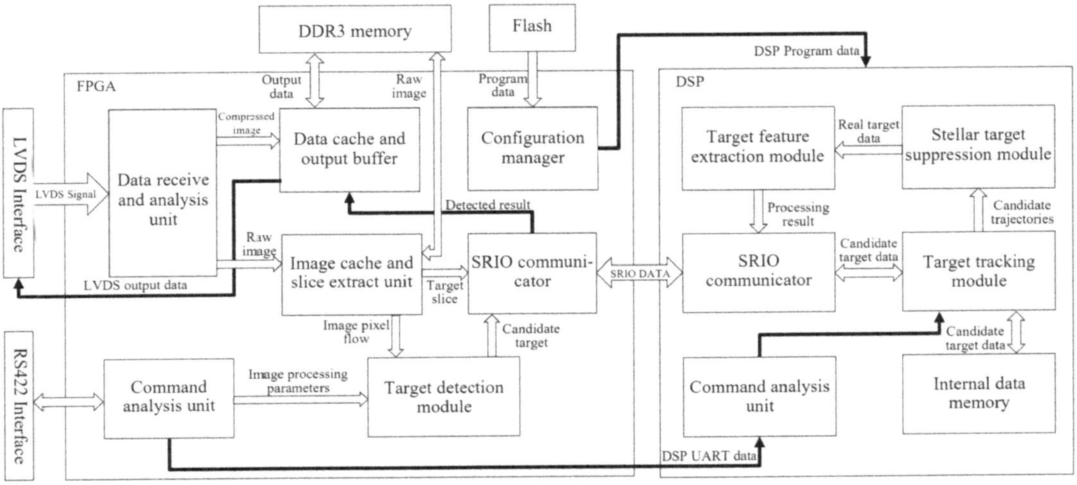

Figure 6. Architecture schematic of system hardware.

The DSP implementation consists of the target tracking module, the SRIO communicator, the stellar target suppression module, the target feature extraction module, and the command analysis unit. The SRIO communicator receives the detected target data and sends them to the target tracking module. The target tracking module adopts the Kalman filter and Hungarian matching algorithm to predict the target state and associated trajectories. The target coordinate data are stored in the internal data memory and read by the target tracking module when updating and associating the target trajectories. The stellar target suppression module uses the real-time satellite attitude data in the auxiliary data package to classify the sidereal and space targets. The feature information of the identified space target are calculated by the target feature extraction module, including the local SNR, the type of targets, and the azimuth and pitch angle. Then, the processing results are packaged and sent to the FPGA chip by the SRIO communicator. The data cache and output buffer in FPGA consolidate the detected results and send them to the integrated control system via the LVDS interface.

4.2. FPGA Implementation

The essential task of the target detection module is to run the LFC algorithm for the optical image sequences. After receiving the image from the space-based image acquisition system through the LVDS interface, FPGA executes the target detection module. The process of the target detection module is illustrated in Figure 7. This module contains five processing stages, including image down-sample, LFC filter, threshold segmentation, connected domain notation, and target centroid extraction. FPGA loads configuration data from Flash by self-starting. After sufficient testing and debugging, software configuration data with default parameters are burned into Flash. During the operation of the system, we can send relevant instructions via UART (universal asynchronous receiver/transmitter) to modify the algorithm parameters.

The hardware architecture of the LFC filter stage is shown in Figure 8. This module will perform the filtering operation in parallel for each pixel during the LFC filter stage. In detail, we split the $n \times m$ filter window from the image pixel stream centered on the point (x, y). The filter window is scanned across the image in Figure 8a from top to bottom and left to right. To quickly calculate the grayscale average value and standard deviation of the filtering region, the filtering window is divided into nine image blocks.

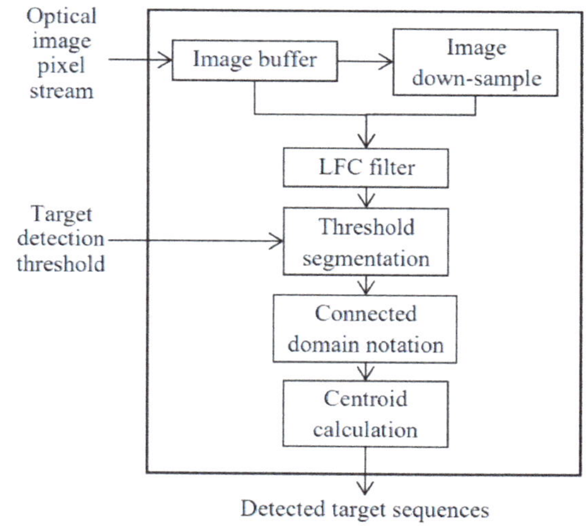

Figure 7. Block diagram of the target detection module.

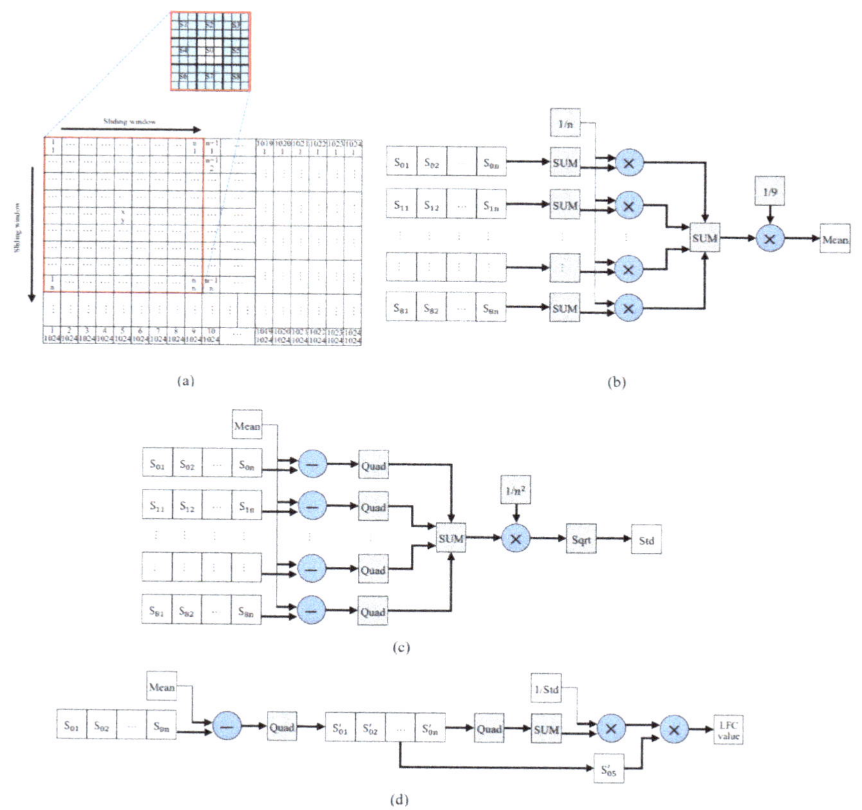

Figure 8. The hardware structure of the LFC filtering: (**a**) the filter window sliding operation; (**b**) the hardware structure of the image block gray mean calculation; (**c**) the hardware structure of the image block gray standard deviation calculation; (**d**) the hardware structure of the LFC value.

This module runs related calculations for the divided image blocks in parallel, as depicted in Figure 8b,c. Then, a series of calculating operations mentioned in Section 3 are also performed for the filtering window. Finally, the LFC value of the center point (x, y) is calculated, as shown in Figure 8d.

In the hardware implementation, the sliding window size of the LFC filter is set to 9×9. In practical scenarios, the projection area of the target on the detector may expand, given the target movement and the change of detector lens parameters. As shown in Figure 9, the uneven grayscale distribution and threshold segmentation processing of large targets may lead to the segmentation of one target into several targets, which increases the difficulty of subsequent target tracking. However, by altering the size of the image filter window, the proposed algorithm might be able to adapt to the target size variation.

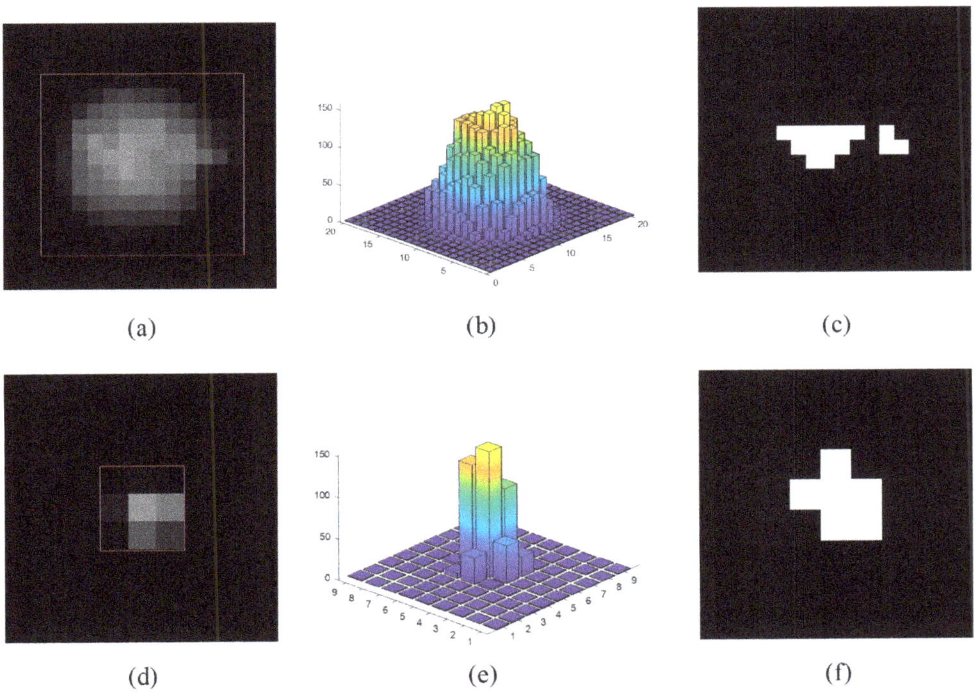

Figure 9. Region of the large target and adjacent neighbor before and after the down-sample operation: (**a**) the region of the large target and adjacent neighbor; (**b**) the 3D plot of the large target; (**c**) the saliency map of the large target; (**d**) the region of the large target and adjacent neighbor after the down-sample operation; (**e**) 3D plot of the large target after the down-sample operation; (**f**) the saliency map of the large target after the down-sample operation.

This will increase the difficulty and cost of the hardware implementation for the detection algorithm. To guarantee the precision and processing speed of target detection, the image downscaling stage executes the down-sample operation on the optical image sequence. We also apply the same scale LFC filter operation to the down-sampled image to extract large-size targets. As shown in Figure 9, the large targets can be accurately detected after the down-sample operation. Since the processing of original and down-sample data is independent, the two LFC filtering procedures for these images are executed concurrently in the FPGA implementation.

After the image LFC filter, the adaptive threshold segmentation is conducted on the local feature contrast image to extract the potential target. The selection of the adaptive segmentation threshold is discussed in Section 5.3. Then, this module applied the scanning line technology to label the connected domain of the target. The distance-weighted centroid method is also adopted to calculate the target centroid coordinate data. Finally, we can obtain the detection target coordinate data based on the original and down-sampled images The target centroid coordinates and the slice data segmented according to the coordinate data are packaged and sent to the DSP processor.

4.3. DSP Implementation

The flowchart of the software in the DSP system is shown in Figure 10. The DSP system mainly performs image processing tasks such as the multi-frame association of detected targets, matching update of candidate trajectories, stellar target suppression, and target feature calculation.

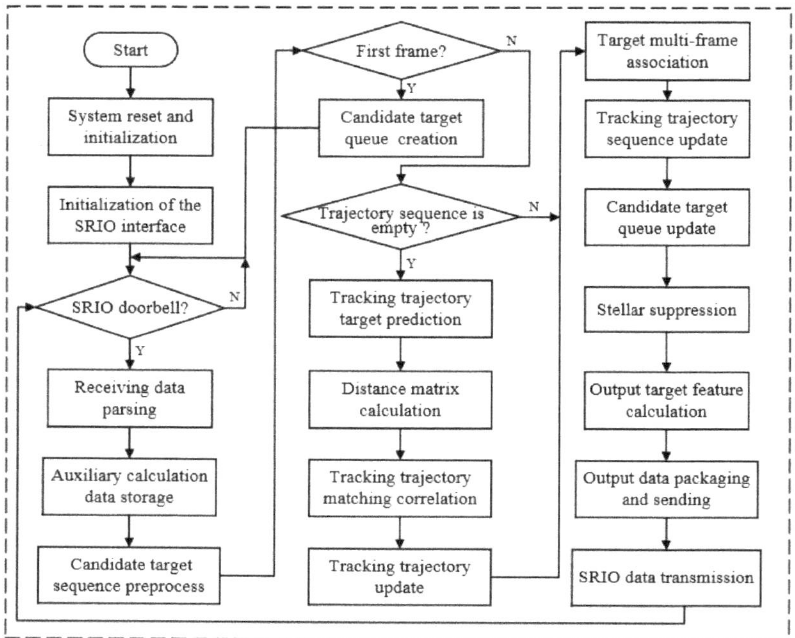

Figure 10. DSP software flow block diagram.

We employ the main Core 0 for software program development and use the kernel's local L2 SDRAM and MSM SDRAM as data storage because the DSP software processing tasks in this system are relatively simple and the amount of data processed is small. The peripheral interfaces of other cores are turned off to reduce system power consumption. The DSP is started by SPI boot mode, and the configuration interface is connected to the FPGA. The FPGA realizes the option of the SPI configuration mode by controlling the high and low levels of the configuration interface. After the DSP is powered up in the prescribed order and the configuration mode is set up, it reads the program data through the SPI interface and loads it into the program storage space in the core and starts to run the program. After sufficient testing and debugging, the DSP software program is written into FLASH, and it comes with a set of default algorithm parameters. During the process of software operation, we can send the relevant execution through the UART interface to change and optimize the algorithm parameters to achieve the optimal processing effect.

When the DSP system completes the program loading and system initialization, it enters the idle state and waits for the SRIO doorbell interrupt. After the FPGA completes the candidate target extraction of a frame image, the detection result data package is written to the memory of the DSP through the SRIO interface, and the doorbell signal is delivered to the DSP after the data transmission is completed by the FPGA. The DSP starts the processing of the candidate target data for the current frame under the trigger of the doorbell interrupts. Firstly, the software parses the candidate target data and auxiliary calculation data in the data packet according to the relevant protocol. The software updates the candidate target data and auxiliary calculation data in the packet to the corresponding storage array. Then, the software calls different functions according to the processing stage to associate the target. In the initial tracking stage, the software performs multi-frame correlation to confirm the real target for multiple consecutive frames of candidate targets. In the middle tracking stage, the software creates the tracking trajectory for the real targets and predicts their motion state in subsequent frames. In the subsequent stable tracking stage, multi-frame association and trajectory association operations are alternated between the candidate target queue and the trajectory sequence, and for the trajectory sequence, the software discriminates the type of the target and suppresses the stellar target to confirm the real space target. Finally, after continuous multi-frame stable tracking correlation and recognition, the feature data of the real target are calculated by the software, packaged, and sent to FPGA. If there is no target in the current frame that is judged to be the true target, the relevant data are not sent. The software enters the idle state after completing the processing of the current frame.

4.3.1. Target Tracking Module

In the DSP implementation, the target tracking module predicts the candidate target state prediction and associates the tracking trajectories with the detected target sequence. Figure 11 illustrates the block diagram of this module. Firstly, this module executes the multi-frame association before the tracking trajectory update to suppress noise points. It calculates the distance between targets in the current frame detected target sequence and the former target sequence in the candidate target queue. When the separation between the targets is below a predetermined threshold, the target is considered a potential target, and the coordinates of the latest frame are updated to the candidate target queue. The target data in the detected target sequence are directly updated to the candidate target queue when the system first receives the target data. This module generates the associated tracking trajectory for subsequent target state prediction and trajectory association when the target in the candidate target queue has more occurrences than a set threshold. In detail, we use the Vision Library, a collection of optimized computer vision algorithm libraries developed by Texas Instruments for digital media processors. It contains the Application Programming Interface for Kalman filter algorithms, which can quickly perform complex function operations in hundreds of machine cycles. The VLIB_kalmanFilter_2 $\times$ 4 is the structural variable type used for the Kalman filter calculation with two-dimensional observation and four-dimensional state vectors. This module creates the corresponding VLIB_kalmanFilter_2 $\times$ 4 structural variable for each tracking trajectory, which is convenient for calling the predict and correct API function to achieve the prediction and state update of the tracking trajectory.

Moreover, this module employs the Hungarian matching algorithm to associate the tracking trajectory queue and the detected target sequence. The distance between the tracking trajectory predicted sequence and the detected target sequence is calculated and integrated into a distance matrix used as the weight of trajectory matching. The Hungarian matching algorithm uses a recursive method to find the path with the maximum expectation value to match the trajectories with the target sequence. The detected target sequence of each frame image is preferentially associated with the tracking trajectories in this module. Moreover, in the multi-frame association stage, the targets in the detected target sequence that successfully matched the tracking trajectory are not associated with the candidate target queue.

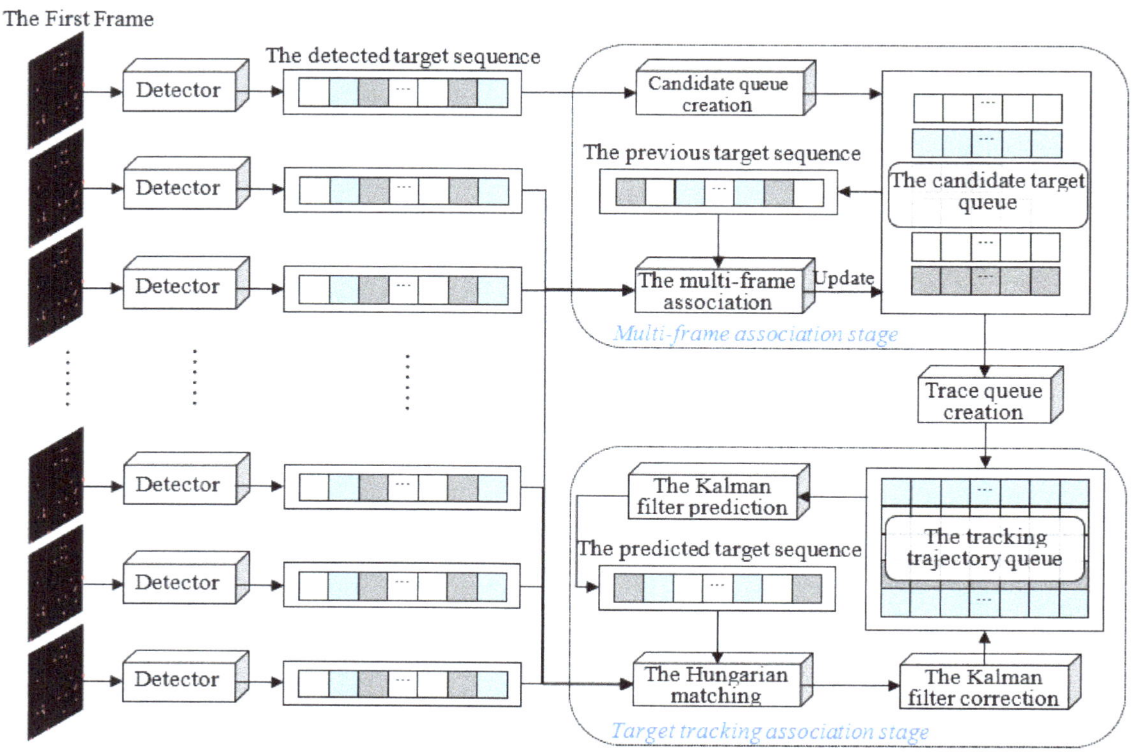

Figure 11. Block diagram of the target tracking module.

4.3.2. Stellar Target Suppression Module

After the latest tracking trajectories of the candidate targets are associated and updated, the potential target type needs to be confirmed, as mentioned in Section 2. The stellar target suppression module, which predicts the target position using the real-time satellite attitude data to classify the stellar targets and space targets, is shown in Figure 12. In the implementation, in order to eliminate the influence of platform jitter on the detection results, the historical coordinate data in the candidate tracking trajectory are utilized to identify the target type. As shown in the diagram, this module computes the target coordinate data relative to the Earth by using the image plane coordinate data and the satellite platform attitude data. Each frame's satellite platform attitude data are included in the auxiliary package transmitted together with the detected target package sent by FPGA. Moreover, the predicted target coordinate data at time n in the image coordinate system could be calculated with the attitude data of the satellite platform at time n. The mean value of the difference between the target actual coordinate and the predicted coordinate calculated is used to judge whether the target is a real space target or not. In detail, this module begins to classify the type of candidate target for each tracking trajectory when its length surpasses a specific value. The threshold value that determines the target type also can be changed by sending instructions from the integrated management system.

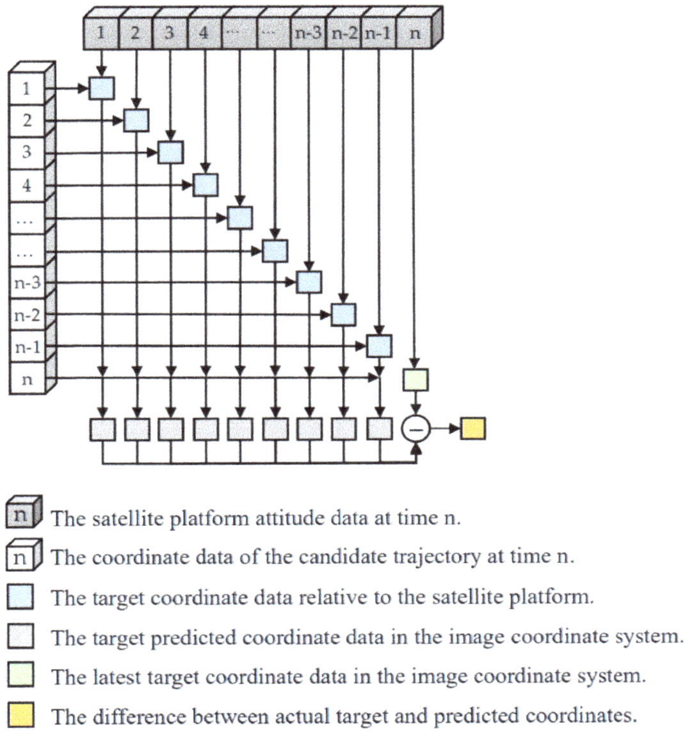

Figure 12. Stellar target suppression module.

5. Experiment

The hardware architecture described in Section 4 was implemented on a dedicated embedded image processing platform. The pictorial diagram of the platform is shown in Figure 13. The architecture was implemented using Verilog and C. To better evaluate the performance of the target detection algorithm, the simulations were conducted using Matlab2018b in the Win10 system, using an i7-10750H CPU with 2.6 GHz and 16 GB of main memory. The remainder of the section is organized as follows. First, the optical image dataset used in the experiment is described. Then, we assess the proposed architecture regarding the target detection rate, the efficiency of the stellar target suppression algorithm, and the target angle calculation accuracy. Finally, we present an evaluation of processing efficiency for the hardware implementation.

5.1. Experimental Dataset

To validate the detection performance of the proposed algorithm, we use the simulated and real image sequences. The simulated image dataset comprises image sequences of the wide and narrow field of view, of which the image size is 1024 × 1024 pixels and the grayscale value is 12 bits. The simulated images whose background is the deep space background contain stellar targets and moving space targets that simulate the space optical image in space-based scenarios. The motion attitude data of the satellite platform are synchronously generated with the image sequences.

Figure 13. Picture of the embedded image processing platform.

Furthermore, the real image data are captured by two cameras of the space-based optical image acquisition system introduced in Section 4 in the ground simulation scenario. The wide and narrow field-of-view camera covers the range of $90^\circ \times 90^\circ$ and $8^\circ \times 8^\circ$ field of view, respectively. The image size and pixel gray level are the same as the simulated image. The real image data are taken on a clear cloudless night, and the camera is placed vertically on the ground, capturing the sidereal points and the unmanned aerial vehicle and civil aviation aircraft targets that simulate the moving space target under the night sky background. The image dataset introduced in detail in Table 1 contains two groups of simulated image sequences and two groups of real image sequences.

Table 1. Details of the space target image sequences.

Sequence	Frame	Field View	Background Details	Target Details
Seq.1	300	Wide field	Simulated deep space background; random noise	Simulated target; 3 × 3
Seq.2	300	Narrow field	Simulated deep space background; random noise	Simulated target; 3 × 3
Seq.3	300	Wide field	Real background; sky	Civil aviation aircraft; 3 × 3
Seq.4	300	Narrow field	Real background; cloud and sky	Unmanned aerial vehicle; 12 × 12

5.2. Target Detection and Tracking Experiment

In this section, we evaluate the performance of the proposed space target detection and tracking algorithm, including the accuracy of target detection, the efficiency of target tracking, and the accuracy of the stellar target suppression algorithm, by using the simulated and real image sequence.

First, we evaluate the proposed target detection algorithm using the four groups of image datasets mentioned in the previous section. To quantitatively assess the performance of the detection algorithm, we use two evaluation criteria: the detection probability and the false alarm rate. The detection probability P_d and false alarm rate P_f are defined as

$$P_d = \frac{N_d}{N_t} \tag{43}$$

$$P_f = \frac{N_f}{N_p} \tag{44}$$

where N_d represents the number of the true detected targets, N_t denotes the number of real targets, N_f indicates the number of false targets, and N_p denotes the total number of pixels in the processed images. Meanwhile, we present the receiver operating characteristic (ROC) curves and calculate the area under the curve (AUC) to intuitively appraise the algorithm's performance. The ROC curve could illustrate the corresponding relationship between the detection probability and the false alarm rate, which is one of the quantitative methods to evaluate detection efficiency. The closer the curve is to the upper left corner, the better the algorithm performs.

Simultaneously, the proposed algorithm is compared with state-of-the-art small target detection algorithms, including the multiscale tri-layer local contrast measure (TLLCM) and the weighted strengthened local contrast measure (WSLCM) [55]. Figure 14 shows the ROC curves of these three algorithms for four groups of space optical image sequences. As shown in Figure 14a, the proposed method obtained a P_d exceeding 95% when P_f reached 10^{-4}. As shown in Figure 14c,d, the proposed and comparative algorithm could reach 95% P_d when $P_f < 10^{-6}$. As shown in Figure 14b, both the proposed algorithm and the comparison algorithm reach 95% P_d when the P_f does not exceed 10^{-4}. Since the target is weak, the performance of the proposed algorithm is slightly inferior to that of WSLCM. The area under the curve (AUC) can further be used to evaluate the performance of the target detection method and provide a more comprehensive comparison. The results of AUC are shown in the figure. The AUC values of the proposed method were 0.9759, 0.9819, 0.9843, and 0.9919. In contrast to Seq.2, the proposed algorithm obtains the maximum AUC value on the other three datasets. The experimental results show that compared with other algorithms, the energy accumulation step in the proposed algorithm enhances the target, achieving a better target detection performance and ensuring the algorithm implementation detection efficiency.

The adaptive segmentation threshold is a significant factor that decides the accuracy of target detection. Figure 15 demonstrates the detection rate and false alarm rate performance of the algorithm on four sequences under different values of the parameter k_1. Figure 15a indicates that when the parameter k_1 is between 30 and 50, the target detection rate exceeds 95%. which is sufficient for practical applications. When the parameter k_1 is less than 35, the P_f is less than 10^{-4}, which satisfies the majority of application requirements, as shown in Figure 15b. In conclusion, we recommend that the parameter k_1 should have a range of values between 35 and 50, so that the detection performance can attain over 95% P_d and $P_f < 10^{-4}$.

After this, we performed a series of experiments using the simulated wide-field and narrow-field image sequences. There are twenty to thirty sidereal points and one space target setting in the two simulated groups of image sequences. The wide- and narrow-field image detection results are shown in Figure 16a,b. It can be seen from the figure that the detection algorithm can accurately detect space targets and stellar points. For the target sequence of each image frame, we created the corresponding tracking trajectory to track the target. When the number of consecutive occurrences of the target is greater than seven, the satellite attitude data generated via simulation are used to classify the space target and the sidereal points of the track trajectories. The results of trajectory tracking and stellar target suppression can be seen in Figure 16b,c, and the efficiency of target tracking after the stellar target suppression is shown in Figure 16e,f. The results show that the algorithm can successfully identify space targets.

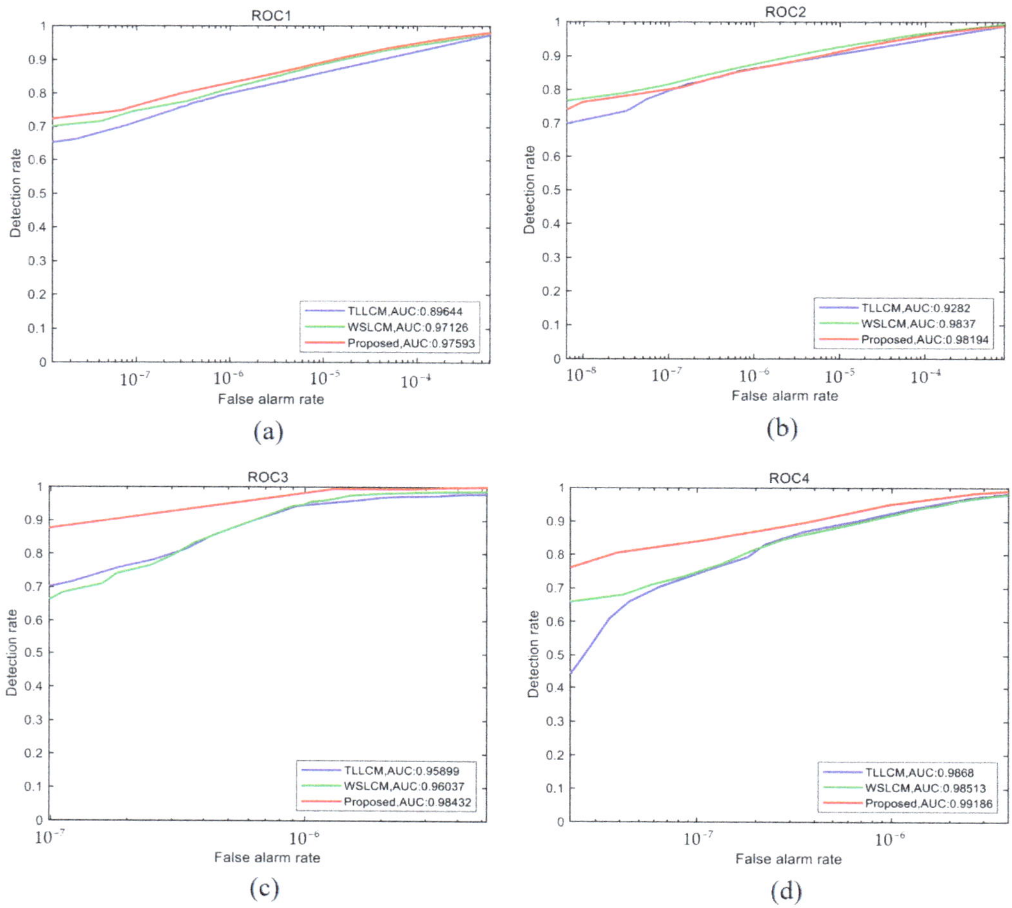

Figure 14. ROC curves of four groups of the image sequences. (**a**) ROC curve of Seq.1; (**b**) ROC curve of Seq.2; (**c**) ROC curve of Seq.3; (**d**) ROC curve of Seq.4.

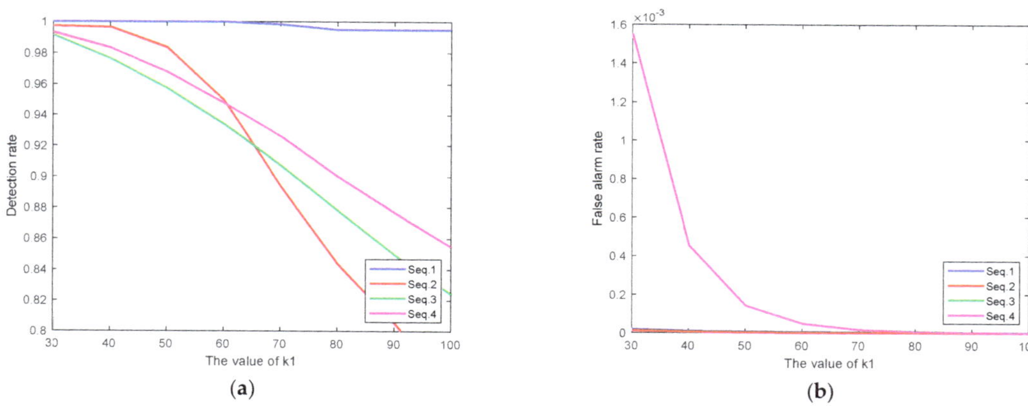

Figure 15. Influence of k_1. (**a**) Relationship between k_1 and P_d; (**b**) relationship between k_1 and P_f.

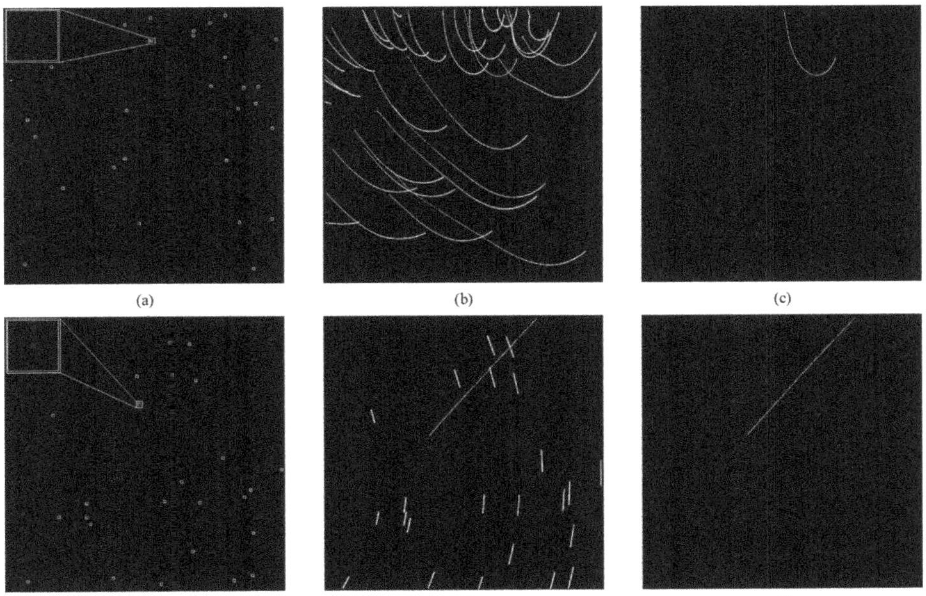

Figure 16. Simulation image detection and tracking results. (**a**) Target detection results of Seq.1 (left upper corner is space target area slice); (**b**) trajectory tracking results of Seq.1 (red is space target trajectory, white is star trajectory); (**c**) tracking trajectory result of Seq.1 after the stellar target suppression; (**d**) target detection results of Seq.2 (upper left corner is target area slice); (**e**) trajectory tracking results of Seq.2; (**f**) tracking trajectory result of Seq.2 after the stellar target suppression.

In addition, to illustrate the effect of the stellar target suppression algorithm, we define the detection probability and false alarm rate based on the whole image sequence to quantitatively evaluate the stellar target suppression algorithm. The detection probability P_t and false alarm rate F_a are defined as

$$P_t = \frac{N_{rd}}{N_{rt}} \tag{45}$$

$$F_a = \frac{N_{rf}}{N_{pf}} \tag{46}$$

where N_{rd} represents the number of detected frames of real space targets, N_{rt} denotes the frame number of real space targets, N_{rf} represents the number of detected frames of false space targets, and N_{pf} denotes the total number of frames in the sequence of images. Table 2 shows the calculation results of this algorithm's detection rate and false alarm rate on two sets of simulated image sequences. The target tracking stage monitors potential targets in multi-frame optical image sequences and suppresses the stellar targets using platform attitude data and historical coordinate data that can reflect target motion differences. The experimental results show that this stage achieves high-precision detection of real space targets.

Table 2. Detection probability and false alarm rate of the stellar target suppression algorithm on different image sequences.

Sequence	P_t	F_a
Seq.1	91.72%	1.33%
Seq.2	97.25%	0%
Seq.3	80.9%	0%
Seq.4	95.67%	0%

We also conducted relative experiments for the real image set, including space target detection, tracking, and stellar target suppression. First, we perform the target detection experiment on the image sequence. Besides the target detection of the original image, we also detect the target in the down-sample real image sequence, which is used to realize the detection of large-size targets. As shown in Figure 17e, when the target is close to the detector, the number of pixels occupied by the target on the detector plane will increase, leading to the algorithm marking one target as two targets. The down-sample processing step of the image reduces the area of the big target to ensure the accuracy of the target detection. Specifically, this step reduces the image size from 1024 × 1024 pixels to 256 × 256 pixels by sampling the original image every four pixels. The target can be down-sampled from 12 × 12 to 3 × 3 by quadrupling this down-sample rate. The hardware implementation of the detection algorithm's filter window can accurately detect targets with a diameter of less than 3, and after down-sampling, the diameter of targets with a diameter of 4 to 12 is reduced to between 3 and 1, allowing our hardware implementation to also accurately detect the target. A target with an area greater than 12 × 12 is not considered in this paper. The detection results of the original image and down-sampled image are shown in Figure 17. We obtain two sets of target sequences after the space detection of the original image and the down-sampled image. A fusion operation is executed to merge the two target sequences.

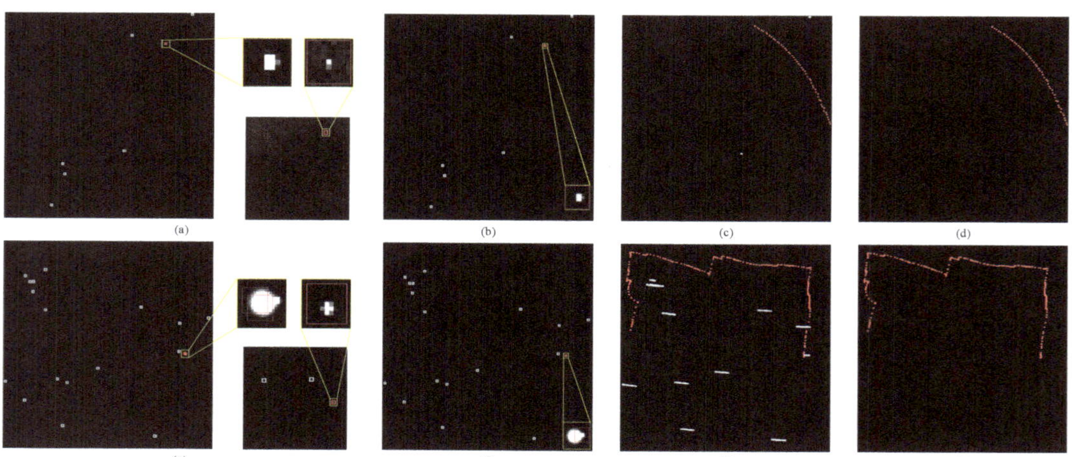

Figure 17. Real image detection and tracking results: (**a**) the detection results of the original image and the down-sampled image of Seq.3; (**b**) the fusion results of the dual-size target detection results of Seq.3; (**c**) trajectory tracking results of Seq.3; (**d**) tracking trajectory results of Seq.3 after the stellar target suppression; (**e**) the detection results of the original image and the down-sampled image of Seq.4; (**f**) the fusion results of the dual-size target detection results of Seq.4; (**g**) trajectory tracking results of Seq.4; (**h**) tracking trajectory results of Seq.4 after the stellar target suppression.

The detection results are shown in Figure 17b,f, and the large target is marked as one target after fusion. We create the corresponding tracking trajectories for the target sequence in the target tracking experiments. When the length of the tracking trajectory is greater than 7, we use the simulated satellite attitude data to classify the space targets and sidereal points in the trajectory sequence. The target tracking results are shown in Figure 17c,g, in which the white trajectories belong to sidereal points and the red one belongs to the simulated space target (UAV and aircraft) trajectory. The results of target tracking after the stellar target suppression are shown in Figure 17d,h. It can be seen from the figure that the stellar points and space targets are precisely distinguished using the difference in motion between them. Table 2 also shows the stellar target suppression results of the proposed algorithm on the real image sequences.

Finally, the target angle information of the space target is calculated. The accuracy of the target angle measurement method will also be calculated in the following experiment. For the wide-field camera with a 90° × 90° field of view, camera distortion correction is required to ensure the accuracy of the angle calculation before performing the target angle calculation. The specific correction scheme is as follows. The camera is mounted on a two-dimensional rotating platform, and a point target is set in front of the rotating platform to simulate a space target. The correction method is shown in Figure 18a. First, we calculate the mounting matrix of the camera, in which the rotating platform rotates to the corresponding angle according to the set angle sequence. The camera acquires 25 target images of the angle near the center point to generate a set of image sequences. We extract the coordinates of 25 target points and fit the camera mounting matrix using these points.

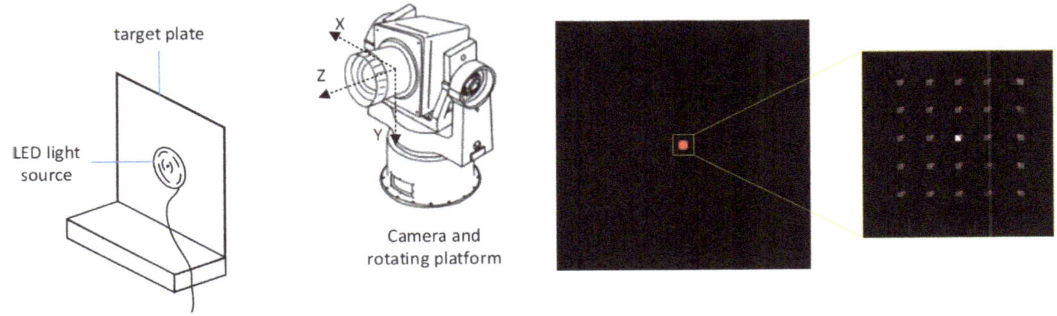

(a) (b)

Figure 18. Wide-field camera distortion correction method. (**a**) Distortion correction method; (**b**) collated sampled point grid image.

Then, the rotating platform also rotates to the corresponding angle according to the set angle sequence, and the camera acquires 225 setting target points images to generate a set of image sequences. We also detect and extract the coordinates of target points for the acquired image sequence, and the collated sampled point grid image is shown in Figure 19a. Due to the barrel distortion of the wide-field camera, the target's actual imaging position is often not in the ideal projection model coordinates. Consequently, we use the target ideal coordinate sequence and the actual imaging coordinate sequence to generate the inverse distortion model. The grid of sampling points corrected by the inverse distortion model is shown in Figure 19b. To verify the accuracy of this inverse distortion model, we take images of 25 test target points. The target points are detected using the proposed algorithm, and the azimuth and pitch angle are also calculated using the inverse distortion model. The measured angle results of the test target point before and after the distortion correction are shown in Figure 19c,d. The red star mark in the figure is the angle value calculated using the imaging position of the target image plane, and the blue is the actual angle value of the target. As shown in the figure, the azimuth and pitch angle of the target are accurately calculated after the aberration correction. The average angle measurement error is 0.1334, and the maximum value of the side angle error is 0.2419. The angle measurement accuracy can reach 99.73%. The formula of the side angle error ϕ and the angle measurement accuracy ε are as follows.

$$\phi = \sqrt{(\theta_i - \theta_r)^2 + (\varphi_i - \varphi_r)^2} \tag{47}$$

$$\varepsilon = 1 - \phi/FOV \tag{48}$$

where (θ_i, φ_i) denotes the actual value of the azimuth and pitch angle of the target point, (θ_r, φ_r) represents the corrected value of the calculated azimuth and pitch angle of the target point after the distortion correction operation, FOV denotes the camera field of view, and ε is the angle measurement error.

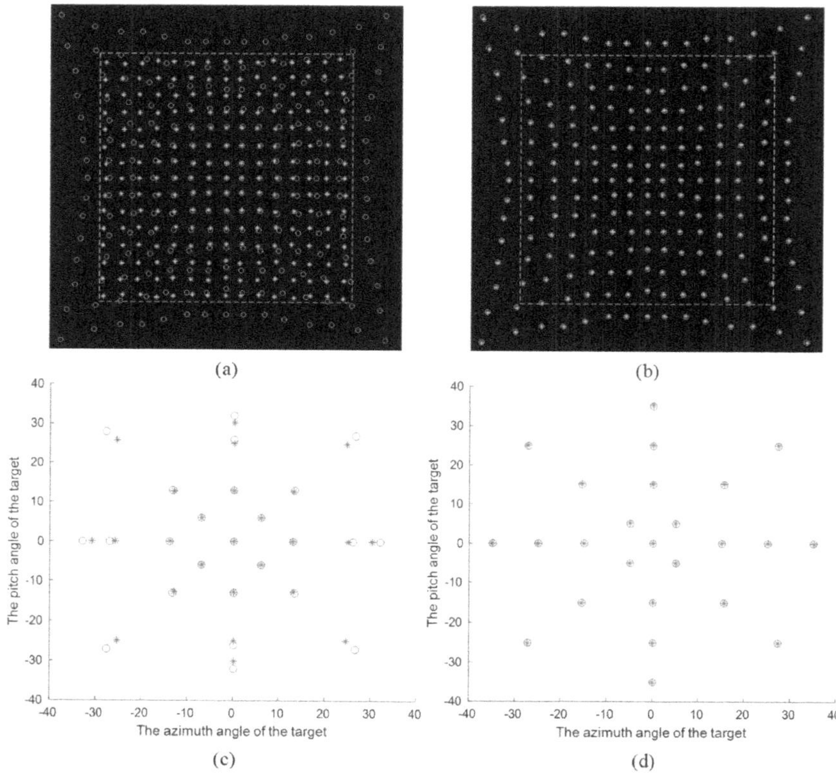

Figure 19. Wide-field camera distortion correction results. (**a**) Distribution of scanned target imaging position and theoretical position before aberration correction; (**b**) distribution of scanned target correction position and theoretical position after aberration correction; (**c**) calculated and actual angles of test target points before distortion correction; (**d**) calculated and actual angles of test target points after distortion correction.

5.3. Hardware System Computational Performance Analysis

In this section, we conduct several experiments to evaluate the proposed implementation's computational performance and operational efficiency. As described in Section 4, the proposed target detection algorithm is implemented on the Xilinx Kintex-7 FPGA with the specific hardware resource consumption rates shown in Table 3. We utilized these resources in the FPGA implementation to optimize the design. Table 4 reports the processing time and power consumption obtained for the hardware implementation of the proposed algorithm on the considered FPGA and DSP architecture. The single-frame image processing time measured in the FPGA is only 22.064 ms. Therefore, the designed space target detection architecture can realize a processing speed of 45 frames per second. In experimental tests, the power consumption of DSP and FPGA is 7.02 W and 6.168 W, respectively, and the overall power consumption can be controlled within 15 W, which satisfies the space-based platform application requirements.

Table 3. Summary of resource utilization for the FPGA implementation of the proposed target detection algorithm.

Component	Number of LUTs	Number of FFs	Number of BRAMs	Number of DSPs	Number of BUFGs
Units	4.9999	6.1274	233	53	22
Percentage	24.53%	15.03%	52.36%	6.31%	68.75%

Table 4. Processing time measured for space detection and tracking method in the hardware system.

Hardware Platform	Processing Time	Clock Period	Hardware Operation Frequency	Hardware Power Consumption
FPGA	22.046 ms	1,102,300	50 MHz	7.02 W
DSP	0.5946 ms	595,460	1000 MHz	6.168 W

Finally, we use the simulated image sequences Seq.1 and Seq.2 to evaluate the performance of the hardware implementation for target detection and tracking. The experiment results of the target detection and stellar target suppression are shown in Table 5. Due to the complexity of implementation, we have not drawn the ROC curve on the FPGA implementation program. The threshold of the adaptive segmentation algorithm is set between 10 and 20 for testing. Finally, a detection rate of 96.36% can be achieved when the false alarm rate is less than 0.4% during the detection stage of the FPGA implementation. The stellar suppression algorithm of the target tracking algorithm module implemented by DSP can accomplish an average detection rate of 87.8% for actual space targets in sequence images. The DSP implementation is inferior to that of the PC platform. The primary reason is that the multi-frame correlation module will confirm the target in the first few frames of the target. The target tracking stage of the DSP implementation will execute the stellar suppression algorithm when the tracking trajectory length exceeds a certain threshold. Consequently, the detection rate will be low in the early stages of tracking and will increase as the trajectory length increases.

Table 5. The result of the hardware implementation for the proposed algorithm.

Hardware Platform	Sequence	Seq.1	Seq.2
FPGA implementation	P_d	97.37%	96.36%
	P_f	0.0332%	0.0335%
DSP implementation	P_t	87.27%	88.33%
	F_a	0%	0%

6. Discussion

In this paper, a space target detection and tracking model is presented with its hardware implementation scheme. A dim small space target detection approach is proposed in the target detection stage, which improves the local contrast method. According to the experimental results on the real and simulated image datasets, as illustrated in Figure 14, its detection performance is stronger than that of TLLCM and WSLCM. In detail, our algorithm can obtain 95% P_d when $P_f < 10^{-4}$ on all sequences. The target detection method is implemented on an FPGA, and Tables 3–5 reveal the resource and time consumption and the experimental results. The target detection software is self-started by the FPGA, and the segmentation parameter will remain constant for a while. The detection performance is significantly impacted by the selection of the adaptive segmentation threshold. On the one hand, we define the default threshold using the results of the PC platform. On the other hand, the detection rate and false alarm rate information are calculated with the output of the detection results from the FPGA, and the best detection effect can be obtained by appropriately modifying the segmentation threshold parameters in accordance with the detection effect. Furthermore, the hardware development could complete the target detection of a single frame image in 22 ms thanks to the parallel processing capabilities of FPGA, which guarantees real-time performance of image processing.

The Kalman filter algorithm and the Hungarian matching algorithm collaborate at the target tracking stage to stabilize the tracking target. The experimental results of the model and the detection effect are displayed in Figure 17 and Table 2. The satellite's attitude data are easily obtained on the space-based platform. Given that attitude data undoubtedly contain errors, we begin the stellar suppression algorithm once the tracking trajectory reaches a particular threshold to avoid the effects of incorrect attitude data on the star suppression effect. We utilize simulated attitude data for the experiment on the PC platform, and the threshold is set at 8 since the simulated data error is minor. On the one hand, calculation errors due to platform differences may have an impact on the detection effect. On the other hand, employing more frames of historical target coordinate data for statistics can eliminate the calculation error caused by attitude data error and ensure the target's detection rate in the actual space-based scene. In order to provide accurate target angle information, we also present a distortion correction scheme for the large field-of view-optical lens. As illustrated in Figure 19, the distortion correction scheme could reduce the angle measurement error to less than 0.3%.

In conclusion, the experimental results validate the efficacy and viability of the model and hardware architecture and confirm that the processing system is capable of real-time space target detection and tracking, thereby meeting the requirements of the space-based platform application.

7. Conclusions

In this paper, a multi-stage joint detection and tracking model is developed to solve the problem of space target detection and tracking in the deep space background and a hardware implementation of this model for space-based surveillance applications is provided. The experiments conducted with the simulated and real image sequences demonstrate that the proposed implementation can lead to improvements in detection accuracy while maintaining real-time processing speed. However, the proposed model may not have a good detection performance for low SNR targets and depends on real-time satellite attitude data. In future work, we will improve the method to address these shortcomings and apply it in other complex scenarios.

Author Contributions: Conceptualization, P.R.; methodology, Y.S. and X.C.; software, Y.S. and G.L.; validation, Y.S., X.C. and C.C.; formal analysis, Y.S.; investigation, Y.S. and G.L.; resources, Y.S. and C.C.; data curation, Y.S. and X.C.; writing—original draft preparation, Y.S.; writing—review and editing, X.C.; visualization, G.L.; supervision, P.R.; project administration, X.C.; funding acquisition, P.R. All authors have read and agreed to the published version of the manuscript.

Funding: This research received no external funding.

Data Availability Statement: Not applicable.

Conflicts of Interest: The authors declare no conflict of interest.

References

1. Li, M.; Yan, C.; Hu, C.; Liu, C.; Xu, L. Space Target Detection in Complicated Situations for Wide-Field Surveillance. *IEEE Access* **2019**, *7*, 123658–123670. [CrossRef]
2. Wang, X.; Chen, Y. Application and Development of Multi-source Information Fusion in Space Situational Awareness. *Spacecr. Recovery Remote Sens.* **2021**, *42*, 11–20. [CrossRef]
3. Chen, L.P.; Zhou, F.Q.; Ye, T. Design and Implementation of Space Target Detection Algorithm. *Appl. Mech. Mater.* **2015**, *738–739*, 319–322. [CrossRef]

4. Barniv, Y. Dynamic programming solution for detecting dim moving targets. *IEEE Trans. Aerosp. Electron. Syst.* **1985**, *AES-21*, 144–156. [CrossRef]
5. Barniv, Y.; Kella, O. Dynamic programming solution for detecting dim moving targets part II: Analysis. *IEEE Trans. Aerosp. Electron. Syst.* **1987**, *AES-23*, 776–788. [CrossRef]
6. Doucet, A.; Gordon, N.J.; Krishnamurthy, V. Particle filters for state estimation of jump Markov linear systems. *IEEE Trans. Signal Process.* **2001**, *49*, 613–624. [CrossRef]
7. Salmond, D.; Birch, H. A particle filter for track-before-detect. In Proceedings of the 2001 American Control Conference (Cat. No. 01CH37148), Arlington, VA, USA, 25–27 June 2001; pp. 3755–3760.
8. Reed, I.S.; Gagliardi, R.M.; Shao, H. Application of three-dimensional filtering to moving target detection. *IEEE Trans. Aerosp. Electron. Syst.* **1983**, *AES-19*, 898–905. [CrossRef]
9. Zhang, C.; Chen, B.; Zhou, X. Small target trace acquisition algorithm for sequence star images with moving background. *Opt. Precision Eng.* **2008**, *16*, 524–530.
10. Cheng, J.; Zhang, W.; Cong, M.; Pan, H. Research of detecting algorithm for space object based on star map recognition. *Opt. Tech.* **2010**, *36*, 439–444.
11. Zhang, J.; Ren, J.-C.; Cheng, S.-C. Space target detection in star image based on motion information. In *International Symposium on Photoelectronic Detection and Imaging 2013: Optical Storage and Display Technology*; SPIE: Bellingham, WA, USA, 2013; pp. 35–44.
12. Xi, X.-L.; Yu, Y.; Zhou, X.-D.; Zhang, J. Algorithm based on star map matching for star images registration. In *International Symposium on Photoelectronic Detection and Imaging 2011: Space Exploration Technologies and Applications*; SPIE: Bellingham, WA, USA, 2011; p. 81961N.
13. Boccignone, G.; Chianese, A.; Picariello, A. Small target detection using wavelets. In Proceedings of the Fourteenth International Conference on Pattern Recognition (Cat. No. 98EX170), Brisbane, QLD, Australia, 20 August 1998; pp. 1776–1778.
14. Jiang, P.; Liu, C.; Yang, W.; Kang, Z.; Li, Z. Automatic Space Debris Extraction Channel Based on Large Field of view Photoelectric Detection System. *Publ. Astron. Soc. Pac.* **2022**, *134*, 024503. [CrossRef]
15. Chen, C.P.; Li, H.; Wei, Y.; Xia, T.; Tang, Y.Y. A local contrast method for small infrared target detection. *IEEE Trans. Geosci. Remote Sens.* **2013**, *52*, 574–581. [CrossRef]
16. Chen, L.; Rao, P.; Chen, X. Infrared dim target detection method based on local feature contrast and energy concentration degree. *Optik* **2021**, *248*, 167651. [CrossRef]
17. Sun, R.-Y.; Zhan, J.-W.; Zhao, C.-Y.; Zhang, X.-X. Algorithms and applications for detecting faint space debris in GEO. *Acta Astronaut.* **2015**, *110*, 9–17. [CrossRef]
18. Deshpande, S.D.; Er, M.H.; Venkateswarlu, R.; Chan, P. Max-mean and max-median filters for detection of small targets. In *Signal and Data Processing of Small Targets 1999*; SPIE: Bellingham, WA, USA, 1999; pp. 74–83.
19. Bai, X.; Zhou, F. Infrared small target enhancement and detection based on modified top-hat transformations. *Comput. Electr. Eng.* **2010**, *36*, 1193–1201. [CrossRef]
20. Lv, P.; Sun, S.; Lin, C.; Liu, G. A method for weak target detection based on human visual contrast mechanism. *IEEE Geosci. Remote Sens. Lett.* **2018**, *16*, 261–265. [CrossRef]
21. Han, J.; Moradi, S.; Faramarzi, I.; Liu, C.; Zhang, H.; Zhao, Q. A local contrast method for infrared small-target detection utilizing a tri-layer window. *IEEE Geosci. Remote Sens. Lett.* **2019**, *17*, 1822–1826. [CrossRef]
22. Lan, Y.; Peng, B.; Wu, X.; Teng, F. Infrared dim and small targets detection via self-attention mechanism and pipeline correlator. *Digit. Signal Process.* **2022**, *130*, 103733. [CrossRef]
23. Shi, F.; Qiu, F.; Li, X.; Tang, Y.; Zhong, R.; Yang, C. A method to detect and track moving airplanes from a satellite video. *Remote Sens.* **2020**, *12*, 2390. [CrossRef]
24. Fujita, K.; Hanada, T.; Kitazawa, Y.; Kawabe, A. A debris image tracking using optical flow algorithm. *Adv. Space Res.* **2012**, *49*, 1007–1018. [CrossRef]
25. Kalman, R.E. A New Approach to Linear Filtering and Prediction Problems. *J. Basic Eng.* **1960**, *82*, 35–45. [CrossRef]
26. La Scala, B.F.; Bitmead, R.R. Design of an extended Kalman filter frequency tracker. *IEEE Trans. Signal Process.* **1996**, *44*, 739–742. [CrossRef]
27. Huang, T.; Xiong, Y.; Li, Z.; Zhou, Y.; Li, Y. Space Target Tracking by Variance Detection. *J. Comput.* **2014**, *9*, 2107–2115. [CrossRef]
28. Hao, L.; Mao, Y.; Yu, Y.; Tang, Z. A method of GEO targets recognition in wide-field opto-electronic telescope observation. *Opto-Electron. Eng.* **2017**, *44*, 418–426.
29. Lin, J.; Ping, X.; Ma, D. Small target detection method in drift-scanning image based on DBT. *Infrared Laser Eng.* **2013**, *42*, 3440–3446.
30. Mehta, D.S.; Chen, S.; Low, K.-S. A rotation-invariant additive vector sequence based star pattern recognition. *IEEE Trans. Aerosp. Electron. Syst.* **2018**, *55*, 689–705. [CrossRef]
31. Yang, L.; Niu, Y.; Zhang, Y.; Lü, J.; Li, J.; Niu, H.; Liu, W.; Zhang, Y. Research on Detection and Recognition of Space Targets Based on Satellite Photoelectric Imaging System. *Laser Optoelectron. Prog.* **2014**, *51*, 121102. [CrossRef]
32. Bo, M. Research on Aerial Infrared Small Target Detection and Hardware Acceleration. Master's Thesis, Beijing University of Technology, Beijing, China, 2016.
33. Zhang, Q. Design and Implementation of Spaceborne Infrared Small Target Detection System Based on FPGA. Master's Thesis, Huazhong University of Science and Technology, Wuhan, China, 2019.

34. Liu, W. Object tracking under complicated background based on DSP+FPGA platform. *Chin. J. Liq. Cryst. Disp.* **2014**, *29*, 1151–1155.
35. Seznec, M.; Gac, N.; Orieux, F.; Naik, A.S. Real-time optical flow processing on embedded GPU: An hardware-aware algorithm to implementation strategy. *J. Real-Time Image Process.* **2022**, *19*, 317–329. [CrossRef]
36. Diprima, F.; Santoni, F.; Piergentili, F.; Fortunato, V.; Abbattista, C.; Amoruso, L. Efficient and automatic image reduction framework for space debris detection based on GPU technology. *Acta Astronaut.* **2018**, *145*, 332–341. [CrossRef]
37. Tian, H.; Guo, S.; Zhao, P.; Gong, M.; Shen, C. Design and Implementation of a Real-Time Multi-Beam Sonar System Based on FPGA and DSP. *Sensors* **2021**, *21*, 1425. [CrossRef]
38. Sun, Q.; Niu, Z.D.; Yao, C. Implementation of Real-time Detection Algorithm for Space Debris Based on Multi-core DSP. *J. Phys. Conf. Ser.* **2019**, *1335*, 012003. [CrossRef]
39. Gyaneshwar, D.; Nidamanuri, R.R. A real-time FPGA accelerated stream processing for hyperspectral image classification. *Geocarto Int.* **2022**, *37*, 52–69. [CrossRef]
40. Han, K.; Pei, H.; Huang, Z.; Huang, T.; Qin, S. Non-cooperative Space Target High-Speed Tracking Measuring Method Based on FPGA. In Proceedings of the 2022 7th International Conference on Image, Vision and Computing (ICIVC), Xi'an, China, 26–28 July 2022; pp. 222–231.
41. Yang, B.; Yang, M.; Plaza, A.; Gao, L.; Zhang, B. Dual-mode FPGA implementation of target and anomaly detection algorithms for real-time hyperspectral imaging. *IEEE J. Sel. Top. Appl. Earth Observ. Remote Sens.* **2015**, *8*, 2950–2961. [CrossRef]
42. Xu, Y.; Zhang, J. Real-time detection algorithm for small space targets based on max-median filter. *J. Inf. Comput. Sci.* **2014**, *11*, 1047–1055. [CrossRef]
43. Han, L.; Tan, C.; Liu, Y.; Song, R. Research on the On-orbit Real-time Space Target Detection Algorithm. *Spacecr. Recovery Remote Sens.* **2021**, *42*, 122–131.
44. Choi, E.-J.; Yoon, J.-C.; Lee, B.-S.; Park, S.-Y.; Choi, K.-H. Onboard orbit determination using GPS observations based on the unscented Kalman filter. *Adv. Space Res.* **2010**, *46*, 1440–1450. [CrossRef]
45. Babu, P.; Parthasarathy, E. FPGA implementation of multi-dimensional Kalman filter for object tracking and motion detection. *Eng. Sci. Technol. Int. J.* **2022**, *33*, 101084. [CrossRef]
46. Zhang, X.; Xiang, J.; Zhang, Y. Space Object Detection in Video Satellite Images Using Motion Information. *Int. J. Aerosp. Eng.* **2017**, *2017*, 1024529. [CrossRef]
47. Li, Q.; Li, R.; Ji, K.; Dai, W. Kalman filter and its application. In Proceedings of the 2015 8th International Conference on Intelligent Networks and Intelligent Systems (ICINIS), Tianjin, China, 1–3 November 2015; pp. 74–77.
48. Kuhn, H.W. The Hungarian method for the assignment problem. *Nav. Res. Logist. Q.* **1955**, *2*, 83–97. [CrossRef]
49. Zhu, H.; Zhou, M. Efficient role transfer based on Kuhn–Munkres algorithm. *IEEE Trans. Syst. Man Cybern. Part A Syst. Hum.* **2011**, *42*, 491–496. [CrossRef]
50. Mirzaeinia, A.; Hassanalian, M. Minimum-cost drone–nest matching through the kuhn–munkres algorithm in smart cities: Energy management and efficiency enhancement. *Aerospace* **2019**, *6*, 125. [CrossRef]
51. Luetteke, F.; Zhang, X.; Franke, J. Implementation of the hungarian method for object tracking on a camera monitored transportation system. In Proceedings of the ROBOTIK 2012: 7th German Conference on Robotics, Munich Germany, 21–22 May 2012; pp. 1–6.
52. Kuipers, J.B. *Quaternions and Rotation Sequences: A Primer with Applications to Orbits, Aerospace, and Virtual Reality*; Princeton University Press: Princeton, NJ, USA, 1999.
53. Tang, Z.; Von Gioi, R.G.; Monasse, P.; Morel, J.-M. A precision analysis of camera distortion models. *IEEE Trans. Image Process.* **2017**, *26*, 2694–2704. [CrossRef]
54. Weng, J.; Cohen, P.; Herniou, M. Camera calibration with distortion models and accuracy evaluation. *IEEE Trans. Pattern Anal. Mach. Intell.* **1992**, *14*, 965–980. [CrossRef]
55. Han, J.; Moradi, S.; Faramarzi, I.; Zhang, H.; Zhao, Q.; Zhang, X.; Li, N. Infrared small target detection based on the weighted strengthened local contrast measure. *IEEE Geosci. Remote Sens. Lett.* **2020**, *18*, 1670–1674. [CrossRef]

remote sensing

MDPI

Article

SE-RRACycleGAN: Unsupervised Single-Image Deraining Using Squeeze-and-Excitation-Based Recurrent Rain-Attentive CycleGAN

Getachew Nadew Wedajew [1] and Sendren Sheng-Dong Xu [1,2,*]

[1] Graduate Institute of Automation and Control, National Taiwan University of Science and Technology, Taipei 106, Taiwan; d11112801@mail.ntust.edu.tw

[2] Advanced Manufacturing Research Center, National Taiwan University of Science and Technology, Taipei 106, Taiwan

* Correspondence: sdxu@mail.ntust.edu.tw

Abstract: In computer vision tasks, the ability to remove rain from a single image is a crucial element to enhance the effectiveness of subsequent high-level tasks in rainy conditions. Recently, numerous data-driven single-image deraining techniques have emerged, primarily relying on paired images (i.e., in a supervised manner). However, when dealing with real deraining tasks, it is common to encounter unpaired images. In such scenarios, removing rain streaks in an unsupervised manner becomes a challenging task, as there are no constraints between images, resulting in suboptimal restoration results. In this paper, we introduce a new unsupervised single-image deraining method called SE-RRACycleGAN, which does not require a paired dataset for training and can effectively leverage the constrained transfer learning capability and cyclic structures inherent in CycleGAN. Since rain removal is closely associated with the analysis of texture features in an input image, we proposed a novel recurrent rain attentive module (RRAM) to enhance rain-related information detection by simultaneously considering both rainy and rain-free images. We also utilize the squeeze-and-excitation enhancement technique to the generator network to effectively capture spatial contextual information among channels. Finally, content loss is introduced to enhance the visual similarity between the input and generated images. Our method excels at removing numerous rain streaks, preserving a smooth background, and closely resembling the ground truth compared to other approaches, based on both quantitative and qualitative results, without the need for paired training images. Extensive experiments on synthetic and real-world datasets demonstrate that our approach shows superiority over most unsupervised state-of-the-art techniques, particularly on the Rain12 dataset (achieving a PSNR of 34.60 and an SSIM of 0.954) and real rainy images (achieving a PSNR of 34.17 and an SSIM of 0.953), and is highly competitive when compared to supervised methods. Moreover, the performance of our model is evaluated using RMSE, FSIM, MAE, and the correlation coefficient, achieving remarkable results that indicate a high degree of accuracy in rain removal and strong preservation of the original image's structural details.

Keywords: content loss; Recurrent Rain-Attentive Module; single-image deraining; squeeze-and-excitation

Citation: Wedajew, G.N.; Xu, S.S.-D. SE-RRACycleGAN: Unsupervised Single-Image Deraining Using Squeeze-and-Excitation-Based Recurrent Rain-Attentive CycleGAN. *Remote Sens.* **2024**, *16*, 2642. https://doi.org/10.3390/rs16142642

Academic Editor: Salah Bourennane

Received: 8 May 2024
Revised: 28 June 2024
Accepted: 6 July 2024
Published: 19 July 2024

1. Introduction

Weather conditions, including snow, haze, rain, and wind, cause low visibility in images and videos. This can considerably impair the performance of outdoor vision tasks, such as facial recognition, pedestrian detection, visual tracking, traffic sign identification, object detection, and intelligent surveillance [1–6]. As a result, it is crucial to remove rain from input rainy images to develop trustworthy computer vision systems. Thus, algorithms that can successfully remove rain from a rainy image are of great interest [7].

To address the restoration of rain degradation, various strategies have been proposed including video deraining [8,9] and single-image deraining (SID) [10–13]. Video deraining uses continuous information between frames to recognize the rain line and restore the background image. This method suffers from poor performance when the camera movements are dynamic. Additionally, as they analyze multiple sequential frames, they require significant computational time, which is critical for some applications, like self-driving cars [10]. On the other hand, SID relies solely on spatial information between adjacent pixels and the visual characteristics of the rain line to remove rain [14]. SID approaches are more difficult since they can only exploit the spatial information in an image, unlike video deraining methods, which can benefit from the dynamics of rainfall and temporal redundancy. In this study, we mainly concentrate on the issue of removing rain streaks from a single image. The goal of single-image rain removal is to eliminate raindrops or streaks from an input image and restore the clean backdrop. Removing rain streaks from single-image is therefore an important research topic that has recently gained a lot of attention in the field of computer vision and pattern recognition [15,16].

SID techniques from the past can be largely split into two categories: model-based methods and data-driven methods [10,12]. The model-based methods focus on incorporating rain's physical characteristics and background scene knowledge into an optimization problem, and they develop logical algorithms to solve it. They often estimate raindrops or rain streaks using various priors or assumptions, such as sparse coding, low rank, and Gaussian. Despite the significant advancement gained by these approaches, their performance is usually limited, especially when the background is messy and contains intricate illuminations. The main reason for this limited performance is that real-world raindrops and rain streaks do not strictly conform to a sparse or Gaussian distribution. Recently, data-driven methods have been developed by creating certain network architectures and pre-collecting pairs of rainy and clean (ground-truth) images to train network parameters in a supervised manner, aiming to achieve sophisticated rain removal functions [10,17]. However, obtaining accurate paired datasets in the real world is challenging due to environmental constraints. Consequently, supervised learning methods rely on synthetic datasets, posing a challenge to generalization due to the disparity between synthetic and real datasets.

Hence, studying unsupervised SID techniques is essential for enhancing rain removal performance on real images, as they can be trained with real rainy images without a ground truth. The unsupervised CycleGAN network [18] is a logical choice for rain removal. While CycleGAN has demonstrated effectiveness in multiple low-level tasks, applying it to remove rain from single images remains challenging due to the asymmetrical domain knowledge between rainy and rain-free images. In particular, a rainy image consists of both background and rain information, whereas the rain-free image only comprises the background. Consequently, directly employing CycleGAN may lead to issues with color and structural distortion, as well as difficulty in completely erasing rain marks (see Figure 1b).

(**a**) (**b**) (**c**) (**d**)

Figure 1. Comparisons of our result with CycleGAN [18] on Rain100L. (**a**) Input, (**b**) CycleGAN [18], (**c**) our approach, and (**d**) GT.

In this study, we introduce a novel unsupervised approach to SID, eliminating the need for aligned image pairings. Our method incorporates a rain-attentive module, allowing it to adapt to any image and leverage the circulatory architecture of CycleGAN. Our model is specifically tailored for unsupervised single-image deraining (SID), with the ability to retain the color and structure of images achieved through the multi-loss constrained rain-attentive module (see Figure 1c). The contribution of our work can be summarized as follows:

1. We propose SE-RRACycleGAN, which can generate high-quality, clean, derained images without supervision in the form of aligned rainy and clean images.

2. We propose a novel rain streak extractor termed the Recurrent Rain-Attentive Module (RRAM), which can detect rain information in both rainy and rain-free images.

3. A squeeze-and-excitation (SE) component is introduced to the generator network, so that it can push the network to learn more useful features, prevent model overfitting, and reinforce the network's generalization ability. We also introduce content loss to generate an image that is visually similar to the input image.

4. Extensive experiments on synthetic and real datasets show that our method gives a competitive result with semi-supervised and supervised methods and outperforms the state-of-the-art unsupervised methods on real rainy images.

The paper is structured as follows: Section 2 describes related works; Section 3 introduces the proposed method, detailing its components and the objective function; experimental results and discussions are presented in Section 4; and the final section summarizes the work presented in this paper.

2. Related Works

In this section, we provide a concise overview of the existing literature on SID and position the proposed method within the appropriate context. As discussed in Section 1, methods for SID can be broadly categorized into two groups: model-based approaches and data-driven approaches.

2.1. Model-Based Approaches

Numerous model-based methods consider the deraining of a single image as an image decomposition problem, where a rainy image is typically represented as the sum of a rain layer and a rain-free background layer. Using a bilateral filter, Kang et al. [19] decomposed a rainy image into low-frequency (LF) and high-frequency (HF) components. They used sparse coding and morphological component analysis (MCA)-based dictionary learning to separate the rain streaks in the HF component. Luo et al. [20] developed an image patch-based discriminative sparse coding framework that distinguished between rain streaks and the rain-free background. Li et al. [21] proposed a patch-based priors for the rain and rain-free background layers. These priors can take into account different rain streak directions and scales since they are based on Gaussian mixture models. Wang et al. [22] proposed an algorithm that takes an advantage of image decomposition and dictionary learning methods. While these model-based techniques, relying on assumptions and priors, perform well in certain scenarios, they face challenges in eliminating complex rain patterns in real-world environments. This is because the assumptions and priors upon which they rely do not always hold, as real-world raindrops and rain streaks do not strictly adhere to a sparse or Gaussian distribution.

2.2. Data-Driven Approaches

2.2.1. Supervised Learning Method

To enhance prediction accuracy, this method employs a network specifically designed to automatically learn rainline properties from extensive paired data. Ahn et al. [10] proposed a two-step rain removal method. The proposed method first predicts rain streaks, including rain density and streak intensity, from an input rainy image. Then, it can effectively remove rain streaks from images taken under diverse rain conditions. Zhang et al. [23] presented a framework termed an image-deraining conditional generative adversarial network (ID-CGAN), which incorporates discriminative, quantitative, and visual performance into the objective function. Yang et al. [24] introduced a rain removal architecture that effectively detects and removes rain streaks, demonstrating superior performance in heavy rain conditions. They utilized a recurrent process to progressively eliminate overlapping rain streaks in diverse forms and directions. Wang et al. [17] proposed a kernel-guided convolutional neural network (KGCNN) and achieved a good result in solving the problem of over- and under-deraining. However, all of the techniques mentioned above rely on paired datasets, which are difficult to obtain in real-world scenarios.

2.2.2. Unsupervised Learning Method

Recently, some researchers have proposed a GAN-based unsupervised learning approach for SID, drawing inspiration from GANs' remarkable success in image-to-image translation [25]. Zhu et al. [18] proposed a CycleGAN for learning image-to-image translation in the absence of paired images. The proposed method was constrained by using adversarial loss and cycle consistency loss to make the translated image indistinguishable from the ground truth. Yang et al. [26] presented an unsupervised end-to-end rain removal network termed Rain Removal-GAN (RR-GAN). By introducing a physical model that explicitly learns recovered images and related rain streaks from a differentiable programming perspective, their network mitigates the paired training constraints. Moreover, to recover the clean image, the multi-scale attention memory generator and multi-scale discriminator, which impose constraints on the clean output image, were employed. However, the attention memory used relies only on the single branch from rainy to rain-free images and is also not constrained to learn more about the rain line in the rainy image, which makes it unstable for an unsupervised model due to its limited detection ability. Guo et al. [13] proposed unsupervised derain attention-guided GAN (Derain Attention GAN), which contains a generator featuring an attention mechanism and a multi-scale discriminator to produce rain-free images and distinguish the generated rain-free images, respectively. Also, perceptual consistency loss and internal feature perceptual loss are presented to reduce the artifact features on the generated images. However, they cannot accurately extract rain streaks from a rainy image solely by considering cycle-consistency loss. This is because they do not consider the constraint on the attention mechanism, resulting in a weak constraint between the rainy and rain-free images. Wei et al. [11] presented an unsupervised framework for single-image rain removal and generation termed DerainCycleGAN. They developed an unsupervised rain-attentive detector (URAD) to improve the detection of rain information in both rainy and rain-free images. Moreover, they generated a rain streak with varied shapes and directions, which is distinct from the previous methods. However, the constraint of URAD is weak, and it detects rain masks from rain-free images.

2.3. Visual Attention

Visual attention models have been utilized to pinpoint specific areas within an image for the purpose of capturing distinctive local features [27]. The idea has been employed for visual identification, categorization, and image classification [28–30]. Likewise, the concept has demonstrated its effectiveness in both the supervised SID method [31] and the unsupervised SID method [12,26], as it enables the network to determine the specific areas where the removal or restoration process should be concentrated and enhance the precision of the SID task. However, it is important to note that the attention module used in [26,31] primarily relies on a single branch, focusing only on a mapping from rain to rain-free. Furthermore, their constraints are relatively weak, and the attention module used in [12] is unconstrained. Thus, in unsupervised mode, they tend to be unstable due to their limited detection capability, as they rely solely on the rain information from rainy images. On the contrary, we introduced RRAM, which detects rain-related information by simultaneously considering both rainy and rain-free images and is constrained by attentive loss to learn without supervision. So the rain information detected by our RRAM is more stable and accurate than the attention modules used previously.

3. The Proposed Method

Our objective is to acquire the ability to eliminate rain streaks from a sole input image without relying on paired training data. Figure 2 illustrates the architecture of our

SE-RRACycleGAN, which consists of three components: (1) RRAM, which focuses on rain-related details in both images with and images without rain, (2) a pair of generators G and F, responsible for generating rain-free and rainy images, respectively, and (3) a pair of discriminators D_G and D_F, designed to distinguish real images from generated ones. In the subsequent sections, we present detailed explanations of each of the three components and the objective function.

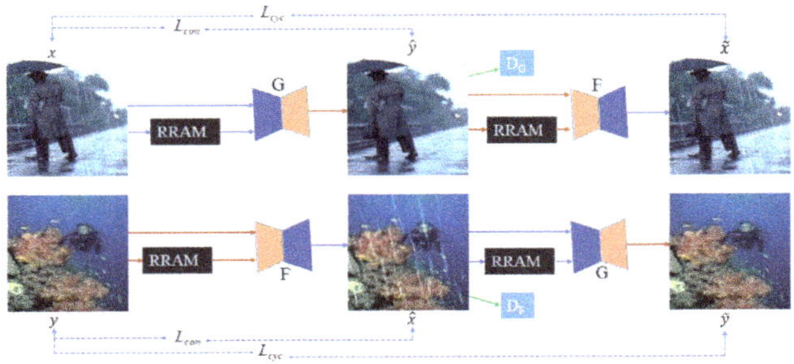

Figure 2. The architecture of SE-RRACycleGAN. The SE-RRACycleGAN process involves two distinct branches. The rainy-to-rainy branch starts with a rainy image and employs a generator G to produce a rain-free image. Then, this rain-free image is utilized to reconstruct a rainy image using generator F. On the other hand, the rain-free to rain-free branch, begins with a rain-free image, which is initially transformed into a rain image by F and then reversed back to a rain-free image using G.

3.1. Recurrent Rain-Attentive Module (RRAM)

In Figure 3, we introduced RRAM to enhance rain-related information detection by simultaneously considering both rainy and rain-free images. To identify rain information within arbitrary images, our RRAM is constrained by attentive loss in both rain-free and rainy images when learning without supervision. As a result, the rain information detected by our RRAM offers a higher level of accuracy compared to the attention modules used previously. RRAM differs from Cycle-Attention-Derain [12] and CBAM [32] in that it not only incorporates channel and spatial attention blocks, but also utilizes an *LSTM* unit [33]. This combination allows RRAM to effectively capture spatial and temporal dependencies in images, facilitating the modeling of intricate patterns of rain streaks interacting with the scene. *LSTM*, with its ability to retain past states [34], effectively models the accumulation of rain streaks and the evolution of rain patterns over time in a single image while mitigating the vanishing gradient problem. Furthermore, LSTM can discern between rain streaks and other image elements, ensuring that the generated rain mask by our RRAM accurately represents the presence of rain compared to those generated by previously used attention modules. Its ability to handle variable-length sequences is crucial in situations where the density and size of rain streaks vary throughout the image.

In our RRAM, each iteration involves *Conv* + *ReLU* operations for feature extraction from the input image and the preceding segment's mask. This is followed by an *LSTM* unit [33], *CSAB*, and a concluding *Conv* layer for generating 2D attention maps. The *LSTM* unit comprises input gate i_t, forget gate f_t, output gate o_t, and cell state c_t. The interactions of these states and gates over time are described as follows:

$$\begin{cases} i_t = \delta(W_{xi} * X_t + W_{hi} * H_{t-1} + b_i) \\ f_t = \delta(W_{xf} * X_t + W_{hf} * H_{t-1} + b_f) \\ g_t = \tanh(W_{xg} * X_t + W_{hg} * H_{t-1} + b_g) \\ c_t = f_t \circ c_{t-1} + i_t \circ g_t \\ o_t = \delta(W_{xo} * X_t + W_{ho} * H_{t-1} + b_o) \\ h_t = o_t \circ \tanh(c_t) \end{cases} \tag{1}$$

where X_t represents the features obtained by $Conv + ReLU$ unit; c_t denotes the cell state that will be fed to the next iteration of the $LSTM$ unit; H_t denotes the output features of the $LSTM$ unit, W and b are convolutional matrix and bias vector, respectively; $*$ denotes the convolution operation; $[\circ]$ represents the concatenate operation; and δ denotes the sigmoid activation function. Each convolution in LSTM uses 32 + 32 input channels, 32 output channels, a kernel size of 3×3, 1 stride, and 1 padding.

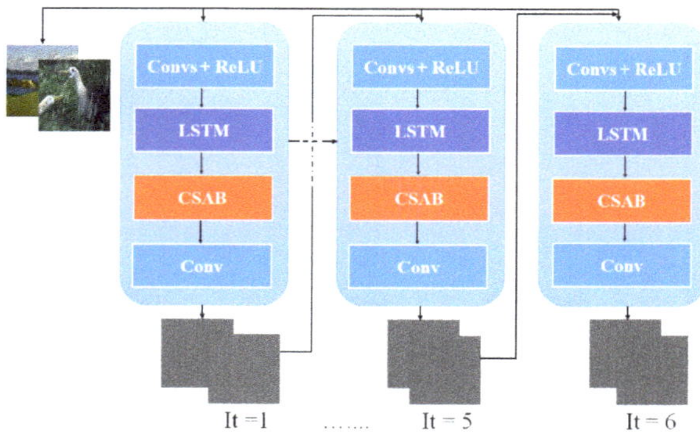

Figure 3. Recurrent rain attentive module architecture.

In our RRAM, *CSAB* denotes a combination of a channel attention block (CAB) and a spatial attention block (SAB), as illustrated in Figure 4. We suggest employing the CAB to distinguish rain streak features from the background and utilizing the SAB to recognize specific attributes such as the sizes, shape, and positioning of the rain streaks. The $LSTM$ unit output, H_t, passes through three convolution layers with kernel size of 3×3, which is represented by *Convs* in Figure 4. The first layer is followed by batch normalization and *ReLU* [35], the second layer is followed by batch normalization, and the third layer produces the intermediate feature map, $Y = R^{C \times H \times W}$. Y is input to the CAB to obtain the channel attention map of $C \times 1 \times 1$, multiplied with Y to yield the feature map of Z_c. Z_c passes through SAB to generate the spatial attention map of $1 \times H \times W$, multiplied by Z_c to yield Z_s, summarized as follows:

$$\begin{cases} Z_c = M_c(Y) \otimes Y \\ Z_s = M_c(Z_c) \otimes Z_c \end{cases} \tag{2}$$

where $\otimes$ denotes element-wise multiplication, $M_c(\cdot)$ represents the CAB, and $M_s(\cdot)$ represents the SAB.

Lastly, the output feature F is obtained by adding H_t to Z_s and passing the result through *ReLU* nonlinearity (Equation (3)). This output is then passed through the last convolution layer of RRAM to generate the rain mask (Equation (4)).

$$F = ReLU(H_t \oplus Z_s) \tag{3}$$

$$Rainmask = Conv(F) \tag{4}$$

where $\oplus$ denotes element-wise addition, ReLU is the ReLU activation function, and *Conv* is the convolution layer with 3×3 kernel size, 1 stride and 1 padding.

The following presents the specifics of the channel attention block and the spatial attention block.

Channel Attention Block (CAB): As each channel within a feature map is considered as a feature detector, channel attention directs its focus toward determining the rain streak of the input rainy image. Thus, to more effectively differentiate rain streak characteristics from background attributes, we have integrated the CAB [32] into RRAM. In our CAB, we employ both average-pooling and max-pooling simultaneously for feature aggregation, for which it is confirmed that it improves the representation power of RRAM rather than using each independently. Spatial information is gathered from the feature map Y by employing average-pooling and max-pooling operations, forming two distinct spatial context features: average-pooled and max-pooled features. Then, both features are fed into the convolutional network, and their outputs are summed together by using element-wise summation to obtain channel attention map $M_c(Y) \epsilon R^{C\times 1\times 1}$ [32]. The convolutional network consists of two convolution layers, as shown in (Equation (5)). In summary, the computation of the channel attention map is as follows:

$$M_c(Y) = \delta(Conv(AvgPool(Y)) + Conv(MaxPool(Y))) \quad (5)$$

where δ represents the sigmoid activation function, $Conv$ denotes two convolutional layer with 1×1 kernel size, 1 stride and 0 padding.

Spatial Attention Block (SAB): Unlike the channel attention map, the spatial attention map focuses on locating rain streaks in rainy images, complementing channel attention. We employ average-pooling and max-pooling along the channel axis, concatenating the results to create an effective feature descriptor for spatial attention computation. The effective highlighting of rain streak regions is achieved through pooling operations along the channel axis [36]. We form a spatial attention map, $M_c(Y) \epsilon R^{1\times H\times W}$, to indicate areas for emphasis or suppression. This is achieved by applying a convolution layer to the concatenated feature descriptor. The process involves creating two 2D maps, $Z_s max \epsilon R^{1\times H\times W}$ (max-pooled features) and $Z_s avg \epsilon R^{1\times H\times W}$ (average-pooled features), through channel information combination using two pooling methods. The final spatial attention map is obtained by concatenating and convolving these features with a typical convolution layer (Equation (6)).

$$M_s(Z_c) = \sigma(Conv(AvgPool(Z_c); MaxPool(Z_c))) \quad (6)$$

where σ represents the sigmoid activation function and $Conv$ denotes convolutional layer with 7×7 kernel size, 3 stride, and 0 padding.

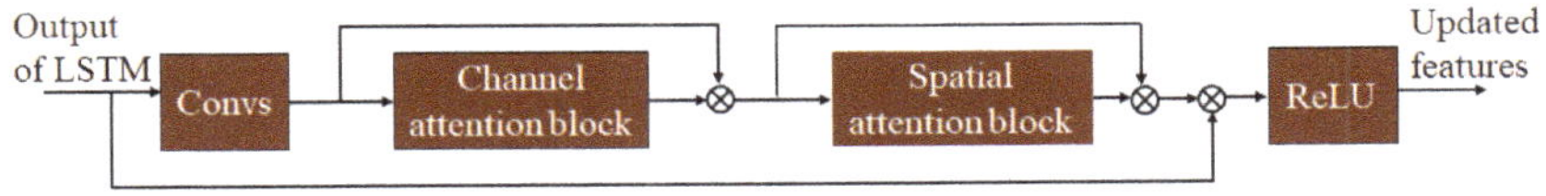

Figure 4. The structure of our CSAB. *Convs* represents three convolution layers with kernel size of 3×3, where the first convolution layer is followed by batch normalization and ReLU, the second convolution layer is followed by batch normalization, and the third convolution layer results in the intermediate feature map.

To validate our RRAM's effectiveness, in Figure 5, we present rain masks detected by RRAM in both rainy and rain-free images. The RRAM input can be rainy images or rain-free images, with the output specifically addressing rain information present in the input image. In Figure 5a, rainy images and their corresponding rain masks identified by RRAM are displayed. It is evident from the illustration that the rain mask becomes clear from iteration to iteration. In Figure 5b, rain-free images are shown, where RRAM correctly identifies the absence of a rain mask from the outset. This distinguishes our RRAM from previous rain attention mechanisms, which erroneously detect rain masks in rain-free images during initial iterations. In order to clearly illustrate the differences between iterations, we made adjustments to the grey color scale of the rain masks detected by our RRAM. Specifically, we shifted the scale from 255 to 176 and reduced the brightness from 0 to −6. This modification enhances the visibility of subtle variations in the rain mask, ensuring they are clearly discernible while maintaining a distinct contrast with the background.

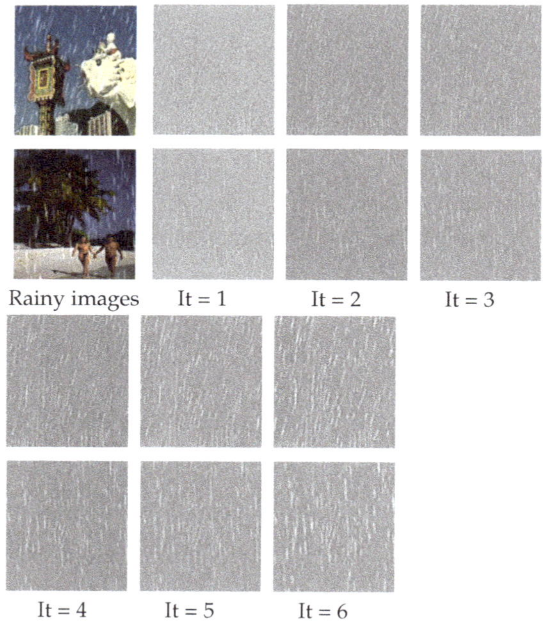

(**a**) The rain mask detected by our RRAM in rainy images.

(**b**) The rain mask detected by our RRAM in rain-free images.

Figure 5. The rain mask detected by RRAM in both (**a**) rainy images and (**b**) rain-free images. The rain mask in rainy images becomes clearer as the number of iterations increases, while there is no rain mask detected in rain-free images. where It denotes iteration.

3.2. Generator

To generate rain-free images from rainy images, we used a generator with a structure similar to the U-net contextual encoder–decoder network [37]. However, our generator differs by incorporating squeeze-and-excitation (SE) blocks [38], which adaptively re-weight feature channels. The encoder module comprises eight Conv-ReLU blocks with strided convolutions, aimed at reducing the spatial dimensions of the feature maps, effectively downsampling the input image to extract hierarchical features. Additionally, one block of SE is included, as depicted in Figure 6a (the SE blocks are highlighted in blue, with the left part for the encoder and the right part for the decoder). The output feature of the fourth Conv-ReLU serves as the input features for the SE block, and the output feature of the SE block is then used as the input feature for the fifth Conv-ReLU. Similarly, the decoder is structured with eight Conv-ReLU blocks and one SE block. In the decoding stage, transposed convolutional layers (also known as deconvolutional layers) are used to increase the spatial dimensions of feature maps (i.e., upsampling). The SE block is inserted into the decoder part of the Conv-ReLU blocks, mirroring its position in the encoder part of the Conv-ReLU blocks. Moreover, two skip connections are utilized to propagate fine-grained information from earlier layers to later layers in the network, aiding in the recovery of spatial details lost during encoding. Our generator takes as input the concatenation of the original image and the rain information identified by our proposed RRAM. Notably, this marks the first instance of incorporating SE in a generator for unsupervised single-image rain streak removal, as far as our knowledge extends. A detailed discussion of SE is given below.

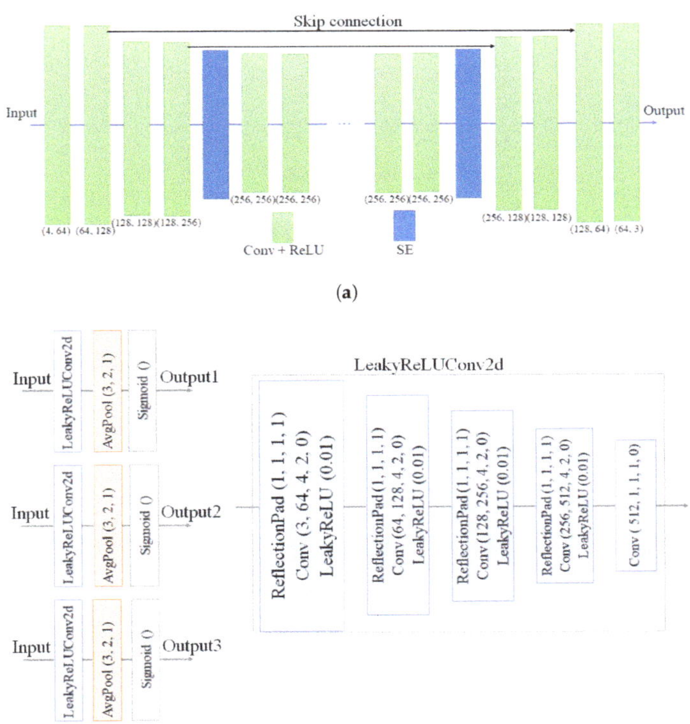

Figure 6. The structure of (**a**) generator and (**b**) discriminator.

Squeeze-and-Excitation Block

Hu et al. [38] introduced SE, a channel relationship representation that adaptively recalibrates channel-wise feature responses by modeling interdependencies between channels. Each channel has a varied capacity to extract rain components, which can have varying densities and directions in SID tasks. It is generally not fair to consider all feature maps identically when extracting the rain component layer. The contributions made by various feature maps to the rain component layer may vary. We thus apply SE enhancement within our generator network, leveraging its ability to capture spatial contextual information effectively among channels, which has been found to be significant for SID tasks. In SID, certain channels may hold essential information for rain streak removal, while others contain noise or irrelevant details. The novelty of our approach lies in leveraging SE blocks to enhance rain-related features, thereby minimizing the impact of noise and leading to more accurate and effective deraining results. By integrating an SE block into the generator, the model is empowered to learn discriminative features, facilitating the capture of complex rain patterns and the representation of the image's underlying structure This capability is particularly significant for deraining tasks that necessitate both local and global feature consideration. Moreover, the SE block contributes to the generator's ability to generalize across diverse rain patterns by dynamically adjusting the weights of features based on the characteristics of the input image. Consequently, this enhances the robustness and effectiveness of the deraining model across various rain scenarios. The SE process is illustrated in Figure 7 and briefly reviewed as follows. We consider $F = [f_1, f_2, \ldots, f_c]$ as the input feature map for the SE block. First, F is operated by global average pooling, producing a vector $v_c \in R^{1\times1\times C}$ with its cth element as shown in Equation (7):

$$v_c = \frac{1}{H \times W} \sum_{i=1}^{H} \sum_{j=1}^{W} f_c(i,j) \tag{7}$$

where $f_c(i,j)$ and v_c represent the value at position (i,j) of the cth channel and the corresponding output, respectively.

To fully capture interdependencies among channels resulting from the aggregated information via global average pooling, we follow it with a second operation. The function must satisfy two requirements in order to achieve this goal: Firstly, it should have the capacity to learn nonlinear interactions among channels. Secondly, it should be capable of learning a relationship that is not mutually exclusive, allowing for the emphasis of multiple channel-wise features rather than a one-hot activation. In order to fulfill these requirements, we choose to utilize a straightforward gating mechanism featuring a sigmoid activation function:

$$\beta = \sigma(W_2 * \delta(W_1 * v_c)) \tag{8}$$

where the *sigmoid* and *ReLU* functions are denoted by $\sigma(\cdot)$ and $\delta(\cdot)$, respectively. $*$ Indicates the convolution operation. W_1 represents the weight set of a convolutional layer, serving as channel downscaling with a reduction ratio of r. Following activation by ReLU, the signal of lower dimensionality is subsequently augmented with a ratio of r by a channel-upscaling layer, characterized by the weight set W_2. Afterward, we acquire the ultimate channel statistics, denoted as β, which is utilized to recalibrate the input.

Finally, the output of SE is expressed as Equation (9):

$$F_{SE} = \beta \otimes F \tag{9}$$

where $\otimes$ denotes channel-wise multiplication for feature channels and corresponding channel weights.

Generally, the SE block is used to assign weight to each channel. This process makes it possible to adaptively recalibrate each feature map's feature response, which makes it easier to capture additional spatial contextual information [39].

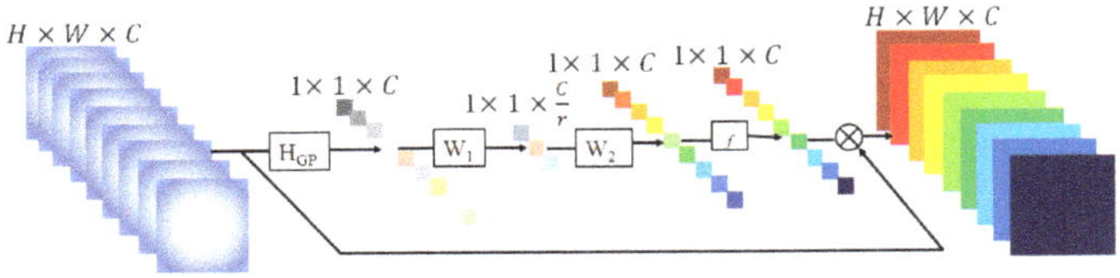

Figure 7. Squeeze-and-excitation [40]: where H_{GP} and $f(\cdot)$ denote the global pooling function and activation function, respectively.

3.3. Discriminator

In the GAN framework, removing rain from a rainy image goes beyond enhancing visual appeal and achieving quantitative comparability with the ground truth. It requires ensuring that the derained output closely resembles the original ground truth image. To achieve this, employing a robust discriminator that captures both local and global information is crucial for distinguishing between real and fake images. We use two discriminators, D_G and D_F, with structures similar to [41], employing a multi-scale structure with feature maps passed through five convolutional layers and supplied into the sigmoid (Figure 6b). The input to the discriminator network is the ground truth image and the generated image by our generator.

3.4. Objective Function

Our objective function contains four types of losses as elaborated below.

Attentive losses: Like in Derain CycleGAN [11], constraining the unsupervised learning of RRAM is essential to identify rain-related details in any given image. Due to the absence of ground truth for rain information, we use a combination of prior knowledge and self-supervision techniques in RRAM training. In particular, we conduct an initial assessment on the mask of the rain-free image, denoted as L_{att_y}, as shown in Equation (10):

$$L_{\text{att}_y} = ||R(y) - Z||_2^2 \tag{10}$$

where $R(y)$ is the mask of the rain-free image identified by RRAM. Z is a distribution comprising zeros with the same shape as the mask. As there is no rain information present in the rain-free image y, the distribution $R(y)$ should closely resemble Z.

Then, we perform self-supervision on the rain image mask, identified as L_{att_y} (Equation (11)).

$$L_{\text{att}_y} = ||(R(x) + \hat{y}) - x||_2^2 \tag{11}$$

where $R(x)$, the mask, identifies the rainy regions in the rainy image as detected by RRAM. $\hat{y}$ and x refer to the derained image and original rainy image, respectively. Engaging in self-supervised learning enables RRAM to focus on rain-related information, as illustrated in Figure 5.

Cycle-consistency losses: Like in CycleGAN [18], we define a cycle-consistency loss with the aim of promoting similarity between the reconstructed image $F(G(x))$ and the original real rain image x and ensuring that $G(F(y))$ matches the input y.

$$L_{cyc} = E_{x \sim P_{\text{data}}(x)}[||\tilde{x} - x||_1] + E_{y \sim P_{\text{data}}(y)}[||\tilde{y} - y||_1] \tag{12}$$

where L_{cyc} is cyclic-consistency losses and $\tilde{x} = F(G(x))$ and $\tilde{y} = G(F(y))$. $P_{data}(y)$ and $P_{data}(x)$ represent the data distributions of locally cropped patches randomly extracted from generated rain-free images and rain images, respectively.

Adversarial losses: The goal of a GAN is to employ a min-max game strategy, aiming to train generators G and F to produce samples resembling the data distribution so that the discriminator cannot distinguish between generated and real samples [42]. Simultaneously, the goal is to train discriminators D_G and D_F to effectively discern generated and real images. To achieve this, the suggested approach iteratively updates both the generators and discriminators following the framework outlined in [43]. $L_{adv}(G, D_G)$ denotes the adversarial relationship between G and D_G, as shown in Equation (13):

$$\begin{aligned} L_{adv}(G, D_G) = & E_{y \sim P_{\text{data}}(y)}[\log D_G(y)] \\ & + E_{x \sim P_{\text{data}}(x)}[\log(1 - D_G(G(R(x), x)))] \end{aligned} \tag{13}$$

where D_G works to maximize the objective function, aiming to differentiate between the generated rain-free image and real ones. Conversely, G minimizes the loss, striving to enhance the realism of the generated rain-free image. Similarly, we can find $L_{adv}(F, D_F))$ by changing the role of G and D_G to F and D_F.

Content losses: The cycle consistency loss minimizes differences between the original image x (or y) and its reconstructed counterpart, *F(G(x))* (or *G(F(y))*). This approach does not consider whether the generated image $G(x)$ (or $F(y)$) visually resembles the original x (or y). Inspired by [44], we decide to include the content loss regularizer in the objective function of SID. We seek to preserve the detailed information of the input image x (or y) while adjusting color for improved visual quality in the resulting image $G(x)$ (or $F(y)$). To achieve this, a *VGG16* pre-trained network is employed to extract feature maps from the *Conv2_3* layer for both input and generated images. We also utilize the 1-norm for assessing content loss, as it demonstrates greater robust to noise and outliers, facilitating a more effective recovery of details of the rainy image. The content loss regularizer is formulated as

$$\begin{aligned} L_{con}(G, F) = & E_{x \sim P_{\text{data}}(x)}[\| V_{GG}(G(x)) - V_{GG}(x) \|_1] \\ & + E_{y \sim P_{\text{data}}(y)}[\| V_{GG}(F(y)) - V_{GG}(y) \|_1] \end{aligned} \tag{14}$$

where $L_{con}(\cdot)$ is content loss, and $V_{GG}(\cdot)$ denotes the *Conv2_3* layer of the *VGG 16* network [45] pre-trained on ImageNet [46]. To our knowledge, this is the first time that content loss has been added to the objective function for single-image rain streak removal.

Total losses: The overall loss function of our proposed network for unsupervised training is expressed as follows:

$$L_{total} = \alpha_1 L_{att} + \alpha_2 L_{adv} + \alpha_3 L_{cyc} + \alpha_4 L_{con}(G, F) \tag{15}$$

where $L_{adv} = L_{adv}(G, D_G) + L_{adv}(F, D_F)$, $L_{att} = L_{\text{att}_x} + L_{\text{att}_y}$ and $\alpha_1, \alpha_2, \ldots, \alpha_4$ are trade-off parameters.

4. Experimental Results and Discussion

In this section, we provide details of the conducted experiments and the quality metrics used to assess the effectiveness of the proposed approach. Additionally, we discuss the dataset and training procedures, followed by a comparison of the proposed method with state-of-the-art approaches, along with the ablation studies.

4.1. Network Training and Parameter Setting

4.1.1. Implementation Details

Our model is trained using the PyTorch 2.2.2 with CUDA 11.8 framework [47] in a Python 3.11.3 environment, leveraging the computational power of an NVIDIA GeForce GTX 2080Ti GPU, manufactured by NVIDIA Corporation, Taipei, Taiwan, with 16 GB of memory. During training, we employ random cropping to extract 256 × 256 image patches from the original input images, augmenting the dataset by including their horizontal flips. We optimize the model's parameters using the Adam optimizer [48] with a mini-batch size of 1, a weight decay of 0.0001, and a momentum of 0.9. The choice of these hyperparameters is based on empirical observations and prior research, where similar values have shown effectiveness in optimizing deep learning models. The training process spans 400 epochs, starting with an initial learning rate of 0.0001, which is annealed using a PyTorch policy after 200 epochs to aid convergence. We select the number of epochs and the learning rate schedule empirically. The parameters $\alpha_1 = 1$, $\alpha_2 = 10$, $\alpha_3 = 10$ and $\alpha_4 = 0.01$ in

Equation (15) are meticulously tuned through a process of trial and error, balancing the contributions of different components in the loss function. We conduct experiments with various values for these parameters and select the ones that yield the best performance on the test set. To obtain the final rain mask, we set the number of iterations, $It = 6$. We employ $SE = 2$ blocks in our generator, considering the trade-off between performance and model complexity. Increasing the number of SE blocks augments the network's complexity, necessitating additional training time and memory resources. The choice of $SE = 2$ blocks is based on empirical observations, where it strikes a good balance between performance and computational cost. Generally, the selection of parameters is guided by empirical observations and experimentation on the available training and test sets, aiming to achieve the best performance on the given task. The training model parameters for the generator and RRAM are shown in Tables 1 and 2, respectively, while the parameters for the discriminator are clearly depicted in Figure 6b.

Table 1. The architecture of the generator and parameter settings.

Layer	Channel (Input, Output)	Kernel Size	Striding	Padding	Dilation
Conv + ReLU	(4, 64)	5 × 5	1 × 1	2 × 2	-
Conv + ReLU	(64, 128)	3 × 3	1 × 1	1 × 1	-
Conv + ReLU	(128, 128)	3 × 3	1× 1	1 × 1	-
Conv + ReLU	(128, 256)	3 × 3	1 × 1	1 × 1	-
AvgPool Conv ReLU	(256, 256/r)	1 × 1	1 × 1	-	-
Conv Sigmoid	(256, 256/r)	1 × 1	1 × 1	-	-
Conv + ReLU	(256, 256)	3 × 3	1 × 1	1 × 1	-
Conv + ReLU	(256, 256)	3 × 3	1 × 1	1 × 1	-
Conv + ReLU	(256, 256)	3 × 3	1 × 1	2 × 2	2
Conv + ReLU	(256, 256)	3 × 3	1 × 1	4 × 4	4
Conv + ReLU	(256, 256)	3 × 3	1 × 1	8 × 8	8
Conv + ReLU	(256, 256)	3 × 3	1 × 1	16 × 16	16
Conv + ReLU	(256, 256)	3 × 3	1 × 1	1 × 1	-
Conv + ReLU	(256, 256)	3 × 3	1 × 1	1 × 1	-
AvgPool Conv ReLU	(256, 256/r)	1 × 1	1 × 1	-	-
Conv Sigmoid	(256, 256/r)	1 × 1	1 × 1	-	-
ConvTranspose AvgPool ReLU	(256, 128)	4 × 4	2 × 2	1 × 1	-
Conv + ReLU	(128, 128)	3 × 3	1 × 1	1 × 1	-
ConvTranspose AvgPool ReLU	(128, 64)	4 × 4	2 × 2	1 × 1	-
Conv + ReLU	(64, 1)	3 × 3	1 × 1	1 × 1	-

Blue color indicates the SE block; r represents the reduction ratio, which is 2 in our case; Conv denotes the convolution; ConvTranspose denotes the transposed convolution; AvgPool denotes the average pooling; and ReLU indicates the ReLU activation function.

Table 2. The architecture of RRAM and parameter settings.

	Layer	Channel (Input, Output)	Kernel Size	Striding	Padding	
	Conv ReLU	(4, 32)	3 × 3	1 × 1	1 × 1	Conv + ReLU
	Conv Sigmoid	(64, 32)	3 × 3	1 × 1	1 × 1	LSTM
	Conv Sigmoid	(64, 32)	3 × 3	1 × 1	1 × 1	
	Conv Tanh	(64, 32)	3 × 3	1 × 1	1 × 1	
	Conv Sigmoid	(64, 32)	3 × 3	1 × 1	1 × 1	
	Conv BatchNorm ReLU	(64, 128)	3 × 3	1 × 1	1 × 1	CSAB
	Conv BatchNorm	(128, 128)	3 × 3	1 × 1	1 × 1	
	Conv	(128, 64)	3 × 3	1 × 1	1 × 1	
Avg out =	AvgPool Conv ReLU Conv	(64, 4) (4, 64)	1 × 1 1 × 1	1 × 1 1 × 1	- -	
Max out =	MaxPool Conv ReLU Conv	(64, 4) (4, 64)	1 × 1 1 × 1	1 × 1 1 × 1	- -	
	Sigmoid (Avg out + Max out)					
	Conv (AvgPool; MaxPool) Sigmoid	(2, 1)	7 × 7	3 × 3	-	
	ReLU					
	Conv	(64, 1)	3 × 3	1 × 1	1 × 1	Conv

Conv denotes the convolution, ReLU indicates the ReLU activation function, Sigmoid indicates the sigmoid activation function, Tahn denotes the tahn activation function, BatchNorm denotes the batch normalization, AvgPool denotes the average pooling, and MaxPool denotes the maximum pooling.

4.1.2. Datasets and Evaluation Metrics

To train and evaluate the proposed SE-RRACycleGAN for single-image rain streak removal, we utilized widely recognized synthetic and real-world image datasets. The synthetic datasets utilized for training and testing include (1) Rain100L [49], comprising 200 rainy image pairs for training and 100 pairs for testing; (2) Rain800 [23], consisting of 700 rainy image pairs for training and 100 pairs for testing; and (3) Rain12 [21], which includes 12 pairs of rain-free and rainy images specifically utilized for testing the model trained on Rain100L. The real-world image datasets employed are (1) SPANet-Data [50], a collection of 1000 rainy images along with corresponding ground-truth images, and (2) SIRR-Data [51], which comprises 147 rainy images without corresponding ground-truth images. These datasets were chosen due to their wide recognition in the research community and their capability to encompass a diverse range of real-world rainy image scenarios. The synthetic datasets provide controlled conditions for training and testing, while the incorporation of real-world image datasets enables the assessment of the model's performance under more challenging and varied environmental conditions.

We evaluate the experimental result of various technique using two widely employed quantitative measures, namely Peak Signal-to-Noise Ratio (PSNR) and Structural Similarity Index (SSIM) [52] for images with ground truth, while visual results are provided for SIRR-Data due to the lack of ground truth. In addition to the above quantitative measures, our model's performance is assessed using Root Mean Square Error (RMSE) [53], Feature Similarity Index (FSIM) [54], Mean Absolute Error (MAE) [55], and Correlation Coefficient (CC) [56]. It is noteworthy that the state-of-the-art methods we compared our model with did not employ these specific quantitative measures, thereby restricting direct comparisons based on these measures.

4.2. Comparisons with State-of-the-Art Methods

We evaluate the performance of our method by comparing it with five supervised networks (i.e., DetailNet [57], Clear [58], RESCAN [59], PReNet [60], SPANet [50]), one semi-supervised technique (SIRR [51]), and five unsupervised techniques (i.e., CycleGAN [18], RR-GAN [26], DerainCycleGAN [11], Derain Attention GAN [13], and Cycle-Attention-Derain [12]). Since our model is unsupervised, the unsupervised methods are primarily compared, although our method is highly competitive with the existing semi-supervised and supervised methods.

4.2.1. Comparisons Using a Synthetic Datasets

We evaluate the proposed method on test images from the synthetic datasets (i.e., Rain100L [49], Rain800 [23], and Rain12 [21]) in the first set of experiments and compare its quantitative and qualitative performance against numerous state-of-the-art methods. For a fair comparison, for certain supervised deep-learning networks, such as DetailNet [57], Clear [58], and SPANet [50], and the semi-supervised deep-learning network SSIR [51], we directly adopt the results reported in [11,61], since the evaluation metrics are the same. Moreover, we utilized the source code provided by the authors in the CycleGAN [18], PReNet [60], Derain CycleGAN [11], and RESCAN [59] papers to train and test on synthetic datasets. Lastly, for the comparison with unsupervised rain removal techniques such as RR-GAN [26], Derain Attention GAN [13], and Cycle-Attention-Derain [12], we directly used the results provided in their respective papers as their code is not publicly available. The quantitative and qualitative comparison results are depicted in Table 3 and Figures 8 and 9, respectively. Table 3 clearly indicates that, compared to unsupervised techniques, the proposed SE-RRAMCycleGAN method achieves the best performance on Rain12 and SSIM on Rain800, and better performance on Rain100L and PSNR on Rain800, following Derain Attention GAN [13] and Cycle-Attention-Derain [12], respectively. The proposed method even outperforms the semi-supervised method SIRR, which can gain an advantage from semi-supervised learning, on Rain12. Moreover, the performance of supervised methods declined rapidly from Rain100L to Rain800, whereas our network remains stable. This stability could be attributed to the ability of the RRAM in our unsupervised network to gain a deeper understanding of rain-related features when handling challenging samples. Furthermore, as shown in Table 4, our model's performance is evaluated using RMSE, FSIM, MAE, and CC. These quantitative metrics collectively assess the model's effectiveness in single-image deraining tasks. The remarkable performance across these metrics underscores our model's capability to effectively remove rain while preserving important image details, positioning it as a promising approach for real-world applications such as improving visibility in rainy conditions.

(a) (b) (c) (d) (e) (f)

Figure 8. Comparisons of qualitative results on Rain100L [49]. (**a–f**) are input rainy images, RESCAN [59], CycleGAN [18], PReNet [60], ours, and ground truth, respectively.

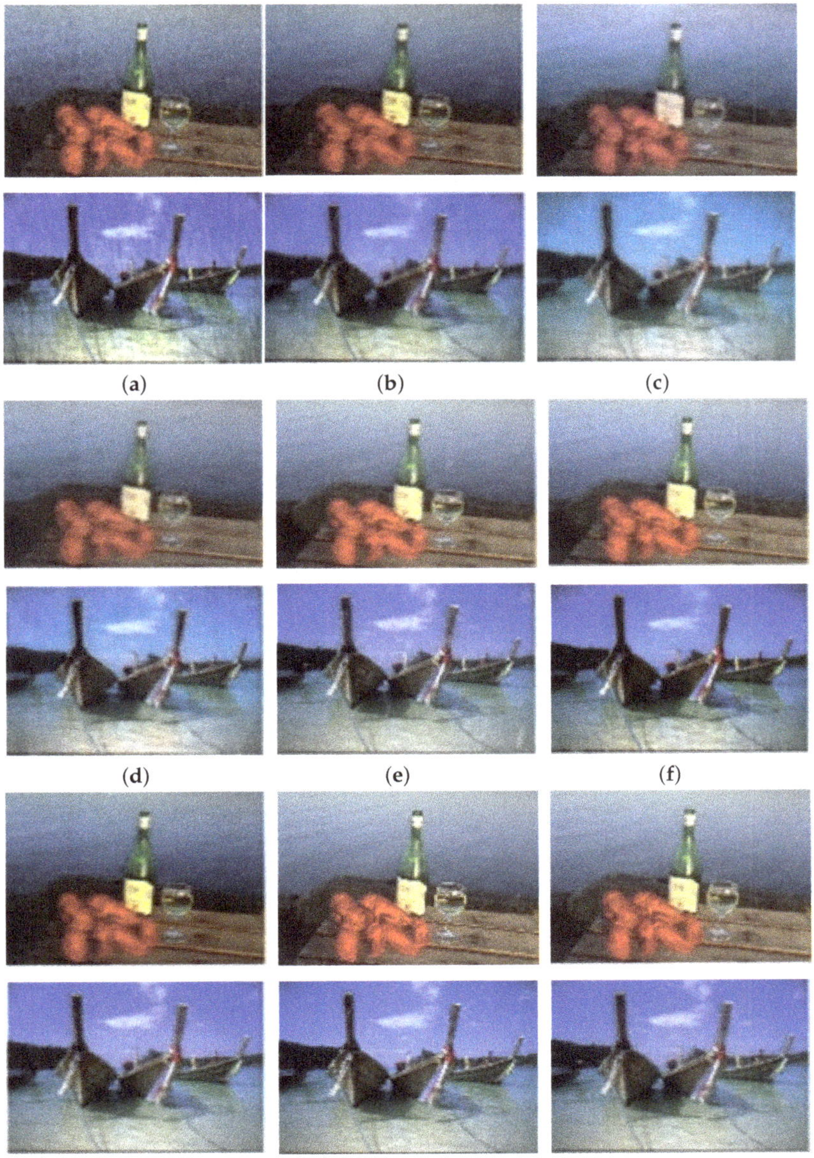

(a) (b) (c) (d) (e) (f) (g) (h) (i)

Figure 9. Comparisons of qualitative results on Rain800 [23], (**a–i**) are Input rainy images, SPANet [50], CycleGAN [18], RR-GAN [26], DerainCycleGAN [11], Derain Attention GAN [13], Cycle-Attention-Derain [12], ours, and ground truth, respectively.

We proceed to qualitatively compare Rain100L [49] and Rain800 [23] in Figures 8 and 9. CycleGAN [18] changes the color of the output image and also leaves rain streaks on the image, and even the supervised method RESCAN [59] leaves some amount of rain streaks (Figure 8). However, our method excels in removing numerous rain streaks, preserving a smooth background, and closely resembling the ground truth compared to other approaches. Figure 9 illustrates that most of the methods left some amount of rain streaks, while others, such as CycleGAN [18] and Cycle-Attention-Derain, exhibit noticeable color shifts. Conversely, our method demonstrates superior performance by eliminating rain streaks while preserving the background color (i.e., similar to the ground truth).

Table 3. Quantitative experiments evaluated on four datasets.

	Rain100L [49]		**Rain12 [21]**		**Rain800 [23]**		**SPANet-Data [50]**	
	PSNR	**SSIM**	**PSNR**	**SSIM**	**PSNR**	**SSIM**	**PSNR**	**SSIM**
DetailNet [57]	32.38	0.926	34.04	0.933	21.16	0.732	34.70	0.926
Clear [58]	30.24	0.934	31.24	0.935	-	-	32.66	0.942
RESCAN [59]	38.52	0.981	36.43	0.952	24.09	0.841	-	-
PReNet [60]	37.45	0.979	36.66	0.961	26.97	0.898	35.06	0.944
SPANet [50]	34.46	0.962	34.63	0.943	24.52	0.51	35.24	0.945
SSIR ** [51]	32.37	0.926	34.02	0.935	-	-	34.85	0.936
CycleGAN * [18]	24.61	0.834	21.56	0.845	23.95	0.819	22.40	0.860
RR-GAN * [26]	-	-	-	-	23.51	0.757	-	-
Derain CycleGAN * [11]	31.49	0.936	*34.44*	*0.952*	24.32	0.842	*34.12*	*0.950*
Derain Attention GAN * [13]	**34.01**	**0.969**	-	-	25.22	0.856	-	-
Cycle-attention-derain * [12]	29.26	0.902	30.77	0.911	**28.48**	*0.874*	33.15	0.921
Ours *	*31.87*	*0.941*	**34.60**	**0.954**	*27.92*	**0.879**	**34.17**	**0.953**

** = semi-supervised, * = unsupervised, **Bold** is for the best value, and *Italic* is for the better value for unsupervised methods.

Table 4. Quantitative experiments evaluated on four datasets using our model.

	Qualitative Measures			
Datasets	**RMSE**	**FSIM**	**MAE**	**CC**
Rain100L [49]	0.167	0.899	0.145	0.918
Rain12 [21]	0.159	0.915	0.138	0.926
Rain800 [23]	0.215	0.847	0.189	0.861
SPANet-Data [50]	0.162	0.908	0.143	0.922

RMSE values are normalized to a range between [0, 1].

4.2.2. Comparisons Using a Real Datasets

We then evaluate each approach on two real rainy datasets: SPANet-Data [50] and SIRR-Data [51]. These datasets are used to test the model that is trained on Rain100L [49] for all methods. Numerical metrics can be employed to evaluate SPANet-Data since it contains ground-truth images. As shown in Table 3, our method obtains the best result among unsupervised methods and is highly competitive with semi-supervised and supervised approaches on SPANet-Data. We also visually compare our method with a supervised and an unsupervised method using SIRR-Data. As illustrated in Figure 10, the supervised PReNet [60] method left the rain streak on the image, and the unsupervised CycleGAN [18] resulted in a color change on the output image. Our unsupervised method demonstrates better performance than even the supervised method. This may be due to the inflexibility of conventional supervised methods in handling real rainy images because the distribution of real rain is very different from the synthetic dataset on which the supervised methods are trained.

Figure 10. Comparisons of deraining results on SIRR-Data [51]. (**a–d**) are rainy images, PReNet [60], CycleGAN [18], and ours, respectively.

4.3. Ablation Study

We conducted ablation experiments on Rain100L [49] to evaluate the effectiveness of each component in our proposed method. The five experiments included CycleGAN as a baseline, CycleGAN + SE, SE-RRACycleGAN without LSTM, SE-RRACycleGAN without SE, and SE-RRACycleGAN. The results, depicted in Figure 11 and Table 5, demonstrate that the introduced components significantly enhanced rain removal performance compared to CycleGAN. All components proved to be crucial for our method, although Figure 11c–e exhibits some undesired artifacts. This experiment underscores the significance of LSTM in our recurrent rain-attention module and SE blocks in the generator. Our method effectively addresses these artifacts, resulting in a smoother background.

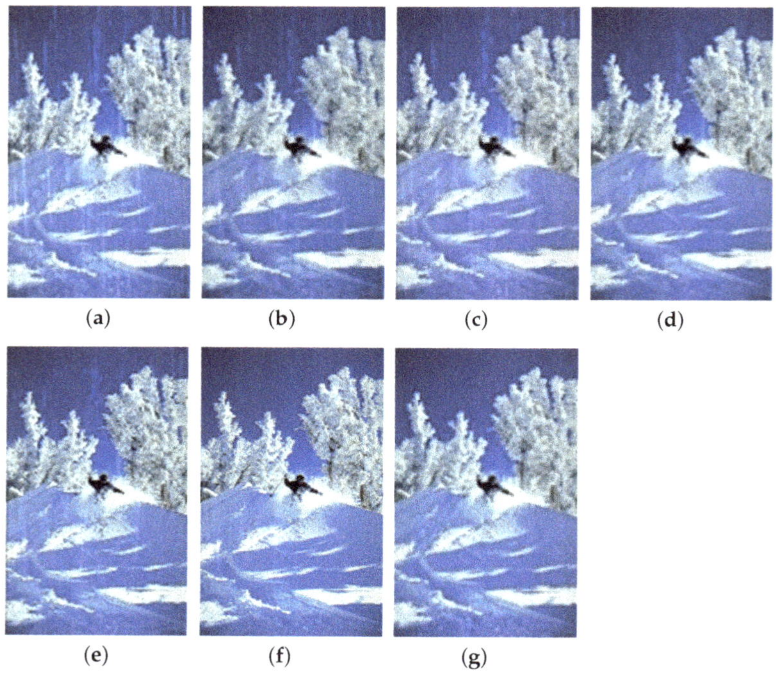

Figure 11. Visual comparisons of components used in our method. (**a**–**g**) are a rainy image, only CycleGAN, CycleGAN + SE, SE-RRACycleGAN without LSTM, SE-RRACycleGAN without SE, our SE-RRACycleGAN, and ground truth, respectively.

Table 5. Ablation study on different components of our method.

Dataset	Metrics	Baseline	Baseline + SE	Ours w/o LSTM	Ours w/o SE	Ours
Rain100L	PSNR	24.61	26.94	28.43	30.25	**31.87**
	SSIM	0.834	0.859	0.878	0.912	**0.941**

Bold indicates best result. Note: w/o represents without, baseline is CycleGAN, and Ours represents SE-RRACycleGAN.

5. Conclusions

In this paper, we introduce an unsupervised SE- RRACycleGAN network for SID. We propose a novel RRAM to enhance the detection of rain-related information by simultaneously analyzing both rainy and rain-free images. In addition, we incorporate the SE into the generator architecture to improve the generator's ability to generalize across diverse rain patterns by dynamically adjusting the weights of features based on the characteristics of the input image. Moreover, we add the content loss into the objective function to enhance the visual similarity of the generated image with the input image. Extensive experiments conducted on both synthetic and real-world datasets reveal that our method surpasses most of the unsupervised state-of-the-art methods both quantitatively and qualitatively, especially on Rain12 and real rainy images, and is highly competitive with supervised techniques. However, our method may not achieve optimal results in situations where rain streaks closely resemble the background texture in the input images. For instance, as shown in the top row of Figure 8, while our method outperforms other techniques in rain removal, it does introduce some blurring in the background regions. Therefore, in the future, we aim to extend our model's capabilities to address this issue and also handle images with intense rain conditions, such as those in the Rain100H dataset, snow and fog, where the background is heavily obscured, making it challenging to accurately restore the images without reference values (i.e., ground truth).

Author Contributions: Conceptualization, methodology, analysis, review, editing, and revision G.N.W. and S.S.-D.X.; coding, simulation, and writing, G.N.W. All authors have read and agreed to the published version of the manuscript.

Funding: This work is partially supported by the Ministry of Science and Technology (MOST), Taiwan, under the Grant MOST 111-2221-E-011-146-MY2.

Data Availability Statement: No new datasets were created by ourselves. All the used datasets in this paper can be found as follows: Rain800 [23] https://github.com/hezhangsprinter/ID-CGAN (accessed on 1 January 2024), Rain100L [49] https://github.com/ZhangXinNan/RainDetectionAndRemoval (accessed on 1 January 2024), Rain12 [21] https://github.com/yu-li/LPDerain (accessed on 1 January 2024), SPANet [50] https://github.com/stevewongv/SPANet (accessed on 1 January 2024), and SIRR-Data [51] https://github.com/wwzjer/Semi-supervised-IRR (accessed on 1 January 2024).

Conflicts of Interest: The authors declare no conflicts of interest.

References

1. Qiao, J.; Song, H.; Zhang, K.; Zhang, X.; Liu, Q. Image super-resolution using conditional generative adversarial network. *IET Image Process.* **2019**, *13*, 2673–2679. [CrossRef]
2. Mao, J.; Xiao, T.; Jiang, Y.; Cao, Z. What can help pedestrian detection? In Proceedings of the IEEE Conference on Computer Vision and Pattern Recognition, Honolulu, HI, USA, 21–26 July 2017; pp. 3127–3136.
3. Song, Y.; Ma, C.; Wu, X.; Gong, L.; Bao, L.; Zuo, W.; Shen, C.; Lau, R.W.; Yang, M.H. Vital: Visual tracking via adversarial learning. In Proceedings of the IEEE Conference on Computer Vision and Pattern Recognition, Salt Lake City, UT, USA, 18–23 June 2018; pp. 8990–8999.
4. Zhu, Z.; Liang, D.; Zhang, S.; Huang, X.; Li, B.; Hu, S. Traffic-sign detection and classification in the wild. In Proceedings of the IEEE Conference on Computer Vision and Pattern Recognition, Las Vegas, NV, USA, 27–30 June 2016; pp. 2110–2118.
5. Tripicchio, P.; Camacho-Gonzalez, G.; D'Avella, S. Welding defect detection: Coping with artifacts in the production line. *Int. J. Adv. Manuf. Technol.* **2020**, *111*, 1659–1669. [CrossRef]
6. Chen, H.; He, X.; Qing, L.; Wu, Y.; Ren, C.; Sheriff, R.E.; Zhu, C. Real-world single image super-resolution: A brief review. *Inf. Fusion* **2022**, *79*, 124–145. [CrossRef]
7. Lian, Q.; Yan, W.; Zhang, X.; Chen, S. Single image rain removal using image decomposition and a dense network. *IEEE/CAA J. Autom. Sin.* **2019**, *6*, 1428–1437. [CrossRef]
8. Liu, J.; Yang, W.; Yang, S.; Guo, Z. D3r-Net: Dynamic routing residue recurrent network for video rain removal. *IEEE Trans. Image Process.* **2018**, *28*, 699–712. [CrossRef] [PubMed]
9. Li, M.; Xie, Q.; Zhao, Q.; Wei, W.; Gu, S.; Tao, J.; Meng, D. Video rain streak removal by multiscale convolutional sparse coding. In Proceedings of the IEEE Conference on Computer Vision and Pattern Recognition, Salt Lake City, UT, USA, 18–23 June 2018; pp. 6644–6653.
10. Ahn, N.; Jo, S.Y.; Kang, S.J. EAGNet: Elementwise attentive gating network-based single image de-raining with rain simplification. *IEEE Trans. Circuits Syst. Video Technol.* **2021**, *32*, 608–620. [CrossRef]
11. Wei, Y.; Zhang, Z.; Wang, Y.; Xu, M.; Yang, Y.; Yan, S.; Wang, M. Deraincyclegan: Rain attentive cyclegan for single-image deraining and rainmaking. *IEEE Trans. Image Process.* **2021**, *30*, 4788–4801. [CrossRef] [PubMed]
12. Chen, M.; Wang, P.; Shang, D.; Wang, P. Cycle-Attention-Derain: Unsupervised rain removal with CycleGAN. *Vis. Comput.* **2023**, *39*, 3727–3739. [CrossRef]
13. Guo, Z.; Hou, M.; Sima, M.; Feng, Z. DerainAttentionGAN: Unsupervised single-image deraining using attention-guided generative adversarial networks. *Signal Image Video Process.* **2022**, *16*, 185–192. [CrossRef]

14. Yang, W.; Tan, R.T.; Wang, S.; Fang, Y.; Liu, J. single-image deraining: From model-based to data-driven and beyond. *IEEE Trans Pattern Anal. Mach. Intell.* **2020**, *43*, 4059–4077. [CrossRef]
15. Yu, X.; Zhang, G.; Tan, F.; Li, F.; Xie, W. Progressive hybrid-modulated network for single-image deraining. *Mathematics* **2023** *11*, 691. [CrossRef]
16. Liu, T.; Zhou, B.; Luo, P.; Zhang, Y.; Niu, L.; Wang, G. Two-Stage and Two-Channel Attention single-image deraining Network for Promoting Ship Detection in Visual Perception System. *Appl. Sci.* **2022**, *12*, 7766. [CrossRef]
17. Wang, Y.T.; Zhao, X.L.; Jiang, T.X.; Deng, L.J.; Chang, Y.; Huang, T.Z. Rain streaks removal for single image via kernel-guided convolutional neural network. *IEEE Trans. Neural Netw. Learn. Syst.* **2020**, *32*, 3664–3676. [CrossRef] [PubMed]
18. Zhu, J.Y.; Park, T.; Isola, P.; Efros, A.A. Unpaired image-to-image translation using cycle-consistent adversarial networks In Proceedings of the IEEE International Conference on Computer Vision, Venice, Italy, 22–29 October 2017; pp. 2223–2232.
19. Kang, L.W.; Lin, C.W.; Fu, Y.H. Automatic single-image-based rain streaks removal via image decomposition. *IEEE Trans. Image Process.* **2011**, *21*, 1742–1755. [CrossRef] [PubMed]
20. Luo, Y.; Xu, Y.; Ji, H. Removing rain from a single image via discriminative sparse coding. In Proceedings of the IEEE International Conference on Computer Vision, Santiago, Chile, 7–13 December 2015; pp. 3397–3405.
21. Li, Y.; Tan, R.T.; Guo, X.; Lu, J.; Brown, M.S. Rain streak removal using layer priors. In Proceedings of the IEEE Conference on Computer Vision and Pattern Recognition, Las Vegas, NV, USA, 27–30 June 2016; pp. 2736–2744.
22. Wang, Y.; Liu, S.; Chen, C.; Zeng, B. A hierarchical approach for rain or snow removing in a single color image. *IEEE Trans. Image Process.* **2017**, *26*, 3936–3950. [CrossRef] [PubMed]
23. Zhang, H.; Sindagi, V.; Patel, V.M. Image de-raining using a conditional generative adversarial network. *IEEE Trans. Circuits Syst. Video Technol.* **2019**, *30*, 3943–3956. [CrossRef]
24. Yang, W.; Tan, R.T.; Feng, J.; Guo, Z.; Yan, S.; Liu, J. Joint rain detection and removal from a single image with contextualized deep networks. *IEEE Trans. Pattern Anal. Mach. Intell.* **2019**, *42*, 1377–1393. [CrossRef] [PubMed]
25. Isola, P.; Zhu, J.Y.; Zhou, T.; Efros, A.A. Image-to-image translation with conditional adversarial networks. In Proceedings of the IEEE Conference on Computer Vision and Pattern Recognition, Honolulu, HI, USA, 21–26 July 2017; pp. 1125–1134.
26. Zhu, H.; Peng, X.; Zhou, J.T.; Yang, S.; Chanderasekh, V.; Li, L.; Lim, J.H. Singe image rain removal with unpaired information: A differentiable programming perspective. In Proceedings of the AAAI Conference on Artificial Intelligence, Honolulu, HI, USA, 27 January–1 February 2019; Volume 33, pp. 9332–9339.
27. Tian, C.; Xu, Y.; Li, Z.; Zuo, W.; Fei, L.; Liu, H. Attention-guided CNN for image denoising. *Neural Netw.* **2020**, *124*, 117–129. [CrossRef] [PubMed]
28. Zhao, B.; Wu, X.; Feng, J.; Peng, Q.; Yan, S. Diversified visual attention networks for fine-grained object classification. *IEEE Trans. Multimed.* **2017**, *19*, 1245–1256. [CrossRef]
29. Mnih, V.; Heess, N.; Graves, A.; Kavukcuoglu, K. Recurrent models of visual attention. *Adv. Neural Inf. Process. Syst.* **2014**, *27*.
30. Zheng, M.; Xu, J.; Shen, Y.; Tian, C.; Li, J.; Fei, L.; Zong, M.; Liu, X. Attention-based CNNs for image classification: A survey. In *Proceedings of the Journal of Physics: Conference Series*; IOP Publishing: Bristol, UK, 2022; Volume 2171, p. 012068.
31. Qian, R.; Tan, R.T.; Yang, W.; Su, J.; Liu, J. Attentive generative adversarial network for raindrop removal from a single image. In Proceedings of the IEEE Conference on Computer Vision and Pattern Recognition, Salt Lake City, UT, USA, 18–23 June 2018; pp. 2482–2491.
32. Woo, S.; Park, J.; Lee, J.Y.; Kweon, I.S. Cbam: Convolutional block attention module. In Proceedings of the European Conference on Computer Vision (ECCV), Munich, Germany, 8–14 September 2018; pp. 3–19.
33. Chao, Z.; Pu, F.; Yin, Y.; Han, B.; Chen, X. Research on real-time local rainfall prediction based on MEMS sensors. *J. Sens.* **2018**, *2018*, 6184713. [CrossRef]
34. Liu, R.W.; Hu, K.; Liang, M.; Li, Y.; Liu, X.; Yang, D. QSD-LSTM: Vessel trajectory prediction using long short-term memory with quaternion ship domain. *Appl. Ocean Res.* **2023**, *136*, 103592. [CrossRef]
35. Brown, M.J.; Hutchinson, L.A.; Rainbow, M.J.; Deluzio, K.J.; De Asha, A.R. A comparison of self-selected walking speeds and walking speed variability when data are collected during repeated discrete trials and during continuous walking. *J. Appl. Biomech.* **2017**, *33*, 384–387. [CrossRef] [PubMed]
36. Zagoruyko, S.; Komodakis, N. Paying more attention to attention: Improving the performance of convolutional neural networks via attention transfer. *arXiv* **2016**, arXiv:1612.03928.
37. Ronneberger, O.; Fischer, P.; Brox, T. U-net: Convolutional networks for biomedical image segmentation. In Proceedings of the Medical Image Computing and Computer-Assisted Intervention: 18th International Conference, Munich, Germany, 5–9 October 2015; pp. 234–241.
38. Hu, J.; Shen, L.; Sun, G. Squeeze-and-excitation networks. In Proceedings of the IEEE Conference on Computer Vision and Pattern Recognition, Salt Lake City, UT, USA, 18–23 June 2018; pp. 7132–7141.
39. Wang, C.; Fan, W.; Zhu, H.; Su, Z. single-image deraining via nonlocal squeeze-and-excitation enhancing network. *Appl. Intell.* **2020**, *50*, 2932–2944. [CrossRef]
40. Zhang, Y.; Li, K.; Li, K.; Wang, L.; Zhong, B.; Fu, Y. Image super-resolution using very deep residual channel attention networks. In Proceedings of the European Conference on Computer Vision (ECCV), Munich, Germany, 8–14 September 2018; pp. 286–301.
41. Lee, H.Y.; Tseng, H.Y.; Huang, J.B.; Singh, M.; Yang, M.H. Diverse image-to-image translation via disentangled representations. In Proceedings of the European Conference on Computer Vision (ECCV), Munich, Germany, 8–14 September 2018; pp. 35–51.

42. Qiao, J.; Song, H.; Zhang, K.; Zhang, X. Conditional generative adversarial network with densely-connected residual learning for single image super-resolution. *Multimed. Tools Appl.* **2021**, *80*, 4383–4397. [CrossRef]
43. Goodfellow, I.; Pouget-Abadie, J.; Mirza, M.; Xu, B.; Warde-Farley, D.; Ozair, S.; Courville, A.; Bengio, Y. Generative adversarial nets. *Adv. Neural Inf. Process. Syst.* **2014**, *27*, 139–144.
44. Du, R.; Li, W.; Chen, S.; Li, C.; Zhang, Y. Unpaired underwater image enhancement based on cyclegan. *Information* **2021**, *13*, 1. [CrossRef]
45. Karen, S. Very deep convolutional networks for large-scale image recognition. *arXiv* **2014**, arXiv:1409.1556.
46. Deng, J.; Dong, W.; Socher, R.; Li, L.J.; Li, K.; Fei-Fei, L. Imagenet: A large-scale hierarchical image database. In Proceedings of the IEEE Conference on Computer Vision and Pattern Recognition, Miami, FL, USA, 20–25 June 2009; pp. 248–255.
47. Paszke, A.; Gross, S.; Chintala, S.; Chanan, G.; Yang, E.; DeVito, Z.; Lin, Z.; Desmaison, A.; Antiga, L.; Lerer, A. Automatic Differentiation in Pytorch. 2017. Available online: https://openreview.net/forum?id=BJJsrmfCZ (accessed on 1 January 2024).
48. Kingma, D.; Ba, J. Adam: A Method for Stochastic Optimization. In Proceedings of the International Conference on Learning Representations (ICLR), San Diego, CA, USA, 7–9 May 2015.
49. Yang, W.; Tan, R.T.; Feng, J.; Liu, J.; Guo, Z.; Yan, S. Deep joint rain detection and removal from a single image. In Proceedings of the IEEE Conference on Computer Vision and Pattern Recognition, Honolulu, HI, USA, 21–26 July 2017; pp. 1357–1366.
50. Wang, T.; Yang, X.; Xu, K.; Chen, S.; Zhang, Q.; Lau, R.W. Spatial attentive single-image deraining with a high quality real rain dataset. In Proceedings of the IEEE/CVF Conference on Computer Vision and Pattern Recognition, Long Beach, CA, USA, 15–20 June 2019; pp. 12270–12279.
51. Wei, W.; Meng, D.; Zhao, Q.; Xu, Z.; Wu, Y. Semi-supervised transfer learning for image rain removal. In Proceedings of the IEEE/CVF Conference on Computer Vision and Pattern Recognition, Long Beach, CA, USA, 15–20 June 2019; pp. 3877–3886.
52. Wang, Z.; Bovik, A.C.; Sheikh, H.R.; Simoncelli, E.P. Image quality assessment: From error visibility to structural similarity. *IEEE Trans. Image Process.* **2004**, *13*, 600–612. [CrossRef]
53. Sara, U.; Akter, M.; Uddin, M.S. Image quality assessment through FSIM, SSIM, MSE and PSNR—A comparative study. *J. Comput. Commun.* **2019**, *7*, 8–18. [CrossRef]
54. Zhang, L.; Zhang, L.; Mou, X.; Zhang, D. FSIM: A feature similarity index for image quality assessment. *IEEE Trans. Image Process.* **2011**, *20*, 2378–2386. [CrossRef] [PubMed]
55. Su, Z.; Zhang, Y.; Shi, J.; Zhang, X.P. A Survey of Single Image Rain Removal Based on Deep Learning. *ACM Comput. Surv.* **2023**, *56*, 1–35. [CrossRef]
56. Ratner, B. The correlation coefficient: Its values range between +1/−1, or do they? *J. Target. Meas. Anal. Mark.* **2009**, *17*, 139–142. [CrossRef]
57. Fu, X.; Huang, J.; Zeng, D.; Huang, Y.; Ding, X.; Paisley, J. Removing rain from single images via a deep detail network. In Proceedings of the IEEE Conference on Computer Vision and Pattern Recognition, Honolulu, HI, USA, 21–26 July 2017; pp. 3855–3863.
58. Fu, X.; Huang, J.; Ding, X.; Liao, Y.; Paisley, J. Clearing the skies: A deep network architecture for single-image rain removal. *IEEE Trans. Image Process.* **2017**, *26*, 2944–2956. [CrossRef] [PubMed]
59. Li, X.; Wu, J.; Lin, Z.; Liu, H.; Zha, H. Recurrent squeeze-and-excitation context aggregation net for single image deraining. In Proceedings of the European Conference on Computer Vision (ECCV), Munich, Germany, 8–14 September 2018; pp. 254–269.
60. Ren, D.; Zuo, W.; Hu, Q.; Zhu, P.; Meng, D. Progressive image deraining networks: A better and simpler baseline. In Proceedings of the IEEE/CVF Conference on Computer Vision and Pattern Recognition, Long Beach, CA, USA, 15–20 June 2019; pp. 3937–3946.
61. Wang, H.; Wu, Y.; Li, M.; Zhao, Q.; Meng, D. Survey on rain removal from videos or a single image. *Sci. China Inf. Sci.* **2022**, *65*, 111101. [CrossRef]

MDPI AG
Grosspeteranlage 5
4052 Basel
Switzerland
Tel.: +41 61 683 77 34

Remote Sensing Editorial Office
E-mail: remotesensing@mdpi.com
www.mdpi.com/journal/remotesensing

www.ingramcontent.com/pod-product-compliance
Lightning Source LLC
Chambersburg PA
CBHW061621220926
51850CB00005B/567
* 9 7 8 3 7 2 5 8 2 1 1 9 8 *